Food SAFETY

Old Habits, New Perspectives

Food SAFETY

Old Habits, New Perspectives

Phyllis Entis, MSc, SM(NRM)

ASM PRESS
Washington, D.C.

ASM Press
American Society for Microbiology
1752 N Street, N.W.
Washington, DC 20036-2804

Library of Congress Cataloging-in-Publication Data

Entis, Phyllis.
Food safety : old habits, new perspectives / by Phyllis Entis.
p. ; cm.
Includes bibliographical references and index.
ISBN-13: 978-1-55581-417-5
ISBN-10: 1-55581-417-4
1. Food—Safety measures. 2. Food contamination. 3. Foodborne diseases. I. Title.
[DNLM: 1. Food Contamination. 2. Food Handling—methods. 3. Food Poisoning—microbiology. 4. Safety. WA 701 E61f 2007]

RA601.E58 2007
363.19′26—dc22

2006036301

10 9 8 7 6 5 4 3 2 1

Address editorial correspondence to: ASM Press, 1752 N St., N.W., Washington, DC 20036-2904, U.S.A.

Send orders to: ASM Press, P.O. Box 605, Herndon, VA 20172, U.S.A.
Phone: 800-546-2416; 703-661-1593
Fax: 703-661-1501
Email: Books@asmusa.org
Online: estore.asm.org

To MPE, with love

Contents

Illustrations

Figures

Tables

Preface

When I was in my teens, one of my favorite writers was Frances Parkinson Keyes. A "romance novelist" who began her writing career as a journalist, Ms. Keyes imbued her stories with a strong sense of place and a believable set of characters, and swept her readers along with a strong and smooth narrative style. Unlike many other authors in the genre, she also prefaced each book with a detailed explanation of its evolution, deftly setting the scene for her "Gentle Readers." Often, her prefaces were almost as interesting as the novels they introduced. While I can't hope to match Ms. Keyes's powers of observation and storytelling, I beg the indulgence of my own Gentle Readers as I sketch in some background to this book.

My involvement in food safety began in mid-1972, when I joined Canada's Health Protection Branch (HPB). I began my career with HPB in the Winnipeg, Manitoba, regional laboratory and, in mid-1974, moved to the agency's Quebec regional lab, based in the Montreal area. In 1975, I took over responsibility for managing HPB's Montreal area microbiology group. The Regent Chocolate *Salmonella* outbreak described in chapter 4 took place while I was working in Winnipeg, and the investigation into the source of repeated contamination of milk powder production plants with *Salmonella* (chapter 2) was carried out while I was in Montreal. Many of the details included in the description of both of those events (those not supported by specific literature reference citations) are based on first-hand information.

In 1979 I left HPB and, with my husband, co-founded QA Laboratories (later QA Life Sciences). The description of the wiener processing facility in chapter 2 is based on first-hand information. I was the consultant hired by the company to determine the source of their ongoing post-process contamination program.

In 2003, I wrote a series of articles on food safety for the *Del Mar Times*, a Del Mar, CA, weekly newspaper. Some of the information and anecdotes that appeared in those articles are scattered through this book. Notably, the story of Kevin Kowalcyk (chapter 6) and portions of the discussion of the BARF (raw food) diet for companion animals (chapter 15) first appeared in those *Del Mar Times* articles.

Gentle Readers should keep in mind that science doesn't stand still. All cumulative totals of outbreaks, cases, etc. are valid as of June 2006. Likewise, statements in the text that relate to ongoing investigations or situations also are effective as of that same date. The continuing saga of U.S.-Japan trade talks aimed at reopening the Japanese market to American beef is an example of a situation that can change from day to day. Also, URLs (web page addresses) cited in the References at the end of each chapter were verified on the dates shown. Given the ephemeral nature of the Internet, these are also subject to change.

What is not—and never should become—subject to change is the responsibility of food producers and processors to put food safety concerns ahead of expediency when making decisions. Choosing to ignore unfavorable or inconvenient test results, opting for the least expensive, most "cost-effective" processing method, and establishing token food safety programs that look good on paper but are ineffective, are not the actions one would wish to associate with major food companies. Yet these choices are made again and again—not just by small food processors, but also by major, multinational food companies. As I write this Preface, Cadbury-Schweppes is facing possible prosecution for its involvement in a U.K. *Salmonella* outbreak traced to chocolate produced in the company's Herefordshire production facility. Cadbury's management waited five months before alerting British health authorities to a leaking pipe that had contaminated some of its chocolate crumb.

A major outbreak of *Escherichia coli* O157:H7 (183 cases and one death as of September 26, 2006) linked to spinach grown and packaged in the Salinas Valley area of California—unfolding as this book goes to press—is an example of what can happen when an industry chooses to stay with its old habits. Between 1995 and 2005, there were 19 outbreaks of *E. coli* O157:H7 tied by epidemiological evidence to spinach or lettuce; at least 8 of the outbreaks were traced to Salinas Valley produce. According to newspaper reports, in 2004 and 2005, the FDA advised farmers in California that their crops could become contaminated with *E. coli* O157:H7. Unfortunately, the growers closed their eyes and ears to the government's warnings, and are now suffering the consequences—as are the 183 (or more) outbreak victims and their families.

Legislators and regulators also bear a responsibility for improving and maintaining food safety. Notwithstanding industry's pleas for voluntary programs and self-regulation, government oversight is an essential part of the food safety mosaic. Just as drivers will push the speed limit when they know that they are not being monitored, so too will food processors push the limits of "voluntary compliance"—not maliciously or with intent to harm the consumer, but simply because it's human nature to do so. Self-regulation is an oxymoron.

The public, too, has an important role to play. All too often, food preparers and consumers engage in risky behavior—eating raw or undercooked meat, poultry, eggs, or seafood, drinking unpasteurized milk or cider, neglecting proper kitchen sanitation practices, or storing food at an incorrect temperature. Lapses on the part of large food companies can result in massive food-borne disease outbreaks, but these occur only occasionally. Far more common are the sporadic cases and small outbreaks of food-borne disease caused by mishandling of food on the part of food service workers and by individual food preparers in the home.

The need for a safe food supply is not debatable, but experts differ on the best ways to achieve and maintain that goal. Irradiation of raw meats and poultry, the role of microbiological testing, and the precise role that regulatory authorities should play are all areas of controversy. While I have received and considered the opinions of my reviewers, I alone am responsible for the accuracy and completeness of the contents of this book and for any opinions expressed therein.

Phyllis Entis

Acknowledgments

To MICHAEL ENTIS, who gave me the original idea for the theme and orientation of this book, and whose constructive critique of the initial chapters set my feet on the correct path,

To Jeff Holtmeier of ASM Press for his enthusiastic support of this project,

To Ellie Tupper, my Production Editor, for skillfully guiding my book through the production process,

To Laura Ledbetter and the rest of the editorial, production, and marketing staff of ASM Press for their invaluable assistance,

To the libraries of the University of California at San Diego, the libraries of the University of Vermont, the San Diego Public Library, and the Stowe Free Library for access to their collections and on-line search services,

To Jeffrey Farber of Health Canada and Richard A. Holley of the University of Manitoba for obtaining on my behalf copies of several older articles that would not otherwise have been available to me,

To all those authors who so kindly responded to my requests for reprints of published research and review articles,

To Wallace H. Andrews, Steven C. Ingram, and Tom Montville for reviewing my book proposal,

To Kathryn J. Boor, Robert E. Brackett, Richard A. Holley, Lee-Ann Jaykus, and Ewen C. D. Todd for reviewing portions of this manuscript and offering their helpful corrections and constructive suggestions, and

To Dean O. Cliver, for his in-depth and extremely helpful review of my entire manuscript, including the Glossary and Appendix,

I extend my thanks and appreciation.

chapter 1

Old Habits Die Hard

TEN-YEAR-OLD DENISE STOOD in the kitchen watching as her mother prepared a pot roast for the oven. Her mother washed and seasoned the meat and then, just before placing it into the roasting pan, cut off one end of the brisket and put it into the pan separately.

"Mom," asked Denise, "why did you cut off the end of the meat before putting it into the pot?"

"Because your Grandma always does it that way and that's how I learned," replied her mother.

"But why does Grandma do it that way?"

"I don't know, but we'll ask the next time we visit her."

A few days later, they found themselves at Grandma's house. "Grandma, when you make a pot roast, why do you always cut off the end of the meat before putting it into the pot?" inquired Denise.

"Because that's how your great-grandmother used to do it," replied Grandma. "Why don't we call and ask her? She should be at home." And she was.

"Hello, Great-grandma. I have a question for you," Denise said breathlessly. "Why did you use to cut the end off of the brisket whenever you made a pot roast? Mom and Grandma and I all want to know!"

"That's easy, sweetie," was the instant reply. "My pot was too small for the roast!"

No doubt, Great-grandma also cooled freshly cooked food on the kitchen counter or window ledge before putting it into the refrigerator. She learned this from her own mother in the days when kitchens had iceboxes instead of refrigerators. Putting a hot dish of food directly into the icebox—an insulated cabinet cooled by a large block of ice—was a recipe for disaster. The heat melted the ice and, unless the ice could be replaced promptly (not an easy task—ice was usually delivered only once a day), the food in the icebox spoiled due to lack of adequate cooling. Hot dishes had to be cooled to room temperature on the kitchen counter or in front of an open window before being put away. And the first refrigerators were not much better.

Salmonella—a Zoonosis for All Seasons

A zoonosis is a disease caused by a human pathogen that can also be hosted by one or more animals, with or without causing disease in those animals. Host animals serve as reservoirs and sometimes as vectors of zoonotic human pathogens (16).

While many food-borne pathogens are zoonotic, their host ranges vary enormously. Hepatitis A virus, for example, only infects primates; *Cryptosporidium parvum*, a protozoan parasite, infects a variety of mammals, but not birds, fish, or reptiles (2, 13–15). *Salmonella*, with a host range that encompasses mammals, other vertebrate and invertebrate animals (including birds, reptiles, and fish), and insects, is one of the most versatile and widespread zoonoses (19). Infected animals—often asymptomatic—transmit the pathogen to other animals and to humans by shedding *Salmonella* in their feces (6, 28).

Members of the genus *Salmonella* have been recognized as both human and animal pathogens for more than a century. *Salmonella enterica* serotype Typhi, one of the few members of this genus that is not zoonotic, was cultured for the first time in 1884 (4). *S. enterica* serotype Choleraesuis was the first representative of this genus to be isolated from a nonhuman host. The cause of hog cholera, it was isolated from a diseased pig in 1885 (8).

Whether or not a pathogen is zoonotic, and how wide a range of animal hosts it can infect, can have a significant impact on attempts to control the pathogen's spread to humans. Microbes with a limited host range—*Vibrio parahaemolyticus*, for example, which is found in seafood harvested from coastal marine waters—may only be a problem for certain sectors of the food industry, in relatively specific geographical areas, and at certain times of the year (7). On the other hand, *Salmonella*, with its broad spectrum of available hosts, must be guarded against year-round by anyone who handles food or animals anywhere in the world.

Unfortunately, many food handlers have never realized that, just like the size of Great-grandma's roasting pan, the cooling capacity of refrigerators has changed. Modern commercial and household refrigerators can easily handle a hot item without endangering the other foods. Allowing a hot dish to cool on the countertop is no longer advisable; in fact, it can be downright dangerous. A group of school kids, teachers, and cafeteria workers found that out in the spring of 1986.

April Fool!

On March 31, 1986, the workers in an Oklahoma school district food service kitchen began preparing chicken to be served in four school cafeterias. They started by allowing the frozen chicken to thaw overnight at room temperature. On April 1, some of the thawed chicken was cooked in pans of water in a 350°F (177°C) oven. The oven was turned off after 2 hours and the chicken was left overnight in the warm oven. The kitchen workers cooked the rest of the chicken in a steam cooker for 2 hours, then readjusted the cooker to the lowest setting and allowed the food to remain in the warm cooker overnight. The chicken was delivered to the four cafeterias on April 2.

The outbreak erupted that same afternoon. Students, teachers, and cafeteria workers—no one was spared. Victims variously reported experiencing nausea, vomiting, cramps, or fever; many suffered from a combination of two or more symptoms. Twenty-two ended up in the hospital. In all, more than 200 people were stricken with *Salmonella* food poisoning courtesy of the chicken. Fortunately, everyone survived (5).

How could this have happened? Shouldn't chicken cooked for 2 hours in a 350°F (177°C) oven be safe? Usually it is, if the food is handled properly in all other respects. But in this case, just about every rule for safe handling was broken. When interviewed by the county and state health authorities, the cafeteria workers had no idea that they had done anything wrong. In fact, they made three major errors.

Their first mistake was to thaw the chicken at room temperature overnight. Of course, the frozen meat didn't thaw uniformly. The outer surface thawed first, followed by the interior. Once the surface warmed, the bacteria on it began to multiply. And the *Salmonella* bacteria—guaranteed to be on at least some of the chicken—generated millions of offspring by morning.

Having succeeded in producing chicken laden with salmonellae, the food handlers made their second mistake. No one thought to verify the cooking procedure or bothered to check the temperature of the cooked chicken with a meat thermometer. The investigation report does not state how tightly the chicken was packed into water-filled pans for the 2-hour cooking at 350°F (177°C). Nor do we know how full the steam cooker was. But it's a pretty good guess that the chicken did not cook evenly. One "cold spot" or undercooked area would have been enough to allow a few salmonellae to survive.

As for their third error, the kitchen staff allowed the cooked chicken to sit in a warm oven and warm steamer overnight. *Salmonella* reproduces best at or near body temperature, 95 to 100°F (35 to 38°C). Under these cozy conditions, it can double its population every 20 to 30 minutes. In 12 hours, a single *Salmonella* organism is able to generate more than 10 million offspring. Thus, even if only a few salmonellae managed to survive the cooking process, the chicken could easily have been swarming with salmonellae by the next day. All told, a perfect recipe for a bacterial picnic.

Once they determined the likely source of the outbreak, health authorities took corrective action. Cafeteria workers suffering from diarrhea were not permitted to return to their jobs until they were symptom free. All of the cafeteria workers were instructed on proper hand washing and personal hygiene. They were also taught to thaw frozen meat in the refrigerator, to always check the internal temperature of cooked meat with a meat thermometer, and to store cooked foods either hot (above 140°F [60°C]) or cold (below 34.5°F [4.5°C]) to minimize bacterial growth. The cafeteria workers should have been taught these simple rules long before the outbreak. Had the workers been properly trained when they were first hired, 200 people most likely would have been spared the agonies of *Salmonella*

Degrees of Confusion

It would be an exceptional microbe that could withstand 350°F (177°C), even for a second or two. So how can a pathogen such as *Salmonella* (which is not especially heat tolerant) survive 2 hours or more in a 350°F (177°C) oven? In fact, the only thing inside an oven that reaches the nominal set temperature is the air. A roast leg of lamb, for example, is considered to be "well done" when its internal temperature hits 170°F (77°C) (12). Even water only reaches 212°F (100°C) before it boils—unless it is high in solutes or is in a pressurized container.

The temperature at which an oven is set is a very poor predictor of the final core temperature of the food. Several additional factors influence the outcome of a cooking procedure, including the reliability of the oven thermostat, the uniformity of heat distribution in the oven, the density and thickness of the food being cooked, the length of time allowed for cooking, the ability of air to circulate around the food, and evaporative cooling. An oven thermostat that is out of calibration by 25 or 50°F (14 or 28°C) will alter noticeably the length of time required to cook a food.

Many ovens, especially older ones, also suffer from uneven heat distribution. The baking element is usually at the bottom of the oven. When the oven thermostat calls for heat, the element comes on and remains on until the thermostat senses that the set temperature has been reached. This produces a temperature gradient inside the oven, with the hottest area on the bottom. Convection ovens, which incorporate a fan to circulate the air, were developed to eliminate the uneven heat distribution. Some convection oven designs achieve a very uniform temperature distribution, whereas others are less effective.

The way in which food is placed inside an oven also affects heat distribution. Large trays or pans that fill entire shelves right to the oven walls impede the circulation of air, even in a convection oven. Lining shelves with aluminum foil also inhibits airflow, resulting in uneven heating. Squeezing as much food as possible into a cooking pan, or covering the food tightly with foil, prevents air from circulating efficiently around the food. Finally, evaporative cooling—the reduction in temperature that takes place when water evaporates—lowers the temperature of the surface of food, slowing the rise in the food's internal temperature.

Ensuring that food has been cooked to a safe internal temperature is more complicated than putting the food into a pan, setting the oven temperature, and cooking the food for a fixed time. The only way to be certain that food has been cooked adequately is to use one or more meat thermometers, placed in the thickest, most dense parts of the food—the areas that are likely to be the last to reach the target temperature. Relying solely on past experience to determine when food should be removed from an oven is truly a recipe for disaster.

food poisoning, students and teachers would not have missed classes. And the school's insurer would not have been out $40,000 in medical expenses.

Unfortunately, lack of proper instruction for food service workers is the rule, rather than the exception. Consequently, outbreaks due to errors in food handling occur with sickening regularity. The customers and owner of Danny's Deli found this out the hard way in 1993.

Danny's Deli

St. Patrick's Day is a major event for a catering delicatessen famous for its corned beef. Danny's Deli was known throughout Cleveland for the quality and flavor of its signature meat. On March 12, 1993, the deli began to prepare and stockpile meat for the anticipated St. Patrick's Day demand.

Danny's cooked the corned beef briskets by boiling them for 3 hours (29). The cooked meat was allowed to cool at room temperature, after which it was refrigerated until the St. Patrick's Day "rush" on March 16 and 17. To prepare the meat for serving, briskets were removed from the refrigerator and placed in a warmer maintained at 120°F (49°C). Each brisket was then sliced and served. Some of the meat was also used on March 17 to prepare sandwiches for catering. These sandwiches were made around 11 a.m. and held at room temperature until they were eaten during the course of the afternoon (29).

On March 18, the Cleveland City Health Department began receiving telephone calls from individuals stricken with food poisoning. In all, 15 calls were made to the health department that day, reporting approximately 150 cases of illness. Health officials responded promptly with a temporary closure of Danny's pending a full inspection of the facilities and a review of its food handling practices. The deli was permitted to reopen for business the following day.

The Ohio Department of Health, which analyzed the suspect meat, reported that it contained a high concentration (more than 100,000 organisms per gram of food) of *Clostridium perfringens*, a bacterium known to cause food poisoning and often associated with this type of outbreak. The bacteria produce heat-resistant spores that could have easily survived the 3-hour boiling process used to prepare the corned beef. But those spores would not have caused a problem had the boiled meat been refrigerated immediately after cooking.

A reporter from *The Plain Dealer*, a Cleveland daily newspaper, interviewed Danny's owner George Georges on the day the deli reopened (22). According to Mr. Georges, the food poisoning was caused by his having refrigerated the meat too soon after cooking. He was reported by the newspaper to have said that "... an inspector who visited the site Thursday told him the problem might have been the result of not allowing cooked corned beef to stand before refrigerating it." The reporter added that, in the owner's opinion, "... about 60 pounds of the meat apparently was refrigerated too soon after cooking. By doing that, the cooking process stopped before all the bacteria were destroyed."

In fact, the true cause was exactly the opposite. Spores are produced by bacteria as a survival mechanism. They are designed to survive harsh conditions such as high temperatures and to germinate and grow once the conditions are favorable. Had the meat been refrigerated immediately after cooking, the spores would have remained inactive and the meat would have been safe to eat. Allowing the cooked corned beef to cool slowly at room temperature ensured that the meat would remain at a favorable bacterial growth temperature for an extended period of time—long enough to produce a dangerous level of *C. perfringens* in the meat. Once the chain of events leading to the outbreak was established, representatives of the City Health Department provided recommendations to the deli on how to improve its handling practices.

Fortunately for the victims and for Danny's Deli, *C. perfringens* is not one of the more dangerous food poisoning bacteria. Its main symptoms are acute diarrhea, abdominal cramps, and vomiting. The illness typically runs its course in about 24 hours. No one died as a result of this outbreak; no one was even hospitalized. However, more than 150 individuals and their families were severely inconvenienced, and many probably lost a day of work. In describing her symptoms, one victim said, "It was pretty bad. I was crawling on the floor saying, 'God, if you just let me live I'll be a better person'" (23).

The most disturbing aspect of this story is that the owner understood the health inspector to state that the food was cooled too soon. Health inspectors are expected to have the correct information and to communicate it clearly. While we can, perhaps, understand the lack of knowledge shown by the owner, the inspector's poor performance was inexcusable. We expect our health professionals to have mastered their profession. But even the health care industry is susceptible to unsafe practices.

Powdered Infant Formula and the Newborn

In April 2001, a 20-day-old baby boy died in a hospital in Tennessee (17). He had been born prematurely by cesarean section and placed in the hospital's neonatal intensive care unit (NICU). At 11 days, he was suffering from a variety of symptoms, including fever and neurological abnormalities. Lab cultures established that he had contracted meningitis caused by a bacterium called *Enterobacter sakazakii*. His doctors tried, unsuccessfully, to treat the infection with antibiotics.

On learning of the *E. sakazakii* infection, hospital personnel screened the other 48 infants under NICU care to find out whether any of them were carrying this same microbe. The bacterium was found in specimens from 9 of the 48 infants tested. But where did it come from? And how did it infect such a large portion of the NICU population? Premature and underweight infants represent one of the most susceptible populations to infection. It was imperative that the source of the infection be uncovered.

The hospital began by comparing the records of the 9 infected infants—including the baby boy who had died—with those of the 40 infants who were resident in the NICU at the same time but showed no sign of *E. sakazakii* in their specimens. After reviewing all possible variables, hospital personnel could find only one thing that the nine infected babies had in common; they had all been fed Portagen, a powdered infant formula product made by Mead Johnson. However, even this was not fully conclusive, since 21 of the 40 unaffected babies had also received Portagen. The hospital continued its investigation.

Everything used to prepare the powdered formula for feeding came under suspicion. Microbiology lab personnel tested the water in which the powder was dissolved. They analyzed samples from opened cans of two different batches of formula that had been in use in the NICU during the time period under study. They also sampled unopened cans from both batches and performed tests on the countertops where the formula had been prepared for feeding. While the lab tests were under way, hospital personnel carried out an intensive review of all infection-control practices in the NICU and all preparation protocols and records for the powdered formula.

The review of practices and procedures turned up nothing. Everything had been done by the book. Formula had been prepared and stored according to the manufacturer's instructions. The infant who died from meningitis had been fed the formula continuously by tube, and the "hang time" (time that a container of formula was allowed to hang at room temperature during feeding) had not exceeded the 8 hours specified in hospital policy (17).

Fortunately, the lab investigation provided answers. One of the two batches of Portagen in use in the NICU during the time of the outbreak was contaminated with *E. sakazakii*. The microbe was present even in unopened cans of the powdered infant formula. All of the environmental samples were negative, as was the water used to prepare the formula. On learning these results, the hospital immediately made several changes to its practices; it switched from powdered formula to a ready-to-use liquid product, limited the use of powdered formula to certain specific situations, and reduced the "hang time" for continuous tube feeding of formula to 4 hours from 8 hours.

Mead Johnson recalled the contaminated batch of Portagen on March 29, 2002, nearly one full year after the initial outbreak (24). As a result of this incident, both the American Dietetic Association (ADA) and the U.S. Food and Drug Administration (FDA) modified their recommendations for preparation, use, and storage of powdered formula (1, 10). In addition, FDA microbiologists developed a recommended method for detecting *E. sakazakii* in powdered infant formula and published the details on their website. At the same time, FDA let it be known that searching for *E. sakazakii* would become part of its standard protocol when inspecting infant formula manufacturers (9).

Very few people outside the microbiology community have heard about *E. sakazakii*. Some might even consider it to be one of the "emerging pathogens" that turn up from time to time and are impossible to predict. But *E. sakazakii*—and the harm it can cause—has been well known for many years, although it used to travel under a different name, "yellow-pigmented *Enterobacter cloacae*." Several researchers in North America and Europe had already made the connection between infant formula, *E. sakazakii*, and meningitis in infants.

The very first reports linking meningitis in infants to this microbe appeared in the early 1960s (20). By 1981, the ability of *E. sakazakii* to cause fatal meningitis was confirmed by researchers with the Indiana University School of Medicine (21). In 1983, a group of Dutch researchers drew the first tentative conclusion linking the infection to infant formula (25). This was corroborated by the results of a detailed investigation of an Icelandic outbreak, carried out with the cooperation of a representative of the Centers for Disease Control and Prevention (CDC). The report of this investigation appeared in the *Journal of Clinical Microbiology* in 1989 (3).

How common is *E. sakazakii*? A 1988 study evaluated 141 samples of powdered formula obtained in 35 countries (26). The researchers found low levels of *E. sakazakii* in 20 of those samples, from 13 countries. None of the results exceeded the standards of the Food and Agricultural Organization of the United Nations in force at that time for powdered infant formula. However, even very low levels of a dangerous bacterium can grow to high numbers if the conditions are right.

At the time of the 2001 Tennessee outbreak, standard practices allowed reconstituted powdered formula to remain at room temperature for up to 8 hours while an infant was being fed continuously by tube. Yet in 1997, Canadian researchers reported that *E. sakazakii* could begin to multiply in reconstituted formula after only 2.7 hours at room temperature (27). In addition, they established that once it began to grow, *E. sakazakii* could double in population under these conditions every 40 minutes. Allowing for the 2.7-hour lag time, a single *E. sakazakii* organism could produce as many as 256 offspring under the conditions of use still recommended in 2001 by the ADA and the FDA. If the 4-hour limit later introduced had been in effect at the time of the 2001 outbreak, that same *E. sakazakii* cell would only have had time to produce four offspring.

One might argue that the results of a single research report would not have been compelling enough to warrant a major policy shift. However, the Canadian research had been triggered by several clinical reports, issued over a period of years, of death or lifelong disability resulting from *E. sakazakii* infections. But these reports and the results of the Canadian research study appear to have passed beneath the radar screen of the ADA, food safety regulators, and infant formula manufacturers both in North America and in Europe. In March 2002, a Belgian baby fell ill and died shortly after being released from the hospital at 5 days of age. The source of the baby's *E. sakazakii* meningitis was traced to a batch of Nestlé's Beba powdered formula. Nestlé recalled two production lots of the formula in May 2002 (18).

Fortunately, the FDA responded to the lessons learned from the Tennessee outbreak by making some policy changes, including increasing surveillance of infant formula manufacturers, with specific emphasis on *E. sakazakii*. This heightened awareness has produced at least one recall of several lots of powdered formula produced by Wyeth Nutritionals and sold under various names. And this time, the FDA caught the problem before any infants were put at risk (11).

All of us—consumers, food handlers, and food safety professionals—are subject to the "old habits" syndrome. But we cannot afford to be complacent. To ensure the safety of our food and water supply, we must always look to our past experiences to teach us the best and safest ways to produce, prepare, and store food. Just because Great-grandma cut off the end of the brisket, that doesn't mean we have to.

References

1. **American Dietetic Association.** 2003. *Guidelines for Preparation of Formula and Breastmilk in Health Care Facilities.* The American Dietetic Association, Chicago, Ill. [Online.] http://www.eatright.org/cps/rde/xchg/ada/hs.xsl/nutrition_1562_ENU_HTML.htm. Accessed 10 May 2006.
2. **Balayan, M. S.** 1992. Natural hosts of hepatitis A virus. *Vaccine.* **10:**S27–S31.
3. **Biering, G., S. Karlsson, N. C. Clark, K. E. Jonsdottir, P. Ludvigsson, and O. Steingrimsson.** 1989. Three cases of neonatal meningitis caused by *Enterobacter sakazakii* in powdered milk. *J. Clin. Microbiol.* **27:**2054–2056.
4. **Brock, T. D.** 1999. *Robert Koch. A Life in Medicine and Bacteriology.* ASM Press, Washington, D.C.
5. **Carr, R., D. O. Coweta, S. Brown, A. Goodall, D. Head, B. Stacy, R. Bryce, T. Hill, and G. Istre.** 1987. Epidemiologic notes and reports. Salmonellosis in a school system—Oklahoma. *Morb. Mortal. Wkly. Rep.* **36:**74–75.
6. **De Jong, B., Y. Andersson, and K. Ekdahl.** 2005. Effect of regulation and education on reptile-associated salmonellosis. *Emerg. Infect. Dis.* **11:**398–403.
7. **DePaola, A., C. A. Kaysner, J. Bowers, and D. W. Cook.** 2000. Environmental investigations of *Vibrio parahaemolyticus* in oysters after outbreaks in Washington, Texas, and New York (1997 and 1998). *Appl. Environ. Microbiol.* **66:**4649–4654.
8. **Doyle, M. P., L. R. Beuchat, and T. J. Montville (ed.).** 1997. *Food Microbiology: Fundamentals and Frontiers.* ASM Press, Washington, D.C.
9. **Food and Drug Administration.** 2002. Isolation and enumeration of *Enterobacter sakazakii* from dehydrated powdered infant formula. U.S. Food and Drug Administration, Center for Food Safety and Applied Nutrition. [Online.] http://www.cfsan.fda.gov/~comm/mmesakaz.html. Accessed 10 May 2006.
10. **Food and Drug Administration.** 10 October 2002. Health professionals letter on *Enterobacter sakazakii* infections associated with use of powdered (dry) infant formulas in neonatal intensive care units. U.S. Food and Drug Administration, Center for Food Safety and Applied Nutrition, Office of Nutritional Products, Labeling and Dietary Supplements. [Online.] http://www.cfsan.fda.gov/~dms/inf-ltr3.html. Accessed 10 May 2006.
11. **Food and Drug Administration.** 1 November 2002. FDA alerts public regarding recall of powdered infant formula. FDA News Release P02-46. U.S. Food and Drug Administration. [Online.] http://www.fda.gov/bbs/topics/NEWS/2002/NEW00849.html. Accessed 10 May 2006.

12. **Food Safety and Inspection Service.** 2006. Roasting those "other" holiday meats. Food Safety and Inspection Service, U.S. Department of Agriculture. [Online.] http://www.fsis.usda.gov/Fact_Sheets/Roasting_Those_Other_Holiday_Meats/index.asp. Accessed 10 May 2006.

13. **Graczyk, T. K., M. R. Cranfield, R. Fayer, and M. S. Anderson.** 1996. Viability and infectivity of *Cryptosporidium parvum* oocysts are retained upon intestinal passage through a refractory avian host. *Appl. Environ. Microbiol.* **62:**3234–3237.

14. **Graczyk, T. K., R. Fayer, and M. R. Cranfield.** 1996. *Cryptosporidium parvum* is not transmissible to fish, amphibians, or reptiles. *J. Parasitol.* **82:**748–751.

15. **Guerrant, R. L.** 1997. Cryptosporidiosis: an emerging, highly infectious threat. *Emerg. Infect. Dis.* **3:**51–57.

16. **Hendriksen, S. W. M., K. Orsel, J. A. Wagenaar, A. Miko, and E. van Duijkeren.** 2004. Animal-to-human transmission of *Salmonella* Typhimurium DT104A variant. *Emerg. Infect. Dis.* **10:**2225–2227.

17. **Himelright, I., E. Harris, V. Lorch, M. Anderson, T. Jones, A. Craig, M. Kuehnert, T. Forster, M. Arduino, B. Jensen, and D. Jernigan.** 2002. *Enterobacter sakazakii* infections associated with the use of powdered infant formula—Tennessee, 2001. *Morb. Mortal. Wkly. Rep.* **51:**298–300.

18. **International Baby Food Action Network.** 10 May 2002. How safe are infant formulas? The death of a one-week-old formula-fed baby in Belgium. IBFAN Press Release. [Online.] http://www.ibfan.org/english/news/press/press10may02.html. Accessed 10 May 2006.

19. **Jay, J. M.** 2000. *Modern Food Microbiology*, 6th ed. Aspen Publishers, Inc., Gaithersburg, Md.

20. **Joker, R. N., T. Norholm, and K. E. Siboni.** 1965. A case of neonatal meningitis caused by a yellow enterobacter. *Dan. Med. Bull.* **12:**128–130.

21. **Kleiman, M. B., S. D. Allen, P. Neal, and J. Reynolds.** 1981. Meningoencephalitis and compartmentalization of the cerebral ventricles caused by *Enterobacter sakazakii*. *J. Clin. Microbiol.* **14:**352–354.

22. **Marrison, B.** 20 March 1993. Deli gets OK from city to reopen. *The Plain Dealer*, Cleveland, Ohio.

23. **Marrison, B., and C. L. Kissling.** 19 March 1993. Tainted food suspected after 150 are sickened. *The Plain Dealer*, Cleveland, Ohio.

24. **Mead Johnson Nutritionals.** 29 March 2002. Portagen powder recall. News Release. [Online.] http://www.fda.gov/oc/po/firmrecalls/meadjohnson03_02.html. Accessed 10 May 2006.

25. **Muytjens, H. L., H. C. Zanen, H. J. Sonderkamp, L. A. Kollee, I. K. Wachsmuth, and J. J. Farmer III.** 1983. Analysis of eight cases of neonatal meningitis and sepsis due to *Enterobacter sakazakii*. *J. Clin. Microbiol.* **18:**115–120.

26. **Muytjens, H. L., H. Roelofs-Willemse, and G. H. J. Jaspar.** 1988. Quality of powdered substitutes for breast milk with regard to members of the family *Enterobacteriaceae*. *J. Clin. Microbiol.* **26:**743–746.

27. **Nazarowec-White, M., and J. M. Farber.** 1997. Incidence, survival and growth of *Enterobacter sakazakii* in infant formula. *J. Food Prot.* **60:**226–230.

28. **Van Immerseel, F., F. Pasmans, J. De Buck, I. Rychlik, H. Hradecka, J.-M. Collard, C. Wildemauwe, M. Heyndrickx, R. Ducatelle, and**

F. Haesebrouck. 2004. Cats as a risk for transmission of antimicrobial drug-resistant *Salmonella*. *Emerg. Infect. Dis.* **10:**2169–2174.

29. **Zimomra, J., T. Wenderoth, A. Snyder, R. Russ, E. D. Peterson, R. French, T. J. Halpin, J. E. Florance, A. Adkins, J. Andrew, M. Burkgren, K. Crisler, T. Fagen, L. Fass, J. M. Galloway, S. Haines, R. H. Hinton, C. Jackson, N. S. Rivera, E. L. Testor, C. Williams, A. A. DiAllo, D. R. Patel, C. W. Armstrong, D. Woolard, and G. B. Miller.** 1994. *Clostridium perfringens* gastroenteritis associated with corned beef served at St. Patrick's Day meals—Ohio and Virginia, 1993. *Morb. Mortal. Wkly. Rep.* **43:**137–138, 143–144.

chapter 2

Engineering Errors

NAPOLEON BONAPARTE DIDN'T WIN BATTLES by neglecting his soldiers. He knew that "an army marches on its stomach." In 1795, in an attempt to keep that collective stomach filled, Napoleon announced a competition. The person who devised a practical method for preserving food so that it could travel with his armies would win a prize of 12,000 francs (8).

This challenge pricked the interest of Nicholas Appert, a cook and confectioner. Appert, along with most of his contemporaries, believed that contact with oxygen caused food to spoil. He filled glass containers with food, sealed them with wax, and placed them in boiling water. As the containers heated up, their contents expanded, driving out the air. Then, as the containers cooled, the volume of air that was left inside the containers shrank, drawing the wax seals tightly against the openings and creating a partial vacuum inside the jars. The food inside the sealed glass containers remained unspoiled even after more than 4 months of storage at sea. A grateful Napoleon awarded Appert his prize in 1809.

Unfortunately for the French soldiers, Appert's invention did not prevent Napoleon's army from starving during the long retreat from Russia late in 1812. Nevertheless, thanks to Bonaparte's stimulus, Appert had set the stage for modern food processing. But it was Louis Pasteur's careful experiments on the spoilage of beer and wine that provided the true explanation for the success of appertization. Pasteur's work illuminated the microbiology behind Appert's successful invention.

Appert had an excellent excuse for not understanding the microbiology behind his own success. After all, Pasteur was born in 1822—fully 13 years after Appert received his prize—and only began working with microbes in 1854. His first microbiology-related research paper, an explanation of the nature of fermentation, saw the light of day in 1857, 50 years after Appert's landmark invention (21).

Unlike Appert, today's food engineers have no excuse to be ignorant of microbiology. Yet some present-day food engineers still turn a blind eye to the microbiological implications of their designs in the name of efficiency—sometimes to the long-term cost of their employer's reputation and business, and to the detriment of the health and safety of consumers.

Keep Your Powder Dry

Dried milk, one of the world's first engineered foods, is as old as the Mongol civilization. More than 700 years ago, in 1295, Marco Polo reported that the Mongolians boiled milk, skimmed off the fat that rose to the top to make butter, and dried the defatted milk in the sun (14). The Mongols learned through experience that dried milk was a safe, convenient, and stable food. It could be stored for long periods, transported over great distances, and reconstituted for drinking whenever water was available.

In some ways, powdered milk production hasn't changed over the last 700 years. We still heat the milk, skim off its fat to make butter, and then use more heat to dry the milk. But these days, the process is highly automated. Our engineering advances have enabled us to manufacture powdered milk under tightly controlled conditions. As a result, it should be an exceptionally safe product. After all, it is made from pasteurized milk and is too dry to support growth of even the hardiest mold. Yet, due to poor plant design, improper maintenance, and lack of focus on microbiology, milk powder has been the cause of several disease outbreaks and uncounted near-misses.

Nearly all of the powdered milk produced these days for human consumption is spray dried. The process of spray drying is itself more than 80 years old, and while equipment design has continued to evolve, the basic approach has not changed much. Production of milk powder requires several preliminary steps, including clarification, fat separation, pasteurization, and concentration of fluid milk (14). Concentrated milk enters the spray drier through an atomizer. Finally, powdered milk is recovered from the spray drier and packaged (Fig. 2.1). For instantized milk powder, an agglomeration step that follows spray drying reintroduces a small amount of moisture under controlled conditions, making the powder easier to dissolve in cold water.

Although a few sporadic cases of food-borne illness were linked to powdered milk as long ago as the 1940s and 1950s (47), it wasn't until 1965–1966 that the U.S. Food and Drug Administration (FDA) fully realized the risks associated with this product. In late 1965 and early 1966, *Salmonella enterica* serotype Newbrunswick sickened 29 people—12 of them under 1 year of age—in 17 U.S. states. Fortunately, no one died. Epidemiologists traced the outbreak to instant nonfat dry milk from a single drying and instantizing plant (15).

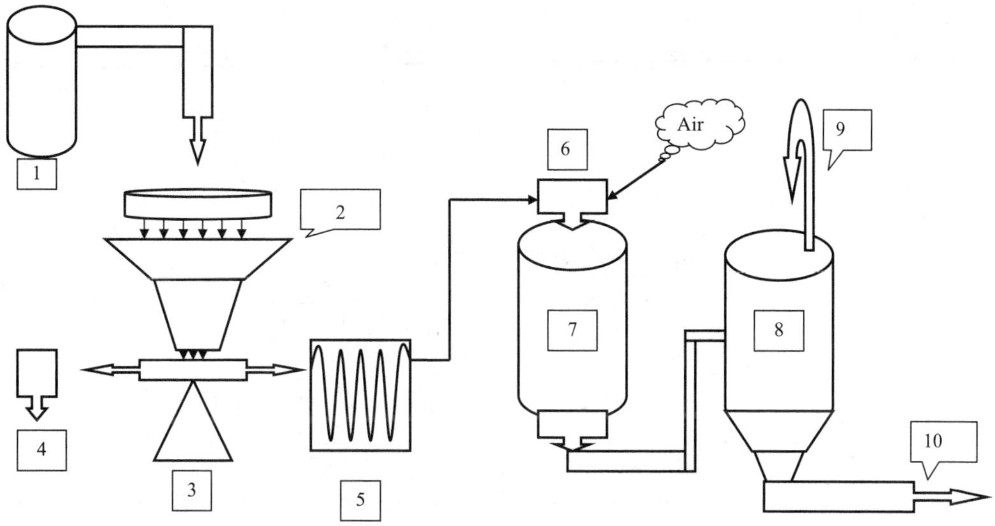

Figure 2.1 Skim milk powder production sequence. (1) Raw milk bulk tank. (2) Clarifier. (3) Separator. (4) Fat (for butter or blending into milk). (5) Pasteurizer. (6) Atomizer. (7) Spray dryer. (8) Cyclone recuperator. (9) Air exhaust. (10) Spray-dried milk to packaging.

Tracing the source of the contaminated milk product was only part of the battle. Two vital questions remained to be answered: "How did the *Salmonella* enter the plant, and where was it lurking?"

Federal investigators descended on the milk powder plant and began to dissect the production equipment and the processing methods. They determined that the processing plant was supplied with milk from as many as 800 farms, any of which—in theory—could have been the original source of contamination. Pasteurization should have eliminated any *Salmonella* organisms that might have been in the raw milk. However, since the plant's pasteurizer was not equipped with either time or temperature controls, there was no way to confirm that the raw milk had been pasteurized effectively. And without proper pasteurization, *Salmonella* could easily have survived to contaminate the spray-dried milk (42).

Investigators also found fault with the other production equipment and its maintenance. Several components of the spray-drying and instantizing systems were difficult to clean. Open seams, crevices, and poorly welded joints in various pieces of equipment were tailor-made refuges for *Salmonella*. The spray dryer was brushed out daily and wet-cleaned only once a month. The agglomerator was wet-cleaned once a week—not often enough to keep the sticky powder from building up inside the equipment.

Air intakes for the dryer and agglomerator also were a problem. The intakes were located inside the room that was being used for sifting and bagging milk powder. Dust produced by these operations clogged the air intake filters, requiring them to be changed frequently. On occasion, filters were removed and replaced while the drying equipment was operating, allowing fine milk dust to be sucked back into the spray dryer and the instantizer.

The owners cleaned and remodeled the plant to correct problems that the investigators had found. And the U.S. Department of Agriculture (USDA) undertook an extensive nationwide testing program in 1966 to determine the extent of *Salmonella* contamination in the U.S. powdered milk supply. To their dismay, they found that 13% of the plants they surveyed had produced at least one batch of finished product containing *Salmonella* (47).

Coincidentally, in 1965, Canada also experienced a *Salmonella* outbreak, traced to powdered milk from a Quebec producer. This time, the culprit was *S. enterica* serotype Newport. Investigators failed to find either the source of the pathogen or its reservoir of contamination inside the plant. They observed that the problem had developed even though the milk had been pasteurized correctly, and despite the air intakes and exhaust vents having been located on the roof to prevent recontamination from within the plant (41). A second outbreak, also involving serotype Newport, occurred in Newfoundland in 1968 (47). Canada responded by implementing a long-term powdered milk surveillance and inspection program.

A strange pattern emerged from some of the inspections that were carried out under this program in the province of Quebec. The investigation team would visit a processing facility and discover *Salmonella* in the equipment, the environment, and in some cases, the finished milk powder. Plant personnel would clean the facilities intensively, and when inspectors returned, they would find little or no evidence of contamination. Months or even years later, a follow-up inspection would determine that the very same *Salmonella* discovered initially in the plant had returned, even while plant sanitation continued to appear good (41). Where was the *Salmonella* hiding, and how did it outwit efforts to keep the plant clean? The answer lay on the roof.

Gerard Lachapelle, a microbiologist, usually accompanied the milk plant inspection teams. Lachapelle's role was to think like *Salmonella*, identify those plant locations that might shelter the microbe from the plant's cleaning and sanitizing procedures, and take samples from each and every suspicious location back to the lab for analysis. When he walked onto the roof of one plant, he was struck by the quantity of powdered milk that had accumulated on it.

The plant had installed a coarsely filtered air intake on the roof to provide fresh air to the spray dryer. An unfiltered air exhaust vent for the cyclone recuperator was also on the roof, and was located upwind of the air intake. Fine milk powder from the cyclone was expelled through the exhaust vent and accumulated on the roof between the air intake and the exhaust vent.

On rainy days, the accumulated powder turned back into fluid milk—an ideal growth medium for bacteria. All that was needed was a source of *Salmonella*—perhaps from a passing bird—to turn the milk into what Lachapelle described as an infected lake. On sunny days, the lake dried up and the milk turned powdery once more. In time, the *Salmonella*-infested milk powder was sucked back into the spray dryer through the coarsely filtered air intake. Lachapelle was able to recover *Salmonella* from inside the spray dryer, inside the cyclones, and on the roof (41).

What was Lachapelle's solution for this and other powdered milk facilities? He proposed that the entire production plant, including the roof, be cleaned intensively, and that the roof be sealed with a waterproof material. He also recommended that manufacturers institute a regular roof cleaning and maintenance program. Simple. And effective.

The Quebec powdered milk industry dodged a bullet. No massive outbreak of *Salmonella* was traced to its products after the 1960s. But milk powder continued to be a source of food-borne illness worldwide (Table 2.1).

Halfway around the world, Australia was also wrestling with *Salmonella* in powdered milk. In 1977, Jennifer Taplin, an alert microbiologist working for the Victorian Commission of Public Health, noticed an unusual rise in the number of *S. enterica* serotype Bredeney isolations (29). Epidemiological investigation pointed to a milk-based powdered infant formula as the source. Whereas three brands of formula were apparently involved, just one company manufactured the powdered milk used in all of the products.

An unannounced inspection of the manufacturing plant revealed that the cone-shaped base of the spray dryer had developed several cracks in the inner wall. These cracks allowed both moisture and powdered milk

Table 2.1 Outbreaks and recalls tied to powdered milk and powdered infant formula

Date	Pathogen[a]	Illnesses (deaths)	Country(ies) affected	Reference(s)
1966	Newbrunswick	29 (0)	United States	15
1977	Bredeney	>80[b] (0)	Australia	29
1977	Adelaide	3[c] (0)	Australia	29
1985	Ealing	76[d] (1)	United Kingdom	61
1993	Tennessee	3[c] (0)	Canada, United States	45
1994	Virchow	48 (0)	Spain	68, 69
1996–1997	Anatum	22 (0)	England, Scotland, France, Belgium	3

[a] *S. enterica* serotype.

[b] Total number of victims unknown. Approximately 80 symptomatic small children in the state of Victoria, including at least 53 under 3 years of age (J. Forsyth, personal communication).

[c] Including three symptomless excreters.

[d] Including 14 symptomless excreters.

to creep into the fiberglass insulation of the dryer. Company officials had been aware of the problem for a year and had tried to weld the cracks. When their welds failed, they simply discarded the portion of powdered milk that fell to the bottom of the cone-shaped dryer and sold the rest. By the time government investigators arrived at the plant, the fiberglass insulation had become extensively contaminated with serotype Bredeney. The microbe easily passed from the insulation back into the spray dryer, contaminating batch after batch of powdered milk destined for use in infant formula.

Sadly, the powdered milk industry is not alone in ignoring the microbiological consequences of its engineering and maintenance decisions. Fluid milk producers can be equally adept at turning a blind eye to bacteria.

Land of Milk and *Salmonella*

Thomas Levins was 44 when he died. Mary Kierzek was 61. Thomas died at home in his sleep. Mary passed away at the end of a 1-week hospital stay. They had only one thing in common—*Salmonella* (70, 74). Both were victims of the largest milk-related food poisoning outbreak in U.S. history.

The first hint of a problem surfaced in August 1984. Two hundred people in northern Illinois developed salmonellosis. The culprit, a strain of *S. enterica* serotype Typhimurium, had a very unusual pattern of resistance to antibiotics. Unfortunately, epidemiologists were never able to trace the source of the illness (62). Had they succeeded, the explosive 1985 outbreak might have been avoided.

On March 28, 1985, an infection-control nurse in northern Illinois came face-to-face with an old acquaintance—serotype Typhimurium. It had the identical antibiotic resistance pattern to the strain that was responsible for the August 1984 outbreak. When that same strain turned up in 31 additional patients the following day, investigators were called in immediately. Local and state health workers began interviewing victims to determine what they had in common. Suspicion quickly focused on pasteurized milk from Jewel Dairy. A single day's production of the 2% milkfat Bluebrook brand with a March 29 expiration date appeared to be the culprit.

News of the outbreak reached the public on April 2 when the *Chicago Tribune* carried a warning issued by Illinois' public health director (73). On April 8, health authorities realized that a second Jewel product—carrying the Hillfarm brand name and an April 8 expiration date—was contaminated with the same microbe. A third contaminated batch that had been produced on April 8 was recalled before it could reach consumers. But the damage had already been done.

In all, more than 16,000 people were diagnosed with salmonellosis after drinking the contaminated milk. *Salmonella* killed Thomas Levins and Mary Kierzek, and might have been a factor in 12 other deaths. Overall, more than 165,000 people were most likely affected—90% of whom never reported their illness to the health authorities.

The magnitude of the outbreak and the perishable nature of the product demanded an immediate and urgent response. Federal, state, university, and industry investigators probed the plant records for clues, dismantled the production equipment—not once, but twice—interviewed victims, and carried out lab analyses. They checked raw milk from Jewel's suppliers, searched for *Salmonella* in the dismantled equipment, and tested unopened containers of milk produced over the entire duration of the outbreak.

Only the pasteurized milk produced on March 20, March 30, and April 8 contained *Salmonella*. Approximately 5.5% of the raw milk samples were positive, but none contained the outbreak strain, nor could serotype Typhimurium be found in the plant equipment or anywhere inside the production facility. And, according to plant records, the milk had been pasteurized correctly.

Investigators were mystified until they unraveled the complexity of the plant design. In most fluid milk plants, pasteurization is one of the last steps before packaging. All of the other processing steps—clarification, separation, blending, and vitamin addition—take place before the milk is pasteurized. This sequence minimizes the risk of a contaminant sneaking back into the product after pasteurization. But Jewel had redesigned its plant in the mid-1970s, shifting the pasteurization step to near the beginning, rather than at the end, of the process (71). This change, made for reasons of production efficiency, exposed the milk needlessly to possible postpasteurization contaminants.

Jewel's plant had another problem as well. Faulty pipe design had created a cross-connection between the raw milk and pasteurized milk lines. The crossed pipes could have permitted raw milk to flow into the pasteurized milk line. In fact, the investigation team concluded that this cross-connection was the most probable source of contamination. Unfortunately, the team's theory could not be confirmed by lab results. The company had thoroughly cleaned and sanitized all of the production equipment and piping the day before the investigators arrived.

Could Jewel have done more to short-circuit the outbreak and to aid the investigation? Robert Riley, the person formerly in charge of the pasteurizer at the dairy, thought so. When testifying during the class action lawsuit brought against Jewel, he stated that the company regularly packaged 2% Bluebrook milk into Hillfarm containers. According to Riley, this was done on March 30, the day after Jewel found out about the initial outbreak (52). The Hillfarm 2% milk that had been produced on March 30 was implicated in the second phase of the *Salmonella* outbreak (62).

Jewel also had a chance to minimize the impact of the second phase of the outbreak—had they conducted *Salmonella* testing on milk produced on and after March 30. In hindsight, although finished product testing for *Salmonella* is not usually carried out for pasteurized milk (nor should it be necessary under normal circumstances), it would have been prudent to test each production lot until the extent of the problem was known. But the independent lab hired to investigate the outbreak on Jewel's behalf received no instructions to test milk produced after March 29. Had the milk produced on March 30 been tested immediately, a preliminary positive result might have been available as early as April 3. Instead, the *Salmonella*-contaminated milk remained on store shelves until its expiration date of April 8 (53).

The cross-connection between the raw and pasteurized milk lines in the Jewel dairy facility was in violation of U.S. and Illinois codes that prohibited dead-end piping (54). As a result of the *Salmonella* outbreak, the state of Illinois also decided to bar milk processors from using the production sequence used by Jewel (72). By engineering for efficiency, Jewel Dairy made tens of thousands of people ill and also engineered the shutdown of its own production plant. And by the company's other actions and omissions, Jewel broadened the scope and extended the duration of the outbreak.

Despite the lessons taught by the Jewel experience, pasteurized milk has continued to be the source of an occasional food poisoning outbreak, both in the United States (1, 55) and elsewhere (4, 19). Nor are the attitudes that led to the Jewel outbreak limited to the dairy industry.

A Dog's Breakfast

"Hot off the grill or cold out of the package, these hot dogs are great," was the message in the radio ads. But the wieners came with an unannounced bonus—coliforms, *Escherichia coli*, and *Listeria*. Not the best garnish for a cold wiener right out of the package!

In the early 1980s, a Canadian poultry processor was looking for a way to add value to the least desirable parts of the chicken—the backs and necks. Initially, the company introduced a mechanical deboning line into its plant and sold the raw, deboned meat. Later on, management decided to use the finely ground deboned meat to produce its own brand of chicken-meat hot dogs.

The company's engineer designed the production line, located in a suite of rooms within the raw poultry processing plant (Fig. 2.2). It seemed like a simple enough operation. Raw, deboned meat was combined with other ingredients in a large mechanical batch mixer and then fed into a continuous-feed sausage stuffer. The long links of sausages traveled on a suspended conveyer line into and through a smokehouse. At the smokehouse exit, the cooked wieners were sprayed with a brine solution to cool them quickly. Then the casings were stripped off and the hot dogs sealed into their retail packages. But somewhere along the way, bacteria were hitching a ride.

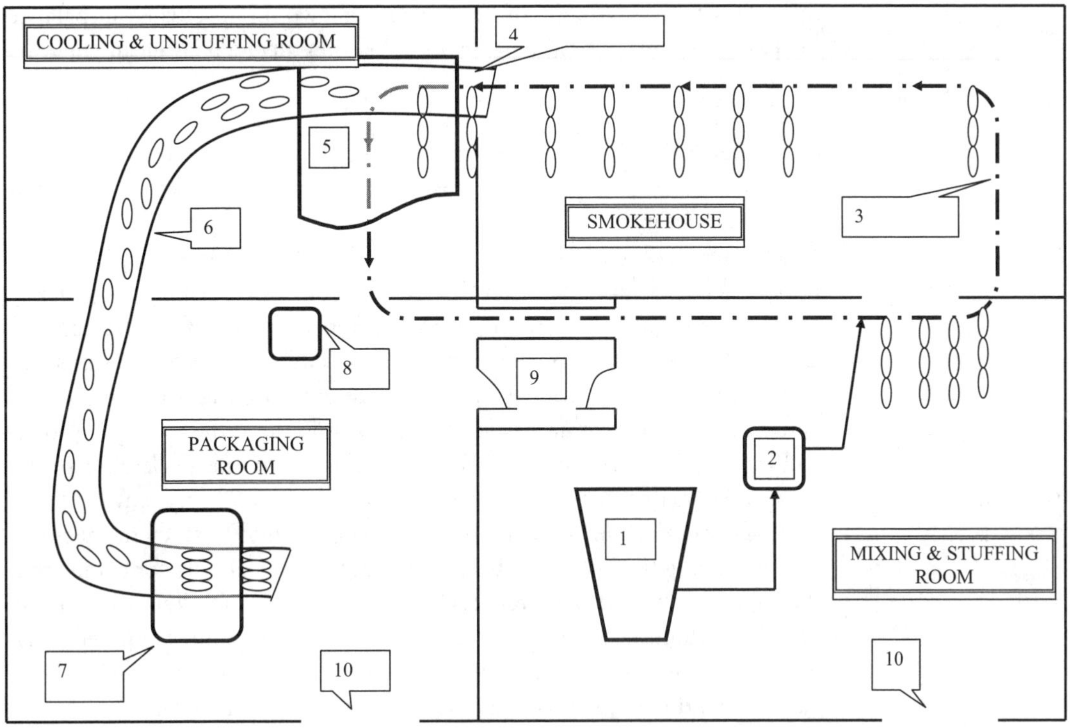

Figure 2.2 Wiener production plant. (1) Mixing bin. (2) Stuffer. (3) Suspended conveyer line. (4) Spray cooler. (5) Casing remover. (6) Wiener conveyer belt. (7) Packaging machine. (8) Brine tank. (9) Passageway between raw and cooked product rooms. (10) Overhead entry doors.

The first clues to the source of the bacteria came from testing packaged hot dogs from each day's production run. It took only a few weeks of testing for a pattern to emerge. Bacterial levels started out low and, over a period of 1 to 2 weeks, got progressively higher. Then the levels dropped all the way back down and the pattern repeated. Something was happening at 1- to 2-week intervals to create the pattern.

The answer—and the contamination—was in the cooling brine. The cold brine reservoir was an open tank located in the packaging room. The brine was pumped from the reservoir to the sprayers at the smokehouse outlet, and was sprayed over the hot dogs to cool them as they emerged from the smokehouse. Brine that dripped off the hot dogs drained down a pipe and back into the brine tank. In time, the brine accumulated bits of broken wieners—and millions of live bacteria.

The tank was cleaned at 1- to 2-week intervals. When the cleaning chart was compared to the bacterial count patterns, the matchup was nearly perfect. Every time the tank was emptied and cleaned, bacterial levels on the hot dogs dropped. The solution appeared to be obvious—to empty and clean the tank daily. But management balked. They feared that discarding a large volume of brine every day would upset the salt balance in their wastewater treatment facility. So they decided to clean the tank twice a week instead.

The change in cleaning routine helped but didn't solve the problem completely. Finally, management called in a consulting microbiologist. She visited the facility while it was operating and found several deficiencies in the design.

The most obvious problem was that equipment and people moved freely back and forth between the stuffing room (where the raw chicken and other ingredients were mixed in an open vat and stuffed into casings) and the packaging room. In addition, there was no air curtain or positive air pressure to keep airborne bacteria from entering the packaging room. The brine tank reservoir was located in the middle of the packaging room and had no lid. And the air conditioning unit for the packaging room, along with its drip pan, was located directly above the brine tank.

Inadvertently, the company had engineered a highly efficient means of contaminating the hot dogs. Results of microbiological tests confirmed that the wieners were virtually sterile when they exited the smokehouse, but the heavily contaminated cooling brine inoculated them almost immediately. And if the hot dogs avoided becoming contaminated during cooling, they had a second chance in the packaging room, where airborne contamination was high.

Management responded to their consultant's findings by modifying personnel traffic patterns and covering the brine tank. But they believed that the cost of relocating the air conditioning unit to the roof and increasing the airflow to create positive pressure in the packaging room would be prohibitive. Instead, the company chose simply to clean the air conditioner drip pan and to monitor the plant environment more closely.

The design of this facility was efficient—from an engineering and production perspective. Unfortunately, no one checked with a microbiologist before implementing the engineering plans. In consequence, the company was stuck with a bacterial time bomb that only constant vigilance and a large dollop of good luck prevented from exploding. However, BilMar Foods, a subsidiary of Sara Lee Corporation, was less fortunate.

It was the 1998 July 4 weekend, but inside the BilMar plant in Zeeland, Mich., a shower of sparks was the only fireworks display employees saw. The plant was shut down for the holiday, and the company had decided to replace an aging and leaky air conditioning unit. Dust from the dismantling operation spread through the area as workers cut the offending unit into pieces small enough to haul away (57).

The obsolete air conditioner had been the source of problems for several months. In November 1997, USDA inspectors shut down the plant because condensation from the unit was dripping onto the hot dog production line.

Repairs helped for a while, but the condensation returned. It was time to replace the unit.

When the plant started back up after the holiday weekend, a new refrigeration unit was cooling the hot dog production plant. But the condensation was still there. And now another problem began to show itself—unusually large numbers of bacteria were turning up in the weekly environmental test samples. In the 6 weeks after the new unit was installed, 11 out of 12 samples tested positive, compared to only 3 positive samples out of 12 in the 6 weeks before the old air conditioning equipment was removed. Management solved that problem quickly. They terminated the environmental testing program (51).

However, doing away with testing did not prevent condensation from dripping onto the production line. USDA inspectors wrote citation after citation—at least 45 by the end of 1998. And the line kept running—20 hours a day, 7 days a week—until December 22.

While BilMar's management concentrated on ignoring the bacteria that were hitching a ride on their hot dogs and deli meats, people began to fall ill. The first case of listeriosis turned up in July, just a few weeks after the refrigeration unit was replaced. But one isolated case wasn't enough to get an outbreak onto the radar screen. In October, the U.S. Centers for Disease Control and Prevention (CDC) began to receive reports and inquiries from state health agencies. Suddenly, an unusual number of people were being stricken with *Listeria monocytogenes* infections. And some of them were dying. By mid-November, CDC knew it had a major outbreak on its hands. By December 2, the case tally had risen to 40 people from 10 states.

The outbreak investigation proceeded along two parallel tracks. Lab investigators, using pulsed-field gel electrophoresis and ribotyping methods, determined that one single strain of *L. monocytogenes* was responsible for all of the illnesses. At the same time, epidemiologists interviewed patients and their families in order to establish the probable source of the microbe. They also interviewed healthy individuals and compared their answers with those of the victims. The results of this case-control study pointed to hot dogs manufactured in BilMar's Zeeland, Mich., plant (9). Discovery of the same strain of *L. monocytogenes* in an unopened package of BilMar hot dogs confirmed the investigators' suspicions (11).

Sara Lee, BilMar's parent company, announced a product recall on December 22. Unfortunately, the USDA didn't follow suit with its own publication of the recall until January 28 (28). By the time the outbreak was over, 21 people were dead—including 6 stillborn or miscarried infants—and BilMar products had caused more than 100 cases of listeriosis (10).

As is so often the case, investigators were unable to prove conclusively that the source of the *Listeria* contamination was triggered by the maintenance work carried out over the July 4 weekend. All the circumstantial

Listeria monocytogenes—Minor Menace or Major Pathogen?

Since it was first reported as a cause of human illness in 1929, *L. monocytogenes* has made a habit of periodically emerging and fading away from medical consciousness, only to emerge once again years or decades later (7, 32, 40). As long ago as the 1950s, this pathogen was known to produce just mild flu-like symptoms in most victims, to cause severe illness—even death—in the elderly, and to provoke miscarriages in pregnant women (32). By the mid-1960s, the first hints appeared that *L. monocytogenes* might be transmitted by ingestion of contaminated raw milk (32). But the microbe, which only appeared to cause occasional, sporadic illnesses and small outbreaks in the 1960s and the 1970s, received relatively little attention until the early 1980s (2, 6, 13, 25, 30, 33, 35, 39, 58, 59).

A 1981–1982 outbreak, traced to coleslaw consumed in the Canadian Maritime Provinces, established unequivocally the food-borne nature of listeriosis (65). This episode was followed by others: a 1983 outbreak in Massachusetts that was linked to contaminated pasteurized milk, and a 1985 Los Angeles outbreak caused by ingestion of contaminated Mexican-style soft cheese (26, 43). Nor was the disease limited to North America. In the decade following the Los Angeles incident, sporadic cases and outbreaks of human listeriosis also cropped up in the United Kingdom, several European countries, and Australia (16, 17, 24, 36, 38, 44, 48, 49, 56, 63, 64).

At first, regulators in several countries, including the United States and the United Kingdom, responded to the risk of food-borne listeriosis by implementing a "zero tolerance" policy for the pathogen (60, 66, 67). Under this policy, the detection of even a single viable *L. monocytogenes* in a 10-g or a 25-g sample of food could trigger an investigation and a recall of the contaminated food. This approach was consistent with regulatory policies that were already in place for dealing with other food-borne pathogens such as *Salmonella* (20).

When *L. monocytogenes* reappeared on the regulatory radar screen in the 1980s, knowledge of the extent of its risk to public health was limited (23, 24). The situation was further complicated by the lack of reliable and rapid methods to test for the presence of *L. monocytogenes* in food and environmental samples (23, 46). In time, researchers developed better analytical methods and learned more about the microbe's pathogenicity, infectivity, virulence, and distribution in foods (12, 22, 31, 37). They found that approximately 6% of the population carried *L. monocytogenes* in their intestines asymptomatically and that the pathogen rarely infected healthy, nonpregnant adults (22). Only the immunocompromised, the elderly, infants, and pregnant women (and their unborn children) were at significant risk of serious illness or death (22, 37).

With access to better information and more reliable testing methods, regulators in Canada and Europe revised their approach to *L. monocytogenes* control. While maintaining a "zero tolerance" for certain higher-risk foods (such as infant foods and special dietary products), pasteurized milk, and aseptically packed foods, Germany established a quantitative limit of 100 *L. monocytogenes* cells per gram for lower-risk ready-to-eat foods (67). Other countries, including Canada and the United Kingdom, adopted a similar approach (34, 60). The United States, thus far, has maintained a policy of "zero tolerance" for *L. monocytogenes*, although the FDA began a review of its regulatory approach in 2004 in response to a petition submitted by a group of 15 food industry trade associations (27).

evidence—the timing of the outbreak, the high numbers of positive bacterial results from environmental samples taken in the weeks following July 4, and the termination of the outbreak once all of the post-July 4 production had been recalled—pointed in the direction of the maintenance work that had been carried out over the July 4 weekend. But some people remained skeptical about the ability of *L. monocytogenes* to survive for long periods on particles of construction dust.

L. monocytogenes, however, has since been proven to be hardier than the skeptics thought. In 2003, researchers at Iowa State University reported that the microbe was able to survive on particles of sand (chosen to mimic construction dust) for up to 151 days, depending on temperature and humidity conditions. And the organism still could attach to and survive on hot dogs and deli meats, even after living on the sand for as long as 1½ months (18). As a result of this research, little doubt remained that the dust stirred up by the removal of the air conditioning unit was the vehicle that spread *L. monocytogenes* through the BilMar plant.

Dust is the inevitable by-product of any construction activity. Company management neglected to plan for extensive and intensive cleaning, sanitizing, and microbiological testing of the entire facility—including those areas not directly affected by the construction work—before restarting production. In addition, the company should have maintained a higher level of microbiological alert during the weeks following startup. Such an approach would have been expensive, but not as expensive as the outbreak, which cost Sara Lee at least $76 million (50) and, tragically, cost 21 people their lives (57).

Engineering for maximum productivity and processing efficiency is not enough. Microbiology cannot be an afterthought or an add-on. Long ago, the food industry recognized the need to design equipment and processing facilities for ease of cleaning and sanitation and, in the 1920s, developed the first set of 3-A Sanitary Standards (5). These industry standards have been updated regularly to meet evolving needs and adapt to new technologies. But standards are of no value unless they are applied. Whether one is designing a new production plant, remodeling an old one, or making maintenance decisions, microbiological considerations must always be an integral part of the process.

References

1. **Ackers, M.-L., S. Schoenfeld, J. Markman, M. G. Smith, M. A. Nicholson, W. De Witt, D. N. Cameron, P. M. Griffin, and L. Slutsker.** 2000. An outbreak of *Yersinia enterocolitica* O:8 infections associated with pasteurized milk. *J. Infect. Dis.* **181:**1834–1837.
2. **Alojipan, L. C., and B. F. Andrews.** 1975. Neonatal sepsis. A survey of eight year's experience at the Louisville General Hospital. *Clin. Pediatr. (Phila).* **14:**181–185.
3. **Anonymous.** 1997. Preliminary report of an international outbreak of *Salmonella anatum* infection linked to an infant formula milk. *Euro. Surveill.* **2:**22–24. [Online.] http://www.eurosurveillance.org/em/v02n03/0203-224.asp. Accessed 14 May 2006.
4. **Anonymous.** 13 May 2004. Outbreak of vero cytotoxin-producing *E. coli* O157 linked to milk in Denmark. *Euro. Surveill. Wkly.* **8.** [Online.] http://www.eurosurveillance.org/ew/2004/040513.asp#2. Accessed 14 May 2006.

5. **Anonymous.** 2006. About 3-A Sanitary Standards, Inc. [Online.] http://www.3-a.org/about/index.htm. Accessed 13 May 2006.

6. **Beck, A., P. K. O'Brien, and V. F. Mackenzie.** 1966. Case of stillbirth due to infection with *Listeria monocytogenes*. *J. Clin. Pathol.* **19:**567–569.

7. **Biegeleisen, J. Z., Jr.** 1964. Immunofluorescence techniques in retrospective diagnosis of human listeriosis. *J. Bacteriol.* **87:**1257–1258.

8. **Can Manufacturers Institute.** 2005. The history of can making. Can Manufacturers Institute. [Online.] http://www.cancentral.com/canc/nontext/history.htm. Accessed 14 May 2006.

9. **Centers for Disease Control and Prevention.** 1998. Multistate outbreak of listeriosis—United States, 1998. *Morb. Mortal. Wkly. Rep.* **47:**1085–1086.

10. **Centers for Disease Control and Prevention.** 17 March 1999. Update: multistate outbreak of listeriosis. Centers for Disease Control and Prevention, Division of Media Relations, Atlanta, Ga. [Online.] http://www.cdc.gov/od/oc/media/pressrel/r990114.htm. Accessed 14 May 2006.

11. **Centers for Disease Control and Prevention.** 1999. Update: multistate outbreak of listeriosis—United States, 1998–1999. *Morb. Mortal. Wkly. Rep.* **47:**1117–1118.

12. **Chen, Y., W. H. Ross, V. N. Scott, and D. E. Gombas.** 2003. *Listeria monocytogenes*: low levels equal low risk. *J. Food Prot.* **66:**570–577.

13. **Clark, R. A.** 1977. Bacterial endocarditis caused by *Listeria monocytogenes*. *West. J. Med.* **126:**403–405.

14. **Clark, W. S., Jr.** 1998. Concentrated and dry milks and wheys, p. 65–80. *In* E. H. Marth and J. L. Steele (ed.), *Applied Dairy Microbiology*. Marcel Dekker, New York, N.Y.

15. **Collins, R. N., M. D. Treger, J. B. Goldsby, J. R. Boring III, D. B. Coohon, and R. N. Barr.** 1968. Interstate outbreak of *Salmonella newbrunswick* infection traced to powdered milk. *JAMA* **203:**838–844.

16. **Corcoran, G. D., J. Flynn, N. Gibbons, and E. Mulvihill.** 1989. Listeriosis: a community problem? *Ir. Med. J.* **82:**70–71.

17. **Cumber, P. M., W. Mumar-Bashi, S. Palmer, and R. D. Hutton.** 1991. *Listeria* meningitis and paté. *J. Clin. Pathol.* **44:**339.

18. **De Roin, M. A., S. C. C. Foong, P. M. Dixon, and J. S. Dickson.** 2003. Survival and recovery of *Listeria monocytogenes* on ready-to-eat meats inoculated with a desiccated and nutritionally depleted dustlike vector. *J. Food Prot.* **66:**962–969.

19. **Djuretic, T., P. G. Wall, and G. Nichols.** 1997. General outbreaks of infectious intestinal disease associated with milk and dairy products in England and Wales: 1992 to 1996. *Commun. Dis. Rep. CDR Rev.* **7:**R41–R45. [Online.] http://www.hpa.org.uk/cdr/archives/CDRreview/1997/cdrr0397.pdf. Accessed 14 May 2006.

20. **Doyle, M. P., L. R. Beuchat, and T. J. Montville (ed.).** 1997. *Food Microbiology: Fundamentals and Frontiers*. ASM Press, Washington, D.C.

21. **Dubos, R.** 1998. *Pasteur and Modern Science*. T. D. Brock (ed.). ASM Press, Washington, D.C.

22. **Farber, J. M.** 2000. Present situation in Canada regarding *Listeria monocytogenes* and ready-to-eat seafood products. *Int. J. Food Microbiol.* **62:**247–251.

23. **Farber, J. M., and J. Z. Losos.** 1988. *Listeria monocytogenes*. A foodborne pathogen. *CMAJ* **138:**413–418.

24. **Farber, J. M., and P. I. Peterkin.** 1991. *Listeria monocytogenes*, a food-borne pathogen. *Microbiol. Rev.* **55:**476–511.

25. **Filice, G. A., H. F. Cantrell, A. B. Smith, P. S. Hayes, J. C. Feeley, and D. W. Fraser.** 1978. *Listeria monocytogenes* infection in neonates: investigation of an epidemic. *J. Infect. Dis.* **138:**17–23.

26. **Fleming, D. W., S. L. Cochi, K. L. MacDonald, J. Brondum, P. S. Hayes, B. D. Plikaytis, M. B. Holmes, A. Audurier, C. V. Broome, and A. L. Reingold.** 1985. Pasteurized milk as a vehicle of infection in an outbreak of listeriosis. *New Engl. J. Med.* **312:**404–407.

27. **Food and Drug Administration.** 24 May 2004. Docket No. 2003P-0574. *Listeria monocytogenes*; petition to establish a regulatory limit. U.S. Food and Drug Administration. [Online.] http://www.fda.gov/ohrms/dockets/98fr/03p-0574-n000001.pdf. Accessed 14 May 2006.

28. **Food Safety and Inspection Service.** 28 January 1999. Bil Mar *Listeria* recall—additional brands sold at retail. Food Safety and Inspection Service, U.S. Department of Agriculture. [Online.] http://www.fsis.usda.gov/OA/recalls/prelease/pr044-98a.htm. Accessed 14 May 2006.

29. **Forsyth, J. R. L., N. M. Bennett, S. Hogben, E. M. S. Hutchinson, G. Rouch, A. Tan, and J. Taplin.** 2003. The year of the *Salmonella* seekers—1977. *Aust. N.Z. J. Public Health* **27:**385–389.

30. **Gantz, N. M., R. L. Myerowitz, A. A. Medeiros, G. F. Carrera, R. E. Wilson, and T. F. O'Brien.** 1975. Listeriosis in immunosuppressed patients. A cluster of eight cases. *Am. J. Med.* **58:**637–643.

31. **Goulet, V., H. de Valk, O. Pierre, F. Stainer, J. Rocourt, V. Vaillant, C. Jacquet, and J.-C. Desenclos.** 2001. Effect of prevention measures on incidence of human listeriosis, France, 1987–1997. *Emerg. Infect. Dis.* **7:**983–989.

32. **Gray, M. L., and A. H. Killinger.** 1966. *Listeria monocytogenes* and listeric infections. *Bacteriol. Rev.* **30:**309–382.

33. **Green, H. T., and M. B. Macaulay.** 1978. Hospital outbreak of *Listeria monocytogenes* septicaemia: a problem of cross infection? *Lancet* **2:**1039–1040.

34. **Health Canada.** 2004. Policy on *Listeria monocytogenes* in ready-to-eat foods. Food Directorate, Health Products and Food Branch, Health Canada. [Online.] http://www.hc-sc.gc.ca/fn-an/legislation/pol/policy_listeria_monocytogenes_politique_toc_e.html. Accessed 14 May 2006.

35. **Houang, E. T., C. J. Williams, and P. F. Wrigley.** 1976. Acute *Listeria monocytogenes* osteomyelitis. *Infection* **4:**113–114.

36. **Jacquet, C., B. Catimel, R. Brosch, C. Buchrieser, P. Dehaumont, V. Goulet, A. Lepoutre, P. Veit, and J. Rocourt.** 1995. Investigations related to the epidemic strain involved in the French listeriosis outbreak in 1992. *Appl. Environ. Microbiol.* **61:**2242–2246.

37. **Jay, J. M.** 2000. *Modern Food Microbiology*, 6th ed. Aspen Publishers, Inc., Gaithersburg, Md.

38. **Jensen, A., W. Frederiksen, and P. Gerner-Smidt.** 1994. Risk factors for listeriosis in Denmark, 1989–1990. *Scand. J. Infect. Dis.* **26:**171–178.

39. **Kalis, P., J. L. Le Frock, W. Smith, and M. Keefe.** 1976. Listeriosis. *Am. J. Med. Sci.* **271:**159–169.

40. **King, E. O., and H. P. R. Seeliger.** 1959. Serological types of *Listeria monocytogenes* occurring in the United States. *J. Bacteriol.* **77:**122–123.

41. **Lachapelle, G.** 1979. *Salmonella* dans les usines de lait en poudre au Québec. *Can. Inst. Food Sci. Technol. J.* **12:**177–179.

42. **Licari, J. J., and N. N. Potter.** 1970. *Salmonella* survival during spray drying and subsequent handling of skimmilk powder. II. Effects of drying conditions. *J. Dairy Sci.* **53:**871–876.

43. **Linnan, M. J., L. Mascola, X. D. Lou, V. Goulet, S. May, C. Salminen, D. W. Hird, M. L. Yonekura, P. Hayes, R. Weaver, A. Audurier, B. D. Plikaytis, S. L. Fannin, A. Kleks, and C. V. Broome.** 1988. Epidemic listeriosis associated with Mexican-style cheese. *New Engl. J. Med.* **319:**823–828.

44. **Loncarevic, S., M.-L. Danielsson-Tham, P. Gerner-Smidt, L. Sahlström, and W. Tham.** 1998. Potential sources of human listeriosis in Sweden. *Food Microbiol.* **15:**65–69.

45. **Louie, K. K., A. M. Paccagnella, W. D. Osei, H. Lior, B. J. Francis, and M. T. Osterholm.** 1993. *Salmonella* serotype Tennessee in powdered milk products and infant formula—Canada and United States, 1993. *Morb. Mortal. Wkly. Rep.* **42:**516–517.

46. **Lovett, J.** 1988. Isolation and enumeration of *Listeria monocytogenes*. *Food Technol.* **42:**172–175.

47. **Marth, E. H.** 1969. Salmonellae and salmonellosis associated with milk and milk products. A review. *J. Dairy Sci.* **52:**283–315.

48. **McLauchlin, J., N. Crofts, and D. M. Campbell.** 1989. A possible outbreak of listeriosis caused by an unusual strain of *Listeria monocytogenes*. *J. Infect.* **18:**179–187.

49. **McLauchlin, J., S. M. Hall, S. K. Velani, and R. J. Gilbert.** 1991. Human listeriosis and paté: a possible association. *Brit. Med. J.* **303:**773–775.

50. **Miller, J. P.** 22 January 1999. Sara Lee emerges mostly unscathed from recent packaged-meats recall. *Wall Street Journal*, New York, N.Y.

51. **Mokhiber, R., and R. Weissman.** 6 August 2001. Hot dog deaths and the corporate death penalty. *The Final Call On-Line Edition.* [Online.] http://www.finalcall.com/perspectives/hotdogs08-06-2001.htm. Accessed 14 May 2006.

52. **Mount, C.** 3 December 1986. Jewel switched labels, witness says. *Chicago Tribune*, Chicago, Ill.

53. **Mount, C.** 8 January 1987. Jewel Milk testing delay told. *Chicago Tribune*, Chicago, Ill.

54. **Mount, C.** 15 January 1987. Jewel Dairy's piping violated federal code, task force found. *Chicago Tribune*, Chicago, Ill.

55. **Olsen, S. J., M. Ying, M. F. Davis, M. Deasy, B. Holland, L. Iampietro, C. M. Baysinger, F. Sassano, L. D. Polk, B. Gormley, M. J. Hung, K. Pilot, M. Orsini, S. Van Duyne, S. Rankin, C. Genese, E. A. Bresnitz, J. Smucker, M. Moll, and J. Sobel.** 2004. Multidrug-resistant *Salmonella typhimurium* infection from milk contaminated after pasteurization. *Emerg. Infect. Dis.* **10:**932–935.

56. **Paul, M. L., D. E. Dwyer, C. Chow, J. Robson, I. Chambers, G. Eagles, and V. Ackerman.** 1994. Listeriosis—a review of eighty-four cases. *Med. J. Aust.* **160:**489–493.

57. **Perl, P.** 16 January 2000. Poisoned package; when 21 people died from eating contaminated meats, it was the nation's most lethal food safety epidemic in 15 years—and one of the quietest. Why didn't the U.S. Department of Agriculture blow the whistle sooner on Sara Lee? *The Washington Post*, Washington, D.C.

58. **Quarles, J. M., Jr., and B. Pittman.** 1966. Unsuccessful attempt to detect *Listeria monocytogenes* in healthy pregnant women. *J. Bacteriol.* **91:**2112–2113.

59. **Relier, J. P., C. Amiel-Tison, J. Krauel, L. Helffer, J. C. Larroche, and A. Minkowski.** 1977. Neonatal listeriosis. Apropos of 53 cases. [Article in French—English abstract.] *J. Gynecol. Obstet. Biol. Reprod. (Paris)* **6:**367–381.

60. **Roberts, D.** 1994. *Listeria monocytogenes* and food: the U.K. approach. *Dairy Food Environ. Sanit.* **14:**198, 200, 202–204.

61. **Rowe, B., N. T. Begg, D. N. Hutchinson, H. C. Dawkins, R. J. Gilbert, M. Jacob, B. H. Hales, F. A. Rae, and M. Jepson.** 1987. *Salmonella ealing* infections associated with consumption of infant dried milk. *Lancet* **2:**900–903.

62. **Ryan, C. A., M. K. Nickels, N. T. Hargrett-Bean, M. E. Potter, T. Endo, L. Mayer, C. W. Langkop, C. Gibson, R. C. McDonald, R. T. Kenney, N. D. Puhr, P. J. McDonnell, R. J. Martin, M. L. Cohen, and P. A. Blake.** 1987. Massive outbreak of antimicrobial-resistant salmonellosis traced to pasteurized milk. *JAMA* **258:**3269–3274.

63. **Salvat, G., M. T. Toquin, Y. Michel, and P. Colin.** 1995. Control of *Listeria monocytogenes* in the delicatessen industries: the lessons of a listeriosis outbreak in France. *Int. J. Food Microbiol.* **25:**75–81.

64. **Samuelsson, S., N. P. Rothgardt, A. Carvajal, and W. Frederiksen.** 1990. Human listeriosis in Denmark 1981–1987 including an outbreak November 1985–March 1987. *J. Infect.* **20:**251–259.

65. **Schlech, W. F., III, P. M. Lavigne, R. A. Bortolussi, A. C. Allen, E. V. Haldane, A. J. Wort, A. W. Hightower, S. E. Johnson, S. H. King, E. S. Nicholls, and C. V. Broome.** 1983. Epidemic listeriosis—evidence for transmission by food. *New Engl. J. Med.* **308:**203–206.

66. **Schuchat, A., B. Swaminathan, and C. V. Broome.** 1991. Epidemiology of human listeriosis. *Clin. Microbiol. Rev.* **4:**169–183.

67. **Teufel, P.** 1994. European perspectives on *Listeria monocytogenes*. *Dairy Food Environ. Sanit.* **14:**212–214.

68. **Usera, M. A., A. Echeita, A. Aladueña, M. C. Blanco, R. Reymundo, M. I. Prieto, O. Tello, R. Cano, D. Herrera, and F. Martinez-Navarro.** 1996. Interregional foodborne salmonellosis outbreak due to powdered infant formula contaminated with lactose-fermenting *Salmonella virchow*. *Eur. J. Epidemiol.* **12:**377–381.

69. **Usera, M. A., A. Rodriguez, A. Echeita, and R. Cano.** 1998. Multiple analysis of a foodborne outbreak caused by infant formula contaminated by an atypical *Salmonella virchow* strain. *Eur. J. Clin. Microbiol. Infect. Dis.* **17:**551–555.

70. **Van, J.** 17 April 1985. *Salmonella* blamed for man's death. *Chicago Tribune*, Chicago, Ill.

71. **Van, J.** 16 May 1985. Jewel Dairy unlikely to reopen. *Chicago Tribune*, Chicago, Ill.

72. **Van, J.** 24 October 1985. Jewel milk process to be barred. *Chicago Tribune*, Chicago, Ill.

73. **Van, J., and R. Davis.** 2 April 1985. Food poison outbreak strikes 200. *Chicago Tribune*, Chicago, Ill.

74. **Van, J., and M. Zambrano.** 9 April 1985. Woman in hospital for *Salmonella* dies. *Chicago Tribune*, Chicago, Ill.

chapter 3

Recipes for Disaster

PITY THE POOR EGG. Once hailed as one of nature's perfect foods, it has fallen on hard times. First, it was ostracized due to our fear of cholesterol. Then in the late 1980s, *Salmonella* further poisoned the egg's reputation. Reports of food-borne *Salmonella* outbreaks—many of them traced to eggs served at food service locations such as cafeterias, hospital kitchens, and restaurants—started flowing into government health department offices in several countries, including the United States.

Have Eggs Instead?

The breakfast buffet was a popular fixture at a Maryland restaurant chain (73). But in 1985, it contained an unexpected and unwelcome ingredient. That August, one employee and three patrons of a restaurant in the chain were stricken with *Salmonella enterica* serotype Enteritidis. But even though county health authorities investigated the restaurant and interviewed the victims and other patrons, they could not determine which food had caused the outbreak. There were just too many individual food items—and not enough victims—to enable investigators to pinpoint the source.

In the first half of September, five customers of a second restaurant belonging to the same chain became ill. And that same week, 113 people who had eaten at a third restaurant in the chain contacted the local health department to complain about diarrhea, cramps, and fever. Seventeen of the 113 ended up in the hospital. Once more, serotype Enteritidis proved to be the uninvited guest at the table. But this time, investigators were able to identify the offending food—scrambled eggs from the breakfast buffet.

All three restaurants used grade A shell eggs to prepare their scrambled eggs. Employees cracked the eggs by hand—as many as 1,800 eggs at a time. The scrambled egg mixture sometimes sat at room temperature for as long as 6 hours before being cooked. And the cooks took pains not to overcook the eggs. Otherwise, they would dry out too much in the breakfast bar's warming tray. Indeed, several of the food poisoning victims at restaurant number 3 reported that the eggs appeared underdone.

At first, it looked like a coincidence that all three locations were hit with a *Salmonella* outbreak in the space of just a few weeks. But an in-depth investigation determined that the three affected restaurants purchased their eggs from the same distributor and that the identical strain of serotype Enteritidis was responsible for all of the cases of gastroenteritis. The combination of contaminated shell eggs and poor handling and cooking procedures made these outbreaks all but inevitable.

State health officials were unable to find *Salmonella* in other batches of eggs from the distributor. As a result, despite strong circumstantial evidence, they could not confirm unequivocally that shell eggs from that distributor were the source of the outbreaks. So the officials contented themselves with pointing out to the restaurant management the correct way to prepare and display scrambled eggs.

The 1985 Maryland outbreaks were early harbingers of a much larger problem. That same year, serotype Enteritidis became the most commonly reported *Salmonella* serotype in three U.S. states—New Hampshire, New Jersey, and New York. The trend of ever-increasing reports of illness due to serotype Enteritidis continued in 1986. Sources of the *Salmonella* outbreaks included a wide variety of foods. Scrambled eggs, pasta (both homemade and commercial), hollandaise sauce, rice balls, protein supplement, and roast beef—all were implicated, but there was no apparent connection to explain the emergence in the U.S. Northeast of serotype Enteritidis as a serotype to be reckoned with (96, 97). And then tragedy hit the Coler Memorial Hospital on Roosevelt Island, New York.

On Friday, July 31, 1987, the headline on page B3 in the *New York Times* read "Tainted Food Possible in Patient's Death." The outbreak had begun on Tuesday of that same week, and by late Thursday more than 175 people had been sickened (105). By the time the outbreak was over, 274 hospital patients were diagnosed with salmonellosis caused by serotype Enteritidis. Nine patients died (107).

Hospital officials worked feverishly to track down the source of the *Salmonella* before the outbreak could spiral out of control. They quickly discovered that the contaminated meal had been served at lunch on July 28. After reviewing the medical records, their attention focused on a tuna-macaroni salad containing hospital-prepared mayonnaise. The mayonnaise was made on July 27 with raw grade A shell eggs. It was mixed with the other ingredients the morning of July 28, and the finished salad sat at room temperature for 5 hours before being served with lunch.

The hospital lab tested remnants of the tuna-macaroni salad and found serotype Enteritidis. They also recovered the same microbe from pooled batches of eggs. And, just to tie up the final loose ends, cultures of hen ovaries from the farm that had produced the eggs were positive for the same strain of serotype Enteritidis that had been found in the patients and in the salad.

Coincidentally with the U.S. outbreaks, serotype Enteritidis was also turning up in the United Kingdom at an alarming rate. Illness due to serotype Enteritidis increased in the United Kingdom from 1,101 to 6,858 cases—more than sixfold—in the period between 1982 and 1987. Whereas some cases of salmonellosis were traced to poultry meat, many others were tied to a variety of foods containing raw or partially cooked eggs (3).

Inexorably, year after year, serotype Enteritidis contamination of shell eggs extended its reach to other countries, as well as to other regions of the United States. Japan was hit in 1989 (106). Serotype Enteritidis was responsible for outbreaks in Canada in 1991, Italy in 1993, and California in 1996 (13, 71, 83). Reports submitted to the World Health Organization between 1979 and 1987 by 35 countries highlighted the magnitude of the problem. The percentage of serotype Enteritidis isolations increased by at least 25% in 21 of the 35 countries, and more than doubled in 15 of them (93). And the number of victims continued to grow. Time and again, serotype Enteritidis outbreaks were traced to foods containing raw or undercooked eggs (Table 3.1).

Hollandaise sauce, Caesar salad, ice cream, cake frostings, meringue pies, cheesecake, scrambled eggs, and mayonnaise—the list of dishes and

Table 3.1 Examples of egg-associated outbreaks of *S. enterica* serotype Enteritidis food-borne illness since 1988

Year	Country	No. of victims (deaths)	Implicated food	Reference
1989	United Kingdom	173 (0)	Egg-based sauces	102
1990	United Kingdom	109 (1)	Beef rissoles	28
1992	United States	74 (0) + 32 (0)[a]	Monte Cristo sandwiches	92
1995	United States	39 (3)	Baked eggs	72
1995	United States	28 (0)	Caesar salad	72
1995	United States	3 (0)	Jamaican malt (beverage)	72
1995	Canada	21 (0)	Mayonnaise-based dip	82
1996	Ireland	65 (0)	Chocolate mousse cake	47
1996	Spain	11 (0)	Omelettes	41
1997	United Kingdom	5 (0)	Fish cakes	29
1997	United Kingdom	8 (0)	Pureed food diet	29
1997	United Kingdom	8 (0)	Tiramisu	29
1997	United States	13 (0)	Cheesecake	90
1997	United States	43 (0)	Lasagna	90
1997	United States	91 (0)	Hollandaise sauce	90
1998	Greece	60 (0)	Mayonnaise	50
1998	Italy	36 (0)	Iced cake	25
1998	Mexico	155 (0)	Egg-covered meat	17
1998	United States	58 (0)	Chiles rellenos	90
1999	Japan	206 (0)	Cream buns	108
1999	Austria	8 (0)	Cake	12

[a] Two separate outbreaks, each caused by a different serotype Enteritidis phage type.

desserts that traditionally use raw or partially cooked eggs goes on and on. And every now and then, the recipe comes with a microscopic bonus—serotype Enteritidis. Microbiologists once thought that *Salmonella* could not penetrate into an egg unless the shell was damaged. But if that was true, how did serotype Enteritidis find its way into eggs? More important, what could be done about it?

To answer these questions, the U.S. Centers for Disease Control and Prevention (CDC) organized a working group in cooperation with state epidemiologists in November 1986 (96). After reviewing 65 outbreaks that had occurred over a 2½-year period beginning in January 1985, the group came to a startling conclusion. Infected hens were contaminating the eggs with *Salmonella* while the immature eggs—not yet covered with their protective shells—were still in the oviduct (103). Not everyone agreed with this assessment at first, but over the next few years, evidence piled up in support of the theory (43, 55, 64, 74, 84) and researchers developed a clearer picture of how and when eggs were becoming infected.

Egg production begins in the ovary, which is filled with immature eggs. As an egg matures, it pops out of the ovary and enters a tube known as the oviduct. At this stage, the egg is a naked yolk. As it travels down the oviduct, the egg increases in volume. Once it reaches full size, the egg yolk is surrounded by the vitelline membrane, around which is deposited a layer of albumen (egg white). Two pieces of fibrous material called chalaza attach the albumen to the vitelline membrane. Finally, the shell membrane and eggshell are deposited around the exterior of the now-complete egg (Fig. 3.1).

An infected hen can carry *Salmonella* in several parts of its body, including—among other organs and tissues—the ovary and oviduct (9, 44). Until an egg receives its protective shell, it is susceptible to *Salmonella* contamination. The microbe can travel either down the oviduct from an infected ovary or up the oviduct from the vagina or cloaca. And if the oviduct itself is contaminated, *Salmonella* can be transferred directly to the egg as it passes through on its way to the cloaca (69).

Knowing how the egg becomes infected is only one part of the battle—and the easiest part, at that. Preventing the infection is far more difficult. The only way to ensure that eggs don't become contaminated with *Salmonella* before they are laid is to start with a *Salmonella*-free flock of laying hens and work to keep it that way.

Which Came First—the Chicken or the Egg?

A *Salmonella*-free breeding hen produces *Salmonella*-free chicks. But that's only the first step. Because newborn chicks are highly susceptible to infection with serotype Enteritidis and other salmonellae, decades of research have gone into strategies to keep chicks free from infection as they mature and develop into laying hens.

In 1971, Finland was hit with a massive outbreak of *S. enterica* serotype Infantis in its broiler flocks. The economic damage was huge, and 277 people were stricken with salmonellosis (89). To prevent a recurrence, Finnish scientists began searching for ways to protect their poultry flocks. They observed that 2- to 3-day-old chicks were highly susceptible to *Salmonella*. But feeding the baby chicks a mixture of the intestinal contents of a healthy adult chicken prevented them from succumbing to infection (78). The researchers theorized that some component of the natural bacterial flora of *Salmonella*-free birds blocked *Salmonella* from colonizing the chickens' guts. They named the phenomenon "competitive exclusion".

A few European countries adopted this approach to *Salmonella* control in the early 1980s. Finland and Sweden, for example, incorporated competitive exclusion into an overall strategy that included using *Salmonella*-free feed and maintaining a clean henhouse environment. The contamination rate of chicken carcasses in Finland dropped to just 5 to 10% in 1992. And in that same year, only 1% of Swedish chickens tested positive for *Salmonella* (60, 111).

Competitive exclusion alone is not enough to eradicate *Salmonella* (34, 106). As the Scandinavians have demonstrated, however, it can be very effective when used in conjunction with other control measures. One such method is vaccination.

The idea of vaccinating chickens against *Salmonella* was first explored as a means of preventing fowl typhoid, a disease of poultry caused by *S. enterica* serotype Gallinarum (10). Early experiments with vaccines made from killed *Salmonella* proved fruitless. Then in 1956, the first successful vaccine—this one using a live, attenuated culture—was announced (100). But, until recently, the use of poultry vaccines was never extended beyond control of serotype Gallinarum.

As regulators struggled in the 1980s and 1990s to contain the sudden eruption of egg-associated serotype Enteritidis outbreaks, poultry scientists revisited vaccination as a control strategy. In 1994 Hassan and Curtiss introduced a live, avirulent vaccine that protected chickens from infection by a variety of salmonellae, including serotype Enteritidis (52). And their vaccine came with a bonus. The baby chicks born from vaccinated hens were more resistant to *Salmonella* infection than chicks from unvaccinated hens (53).

In spite of the promising results obtained from vaccination and competitive exclusion research, egg farmers can't rely on these approaches exclusively to keep their poultry *Salmonella*-free. Several environmental factors conspire to expose laying hens to *Salmonella* infection throughout their lifespan. Often, feed is already contaminated with *Salmonella* when it arrives at the farm (83, 99). And even when the feed is clean to start with, mice carry serotype Enteritidis from henhouse to henhouse, spreading contamination with their droppings (48, 56). In addition, *Salmonella* can attach to dust in the henhouse and circulate through the air, especially in

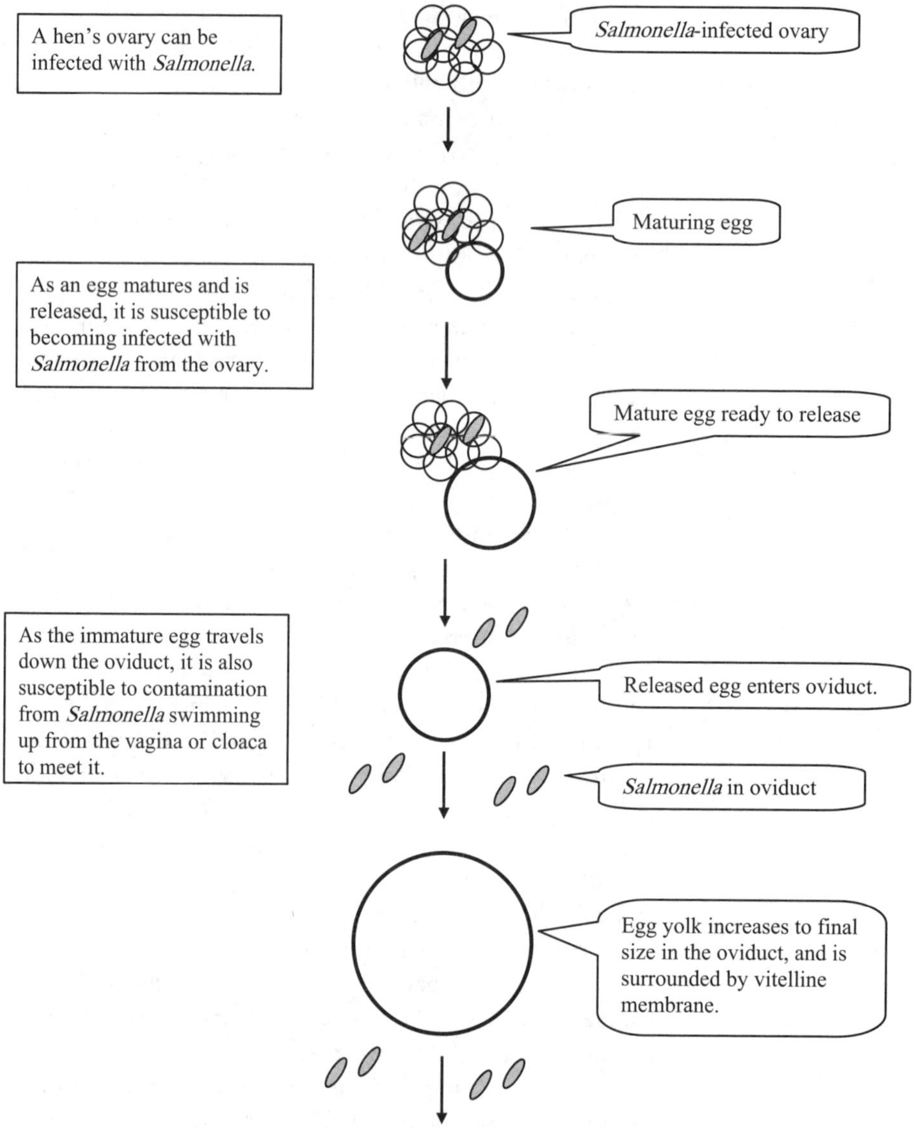

Figure 3.1 Stages in the evolution of a shell egg.

windowless, environmentally controlled facilities (106). Therefore, rodent control and careful attention to sanitation and disinfection must be part of the poultry industry's anti-*Salmonella* arsenal.

Nor is it enough to focus exclusively on the laying hens while ignoring the eggs. Hens expel their eggs through the cloaca, the same opening through which they eliminate feces. Thus, even at the instant a mature egg is laid, it is exposed to the risk of *Salmonella* contamination (95). And every

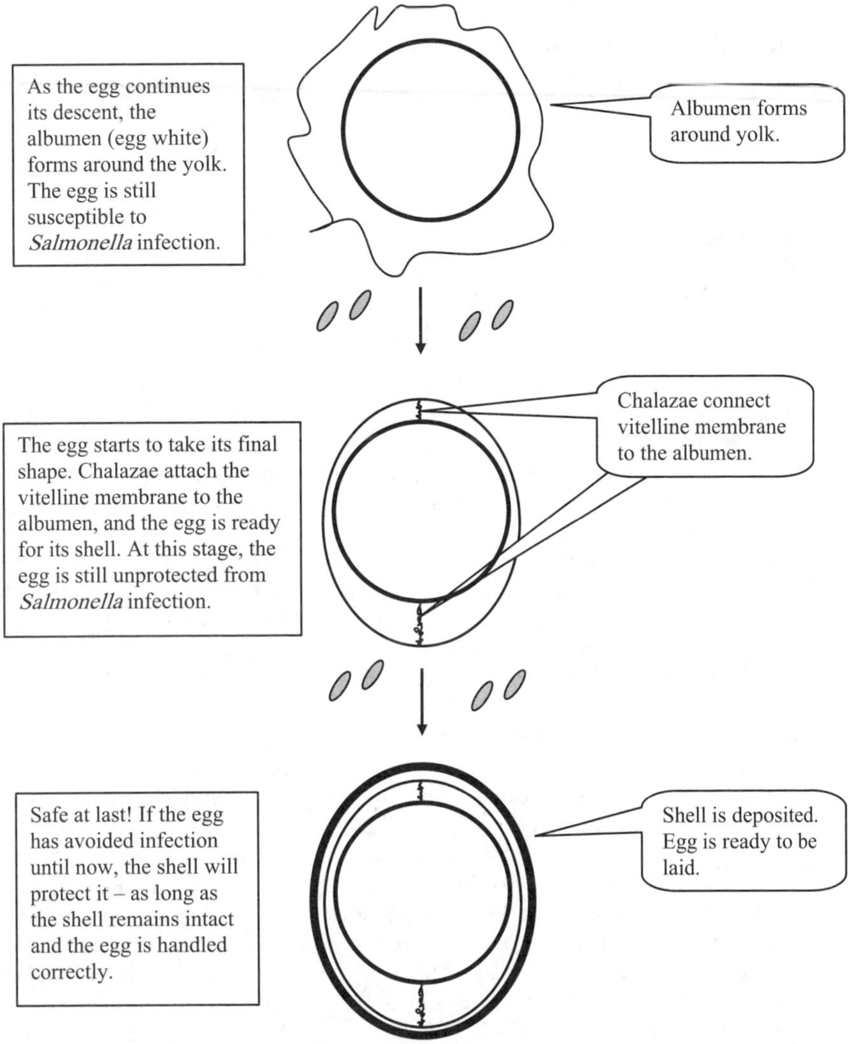

Figure 3.1 *Continued.*

step in its voyage to the breakfast plate provides another opportunity for serotype Enteritidis and other salmonellae to hitch a ride.

Modern egg farms are highly automated and comprise multiple henhouses. When a hen lays her egg, a conveyer belt carries it out of the henhouse. Often, the same belt passes through each house in sequence, accumulating eggs, debris, feces, and sometimes *Salmonella* along the way (76). Being fragile, some eggs may crack or even break open during the collection process. And although cracked or broken eggs are either discarded or diverted, egg yolk and albumen leaking from the damaged eggs can coat intact eggs with protein and act like flypaper to trap stray *Salmonella*.

Competitive Exclusion—Probiotics for Poultry

Infection of poultry flocks with *Salmonella enterica* serotype Gallinarum (including Pullorum variant) was a major economic problem for poultry farmers in the Western Hemisphere and in Europe during the early and middle years of the 20th century. Poultry producers in the affected regions mobilized to eradicate this avian pathogen and, by the 1970s, flocks in the United States, England, and Wales were free from infection with serotype Gallinarum. But according to some, this success story had an unexpected consequence. The elimination of serotype Gallinarum might have cleared the way for *S. enterica* serotype Enteritidis to gain a foothold in commercial poultry flocks (88).

In 1973, more than a decade before *S. enterica* serotype Enteritidis surprised the scientific world by its ability to infect intact eggs, two Finnish researchers proposed a novel way to prevent *Salmonella* from colonizing newborn chicks (89). Their approach, first referred to as the "Nurmi concept" and later called "competitive exclusion," relied on the tendency of the natural intestinal flora of poultry to prevent *Salmonella* from gaining a foothold (101).

The earliest studies were carried out using undefined culture mixtures produced by culturing the intestinal contents of healthy adult chickens. These cultures were fed to 40 1-day-old chicks, and a separate group of 40 chicks served as untreated controls. The next day, both treated and untreated chicks (now 2 days old) were inoculated with *Salmonella*. Several days later (at 8 and 11 days of age), the cecal contents of the chicks were tested for the pathogen. The results were unequivocal. All of the treated chicks were *Salmonella* free, whereas every untreated chick had become colonized with large numbers ($>10^7$/g of cecal content) of *Salmonella* organisms (89).

Understandably, the poultry industry and regulatory agencies worldwide were concerned about using an undefined culture for competitive exclusion therapy (103). After all, undefined cecal cultures could potentially contain pathogens such as *Salmonella* or *Campylobacter* and produce more harm than good (42). So researchers focused on identifying the microbes in the mix that conferred anti-*Salmonella* protection on the baby chicks, in the hope of developing a defined culture mix that would replicate the success of the early Finnish work (22, 23, 49, 58, 81, 98).

E. V. Nurmi was awarded a U.S. patent for his competitive exclusion culture mix in 1987 (79). The first commercial (though still undefined) commercial preparation, Broilact, was introduced in Scandinavia that same year (94). Since then, other experimental and commercial preparations have been introduced and tested, including Avigard and Preempt, which was based on technology developed by researchers at the USDA and was approved for use by the FDA in 1998 (8, 16, 20, 62, 75, 77, 86, 109). And researchers are working to extend the use of competitive exclusion techniques to other pathogens, such as enterohemorrhagic *Escherichia coli* and *Campylobacter*, and to other livestock, including cattle and swine (14, 18, 46, 63, 77, 87, 113).

Although competitive exclusion has been used successfully in Finland and Sweden for many years, it is not a substitute for following good sanitary and flock management practices (8, 60, 111). It does not flush out *Salmonella* from chicks that have already become colonized, and there have been conflicting reports over whether the culture mixtures can retain their effectiveness under normal storage conditions or with repeated subculture (1, 6, 8, 101). But it is a potentially useful component of a *Salmonella*-control toolbox that includes vaccination, sanitation, feed supplementation, and judicious use of antibiotics (19, 21, 40, 59, 70, 91).

Whereas studies have confirmed that caged hens in high-population-density facilities produce *Salmonella*-contaminated eggs, free-range and cage-free hens are not exempt from this problem. Free-range hens have access to the outdoors for at least part of the day; cage-free hens, while not housed in cages, do not necessarily have outdoor access. Both spend at least part of their time indoors. And free-range hens are just as susceptible to *Salmonella* infection as their caged cousins, especially when the poultry farm is located near a sewage runoff creek (7, 70, 71).

Once intact shell eggs arrive at a U.S. egg processing facility, they are washed with a sanitizing solution and coated with a light mineral oil to seal the pores of the shell (39). Washing and sanitizing remove visible soil and reduce the bacterial population on the outside of the shell. But detergent strips the natural protective coating from the shell, potentially enabling *Salmonella* to penetrate into the albumen (57). Because of this perceived risk, several countries—including the United Kingdom and many other members of the European Union—do not wash eggs (5, 19, 68).

Temperature change is also an important factor in egg contamination. When an egg cools—either after having been laid or after washing—the contents shrink slightly. This creates a vacuum that can suck into the egg any bacteria that may be present on the outside of the shell (68).

Once inside an egg, all that *Salmonella* needs in order to multiply is enough time and the right temperature. *S. enterica* serotype Enteritidis barely grows at refrigeration temperature but multiplies very readily in eggs that are stored at room temperature (45, 51). And the older the egg, the better able it is to support the growth of *Salmonella*. Furthermore, the aging effect is hastened by storing eggs without refrigeration (67).

Unlike several other developed countries—such as Australia, Switzerland, and the United Kingdom—the United States requires that eggs be refrigerated at all stages of commercial handling, including while they are in transit and while in storage or on display in retail stores (35, 112). Even so, it's still possible to walk through some retail markets and find a room-temperature egg display.

Notwithstanding all of the things that can go wrong at each step in the egg production, distribution, and marketing system, the risk of finding a contaminated egg is small. In the United States, approximately one egg in 20,000 is likely to contain serotype Enteritidis (27). But the numbers of live bacteria in a contaminated egg can become dangerously elevated if the egg has been mishandled—high enough that traditional cooking methods may not kill all of the *Salmonella* organisms (65).

Salmonella-positive eggs present another risk of which many food handlers are unaware—cross-contamination. Cracking, separating, or beating an egg can spread *Salmonella* over a wide area in the kitchen, contaminating work surfaces, utensils, mixing bowls, and hands (66). It's just as important to clean up the kitchen thoroughly after working with eggs as after handling raw chicken.

But food handlers—even those who work in hospitals—sometimes forget the risks associated with using raw eggs. In 1993, a group of seven adults and seven children who attended a cookout at a psychiatric hospital in Florida learned about the connection between serotype Enteritidis and eggs the hard way. Five of the seven children and all seven adults developed gastroenteritis within 24 hours of attending the party. Eleven of the 12 victims had eaten the homemade ice cream made especially for the cookout (15).

Investigators found serotype Enteritidis in the leftover ice cream and also recovered the microbe from three of the victims. The ice cream had been prepared the morning of the cookout using six raw eggs. There were no obvious preparation, sanitation, or temperature control errors. Raw eggs represented the one fatal flaw in the recipe.

The U.S. Department of Agriculture (USDA) tried to trace the source of the eggs, but to no avail. The hospital was able to point investigators to the distributor from which they had obtained the eggs, but the distributor had purchased eggs from many different suppliers. And the agency had no authority to test poultry flocks for *Salmonella* unless the evidence clearly implicated one specific flock as being the source of an outbreak.

As long as people want to enjoy homemade ice cream, hollandaise sauce, sunny-side-up eggs, and Caesar salad, outbreaks of serotype Enteritidis gastroenteritis will continue to pop up like thunderstorms on hot, humid days. But, unlike the weather, we should be able to control *Salmonella*—if government agencies, the egg industry, and food handlers are prepared to make the effort.

This effort is being made, and it has started to bear fruit. In the United Kingdom, reported cases of serotype Enteritidis infection have been on the decline in recent years—from a peak of nearly 30,000 cases in 1997 to fewer than 16,500 cases in 2001. This improvement follows on the heels of new codes of practice for infection control and hygiene. The British government has mandated *Salmonella* testing of all breeder flocks. And more than 80% of retail shell eggs in the United Kingdom now come from flocks that have been vaccinated against serotype Enteritidis (19).

The United States has taken a somewhat different approach, emphasizing the egg as well as the chicken. The President's Council on Food Safety issued its Egg Safety Action Plan in December 1999 (85). The Council established an interim goal of reducing serotype Enteritidis illness linked to eggs by 50% within 5 years—that is, by 2005—and to eliminate it completely by 2010.

The Action Plan identified the stages of egg production for which improvements were needed, beginning with breeder flocks and poultry feed, and including shell egg processing procedures, improved package labeling, proper storage, shipping, and display temperatures, and better education of food handlers—whether commercial, institutional, or in the home. The Action Plan also highlighted the need for improved outbreak surveillance and for a national system to enable investigators to trace a contaminated batch of eggs through the distribution system back to the flock that produced it. Finally, the Action Plan called for additional research into all aspects of serotype Enteritidis and its association with poultry and eggs, including development of improved vaccines and competitive exclusion products and in-shell pasteurization of eggs.

In 2000, the U.S. Food and Drug Administration (FDA) published a Final Rule requiring that all consumer packages of shell eggs be labeled with instructions for safe handling and storage of the eggs (30). Then, in 2004, FDA took a major step toward achieving the goals set forward in the 1999 Action Plan by issuing a new set of Proposed Rules. The FDA's proposal—intended to cover all egg producers with 3,000 or more laying hens—encompassed rodent and pest control, biosecurity concerns, procurement of chicks and pullets, cleaning and disinfection of poultry houses, and refrigeration of eggs at the farm, among other issues (31, 38). The comment period for the proposal was extended twice in 2005, and as of mid-2006, the FDA's Proposed Rules have not been finalized (32, 33).

The United States still has a long way to go to meet its goal of eliminating serotype Enteritidis from shell eggs by 2010. It will take hard work, ongoing government and industry commitment, and a large helping of good luck to get there. But even if everything goes according to plan, eliminating serotype Enteritidis from eggs will not make food poisoning disappear. Eggs represent only one of many items used in recipes that call for little or no cooking. And some of those ingredients come supplied with a far more deadly contaminant—*Clostridium botulinum*.

The Fatal Error

Botox is in. Whether for eliminating frown lines (36), reducing excessive underarm sweating (37), or treating muscle spasms (54), Botox has become a bandwagon. But this potent neurotoxin and the microbe that produces it can also wreak havoc in our food supply—when we least expect it.

Hazelnut was a common yogurt flavor in the United Kingdom in the 1980s. Hazelnut conserve was prepared from a puree of roasted hazelnuts, water, starch, and sugar. The ingredients were mixed, heated together in a large vat, and pumped into cans. The cans were then sealed and placed in a boiling water retort for 20 minutes, after which they were cooled and stored at room temperature. Yogurt manufacturers purchased the conserve, mixed it with plain yogurt, and then immediately packaged the product into retail containers. All went well until consumers demanded a sugar-free variety.

The supplier of hazelnut conserve responded by substituting aspartame for sugar in his formula. Little did he realize that in doing so, he had opened Pandora's box. In May and June 1989, 27 people living in northwest England and north Wales developed symptoms of botulism—weak limbs, impaired speech, double vision, difficulty swallowing, and weak respiratory muscles. One person died from aspiration pneumonia. Twenty-five others were hospitalized, but survived. Of the 27 victims, 25 had eaten the same brand of hazelnut yogurt (80).

Investigators found *C. botulinum* toxin in unopened packages of the yogurt. But after having inspected the production plant, they concluded that the yogurt manufacturer was maintaining an adequate level of hygiene. Attention turned to the conserve. There had been complaints of swollen cans the previous year. In response, the manufacturer added potassium sorbate to his formula in an attempt to control yeast, which he assumed to be the cause of the swelling. But, as it turned out, yeast contamination wasn't the problem. *C. botulinum* was.

When investigators tested a badly swollen can of the conserve, they found that it contained the same toxin that was present in the yogurt. On reviewing the manufacturer's production process, it quickly became clear that the heating step was not severe enough to kill *C. botulinum* spores. And the pH of the finished product was well within the range that the microbe could tolerate.

By switching from sugar to aspartame without making any other adjustments to his process, the conserve manufacturer had set himself up for disaster. The sugar content of the original formula had prevented *C. botulinum* from growing. Once the sugar was removed, so was the only barrier to a serious outbreak. The manufacturer forgot—or never knew—that he was walking a food safety tightrope.

Long associated with inadequately processed canned food, *C. botulinum* has cropped up in a variety of new places in recent decades. Uneviscerated salted fish (11, 110), fermented beaver tails (61), and oil-packed condiments such as chopped garlic, roasted eggplant, and sautéed onions (24, 26, 104) have all been linked to botulism. And foil-wrapped baked potatoes were responsible for the third-largest outbreak of botulism in U.S. history.

The story began to unfold early on the morning of April 10, 1994, when a father and his teenage son turned up in an El Paso, Tex., hospital complaining of botulism-like symptoms (2). In the days before their arrival at the hospital, they had shared only one meal—at Tassos, a local Greek restaurant (4). The hospital notified local health authorities who, in turn, alerted the other six area hospitals. Within hours, four additional victims were identified. All of them had recently eaten at Tassos.

Faced with an incipient *C. botulinum* outbreak, the El Paso health department acted promptly to prevent the restaurant from opening for business that day. And they began at once to determine which food items the victims had in common. All of the people who showed symptoms of botulism had eaten at the restaurant on either April 8 or 9. And most of them had consumed one of two appetizers made from baked potatoes—skordalia (a potato dip) or melitzanosalata (a potato/eggplant dip). Four of the victims who hadn't eaten either dip were restaurant employees who had handled both appetizers while on the job April 8 and 9.

Investigators tested leftover potato dip and confirmed that it contained high levels of botulinum toxin. The restaurant had no eggplant dip left over, but some was found discarded in a garbage can at the home of one of the patients. It, too, was positive for the same toxin. The pH of the potato dip was 3.7—too acidic for *C. botulinum* to grow. So the toxin must have come from one of the ingredients.

Six ingredients were common to both dips—oil, vinegar, raw onion, raw garlic, feta cheese, and baked potato. In addition, the potato dip contained French bread and the eggplant dip contained baked eggplant. While there was no leftover baked potato or eggplant available for testing, all of the other dip ingredients were tested and contained no toxin.

The potato dip was prepared using two potatoes that had been wrapped in aluminum foil and baked (at 450°F [250°C] for about 2 hours) on April 7. The baked potatoes were stored—still wrapped in foil—for 18 hours at room temperature. The eggplant dip was made on April 5 using one potato and some eggplant baked earlier that evening. It was served to diners on April 6 or 7, with no apparent ill effect. But five people who ate that same batch of dip on April 8 became ill.

Lab investigations confirmed that *C. botulinum* spores could survive baking inside a foil-wrapped potato and could germinate, multiply, and produce toxin in the baked potato at room temperature. Nor was the toxin damaged or inactivated by the acidic environment of the potato dip. The two implicated appetizers had been stored side-by-side in the restaurant's refrigerator, and the same utensils were used for both dishes. Investigators concluded that a serving utensil had transferred toxin from the potato dip to the eggplant dip.

The foil wrapping on the baked potatoes was the key to this outbreak. In a trial baking study, the internal temperature of foil-wrapped baked potatoes only reached 205 to 207°F (96 to 97°C)—a temperature that *C. botulinum* spores could easily have survived. Furthermore, the foil acted as an oxygen barrier, creating near-perfect anaerobic conditions under which the organism could grow and produce its deadly toxin. The chef's decision to store the baked potatoes at room temperature for 18 h was the final ingredient in the *C. botulinum* recipe. Had he remembered that perishable foods must be kept cold, there would have been no outbreak.

Whether it's serotype Enteritidis in eggs or *C. botulinum* in eggplant, the challenge is the same. Recipes that do not include an adequate final cooking step have become increasingly popular with consumers and can be a significant source of food-borne illness. The health risks associated with these recipes are magnified when ingredients are not chosen wisely or are mishandled. Food handlers, whether manufacturers, food service workers, or consumers, must adapt their techniques in the face of newly recognized pathogens—or old familiar pathogens in new settings—in order to ensure the safety of the food they prepare.

References

1. **Almeida, W. A. F., A. Berchieri, Jr., and P. A. Barrow.** 2002. The effect of serial culture and storage on the protective potential of a competitive exclusion preparation. *Brazil. J. Poult. Sci.* **4:**163–167.

2. **Angulo, F. J., J. Getz, J. P. Taylor, K. A. Hendricks, C. L. Hatheway, S. S. Barth, H. M. Solomon, A. E. Larson, E. A. Johnson, L. N. Nickey, and A. A. Ries.** 1998. A large outbreak of botulism: the hazardous baked potato. *J. Infect. Dis.* **178:**172–177.
3. **Anonymous.** 1988. *Salmonella enteritidis* phage type 4: chicken and egg. *Lancet* **ii:**720–722.
4. **Anonymous.** 14 April 1997. Investigators discover source of El Paso botulism outbreak. *Houston Chronicle*, Houston, Tex.
5. **Anonymous.** 2005. Opinion of the scientific panel on biological hazards on the request from the commission related to the microbiological risks on washing of table eggs. *EFSA J.* **269:**1–39. [Online.] http://www.efsa.eu.int/science/biohaz/biohaz_opinions/1196/biohaz_op_ej269_washing_eggs_en1.pdf. Accessed 24 May 2006.
6. **Audisio, M. C., G. Oliver, and M. C. Apella.** 2000. Protective effect of *Enterococcus faecium* J96, a potential probiotic strain, on chicks infected with *Salmonella* Pullorum. *J. Food Prot.* **63:**1333–1337.
7. **Bailey, J. S., and D. E. Cosby.** 2005. *Salmonella* prevalence in free-range and certified organic chickens. *J. Food Prot.* **68:**2451–2453.
8. **Bailey, J. S., N. J. Stern, and N. A. Cox.** 2000. Commercial field trial evaluation of mucosal starter culture to reduce *Salmonella* incidence in processed broiler carcasses. *J. Food Prot.* **63:**867–870.
9. **Barnhart, H. M., D. W. Dreesen, and J. L. Burke.** 1993. Isolation of *Salmonella* from ovaries and oviducts from whole carcasses of spent hens. *Avian Dis.* **37:**977–980.
10. **Barrow, P. A.** 1990. Immunity to experimental fowl typhoid in chickens induced by a virulence plasmid-cured derivative of *Salmonella gallinarum*. *Infect. Immun.* **58:**2283–2288.
11. **Bell, E., P. Bennett, S. Friedman, C. Riceberg, H. Baskind, M. Beim, C. McGiven, M. Moynihan, S. Shahidi, and D. Sencer.** 1985. Botulism associated with commercially distributed kapchunka—New York City. *Morb. Mortal. Wkly. Rep.* **34:**546–547.
12. **Berghold, C., C. Kornschober, and S. Weber.** 2003. A regional outbreak of *S. enteritidis* phage type 5, traced back to the flocks of an egg producer, Austria. *Euro. Surveill.* **8:**195–198. [Online.] http://www.eurosurveillance.org/em/v08n10/0810-222.asp. Accessed 25 May 2006.
13. **Binkin, N., G. Scuderi, F. Novaco, G. L. Giovanardi, G. Paganelli, G. Ferrari, O. Cappelli, L. Ravaglia, F. Zilioli, V. Amadei, W. Magliani, I. Viani, D. Ricco, B. Borrini, M. Magri, A. Alessandrini, G. Bursi, G. Barigazzi, M. Fantasia, E. Filetici, and S. Salmaso.** 1993. Egg-related *Salmonella enteritidis*, Italy, 1991. *Epidemiol. Infect.* **110:**227–237.
14. **Brashears, M. M., D. Jaroni, and J. Trimble.** 2003. Isolation, selection, and characterization of lactic acid bacteria for a competitive exclusion product to reduce shedding of *Escherichia coli* O157:H7 in cattle. *J. Food Prot.* **66:**355–363.
15. **Buckner, P., D. Ferguson, F. Anzalone, D. Anzalone, J. Taylor, W. G. Hlady, and R. S. Hopkins.** 1994. Outbreak of *Salmonella enteritidis* associated with homemade ice cream—Florida, 1993. *Morb. Mortal. Wkly. Rep.* **43:**669–671.
16. **Center for Veterinary Medicine.** 1998. NADA 141-101. Freedom of information summary. U.S. Food and Drug Administration, Center for Veterinary

Medicine. [Online.] http://www.fda.gov/cvm/FOI/886.htm. Accessed 18 May 2006.

17. **Chàvez-de la Peña, M. E., A. L. Higuera-Iglesias, M. A. Huertas-Jiménez, R. Báez-Martinez, J. Morales-de León, F. Arteaga-Cabello, M. S. Rangel-Frausto, and S. Ponce de León-Rosales.** 2001. Brote por *Salmonella enteritidis* en trabajadores de un hospital (An outbreak of *Salmonella enteritidis* infection among hospital workers in Mexico). *Salud Pública Méx.* **43:**211–216.

18. **Chen, H.-C., and N. J. Stern.** 2001. Competitive exclusion of heterologous *Campylobacter* spp. in chicks. *Appl. Environ. Microbiol.* **67:**848–851.

19. **Cogan, T. A., and T. J. Humphrey.** 2003. The rise and fall of *Salmonella enteritidis* in the UK. *J. Appl. Microbiol.* **94:**114S–119S.

20. **Corrier, D. E., D. J. Nisbet, J. A. Byrd II, B. M. Hargis, N. K. Keith, M. Peterson, and J. R. DeLoach.** 1998. Dosage titration of a characterized competitive exclusion culture to inhibit *Salmonella* colonization in broiler chickens during growout. *J. Food Prot.* **61:**796–801.

21. **Corrier, D. E., D. J. Nisbet, B. M. Hargis, P. S. Holt, and J. R. DeLoach.** 1997. Provision of lactose to molting hens enhances resistance to *Salmonella enteritidis* colonization. *J. Food Prot.* **60:**10–15.

22. **Corrier, D. E., D. J. Nisbet, C. M. Scanlan, G. Tellez, B. M. Hargis, and J. R. DeLoach.** 1994. Inhibition of *Salmonella enteritidis* cecal and organ colonization in leghorn chicks by a defined culture of cecal bacteria and dietary lactose. *J. Food Prot.* **56:**377–381.

23. **Cosby, D. E., S. E. Craven, M. A. Harrison, and N. A. Cox.** 1997. Bacterial isolates from the chicken gizzard and ceca with in vitro inhibitory activity against *Salmonella typhimurium*. *J. Food Prot.* **60:**120–124.

24. **D'Argenio, P., F. Palumbo, R. Ortolani, R. Pizzuti, M. Russo, R. Carducci, M. Soscia, P. Aureli, L. Fenicia, G. Franciosa, A. Parella, and V. Scala.** 1995. Type B botulism associated with roasted eggplant in oil—Italy, 1993. *Morb. Mortal. Wkly. Rep.* **44:**33–36.

25. **D'Argenio, P., A. Romano, and F. Autorino.** 1999. An outbreak of *Salmonella enteritidis* infection associated with iced cake. *Euro. Surveill.* **9:**24–26. [Online.] http://www.eurosurveillance.org/em/v04n02/0402-224.asp. Accessed 25 May 2006.

26. **Doughty, S. C., R. P. O'Connor, J. Alexander, G. J. Sidler, S. Churchill, J. W. Parker, R. Tarter, T. Woods, T. F. Jackamore, Jr., J. C. Bhalerao, E. J. Menamin, P. Hays, M. McVay, C. Gibson, C. Langkop, R. J. Martin, B. J. Francis, M. A. Malik, and J. Damare.** 1984. Foodborne botulism—Illinois. *Morb. Mortal. Wkly. Rep.* **33:**22–23.

27. **Ebel, E., and W. Schlosser.** 2000. Estimating the annual fraction of eggs contaminated with *Salmonella enteritidis* in the United States. *Int. J. Food Microbiol.* **61:**51–62.

28. **Evans, M. R., P. G. Hutchings, C. D. Ribeiro, and D. Westmoreland.** 1996. A hospital outbreak of *Salmonella* food poisoning due to inadequate deep-fat frying. *Epidemiol. Infect.* **116:**155–160.

29. **Evans, M. R., W. Lane, and C. D. Ribeiro.** 1998. *Salmonella enteritidis* PT6: another egg-associated salmonellosis? *Emerg. Infect. Dis.* **4:**667–669.

30. **Federal Register.** 2000. Food labeling, safe handling statements, labeling of shell eggs; refrigeration of shell eggs held for retail distribution; Final Rule. *Fed. Regist.* **65:**76092–76114.

31. **Federal Register.** 2004. Prevention of *Salmonella* Enteritidis in shell eggs during production; Proposed Rule. *Fed. Regist.* **69:**56824–56906.

32. **Federal Register.** 2005. Prevention of *Salmonella* Enteritidis in shell eggs during production; reopening of comment period. *Fed. Regist.* **70:**24490–24491.

33. **Federal Register.** 2005. Prevention of *Salmonella* Enteritidis in shell eggs during production; extension of comment period. *Fed. Regist.* **70:**33404–33405.

34. **Ferreira, A. J. P., C. S. A. Ferreira, T. Knobl, A. M. Moreno, M. R. Bacarro, M. Chen, M. Robach, and G. C. Mead.** 2003. Comparison of three commercial competitive-exclusion products for controlling *Salmonella* colonization of broilers in Brazil. *J. Food Prot.* **66:**490–492.

35. **Food and Drug Administration.** 1 July 1999. New egg safety steps announced, safe handling labels and refrigeration will be required. U.S. Department of Health and Human Services, Food and Drug Administration. [Online.] http://www.cfsan.fda.gov/~lrd/hhseggs.html. Accessed 24 May 2006.

36. **Food and Drug Administration.** 15 April 2002. FDA approves Botox to treat frown lines. U.S. Food and Drug Administration. [Online.] http://www.fda.gov/bbs/topics/ANSWERS/2002/ANS01147.html. Accessed 24 May 2006.

37. **Food and Drug Administration.** 20 July 2004. FDA approves Botox to treat severe underarm sweating. U.S. Food and Drug Administration. [Online.] http://www.fda.gov/bbs/topics/answers/2004/ANS01301.html. Accessed 24 May 2006.

38. **Food and Drug Administration.** 20 September 2004. Fact sheet on FDA's proposed regulation: prevention of *Salmonella* Enteritidis in shell eggs during production. U.S. Food and Drug Administration, Center for Food Safety and Applied Nutrition. [Online.] http://www.cfsan.fda.gov/~dms/fs-eggs6.html. Accessed 24 May 2006.

39. **Food Safety and Inspection Service.** 2006. Shell eggs from farm to table. U.S. Department of Agriculture, Food Safety and Inspection Service. [Online.] http://www.fsis.usda.gov/Fact_Sheets/Focus_On_Shell_Eggs/index.asp. Accessed 24 May 2006.

40. **Fukata, T., K. Sasai, T. Miyamoto, and E. Baba.** 1999. Inhibitory effects of competitive exclusion and fructooligosaccharide, singly and in combination, on *Salmonella* colonization of chicks. *J. Food Prot.* **62:**229–233.

41. **Furtado, C., S. Crespi, L. Ward, and P. Wall.** 1997. Outbreak of *Salmonella enteritidis* phage type 1 infection in British tourists visiting Mallorca, June 1996. *Euro. Surveill.* **2:**6–7. [Online.] http://www.eurosurveillance.org/em/v02n01/0201-224.asp. Accessed 25 May 2006.

42. **Garriga, M., M. Pascual, J. M. Monfort, and M. Hugas.** 1998. Selection of lactobacilli for chicken probiotic adjuncts. *J. Appl. Microbiol.* **84:**125–132.

43. **Gast, R. K., and C. W. Beard.** 1990. Production of *Salmonella enteritidis*-contaminated eggs by experimentally infected hens. *Avian Dis.* **34:**438–446.

44. **Gast, R. K., and C. W. Beard.** 1990. Isolation of *Salmonella enteritidis* from internal organs of experimentally infected hens. *Avian Dis.* **34:**991–993.

45. **Gast, R. K., and P. S. Holt.** 2001. Multiplication in egg yolk and survival in egg albumen of *Salmonella enterica* serotype Enteritidis strains of phage types 4, 8, 13a, and 14b. *J. Food Prot.* **64:**865–868.

46. **Genovese, K. J., R. C. Anderson, R. B. Harvey, T. R. Callaway, T. L. Poole, T. S. Edrington, P. J. Fedorka-Cray, and D. J. Nisbet.** 2003. Competitive

exclusion of *Salmonella* from the gut of neonatal and weaned pigs. *J. Food Prot.* **66:**1353–1359.

47. **Grein, T., D. O'Flanagan, T. McCarthy, and T. Prendergast.** 1997. An outbreak of *Salmonella enteritidis* food poisoning in a psychiatric hospital in Dublin, Ireland. *Euro. Surveill.* **2:**84–86. [Online.] http://www.eurosurveillance.org/em/v02n11/0211-222.asp. Accessed 25 May 2006.
48. **Guard-Petter, J., D. J. Henzler, M. M. Rahman, and R. W. Carlson.** 1997. On-farm monitoring of mouse-invasive *Salmonella enterica* serovar Enteritidis and a model for its association with the production of contaminated eggs. *Appl. Environ. Microbiol.* **63:**1588–1593.
49. **Gusils, C., A. P. Chaia, S. González, and G. Oliver.** 1999. Lactobacilli isolated from chicken intestines: potential use as probiotics. *J. Food Prot.* **62:**252–256.
50. **Hadjichristodoulou, C., E. Nikolakopoulou, K. Karabinis, E. Karakou, A. Markogiannakis, C. Panoulis, M. Lampiri, and P. Tassios.** 1999. Outbreak of *Salmonella* gastroenteritis among attendees of a restaurant opening ceremony in Greece, June 1998. *Euro. Surveill.* **4:**72–75. [Online.] http://www.eurosurveillance.org/em/v04n06/0406-224.asp. Accessed 25 May 2006.
51. **Hammack, T. S., P. S. Sherrod, V. R. Bruce, G. A. June, F. B. Satchell, and W. H. Andrews.** 1993. Research note: growth of *Salmonella enteritidis* in grade A eggs during prolonged storage. *Poult. Sci.* **72:**373–377.
52. **Hassan, J. O., and R. Curtiss III.** 1994. Development and evaluation of an experimental vaccination program using a live avirulent *Salmonella typhimurium* strain to protect immunized chickens against challenge with homologous and heterologous *Salmonella* serotypes. *Infect. Immun.* **62:**5519–5527.
53. **Hassan, J. O., and R. Curtiss III.** 1996. Effect of vaccination of hens with an avirulent strain of *Salmonella typhimurium* on immunity of progeny challenged with wild-type *Salmonella* strains. *Infect. Immun.* **64:**938–944.
54. **Henkel, J.** 1995. Orphan products. New hope for people with rare disorders. *FDA Consumer Special Report.* U.S. Food and Drug Administration. [Online.] http://www.fda.gov/fdac/special/newdrug/orphan.html. Accessed 24 May 2006.
55. **Henzler, D. J., E. Ebel, J. Sanders, D. Kradel, and J. Mason.** 1994. *Salmonella enteritidis* in eggs from commercial chicken layer flocks implicated in human outbreaks. *Avian Dis.* **38:**37–43.
56. **Henzler, D. J., and H. M. Opitz.** 1992. The role of mice in the epizootiology of *Salmonella enteritidis* infection on chicken layer farms. *Avian Dis.* **36:**625–631.
57. **Himathongkham, S., H. Riemann, and R. Ernst.** 1999. Efficacy of disinfection of shell eggs externally contaminated with *Salmonella enteritidis*. Implications for egg testing. *Int. J. Food Microbiol.* **49:**161–167.
58. **Hinton, A., Jr., D. E. Corrier, and J. R. DeLoach.** 1992. In vitro inhibition of *Salmonella typhimurium* and *Escherichia coli* O157:H7 by an anaerobic gram-positive coccus isolated from the cecal contents of adult chickens. *J. Food Prot.* **55:**162–166.
59. **Hinton, M., G. C. Mead, and C. S. Impey.** 1991. Protection of chicks against environmental challenge with *Salmonella enteritidis* by 'competitive exclusion' and acid-treated feed. *Lett. Appl. Microbiol.* **12:**69–71.
60. **Hirn, J., E. Nurmi, T. Johansson, and L. Nuotio.** 1992. Long-term experience with competitive exclusion and salmonellas in Finland. *Int. J. Food Microbiol.* **15:**281–285.

61. **Horn, A., K. Stamper, D. Dahlberg, J. McCabe, M. Beller, and J. P. Middaugh.** 2001. Botulism outbreak associated with eating fermented food—Alaska, 2001. *Morb. Mortal. Wkly. Rep.* **50:**680–682.

62. **Hume, M. E., D. E. Corrier, D. J. Nisbet, and J. R. DeLoach.** 1998. Early *Salmonella* challenge time and reduction in chick cecal colonization following treatment with a characterized competitive exclusion culture. *J. Food Prot.* **61:**673–676.

63. **Hume, M. E., D. J. Nisbet, S. A. Buckley, R. L. Ziprin, R. C. Anderson, and L. H. Stanker.** 2001. Inhibition of in vitro *Salmonella* Typhimurium colonization in porcine cecal bacteria continuous-flow competitive exclusion cultures. *J. Food Prot.* **64:**17–22.

64. **Humphrey, T. J., A. Baskerville, S. Mawer, B. Rowe, and S. Hopper.** 1989. *Salmonella enteritidis* phage type 4 from the contents of intact eggs: a study involving naturally infected hens. *Epidemiol. Infect.* **103:**415–423.

65. **Humphrey, T. J., M. Greenwood, R. J. Gilbert, B. Rowe, and P. A. Chapman.** 1989. The survival of salmonellas in shell eggs cooked under simulated domestic conditions. *Epidemiol. Infect.* **103:**35–45.

66. **Humphrey, T. J., K. W. Martin, and A. Whitehead.** 1994. Contamination of hands and work surfaces with *Salmonella enteritidis* PT4 during the preparation of egg dishes. *Epidemiol. Infect.* **113:**403–409.

67. **Humphrey, T. J., and A. Whitehead.** 1993. Egg age and the growth of *Salmonella enteritidis* PT4 in egg contents. *Epidemiol. Infect.* **111:**209–219.

68. **Hutchison, M. L., J. Gittins, A. Walker, N. Sparks, T. J. Humphrey, C. Burton, and A. Moore.** 2004. An assessment of the microbiological risks involved with egg washing under commercial conditions. *J. Food Prot.* **67:**4–11.

69. **Keller, L. H., C. E. Benson, K. Krotec, and R. J. Eckroade.** 1995. *Salmonella enteritidis* colonization of the reproductive tract and forming and freshly laid eggs of chickens. *Infect. Immun.* **63:**2443–2449.

70. **Kinde, H., D. H. Read, A. Ardans, R. E. Breitmeyer, D. Willoughby, H. E. Little, D. Kerr, R. Gireesh, and K. V. Nagaraja.** 1996. Sewage effluent: likely source of *Salmonella enteritidis*, phage type 4 infection in a commercial chicken layer flock in southern California. *Avian Dis.* **40:**672–676.

71. **Kinde, H., D. H. Read, R. P. Chin, A. A. Bickford, R. L. Walker, A. Ardans, R. E. Breitmeyer, D. Willoughby, H. E. Little, D. Kerr, and I. A. Gardner.** 1996. *Salmonella enteritidis*, phage type 4 infection in a commercial layer flock in Southern California: bacteriologic and epidemiologic findings. *Avian Dis.* **40:**665–671.

72. **Levy, M., M. Fletcher, M. Moody, D. Cory, W. Corbitt, C. Borowiecki, D. Gries, J. Heidingsfelder, A. Oglesby, J. Butwin, D. Ewert, D. Bixler, B. Barrett, K. Laurie, E. Muniz, G. Steele, A. Baldonti, B. Williamson, M. Layton, E. Griffin, M. Cambridge, N. Fogg, J. Guzewich, T. Root, D. Morse, J. Wagoner, M. Deasey, and K. Miller.** 1996. Outbreaks of *Salmonella* serotype *enteritidis* infection associated with consumption of raw shell eggs—United States, 1994–1995. *Morb. Mortal. Wkly. Rep.* **45:**737–742.

73. **Lin, F.-Y. C., J. G. Morris, Jr., D. Trump, D. Tilghman, P. K. Wood, N. Jackman, E. Israel, and J. P. Libonati.** 1988. Investigation of an outbreak of *Salmonella enteritidis* gastroenteritis associated with consumption of eggs in a restaurant chain in Maryland. *Am. J. Epidemiol.* **128:**839–844.

74. **Miyamoto, T., E. Baba, T. Tanaka, K. Sasai, T. Fukata, and A. Arakawa.** 1997. *Salmonella enteritidis* contamination of eggs from hens inoculated by vaginal, cloacal, and intravenous routes. *Avian Dis.* **41:**296–303.

75. **Morner, A., R. Froyman, and B. Gautrais.** 1999. Aviguard, a competitive exclusion product: a novel approach for increased food safety in poultry production. Ontario Ministry of Agriculture, Food and Rural Affairs. [Online.] http://www.omafra.gov.on.ca/english/livestock/animalcare/amr/facts/morner.htm. Accessed 24 May 2006.

76. **Murase, T., K. Senjyu, T. Maeda, M. Tanaka, H. Sakae, Y. Matsumoto, Y. Kaneda, T. Ito, and K. Otsuki.** 2001. Monitoring of chicken houses and an attached egg-processing facility in a laying farm for *Salmonella* contamination between 1994 and 1998. *J. Food Prot.* **64:**1912–1916.

77. **Nisbet, D.** 2002. Defined competitive exclusion cultures in the prevention of enteropathogen colonisation in poultry and swine. *Antonie Van Leeuwenhoek* **81:**481–486.

78. **Nurmi, E., and M. Rantala.** 1973. New aspects of *Salmonella* infection in broiler production. *Nature* **241:**210–211.

79. **Nurmi, E. V., J. E. Schneitz, and P. H. Makela.** 1987. Process for the production of a bacterial preparation for the prophylaxis of intestinal disturbances in poultry. U.S. patent number 4689226. U.S. Patent and Trademark Office. [Online.] http://patft1.uspto.gov/netahtml/PTO/srchnum.htm. Accessed 24 May 2006.

80. **O'Mahony, M., E. Mitchell, R. J. Gilbert, D. N. Hutchinson, N. T. Begg, J. C. Rodhouse, and J. E. Morris.** 1990. An outbreak of foodborne botulism associated with contaminated hazelnut yoghurt. *Epidemiol. Infect.* **104:**389–395.

81. **Pascual, M., M. Hugas, J. I. Badiola, J. M. Monfort, and M. Garriga.** 1999. *Lactobacillus salivarius* CTC2197 prevents *Salmonella enteritidis* colonization in chickens. *Appl. Environ. Microbiol.* **65:**4981–4986.

82. **Pilon, P. A., and M. Laurin.** 1997. Outbreak of *Salmonella enteritidis* phage type 8 in a Montreal hotel. *Can. Communic. Dis. Rep.* **23:**148–150. [Online.] http://www.phac-aspc.gc.ca/publicat/ccdr-rmtc/97vol23/dr2319eb.html. Accessed 25 May 2006.

83. **Poppe, C., R. J. Irwin, C. M. Forsberg, R. C. Clarke, and J. Oggel.** 1991. The prevalence of *Salmonella enteritidis* and other *Salmonella* spp. among Canadian registered commercial layer flocks. *Epidemiol. Infect.* **106:**259–270.

84. **Pratt, S.** 14 April 1988. Officials issue warning of possible *Salmonella* contamination in eggs. *Chicago Tribune*, Chicago, Ill.

85. **President's Council on Food Safety.** 10 December 1999. Egg safety from production to consumption: an action plan to eliminate *Salmonella* Enteritidis illnesses due to eggs. President's Council on Food Safety. [Online.] http://www.foodsafety.gov/~fsg/ceggs.html#actionplan. Accessed 24 May 2006.

86. **Promsopone, B., T. Y. Morishita, P. P. Aye, C. W. Cobb, A. Veldkamp, and J. R. Clifford.** 1998. Evaluation of an avian-specific probiotic and *Salmonella typhimurium*-specific antibodies on the colonization of *Salmonella typhimurium* in broilers. *J. Food Prot.* **61:**176–180.

87. **Prukner-Radovcic, E., and I. C. Grozdanic.** 2003. Competitive exclusion against *Salmonella enterica* subspecies *enterica* serovar Enteritidis infection in chicken. *Vet. Arhiv* **73:**141–152.

88. **Rabsch, W., B. M. Hargis, R. M. Tsolis, R. A. Kingsley, K.-H. Hinz, H. Tschäpe, and A. J. Bäumler.** 2000. Competitive exclusion of *Salmonella enteritidis* by *Salmonella gallinarum* in poultry. *Emerg. Infect. Dis.* **6:**443–448.

89. **Rantala, M., and E. Nurmi.** 1973. Prevention of the growth of *Salmonella infantis* in chicks by the flora of the alimentary tract of chickens. *Br. Poult. Sci.* **14:**627–630.

90. **Reporter, R., L. Mascola, L. Kilman, A. Medina, J. Mohle-Boetani, J. Farrar, D. Vugia, M. Fletcher, M. Levy, O. Ravenholt, L. Empey, D. Maxson, P. Klouse, A. Bryant, R. Todd, M. Williams, G. Cage, and L. Bland.** 2000. Outbreaks of *Salmonella* serotype Enteritidis infection associated with eating raw or undercooked shell eggs—United States, 1996–1998. *Morb. Mortal. Wkly. Rep.* **49:**73–79.

91. **Reynolds, D. J., R. H. Davies, M. Richards, and C. Wray.** 1997. Evaluation of combined antibiotic and competitive exclusion treatment in broiler breeder flocks infected with *Salmonella enterica* serovar Enteritidis. *Avian Pathol.* **26:**83–95.

92. **Ridzon, R., P. Kludt, J. Peppe, K. Sharifzadeh, and S. Lett.** 1997. Two outbreaks of *Salmonella enteritidis* associated with Monte Cristo sandwiches. *J. Food Prot.* **60:**1568–1570.

93. **Rodrigue, D. C., R. V. Tauxe, and B. Rowe.** 1990. International increase in *Salmonella enteritidis*: a new pandemic? *Epidemiol. Infect.* **105:**21–27.

94. **Schneitz, C.** 2006. The competitive exclusion product Broilact®. Orion Corporation, Turku, Finland. [Online.] http://www.orion.fi/english/business_divisions/alasivu.shtml/a03?20509. Accessed 24 May 2006.

95. **Schoeni, J. L., K. A. Glass, J. L. McDermott, and A. C. L. Wong.** 1995. Growth and penetration of *Salmonella enteritidis*, *Salmonella heidelberg* and *Salmonella typhimurium* in eggs. *Int. J. Food Microbiol.* **24:**385–396.

96. **Schultz, S., D. Morse, W. Parkin, G. F. Grady, E. J. Witte, J. L. Hadler, R. L. Vogt, E. Schwartz, K. F. Gensheimer, P. R. Silverman, and E. Israel.** 1987. Increasing rate of *Salmonella enteritidis* infections in the northeastern United States. *Morb. Mortal. Wkly. Rep.* **36:**10–11.

97. **Schultz, S., D. Morse, W. Parkin, G. F. Grady, E. J. Witte, J. L. Hadler, R. L. Vogt, E. Schwartz, K. F. Gensheimer, P. R. Silverman, and E. Israel.** 1987. *Salmonella enteritidis* infections in the northeastern United States. *Morb. Mortal. Wkly. Rep.* **36:**204–205.

98. **Sheffield, C., K. Andrews, R. Harvey, T. Crippen, and D. Nisbet.** 2006. Dereplication by automated ribotyping of a competitive exclusion culture bacterial isolate library. *J. Food Prot.* **69:**228–232.

99. **Shirota, K., H. Katoh, T. Murase, T. Ito, and K. Otsuki.** 2001. Monitoring of layer feed and eggs for *Salmonella* in eastern Japan between 1993 and 1998. *J. Food Prot.* **64:**734–737.

100. **Smith, H. W.** 1956. The use of live vaccines in experimental *Salmonella gallinarum* infection in chickens with observations on their interference effect. *J. Hyg.* **54:**419–432.

101. **Stavric, S., and J.-Y. D'Aoust.** 1993. Undefined and defined bacterial preparations for the competitive exclusion of *Salmonella* in poultry—a review. *J. Food Prot.* **56:**173–180.

102. **Stevens, A., C. Joseph, J. Bruce, D. Fenton, M. O'Mahony, D. Cunningham, B. O'Connor, and B. Rowe.** 1989. A large outbreak of *Salmonella enteritidis* phage type 4 associated with eggs from overseas. *Epidemiol. Infect.* **103**:425–433.

103. **St. Louis, M. E., D. L. Morse, M. E. Potter, T. M. DeMelfi, J. J. Guzewich, R. V. Tauxe, P. A. Blake, and the *Salmonella enteritidis* Working Group.** 1988. The emergence of grade A eggs as a major source of *Salmonella enteritidis* infections. *JAMA* **259**:2103–2107.

104. **St. Louis, M. E., S. H. S. Peck, D. Bowering, G. B. Morgan, J. Blatherwick, S. Banerjee, G. D. M. Kettyls, W. A. Black, M. E. Milling, A. H. W. Hauschild, R. V. Tauxe, and P. A. Blake.** 1988. Botulism from chopped garlic: delayed recognition of a major outbreak. *Ann. Intern. Med.* **108**:363–368.

105. **Sullivan, R.** 31 July 1987. Tainted food possible in patient's death. *New York Times*, New York, N.Y.

106. **Sunagawa, H., T. Ikeda, K. Takeshi, T. Takada, K. Tsukamoto, M. Fujii, M. Kurokawa, K. Watabe, Y. Yamane, and H. Ohta.** 1997. A survey of *Salmonella enteritidis* in spent hens and its relation to farming style in Hokkaido, Japan. *Int. J. Food Microbiol.* **38**:95–102.

107. **Telzak, E. E., L. D. Budnick, M. S. Z. Greenberg, S. Blum, M. Shayegani, C. E. Benson, and S. Schultz.** 1990. A nosocomial outbreak of *Salmonella enteritidis* infection due to the consumption of raw eggs. *N. Engl. J. Med.* **323**:394–397.

108. **Tsuji, H., K. Shimada, K. Hamada, and H. Nakajima.** 2000. Outbreak of *Salmonella enteritidis* caused by contaminated buns peddled by a producer using traveling cars in Hyogo and neighboring prefectures in 1999: an epidemiological study using pulsed-field gel electrophoresis. *Jpn. J. Infect. Dis.* **53**:23–24.

109. **Wagner, R. D., M. Holland, and C. E. Cerniglia.** 2002. An in vitro assay to evaluate competitive exclusion products for poultry. *J. Food Prot.* **65**:746–751.

110. **Weber, J. T., R. G. Hibbs, Jr., A. Darwish, B. Mishu, A. L. Corwin, M. Rakha, C. L. Hatheway, S. el Sharkawy, S. A. el-Rahim, M. F. al-Hamd, J. E. Sarn, P. A. Blake, and R. V. Tauxe.** 1993. A massive outbreak of type E botulism associated with traditional salted fish in Cairo. *J. Infect. Dis.* **167**:451–454.

111. **Wierup, M., H. Wahlstrom, and B. Engstrom.** 1992. Experience of a 10-year use of competitive exclusion treatment as part of the *Salmonella* control programme in Sweden. *Int. J. Food Microbiol.* **15**:287–291.

112. **Wilks, C., G. Parkinson, and P. Young.** 2000. RIRDC Publication no. 00/145. International review of *Salmonella* Enteritidis (SE) epidemiology and control policies. Rural Industries Research and Development Corporation, Australia. [Online.] http://www.rirdc.gov.au/reports/EGGS/00-145.pdf. Accessed 23 May 2006.

113. **Zhao, T., S. Tkalcic, M. P. Doyle, B. G. Harmon, C. A. Brown, and P. Zhao.** 2003. Pathogenicity of enterohemorrhagic *Escherichia coli* in neonatal calves and evaluation of fecal shedding by treatment with probiotic *Escherichia coli*. *J. Food Prot.* **66**:924–930.

chapter 4

See No Evil

IT WAS A CHOCOLATE LOVER'S NIGHTMARE. The lab was stacked to its acoustical tile ceiling with cases of Santas, Easter eggs, and Easter bunnies. And we couldn't so much as nibble an ear; even the broken crumbs were off limits. Nor were we tempted to indulge. The novelties had been manufactured by Regent Chocolate, a Canadian company, and were suspected of containing *Salmonella enterica* serotype Eastbourne.

I was working in the Winnipeg, Manitoba, district lab of Canada's Health Protection Branch (HPB) when news of the *Salmonella* outbreak broke in January 1974 (45). One week later, on February 1, the U.S. Food and Drug Administration (FDA) announced a voluntary recall of Regent's foil-wrapped chocolate balls (49). Because the manufacturer was located in a different province (Quebec), we didn't expect to become involved in the investigation. But a dispute arose between company management and government investigators over how much of the chocolate was contaminated. As a result, HPB decided to test every single batch of chocolate produced by Regent and stored in warehouses across Canada.

Our lab was responsible for checking all of the production lots of suspect chocolate stored in the Canadian provinces of Manitoba and Saskatchewan. We didn't find *Salmonella* in any of the samples that we tested, but some of the other district labs struck paydirt. As a result, the company was forced to recall its entire outstanding inventory, a total value of $6 million. The cost of the recall—and the damage to the company's reputation—was a body blow that broke Regent's financial back and forced it into bankruptcy protection.

In the Beginning

Prior to the outbreak, Regent Chocolate was a large Canadian manufacturer of novelty chocolate items. Fourteen million pounds of chocolate passed through its hands in 1973—over half of it destined for the United States, and the rest remaining in Canada. Regent manufactured chocolate

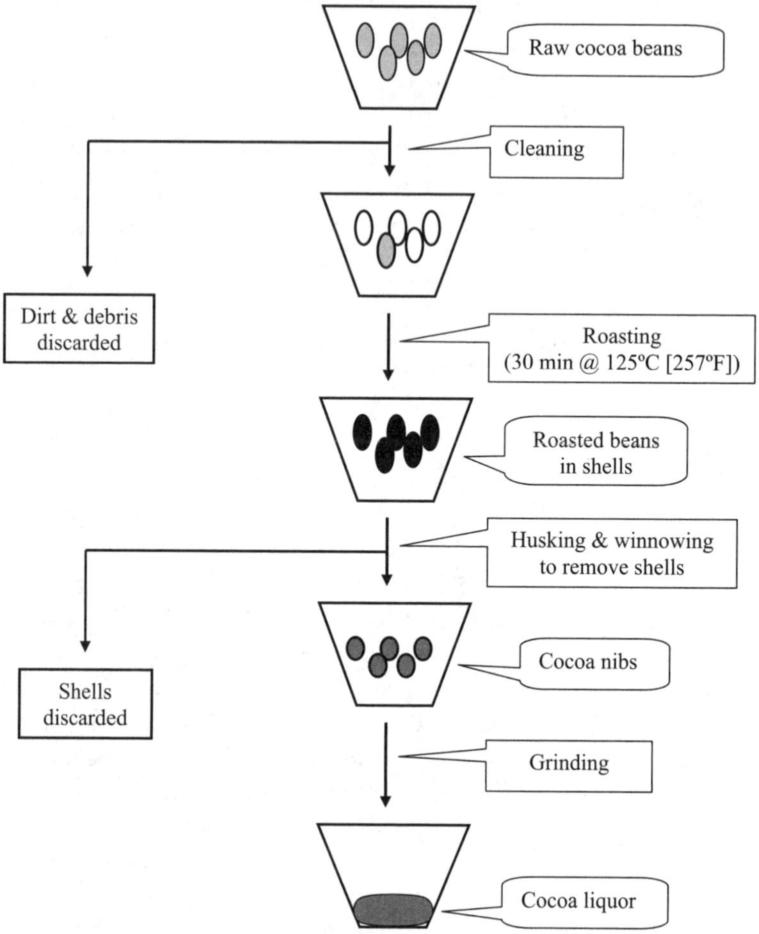

Figure 4.1 Producing cocoa liquor from raw cocoa beans.

for sale under its own brand names and also produced items under contract for other companies (8, 12).

Regent purchased its raw cocoa beans from Nigeria, Ghana, Ecuador, and Brazil. The raw beans were cleaned, roasted, ground, and processed into cocoa liquor, an intermediate step in making chocolate (Fig. 4.1). The company used the liquor for its own chocolate production and also sold some to other manufacturers. Cocoa liquor was combined with a number of other ingredients, including sugar, vanillin, chocolate crumb (a dry mixture of 85% whole milk powder, 5% cocoa powder, and 10% sugar), lecithin, and cocoa butter. Then it was refined, conched, tempered to 104°F (40°C) in a holding tank, and molded into the desired shapes (Fig. 4.2). Once set, the molded chocolate was wrapped either in foil or in cellophane.

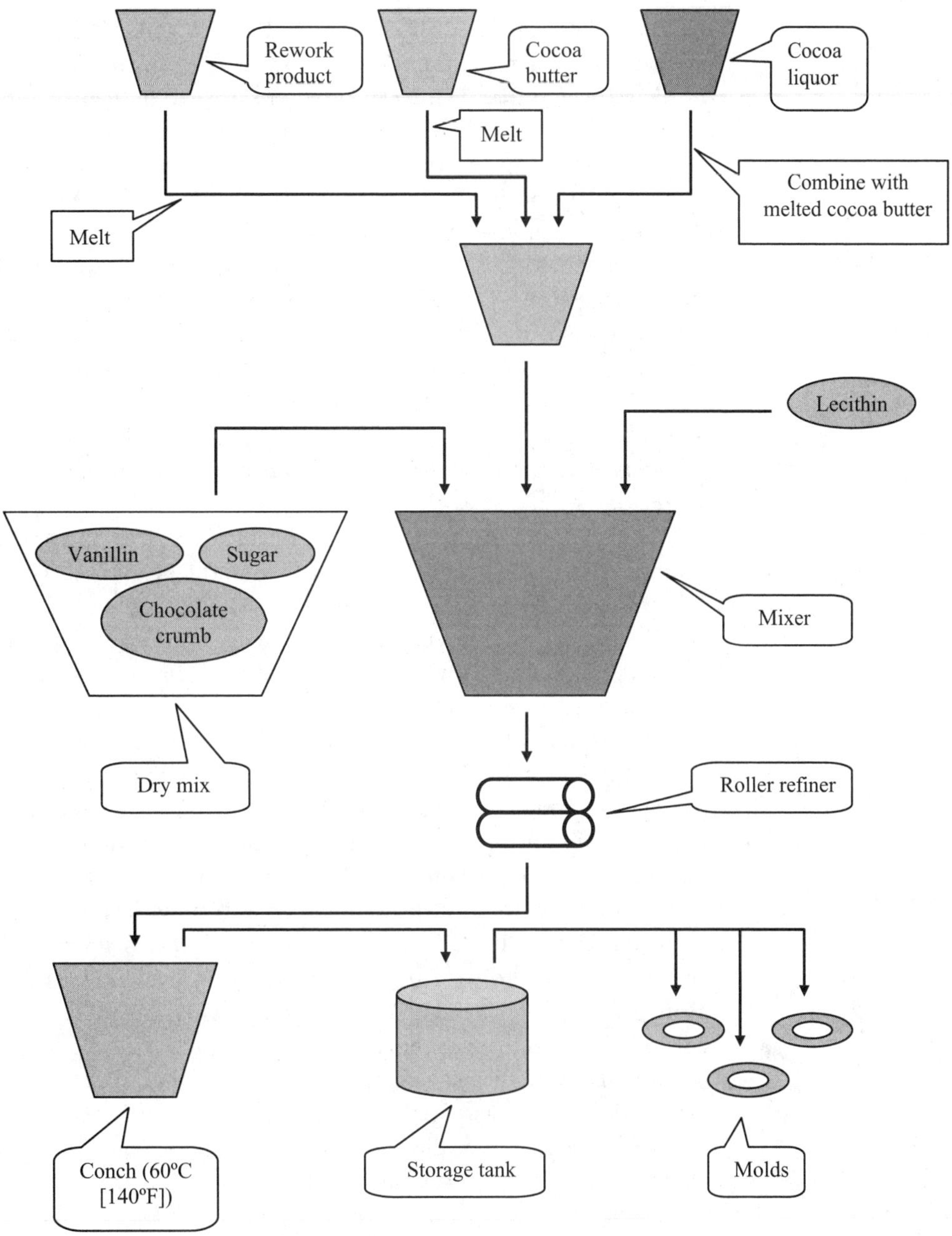

Figure 4.2 Finished chocolate production.

Everything appeared to be running smoothly—until disaster struck late in 1973. On December 4, the U.S. Center for Disease Control (CDC) began receiving reports of salmonellosis due to a previously rare *Salmonella* serotype—*S. enterica* serotype Eastbourne (45). By the end of

February 1974, at least 80 persons in 23 U.S. states and 85 individuals in seven Canadian provinces were stricken (8, 12). And before the outbreak was over, more than 200 people—nearly half of them children under 4 years of age—had been victimized by serotype Eastbourne (11). The results of telephone interviews with victims and their families, and of epidemiological tracking, pointed to Regent's chocolate novelty products as the probable source of the outbreak (3).

Canadian and U.S. investigators descended on the Regent Chocolate facility on February 4, 1974, and spent more than a week combing the production and warehouse operations for clues. They collected hundreds of samples from all parts of the plant, including storage areas, production rooms, the bean roasting room, production line equipment, and all available ingredients. And they began to sample a wide variety of finished products. Everything possible was scrutinized in their effort to find the source of the *Salmonella*.

The only locations in the plant from which the investigators recovered *Salmonella* were the raw bean cleaning room, the bean roasting room, and one of the molding rooms—the first and last links in the production chain (3). Everything in between was negative, as were all of the ingredients stored in the company's warehouse at the time of the investigation— except for a single sample of freshly roasted cocoa beans (12).

How would *Salmonella* have found its way from the raw beans to the finished chocolates? Clearly, since it was present on the roasted beans, the microbe was able to survive roasting. The only subsequent heating step occurred during conching, a special mixing and milling process. But Regent's nominal conching conditions of 12 to 24 hours at 140°F (60°C) varied in reality from lot to lot. In 9 of the 14 months immediately prior to the outbreak, fully 25% of the batches were processed for just 10 hours or less. Four of those 10 hours were taken up with filling the conches, heating and cooling the chocolate, and emptying the equipment, leaving only 6 hours of actual conching time at 140°F (60°C). And 6 hours was not enough to ensure that all salmonellae would be killed (30).

In 1968, more than 5 years before the onset of the Regent Chocolate outbreak, researchers at the University of Wisconsin published the results of their study into heat resistance of *Salmonella* in chocolate (30). They had inoculated two serotypes of *Salmonella*—*S. enterica* serotype Typhimurium and *S. enterica* serotype Senftenberg—into milk chocolate and then exposed the contaminated chocolate to three temperatures. Periodically, the researchers tested the chocolate to determine how many salmonellae were left alive. At least 6 hours of exposure to 158°F (70°C) was needed to kill off 90% of the serotype Senftenberg that had been inoculated into the chocolate. *S. enterica* serotype Typhimurium was even hardier; 10% of its population survived 12 or more hours at 158°F (70°C). And Regent's conching temperature was only 140°F (60°C).

Nor were Regent's processing and quality control procedures all that they should have been. The government inspection of the company's manufacturing facilities and practices brought several problem areas to light. In response, management shut down production and restructured its plant operations. They improved their handling, control, and storage of raw materials; increased conching temperature; changed the procedure used to recycle imperfectly molded chocolate in order to maintain finished product lot identity; began to date-code all packaged products; and expanded the company's bacteriology lab and quality control program.

Regent's response to the investigation report following the serotype Eastbourne outbreak was prompt, positive, and productive—and could serve as a model for how a company can learn from its mistakes. Regent's operations were inspected regularly in the years following the outbreak, and the company came through with flying colors every time.

Some aspects of the Regent Chocolate *Salmonella* outbreak were surprising to investigators and food safety specialists. Notably, based on the level of contamination in the chocolate, fewer than 100 serotype Eastbourne cells were all that had been needed to cause disease—one ten-thousandth of what previously had been thought necessary (12). And in experiments carried out in the years following the outbreak, serotype Eastbourne proved to be much hardier than serotype Typhimurium, a far more common serotype on which most of the infectious dose research had been carried out (43, 44). Suddenly, the presence of small quantities of *Salmonella* seemed more dangerous than before.

In light of these findings, it would be easy to jump to the conclusion that Regent was the victim of an "act of God"; that the outbreak had occurred because serotype Eastbourne was special—extremely heat resistant and highly virulent. But *Salmonella*-contaminated chocolate has been the source of at least four more outbreaks since 1974, each one involving a different serotype (Table 4.1). And in each outbreak for which data are available, just a few *Salmonella* cells were enough to cause illness (12, 31, 33, 34).

The Regent incident ought to have sounded an alarm to the chocolate industry. An outbreak that sent a successful company into bankruptcy should have prompted other manufacturers to revamp their manufacturing, quality control, and testing programs. But complacency and inertia are difficult to overcome. In 1982, another major outbreak—this time due to *S. enterica* serotype Napoli in Italian chocolate—took place in England and Wales (29). And, in 1985–1986, *S. enterica* serotype Nima caused yet another outbreak in North America. The Belgian company that manufactured the serotype Nima-contaminated chocolates had chosen, in spite of the prior Canadian and Italian experiences, not to include *Salmonella* testing in its quality control protocols (33). At least 33 people in Canada

Strange Bedfellows: *Salmonella* and Chocolate

Salmonella does not grow in chocolate. The low moisture, high fat content, and presence of anthocyanins make chocolate an inhospitable environment for the pathogen (1, 2, 28). In spite of this, *Salmonella*-contaminated chocolate has been fingered as the cause of several outbreaks of food-borne illness (8, 11, 29, 31, 33, 34, 48). Ironically, the same conditions that prevent *Salmonella* from growing in chocolate conspire to help the microbe survive.

Fat, protein, and lack of moisture all help protect *Salmonella* from the effects of heating and, to some extent, from other sterilizing or sanitizing agents such as alcohol or organic solvents (13, 30, 47). The low moisture content and the presence of fat in chocolate allow the pathogen to survive for several months—even more than a year—in both milk chocolate and bittersweet chocolate (43, 44, 48). And whereas a portion of the microbes die off at first, enough survive to cause illness if other conditions are right.

Outbreaks of salmonellosis due to eating contaminated chocolate have a common element. Regardless of the serotype involved in the outbreak, the estimated infective dose has been low. Based on the numbers of salmonellae recovered from the chocolate implicated in the various outbreaks, ingesting just 100 cells or fewer usually has been enough to cause disease (see Table 4.1). There are at least two possible explanations for the low infective dose.

The first explanation is that the majority of the outbreak victims were children (8, 29, 33, 48). Although this could be due simply to the relative numbers of children and adults who consumed the contaminated chocolate, it's also possible that the children were more susceptible to *Salmonella* infection than the adults and, as a result, were victimized by a lower dose of the microbe (8). Alternatively, researchers have suggested that the fat content of the chocolate protects *Salmonella* by moderating stomach acidity, allowing most of the bacterial cells to survive and enter the intestine (8, 11, 31).

If stomach acidity was, indeed, moderated by the chocolate, one also should expect low levels of *Salmonella* in other foods to produce illness. In fact, very low infective doses (<1,000 cells) of *Salmonella* have been documented in illnesses linked to contaminated cheddar cheese and hamburger as well as chocolate (46). Furthermore, Waterman and Small reported in 1998 that high-protein foods and high-fat foods (but not high-carbohydrate foods) protected *Salmonella*, *Shigella*, *Vibrio cholerae*, and *Campylobacter jejuni* from acid pH (47). Some elderly individuals also have reduced stomach acidity due to underlying disease or as a result of taking certain medications (42). It is evident, therefore, that determining infective dose is not a simple matter; it depends on the context—the nature of the food and the age and immune status of the victims—not just the virulence of the strain (11, 31, 47).

and the United States—most of them young children—paid dearly for the Belgian manufacturer's shortsighted decision.

Many companies regard microbiological testing of raw ingredients, environmental samples, and finished products as a nonproductive expense. In fact, a well-conceived and correctly managed testing program is an insurance policy—as important to a company's financial health as fire, theft, or liability insurance. However, just as is true for other forms of insurance, a testing program has to be reviewed periodically to ensure that it continues to meet a company's needs. And the results that issue from the program must be heeded and acted on. Unfortunately, in 2002, the management of Pilgrim's Pride chose to close its collective ears to the warnings emanating from its microbiological testing program.

Table 4.1 Food-borne disease outbreaks associated with *Salmonella*-contaminated chocolate

Serotype[a]	Year	Country of manufacture	Location(s) of outbreak	No. of cases	Approximate infective dose	Reference(s)
Eastbourne	1973–74	Canada	Canada, United States	212	<100	8, 12
Napoli	1982	Italy	England, Wales	245	50	29
Nima	1986	Belgium	Canada, United States	33	<10	33
Typhimurium	1987	Norway	Norway, Finland	361	<10	34, 35
Oranienburg	2001	Germany	Germany, Denmark, Sweden, Netherlands, Austria, Belgium, Finland	439	Low[b]	48

[a] *S. enterica* serotype.
[b] Chocolate reported to contain 1.1 to 2.8 *S. enterica* serotype Oranienburg cells per gram.

Hear No Evil

The first reports of illness reached state health officials in Pennsylvania in July 2002. By the end of August, 20 Pennsylvanians had been stricken with *Listeria monocytogenes*—two of them fatally (40). Over the next couple of weeks, increasing numbers of reports of illnesses found their way to the CDC in Atlanta. On September 18, the CDC announced that at least 26 patients from five U.S. states had been infected with the identical strain of *L. monocytogenes* (4). A major food-borne outbreak was in progress, but no one knew which food was spreading the disease.

Listeriosis, the most common disease syndrome caused by *L. monocytogenes*, has a long incubation period. It can take anywhere from a few days to 2 months or more for symptoms to appear. As the weeks passed, reports of illnesses continued to trickle into CDC's offices. By October 4, the confirmed tally had risen to 40 ill, including seven deaths and three miscarriages or stillbirths (5). And while the implicated food appeared to be sliced turkey deli meat, authorities did not yet know the source. Finally, on October 9, the U.S. Department of Agriculture (USDA) announced that Pilgrim's Pride was recalling 295,000 pounds of chicken and turkey deli meats (22). Three days later, on October 12, the company expanded the recall to include everything that had been produced between May 1 and October 11, 2002, at its Franconia, Pa., facility—a total of 27.4 million pounds of meat. And the company announced that it had suspended production at the Franconia plant (23).

Yet that wasn't the last of the recalls. On November 2, Jack Lambersky Poultry Company Inc. (JL Foods) of New Jersey recalled 200,000 pounds of poultry products that were potentially contaminated with *L. monocytogenes* (24). The JL Foods recall was expanded to encompass 4.2 million pounds of meat—mostly cooked chicken and turkey breasts—on November 20 (25).

But which company was responsible for the outbreak? Investigators found multiple strains of *L. monocytogenes* in 25 environmental samples and one finished product from the Pilgrim's Pride facility (7). Two of those samples—both of them from floor drains—yielded the outbreak strain. And the identical microbe was detected in samples of ready-to-eat poultry products from JL Foods (39). Was it just a tragic coincidence that two manufacturing plants—each located in a different state and belonging to a different corporation—were both contaminated with the identical strain of *L. monocytogenes*?

In 1996, USDA opened its Microbial Outbreaks and Special Projects Laboratory (MOSPL) in Athens, Ga. (26). MOSPL's primary responsibility is to analyze any USDA-regulated products that are implicated in foodborne illnesses. Therefore, this lab would have handled the environmental and finished product samples from both the Pilgrim's Pride and the JL Foods plants. Could lab error have played a role?

For lab error to be a credible explanation, samples from both Pilgrim's Pride and JL Foods would need to have been under investigation at the same time. But most of the Pilgrim's Pride samples were obtained and analyzed prior to October 12, when the company announced its expanded recall and plant shutdown. The first *Listeria*-positive result obtained from a JL Foods sample was only revealed on November 2. And the subsequent positive sample was collected for submission to the lab on November 14—well after the Pilgrim's Pride recall. Thus, it would seem that the degree of overlap between the Pilgrim's Pride and JL Foods lab investigations was probably minimal, making lab error an unlikely explanation for the presence of the outbreak strain of *L. monocytogenes* in both production plants. The fact that multiple isolations of the *L. monocytogenes* outbreak strain were reported from both facilities (7, 39) also makes lab error an unlikely explanation.

Another possible explanation is the "Typhoid Mary" scenario. It has been known for many years that people can carry *L. monocytogenes* without exhibiting any symptoms; some carriers have excreted the same serotype for 2 months or more (15, 16). Perhaps an employee of JL Foods ate some of the contaminated Pilgrim's Pride meat, became an asymptomatic carrier, and introduced the contamination—in all innocence—into the JL Foods plant. Only about 30 miles separated the two processing facilities, and Pilgrim's Pride meat was distributed in New Jersey (23, 27). This scenario could explain both the sequential nature of the implication of the two facilities and the isolation of the identical outbreak strain from both plants.

On the other hand, the presence of the outbreak strain in both production plants might have been just a coincidence or the result of both plants having purchased raw poultry from the same supplier. Both Pilgrim's Pride and JL Foods had a prior history of *Listeria* contamination (32, 39). And there have been other examples of the identical strain of *L. monocytogenes* turning up in two or more production facilities at about the same time (37, 41). Sometimes, the same *L. monocytogenes* strain has even been found in two or more countries. A group of Japanese scientists discovered that isolates of an *L. monocytogenes* strain from patients in Japan were of the same genetic type as the strain responsible for two prior outbreaks in California and Switzerland in the 1980s (38).

L. monocytogenes also can be transmitted from one processing plant to another through the purchase and installation of used equipment. In 1998 and 1999, a strain of *L. monocytogenes* traveled sequentially in Finland from plant to plant to plant through the transfer of a contaminated piece of equipment—a dicing machine (36). In each case, the same strain of *L. monocytogenes* was detected for the first time downstream of the contaminated dicing machine approximately 4 weeks after the equipment was installed. The contaminant persisted in the processing line until either the dicing machine was removed (plants A and B) or was subjected to an exhaustive cleaning and sanitation (plant C). Once the offending source of contamination was eliminated, that strain of *L. monocytogenes* disappeared from the plant environment.

It is highly unlikely that we shall ever know the true explanation for how the identical strain of *L. monocytogenes* contaminated two geographically and corporately independent manufacturing plants at the same time. Nor has anyone found definitive proof that either or both processing facilities were responsible for the outbreak—lawsuits are still in progress. Nevertheless, neither company was an innocent victim of circumstances. Both JL Foods and Pilgrim's Pride had ample opportunity to discover and correct problems in their processing plants that could have resulted in this outbreak.

The JL Foods facility had a long history of contamination with *L. monocytogenes*. In 1990, and again in 1994, the USDA found *L. monocytogenes* in a sample of product from that plant. Yet when company management reevaluated its Hazard Analysis and Critical Control Points (HACCP) program in 1999, it chose to ignore the potential for *L. monocytogenes* contamination. As a result, the company made no provision in JL Foods' quality control programs for *Listeria* testing of any sort (39). And—just like Regent Chocolate, some 25 years before—JL Foods learned the hard way that ignorance is definitely not bliss.

The Gift That Keeps on Giving

The ramifications of a food-borne outbreak don't end when the outbreak does—either for the victims or for the company whose food was the source of the outbreak. Some victims pay a lifelong price in the form of chronic disease or disability, and the offending company often is confronted with litigation or even with criminal charges as a result of its sins of commission or omission. Pilgrim's Pride is the subject of a federal criminal investigation stemming from the 2002 outbreak. As of May 2004, that investigation was still under way (27).

Several victims of the 2002 *L. monocytogenes* outbreak sued both Pilgrim's Pride and JL Foods. In May 2006, Pilgrim's Pride settled four wrongful death and injury lawsuits out of court (9). The two companies agreed to pay a combined $3 million to Shakandra Hampton, whose son was born prematurely—and with disabilities—after she contracted an infection with *L. monocytogenes* during the outbreak (10). The details of the other settlements were not revealed (9). Several other lawsuits are still outstanding (9, 10).

Pilgrim's Pride, on the other hand, had been carrying out regular testing for *Listeria* spp. since 1999, though they were not looking specifically for *L. monocytogenes*. But the company chose to ignore the results of its testing program. And, according to a government official, they also neglected to advise the USDA of their testing results. A company spokesperson, however, denied having hidden any information, since "... the company routinely filed its listeria test results in a drawer marked 'U.S.D.A.,' where inspectors could review them" (32). It would appear that, as far as Pilgrim's Pride was concerned, the onus was on the USDA inspector to check the contents of the drawer.

Speak No Evil

When USDA established its HACCP-based regulatory system in 1996, the agency acknowledged *L. monocytogenes* as a "... significant food safety hazard ..." but did not require meat packers and processors to test either their plant environment or their finished products for the presence of this pathogen (17). However, the 1998 BilMar Foods outbreak (see chapter 2), which was also due to *L. monocytogenes*, prompted USDA to rethink its policies.

In February 1999, USDA instructed meat processors to review their HACCP plans in order to determine whether *L. monocytogenes* represented a "... food safety hazard reasonably likely to occur ..." in their production process (18). The notice provided guidance on how to evaluate the likelihood of contamination. Production plants at risk for *L. monocytogenes* contamination were required to update their HACCP plans to address that risk.

Two more years went by before the government made its next move. In January 2001, USDA joined with the FDA to publish a Draft Risk Assessment for *L. monocytogenes* (20). In February 2001—1 month after the Draft Risk Assessment was released and several months before the Pilgrim's Pride/JL Foods outbreak—the agency issued a Proposed Rule for *L. monocytogenes* (19). The rule would have required even those companies that did not believe themselves to be at significant risk for *L. monocytogenes* contamination to implement an environmental monitoring program for *Listeria* spp. Companies that had already incorporated *L. monocytogenes* into their HACCP programs were exempt from the *Listeria* spp. environmental testing requirement.

On paper, USDA appeared to be addressing the gaps in its original HACCP-based regulatory program, but appearances were deceiving. The Proposed Rule was not finalized until June 2003, almost one full year after the Pilgrims Pride/JL Foods outbreak (21). And, in the interim, the agency did not even apply the rules that were already on the books. In neglecting to use all of the enforcement tools at its disposal, USDA failed the consumers it was duty-bound to protect.

JL Foods had a prior history of *L. monocytogenes* in its finished product. On that basis alone, company management was required to incorporate

L. monocytogenes into its HACCP program under the 1999 review notice. Yet JL Foods ignored its past experience and claimed that it was not at significant risk for *L. monocytogenes* contamination. And, even though it knew the company's history of *L. monocytogenes*-positive samples, USDA silently acquiesced.

USDA inspectors also remained silent about multiple incidents of noncompliance at the JL Foods plant. The same sanitation issues recurred with unrelenting regularity. For example, cooked finished product was transported routinely through the raw product area on its way to the storage cooler, exposing the cooked product to potential cross-contamination. Yet these problems were rarely documented in official reports. Instead, inspectors often simply notified the company verbally about the deficiencies that they observed. The USDA District Office responsible for JL Foods noted the paucity of noncompliance reports but took no immediate effective action (39).

The situation at Pilgrim's Pride was just as bad. The company began to test its plant environment for *Listeria* spp. in the spring of 2000, but USDA professed to be unaware of a spike in positive samples that occurred in July and August 2002 (32). The plant also had chronic problems with backed-up floor drains, condensation dripping onto processing tables, and mold and algae on walls. A USDA inspector recommended in August 2002—shortly after the first reports of a food-borne outbreak had already begun to reach health authorities—that the plant be forced to close for repairs. He was overruled. USDA allowed Pilgrim's Pride to continue production in its Franconia, Pa., plant until October 13, when the presence of *L. monocytogenes* was confirmed from a drain sample (14).

Silence is never golden where food safety is concerned, and ignorance is never bliss. Fifty-three people were made ill—and eight of them died—because Pilgrim's Pride and JL Foods flouted food safety rules. And USDA, which is both the promoter and the policeman of the U.S. agricultural industry, abrogated its responsibility to enforce those rules by closing its collective eyes, ears, and mouth to long-standing problems at both plants (6).

References

1. **Barrile, J. C., and J. F. Cone.** 1970. Effect of added moisture on the heat resistance of *Salmonella anatum* in milk chocolate. *Appl. Microbiol.* **19:**177–178.
2. **Busta, F. F., and M. L. Speck.** 1968. Antimicrobial effect of cocoa on salmonellae. *Appl. Microbiol.* **16:**424–425.
3. **Center for Disease Control.** 1974. Follow-up on *Salmonella eastbourne* outbreak—United States, Canada. *Morb. Mortal. Wkly. Rep.* **23:**85–86.
4. **Centers for Disease Control and Prevention.** 18 September 2002. Update: listeriosis outbreak investigation. Centers for Disease Control and Prevention, Media Relations. [Online.] http://www.cdc.gov/od/oc/media/pressrel/r020918b.htm. Accessed 28 May 2006.

5. **Centers for Disease Control and Prevention.** 4 October 2002. Update: listeriosis outbreak investigation. Centers for Disease Control and Prevention, Media Relations. [Online.] http://www.cdc.gov/od/oc/media/pressrel/r021004a.htm. Accessed 28 May 2006.
6. **Centers for Disease Control and Prevention.** 21 November 2002. Update: listeriosis outbreak investigation. Centers for Disease Control and Prevention, Media Relations. [Online.] http://www.cdc.gov/od/oc/media/pressrel/r021121.htm. Accessed 28 May 2006.
7. **Centers for Disease Control and Prevention.** 2002. Public health dispatch: outbreak of listeriosis—Northeastern United States, 2002. *Morb. Mortal. Wkly. Rep.* **51:**950–951.
8. **Craven, P. C., D. C. Mackel, W. B. Baine, W. H. Barker, E. J. Gangarosa, M. Goldfield, H. Rosenfeld, R. Altman, G. Lachapelle, J. W. Davies, and R. C. Swanson.** 1975. International outbreak of *Salmonella eastbourne* infection traced to contaminated chocolate. *Lancet* **i:**788–793.
9. **Dale, M.** 7 May 2006. Pa. meat company settles *Listeria* suits. *Insurance Journal,* San Diego, Calif.
10. **Dale, M.** 12 May 2006. Woman wins $3 million in *Listeria* lawsuit. *The York Dispatch,* York, Pa.
11. **D'Aoust, J. Y.** 1977. *Salmonella* and the chocolate industry. A review. *J. Food Prot.* **40:**718–727.
12. **D'Aoust, J. Y., B. J. Aris, P. Thisdele, A. Durante, N. Brisson, D. Dragon, G. Lachapelle, M. Johnston, and R. Laidley.** 1975. *Salmonella eastbourne* outbreak associated with chocolate. *Can. Inst. Food Sci. Technol. J.* **8:**181–184.
13. **De Fiebre, C. W., K. T. Burck, and D. Feldman.** 1969. Elimination of salmonellae from animal glandular products. *Appl. Microbiol.* **17:**344–346.
14. **Drew, C., and E. Becker.** 11 December 2002. Plant's sanitation may have link to deadly bacteria. *New York Times,* New York, N.Y.
15. **El-Shenawy, M. A.** 1998. Sources of *Listeria* spp. in domestic food processing environment. *Int. J. Environ. Health Res.* **8:**241–251.
16. **Farber, J. M., and P. I. Peterkin.** 1991. *Listeria monocytogenes,* a food-borne pathogen. *Microbiol. Rev.* **55:**476–511.
17. **Federal Register.** 1996. Pathogen reduction; hazard analysis and critical control point (HACCP) systems; Final Rule with request for comments. *Fed. Regist.* **61:**38806–38855.
18. **Federal Register.** 1999. *Listeria monocytogenes* contamination of ready-to-eat products; compliance with the HACCP system regulations and request for comment. *Fed. Regist.* **64:**28351–28353.
19. **Federal Register.** 2001. Performance standards for the production of processed meat and poultry products; Proposed Rule. *Fed. Regist.* **66:** 12590–12636.
20. **Federal Register.** 2001. Relative risk to public health from foodborne *Listeria monocytogenes* among selected categories of ready-to-eat foods; draft risk assessment document and risk management action plan; availability; notice. *Fed. Regist.* **66:**5515–5517.
21. **Federal Register.** 2003. Control of *Listeria monocytogenes* in ready-to-eat meat and poultry products; interim Final Rule. *Fed. Regist.* **68:**34208–34254.

22. **Food Safety and Inspection Service.** 9 October 2002. Pennsylvania firm recalls turkey and chicken products for possible *Listeria* contamination. Food Safety and Inspection Service, U.S. Department of Agriculture. [Online.] http://www.fsis.usda.gov/oa/recalls/prelease/pr090-2002a.htm. Accessed 28 May 2006.

23. **Food Safety and Inspection Service.** 12 October 2002. Pennsylvania firm expands recall of turkey and chicken products for possible *Listeria* contamination. Food Safety and Inspection Service, U.S. Department of Agriculture. [Online.] http://www.fsis.usda.gov/oa/recalls/prelease/pr090-2002.htm. Accessed 28 May 2006.

24. **Food Safety and Inspection Service.** 2 November 2002. New Jersey firm recalls poultry products for possible *Listeria* contamination. Food Safety and Inspection Service, U.S. Department of Agriculture. [Online.] http://www.fsis.usda.gov/OA/recalls/prelease/pr098-2002a.htm. Accessed 28 May 2006.

25. **Food Safety and Inspection Service.** 20 November 2002. New Jersey firm expands recall of poultry products for possible *Listeria* contamination. Food Safety and Inspection Service, U.S. Department of Agriculture. [Online.] http://www.fsis.usda.gov/oa/recalls/prelease/pr098-2002.htm. Accessed 28 May 2006.

26. **Food Safety and Inspection Service.** 2005. FSIS laboratories. Food Safety and Inspection Service, U.S. Department of Agriculture. [Online.] http://www.fsis.usda.gov/science/FSIS_laboratories/index.asp. Accessed 28 May 2006.

27. **Foreman, C. T., and C. Waldrop.** 2004. Not "ready-to-eat." How the meat and poultry industry weakened efforts to reduce *Listeria* food-poisoning. The Consumer Federation of America. [Online.] http://www.agribusinessaccountability.org/pdfs//304_CFA-Not-Ready-to-Eat.pdf. Accessed 26 May 2006.

28. **Forsyth, W. G. C., and V. C. Quesnel.** 1957. Cacao polyphenolic substances. 4. The anthocyanin pigments. *Biochem. J.* **65:**177–179.

29. **Gill, O. N., P. N. Sockett, C. L. R. Bartlett, M. S. B. Vaile, B. Rowe, R. J. Gilbert, C. Dulake, H. C. Murrell, and S. Salmaso.** 1983. Outbreak of *Salmonella napoli* infection caused by contaminated chocolate bars. *Lancet* **i:**574–577.

30. **Goepfert, J. M., and R. A. Biggie.** 1968. Heat resistance of *Salmonella typhimurium* and *Salmonella senftenberg* 775W in milk chocolate. *Appl. Microbiol.* **16:**1939–1940.

31. **Greenwood, M. H., and W. L. Hooper.** 1983. Chocolate bars contaminated with *Salmonella napoli*: an infectivity study. *Br. Med. J.* **286:**1394.

32. **Hazelkorn, B.** 21 December 2002. Poultry plant slow to report sharp increase in bacteria. *New York Times*, New York, N.Y.

33. **Hockin, J. C., J.-Y. D'Aoust, D. Bowering, J. H. Jessop, B. Khanna, H. Lior, and M. E. Milling.** 1989. An international outbreak of *Salmonella nima* from imported chocolate. *J. Food Prot.* **52:**51–54.

34. **Kapperud, G., S. Gustavsen, I. Hellesnes, A. H. Hansen, J. Lassen, J. Hirn, M. Jahkola, M. A. Montenegro, and R. Helmuth.** 1990. Outbreak of *Salmonella typhimurium* infection traced to contaminated chocolate and caused by a strain lacking the 60-megadalton virulence plasmid. *J. Clin. Microbiol.* **28:**2597–2601.

35. **Kapperud, G., J. Lassen, K. Dommarsnes, B.-E. Kristiansen, D. A. Caugant, E. Ask, and M. Jahkola.** 1989. Comparison of epidemiological marker methods for identification of *Salmonella typhimurium* isolates from an outbreak caused by contaminated chocolate. *J. Clin. Microbiol.* **27:**2019–2024.

36. **Lundén, J. M., T. J. Autio, and H. J. Korkeala.** 2002. Transfer of persistent *Listeria monocytogenes* contamination between food-processing plants associated with a dicing machine. *J. Food Prot.* **65:**1129–1133.

37. **Lundén, J. M., T. J. Autio, A.-M. Sjöberg, and H. J. Korkeala.** 2003. Persistent and nonpersistent *Listeria monocytogenes* contamination in meat and poultry processing plants. *J. Food Prot.* **66:**2062–2069.

38. **Nakama, A., M. Terao, Y. Kokubo, T. Itoh, T. Maruyama, C. Kaneuchi, and J. McLauchlin.** 1998. A comparison of *Listeria monocytogenes* serovar 4b isolates of clinical and food origin in Japan by pulsed-field gel electrophoresis. *Int. J. Food Microbiol.* **42:**201–206.

39. **Office of Inspector General.** 2004. Audit Report. Food Safety and Inspection Service oversight of the *Listeria* outbreak in the Northeastern United States. U.S. Department of Agriculture, Office of Inspector General, Northeast Region. Report No. 24601-02-Hy. June 2004. [Online.] http://www.usda.gov/oig/rptsauditsfsis.htm. Accessed 28 May 2006.

40. **Pennsylvania Department of Health.** 30 August 2002. Health department announces investigation of *Listeria* infections. Commonwealth of Pennsylvania, Department of Health. [Online.] http://www.dsf.health.state.pa.us/health/CWP/view.asp?A=190&QUESTION_ID=232137. Accessed 28 May 2006.

41. **Sauders, B. D., K. Mangione, C. Vincent, J. Schermerhorn, C. M. Farchione, N. B. Dumas, D. Bopp, L. Kornstein, E. D. Fortes, K. Windham, and M. Wiedmann.** 2004. Distribution of *Listeria monocytogenes* molecular subtypes among human and food isolates from New York State shows persistence of human disease-associated *Listeria monocytogenes* strains in retail environments. *J. Food Prot.* **67:**1417–1428.

42. **Smith, J. L.** 1998. Foodborne illness in the elderly. *J. Food Prot.* **61:** 1229–1239.

43. **Tamminga, S. K., R. R. Beumer, E. H. Kampelmacher, and F. M. van Leusden.** 1976. Survival of *Salmonella eastbourne* and *Salmonella typhimurium* in chocolate. *J. Hyg. Camb.* **76:**41–47.

44. **Tamminga, S. K., R. R. Beumer, E. H. Kampelmacher, and F. M. van Leusden.** 1977. Survival of *Salmonella eastbourne* and *Salmonella typhimurium* in milk chocolate prepared with artificially contaminated milk powder. *J. Hyg. Camb.* **79:**333–337.

45. **Vernon, T. M., C. W. Langkop, R. J. Martin, B. J. Francis, N. J. Fiumara, G. E. Waterman, D. Coohon, N. S. Hayner, D. S. Fleming, V. Kaupas, G. H. Hauser, C. T. Caraway, H. Rosenfeld, R. Altman, D. Nathan, A. R. Hinman, R. S. Westaby, P. K. Mayville, H. G. Skinner, and R. Laidley.** 1974. *Salmonella eastbourne* infections—Colorado, Illinois, Louisiana, Massachusetts, Michigan, Minnesota, New Hampshire, New Jersey, New York, South Dakota, Wisconsin, Canada. *Morb. Mortal. Wkly. Rep.* **23:**35–36.

46. **Warburton, D. W., J. Harwig, and B. Bowen.** 1993. The survival of salmonellae in homemade chocolate and egg liqueur. *Food Microbiol.* **10:**405–410.

47. **Waterman, S. R., and P. L. C. Small.** 1998. Acid-sensitive enteric pathogens are protected from killing under extremely acidic conditions of pH 2.5 when they are inoculated onto certain solid food sources. *Appl. Environ. Microbiol.* **64:**3882–3886.

48. **Werber, D., J. Dreesman, F. Feil, U. van Treeck, G. Fell, S. Ethelberg, A. M. Hauri, P. Roggentin, R. Prager, I. S. T. Fisher, S. C. Behnke, E. Bartelt, E. Weise, A. Ellis, A. Siitonen, Y. Andersson, H. Tschäpe, M. H. Kramer, and A. Ammon.** 2005. International outbreak of *Salmonella* Oranienburg due to German chocolate. *BMC Infect. Dis.* **5:**7. [Online.] http://www.biomedcentral.com/1471-2334/5/7. Accessed 28 May 2006.

49. **Werner, S. B., J. Chin, J. C. Hart, C. W. Langkop, R. J. Martin, B. J. Francis, N. J. Fiumara, G. E. Waterman, D. Coohon, N. S. Hayner, H. Rosenfeld, R. Altman, M. Goldfield, P. Steele, K. Mosser, F. Bradshaw, J. A. Ackerman, and J. W. Davies.** 1974. Follow-up on *Salmonella eastbourne* epidemic—United States, Canada. *Morb. Mortal. Wkly. Rep.* **23:**37–38.

chapter 5

Cross-Contamination

THE LARGEST FOOD-BORNE *Salmonella* outbreak in U.S. history was almost over before most people even knew it had begun. By the time the media broke the story, the Minnesota Department of Health, working together with the Centers for Disease Control and Prevention (CDC), had already traced the problem to ice cream produced by Schwan's Sales Enterprises of Marshall, Minn. Investigators identified 593 confirmed victims of the outbreak, scattered across 41 U.S. states. But epidemiological and laboratory studies put the toll much higher. An estimated 224,000 people were infected by the *Salmonella*-contaminated ice cream, breaking the 10-year-old record of 197,000 victims that had been set by the Jewel Dairy outbreak of 1984 (48, 87).

Case History

In late September 1994, epidemiologists with the Minnesota Department of Health noticed that the state's Public Health Laboratory had reported an unusually high number of isolations of *Salmonella enterica* serotype Enteritidis from patients living in the southeastern part of the state (18). The Department of Health conducted a telephone survey on October 5 and 6 in order to pinpoint the source of the *Salmonella* outbreak. When the results of that survey identified Schwan's ice cream as the likely culprit, the state immediately notified both the U.S. Food and Drug Administration (FDA) and the company. On October 9, Schwan's announced a nationwide recall of all ice cream that had been produced at its Marshall plant.

The FDA and Minnesota Department of Agriculture labs began testing the returned ice cream for *Salmonella*. They reported their first positive sample on October 17, just over a week after the recall had begun. In all, the agencies examined 266 unopened containers of ice cream. Nine samples—more than 3% of the containers tested—were *Salmonella* positive; eight contained serotype Enteritidis and one was contaminated with *S. enterica* serotype Thompson. The probable link between Schwan's and the outbreak was now a certainty, but the way in which the ice cream had become contaminated was still unknown.

State and federal investigators wasted no time. On October 7—2 days before the recall was announced—inspectors from the FDA and the Minnesota Department of Agriculture began a 3½-week examination of the manufacturing plant and the entire production process. The investigators could not find a source of *Salmonella* within the Schwan's plant. Production equipment, sanitation, and quality control procedures were all functioning properly. Schwan's routinely tested its chocolate flavorings and batches of French vanilla premix for *Salmonella*, with uniformly negative results. The contamination appeared to have originated from outside the plant—possibly from a supplier of the ice cream premix.

The company purchased pasteurized premix for its ice creams from two suppliers and then customized the premix to produce the various flavors and specialty ice cream products. The premix was shipped by tanker truck from the suppliers' facilities to Schwan's Marshall, Minn., location. When government inspectors visited both premix production facilities and gave them a clean bill of health, attention then turned to the trucking company responsible for transporting premix to Schwan's ice cream manufacturing plant.

The tanker trucks that transported pasteurized ice cream premix also carried other loads, including bulk quantities of oils, molasses, corn syrup, and raw liquid bulk egg mixtures. Schwan's was aware of this. The contracts it had negotiated with both premix suppliers and the trucking company specified that the tanker trucks were to be cleaned, sanitized, and inspected by the suppliers before being loaded with pasteurized ice cream mix. And the trucking company had also agreed, as part of its contract, "... to properly maintain and inspect the trailers and to refrain from hauling any products which might cause contamination or adulteration of the ice cream mix intended to be sold at Schwan's" (83). But someone dropped the ball.

Upon investigating the trucking company, FDA and state inspectors found several glaring flaws. Dirty outlet valve gaskets, egg residue, poor record-keeping, lack of routine inspection of tanker trailer interiors, lack of documentation, and cracks in the liners of five tankers—including a tanker that had delivered one of the batches of premix implicated in the outbreak—were among the problems they encountered. The investigators also found out that at least three of the tankers had hauled a load of unpasteurized liquid egg for another customer immediately before transporting ice cream premix to Schwan's. And although they couldn't detect *Salmonella* in any of the tanker trailers, investigators found more than one strain of serotype Enteritidis in samples of unpasteurized liquid egg from the three egg production facilities serviced by the trucking firm.

The investigation team concluded that the Schwan's ice cream outbreak was due to cross-contamination of ice cream premix by

Salmonella-contaminated liquid egg (48). They stated that the pasteurized premix was almost certainly *Salmonella* free when it was loaded into the tanker trucks, but—because of improper maintenance, cleaning, and sanitation—became contaminated by *Salmonella* while in the tankers. Schwan's, however, was not entirely an innocent victim of circumstances.

Most ice cream manufacturers pasteurize their products shortly before freezing and packaging in order to minimize the risk of cross-contamination. But Schwan's production plant was not equipped with a pasteurizer. Company management relied on its suppliers to pasteurize the premix and depended on its trucking company to handle the premix safely and with full attention to proper sanitation (49). Also, ingredients added to the premix by Schwan's—flavors, colors, chocolate, fruit, nuts, and stabilizers—were not pasteurized at all (48).

The company could have done more to prevent this outbreak. Had Schwan's pasteurized the finished ice cream mix prior to freezing, the *Salmonella* would have been destroyed. If in-house pasteurization was not feasible, management at least should have implemented a thorough microbiological testing program, especially including an analysis of every batch of ice cream premix, as well as routine testing of environmental samples and meaningful testing of every batch of finished product.

Though apparently legal at the time, Schwan's procedure was risky. The investigation team recommended that, in the future, any bulk food products that would not be repasteurized at their destination should be transported only in dedicated containers (6, 47, 48). To Schwan's credit, once it was advised of the link between its ice cream and the *Salmonella* outbreak, the company moved quickly to improve its controls and implement the improvements recommended by the FDA. Management began construction of a repasteurization facility almost immediately after the recall was announced, even before the October 17 confirmation that its ice cream was contaminated with serotype Enteritidis. The company now repasteurizes the ice cream mix when it arrives at its production facility (93). In addition, Schwan's initiated a program of testing its finished product for *Salmonella* (34).

The Schwan's ice cream outbreak is an extreme example of the harm that can be caused by cross-contamination. Most food-borne illnesses that are traced to cross-contamination are relatively small and are due to mishandling of foods—especially raw meats and poultry—in a household kitchen or a food service facility. A momentary lapse of attention to good food handling practices in the kitchen can expose food handlers and consumers to pathogens such as *Salmonella*, *Campylobacter*, or *Escherichia coli* O157:H7. And, unfortunately, cross-contamination of raw beef, pork, and poultry in slaughterhouses and packing plants has made the kitchen a high-risk area for food-borne illness.

Spreading the Wealth

Salmonella and *E. coli* O157:H7 in beef; *Salmonella* and *Campylobacter* in poultry and pork—how has our meat and poultry supply become so heavily contaminated with pathogens? Whereas only a small portion of live animals arriving at abattoirs may already be infected with *Salmonella*, *Campylobacter*, or *E. coli* O157:H7, cross-contamination during and after slaughter spreads these bacteria over a substantial percentage of our raw meat and poultry supply.

In the case of pork, the problem begins in the lairage, the area where pigs are held while awaiting slaughter. In a study of two Dutch slaughterhouses, researchers found that the environment and water in the lairages were highly contaminated with *Salmonella*—89% of environmental swab samples and 95% of water samples in one slaughterhouse, and 79% and 62%, respectively, in the other. And the normal weekly cleaning and disinfection procedure did not completely eliminate *Salmonella* (100).

The Dutch researchers concluded that holding time in a contaminated lairage could have a significant impact on the number of *Salmonella*-infected pigs at slaughter. *Salmonella* infection can spread rapidly throughout a pig's body, especially when it's introduced through the nose. The pathogen was recovered from the colon and cecum of pigs as soon as 3 hours after the animals received an intranasal dose in one study (33). A different group of researchers, working within a naturally contaminated environment, found that transport or holding times as short as 2 hours could result in infection of pigs prior to slaughter (54, 55). Yet, pigs may be held in a lairage for 6 hours or more before slaughter—more than enough time for an infection to develop (11, 55, 85). And after slaughter, the chances of cross-contamination multiply, especially during evisceration.

Before a pig carcass can be eviscerated, the hide first must be dehaired. This process consists of four steps: scalding, scraping, singeing, and polishing. First, the carcass is scalded by immersion in water at 156°F (69°C) for 5 minutes to loosen the hair. Then a scraper removes gross organic material and loose hair from the hide. Once scraped, the carcass is singed and polished to remove any residual hair (108). The initial scalding destroys most of the bacteria present on the skin, but the subsequent steps in the dehairing process contaminate the carcass anew with spoilage bacteria and pathogens (37, 42).

Once a carcass has been dehaired, it is eviscerated and the head is removed. Then it is split longitudinally in half, washed, and chilled. Removal of the head and evisceration are major sources of carcass contamination with a variety of bacteria, including coliforms, *E. coli*, *Campylobacter*, and *Salmonella* (13, 66). And the handling involved in splitting and washing the carcass spreads those contaminants throughout the production plant via contact with butchers' hands, cutting tools, and environmental surfaces

(66, 108). Cross-contamination at the various stages of pork slaughter and processing accounted for 29% of the *Salmonella*-contaminated carcasses in a recent Belgian study (12).

The situation in the beef packing industry is similar. *E. coli* O157:H7 was present in the feces of 28% of cattle and on 11% of the hides before slaughter in one recent U.S. study (28). And the level of contamination increased after slaughter, to 43% of uneviscerated beef carcasses. This study also demonstrated a significant correlation between fecal contamination and *E. coli*-positive carcasses. Fortunately, subsequent processing steps at the packing plant reduced the carcass contamination level to just 1.8%. In a follow-up study, researchers concluded that most carcass contamination occurred before there was direct contact between carcasses in the processing area. Once a carcass entered the processing area, contact with contaminated equipment and manipulation of the carcass by plant workers could result in cross-contamination (9).

The potential for spread of pathogens such as *Salmonella* and *E. coli* O157:H7 from carcass to carcass in a processing plant is not limited to the U.S. beef industry. For more than 25 years, researchers on at least four continents have wrestled with the need to define and control the extent of the contamination problem (40, 41, 43, 70, 89, 94, 105). Most attempts to reduce the pathogen population of raw beef have focused on methods for decontaminating carcasses. Trimming and washing, hot water or steam pasteurization, acidified sodium chlorite treatment, rinsing with organic acid, and treatment with lactoperoxidase—an antimicrobial enzyme system—are some of the strategies that have been tried (7, 17, 24, 26, 29, 36, 38, 39, 68). Nevertheless, residual amounts of *Salmonella* and *E. coli* O157:H7 survive decontamination and are dispersed even more widely when beef is ground for hamburger.

Until the meat industry gets its house in order, pork, beef, and other red meats will be an ongoing source of contamination in food service facilities and in the home kitchen. But the risks to public health posed by massive contamination of raw poultry with *Salmonella* and *Campylobacter* far outweigh the problems caused by contaminated pork and beef. And cross-contamination plays a significant role in disseminating those pathogens in poultry meat.

The main steps involved in poultry packing operations are slaughter, scalding, defeathering, evisceration, and chilling. Of these activities, defeathering, evisceration, and chilling carry a high risk of spreading fecal contamination onto the skin of a carcass, between carcasses of a production lot, and even between production lots (19, 50, 76). The spread of bacteria to the outer surface of a carcass actually increases the ability of pathogens to survive decontamination. *Salmonella* and *Campylobacter* that become attached to the skin of a poultry carcass are protected, to some extent, from chlorine used in chilling water in some countries. This protection is due, most likely, to the reduction in free chlorine levels that takes place when dissolved chlorine interacts with organic material and to the presence of an oily layer on the surface of the skin that can act as a barrier between the chlorine and the bacteria that are attached to the skin (116).

As a result of widespread asymptomatic infection of live poultry and the propensity for cross-contamination of carcasses during processing, up to 55% of raw poultry sold at retail may be contaminated with *Salmonella* (15, 91, 92) and as much as 70% with *Campylobacter* (71, 77, 97). As long as the meat and poultry industry continues to engage in practices that encourage further spread of contaminants throughout production lots, consumers will continue to be exposed to a high risk of cross-contamination both in the home and when eating out.

Spreading the Word

The presence of pathogens such as *Salmonella*, *Campylobacter*, and *E. coli* O157:H7 in raw meat, poultry, and eggs places the burden of ensuring food safety on the shoulders—and into the hands—of food preparers. Whether paid food service workers or individuals preparing meals at home, food handlers are expected to know how to handle and cook raw foods correctly, how to avoid cross-contamination of ready-to-eat foods with pathogens, and how to clean and disinfect all work surfaces and utensils. Government agencies try to educate the public on safe food handling practices, and training courses are available to workers in the food service industry, but the message doesn't always get through—or is ignored.

In 1993, the FDA sponsored a telephone survey of food preparation behaviors. One-third of the people questioned admitted to at least one unsafe food preparation practice that could lead to cross-contamination; 25% of those polled did not routinely clean their cutting boards after cutting up raw meat or chicken (4, 63). Three years later, a much larger survey determined that 19.5% of U.S. consumers—nearly one person in five—were still reusing unwashed cutting boards, and 18.5% didn't wash their hands after handling raw meat or poultry (117).

The performance of food preparers in other countries was no better. At least one in three Australians surveyed did not wash their cutting boards at all—or simply wiped the work surface with a damp cloth—between using the boards for raw and cooked meats (60, 73). Nearly one-third of Argentineans surveyed also admitted to the same hazardous practices, as did 25% of people living in the south Wales region of the United Kingdom (14, 114).

Clearly, the home kitchen can be a dangerous place. Poor attention to hand washing, inadequate cleaning of utensils and work surfaces, and use of contaminated sponges or dishrags are all lapses in safe food handling that contribute to cross-contamination. Whether in Europe, North America,

Cutting Boards and Other Kitchen Culprits

Routine kitchen procedures can spread contaminants in the kitchen very efficiently (23). Preparing raw chicken has long been recognized as a key source of *Salmonella* and *Campylobacter* in household kitchens, but the ease and extent of their spread can be surprising. Handling (rinsing, cutting, and seasoning) contaminated raw chicken disperses pathogens throughout the kitchen, not just onto work surfaces and hands (20). Raw eggs are also a source of contamination in the kitchen. *Salmonella* from an infected egg can attach to fingers when the egg is broken by hand and be carried to various places in the kitchen (53, 90).

Once released into the kitchen, pathogens can be spread by aerosol or by direct contact to utensils such as spoons, whisks, mixing bowls, and oven mitts (20, 53, 110). *Salmonella*, *Campylobacter*, and other bacteria can be transferred from contaminated hands and utensils to almost any area in a kitchen, including countertops, sinks, doorknobs, refrigerator handles, cupboards, and dishwashers, and even into spices and condiments used to prepare food (20, 23, 51, 98, 103, 111).

Of all the kitchen utensils that are susceptible to contamination, the one that has probably received the greatest attention is the cutting board. Traditionally, wood has been the preferred surface on which to cut meats, bread, and other food in the kitchen (2). Unlike ceramic, stainless steel, or granite cutting surfaces, cutting on wood does not dull the edge of a sharp knife. But the porous nature of wood led to concerns about the potential for cross-contamination if the same board was used to cut both raw meat and ready-to-eat food. Sanitarians questioned whether a wooden cutting board could be cleaned and sanitized as effectively as a nonporous cutting surface such as plastic, and in 1993, the USDA recommended that consumers replace their wooden cutting boards with plastic ones (2).

Surprisingly, there has been little interest on the part of food safety researchers in determining the relative risks and benefits of wood and plastic boards. Carpentier found only a dozen publications on the subject—most of them supporting the acceptability of wood—in preparing a 1997 review of chopping board safety (16). A 1994 study that was a collaborative effort of Michigan State University (MSU), NSF International, and the FDA demonstrated that dry, new wood surfaces were likely to absorb bacteria, whereas wet, conditioned (with oil), or used surfaces did not promote bacterial adherence (1). Using a direct viable count procedure coupled with scanning electron microscopy, they found that bacteria absorbed into the pores of dry wood cutting boards remained viable, although the bacteria could not be recovered by rinsing (1).

Cliver's research groups at the Food Research Institute in Wisconsin and University of California–Davis have pursued an extensive ongoing study of cutting board safety. Using an experimental design developed to simulate conditions in a home kitchen, they found results somewhat different from those of the MSU team. Whereas bacteria inoculated onto plastic surfaces could be recovered even after 12 hours, the same bacteria "disappeared" from wood surfaces after just a few minutes (1). The rate of bacteria absorption varied depending on the species of wood and the type of grain (end grain versus longitudinal grain), but—unlike the results reported in the MSU study—absorption was not affected by whether or not the wood had been oiled or by the status of the surface (new versus old boards). The researchers suspected, but were unable to confirm, that some species of wood, notably oak, might have antibacterial properties (3). On further study, though, the effect of oak proved to be physical rather than chemical (35).

Whether the contaminated item is a cutting board (wood or plastic), countertop, knife handle, sink drain, or sponge, effective cleaning and sanitation are the key to avoiding cross-contamination in a household or commercial kitchen and, therefore, to reducing the transmission of food-borne illness in the home (27). A significant number of contaminants can be removed from many surfaces by simple rinsing or by washing with soap and water, but correct use of an antimicrobial agent can provide an added measure of safety (21). Sodium hypochlorite (household bleach) is an inexpensive and readily available agent for sanitizing work surfaces, sinks, sponges, and dishcloths, even though *Salmonella* and *Staphylococcus aureus* sometimes can develop

Cutting Boards and Other Kitchen Culprits

resistance to this sanitizer (64, 86). An antibacterial dishwashing liquid was effective against several pathogens when tested in culture suspensions, but performed poorly when used in conjunction with household sponges (65). One of the simplest and most effective ways to decontaminate sponges, kitchen rags, and dishcloths is by exposing them to microwaves for about 60 seconds (79). Even cutting boards—wood, not plastic—can be disinfected safely in a household microwave (78).

or Australia, mishandling of food in the home has been—and continues to be—the source of a significant percentage of reported food-borne illness (69, 82, 104). Twenty-one percent of U.S. food-borne disease outbreaks between 1973 and 1987 were traced to food prepared in the home (10). Up to 50% of the cases in New Zealand in 1997 were caused, at least in part, by poor handling practices in home kitchens (82). And the true proportion of outbreaks due to food preparation errors in the home is probably higher than these estimates, since most cases of sporadic food-borne illness go unreported (101).

One can, perhaps, understand the lack of attention of home food preparers to details of safe food handling. Most have learned their way around a kitchen by watching a parent, a grandparent, or an older sibling. And government efforts to educate consumers about the risks associated with improper kitchen practices have been only partially successful, at best. Old habits are hard to change.

Service with a Smile?

Outdated or inappropriate food handling techniques may be risky at home, but their consequences are magnified in a food service environment. Food-borne illness can strike the patrons of a small catering kitchen, a hospital cafeteria, a local market, or a large banquet hall with equal ease, resulting, in some cases, in hundreds of illnesses (Table 5.1). And, just as in politics, cross-contamination can make strange bedfellows, with pathogens more typically associated with raw meat and poultry finding their way into unusual places such as salads. Such was the case in Oklahoma in 1996.

The Jackson County, Okla., health department started receiving an unusual number of calls from residents complaining of diarrhea and vomiting in mid-August 1996. In all, 14 people fell ill between August 16 and 20 of that year; all were diagnosed with *Campylobacter jejuni* infection. And all had eaten at the same restaurant on the same day. Officials of Jackson County notified the state health authorities of the problem on August 29. State and county officials, working together, immediately began to conduct in-depth interviews of the victims as well as of other restaurant patrons who had not

Table 5.1 Summary of selected food-borne illness outbreaks confirmed or suspected to have been due to cross-contamination in a food service, restaurant, or retail food setting

Pathogen	Setting	Country	Year	Cases (deaths)	Reference(s)
Campylobacter sp.	Senior center	United States	1997	16	113
C. jejuni	Restaurant	United States	1996	14	45
C. jejuni	Summer camp	United States	1998	79	84
E. coli O157	Retail meat shop	United Kingdom	1995	10	99
E. coli O157	Retail store	United Kingdom	2001	30	81
E. coli O157:H7	Restaurant	United States	1993	39	57
E. coli O157:H7	Supermarket meat department	United States	1994	21	8
E. coli O157:H7	Restaurant	United States	1999	72	106
L. monocytogenes	Delicatessen	France	1992	279 (63)	58, 88
S. enterica serotype Agona	Restaurant	United States	1995	7	102
S. enterica serotype Enteritidis	Hospital	Germany	1995	102	62
S. enterica serotype Enteritidis	Retail store	United Kingdom	1995	32	112
S. enterica serotype Enteritidis	Restaurant	United Kingdom	1996	49	52
S. enterica serotype Senftenberg	Hospital	United States	1993–94	22	67
Staphylococcus aureus	Institutional kitchen	United States	1997	18	107

fallen ill. Suspicion quickly fell on the lettuce, which all 14 victims and only 4 unaffected customers had consumed on the 15th of August (45).

Unfortunately, it was too late to recover any of the suspect food for lab analysis. Instead, investigators were limited to inspecting the restaurant in an effort to determine how the lettuce might have become contaminated. When they visited the restaurant kitchen, they were struck by the size of the kitchen countertop. It was so small that the staff would not have been able to keep raw poultry segregated from ready-to-eat foods. In fact, the cook admitted to having cut up raw chicken before using the same counter to prepare sandwiches, salad, and lasagna. And although a bleach solution was available to disinfect the countertop between uses, investigators couldn't be sure that the cook had sanitized the counter after cutting up the raw chicken. Nor was it clear whether—or how carefully—the cook had washed her hands or the utensils after handling the raw poultry.

Although *Campylobacter* is not a common contaminant of produce, it is present in a high percentage of raw poultry meat (75). When mishandling creates conditions that open the door to cross-contamination, *Campylobacter* can walk right through. Nor does it take many live cells to cause illness; as few as 500 organisms can do the job (45). Just a few drops of contaminated raw chicken juice were probably all that were needed to sicken those 14 restaurant patrons.

Campylobacter is not the only pathogen that can be transmitted by cross-contamination. *Salmonella*, *Listeria monocytogenes*, viruses, and pathogenic strains of *E. coli* can be transferred from raw products, work surfaces, utensils, or hands to ready-to-eat foods such as cooked meats or salads. That's what happened at four separate Sizzler restaurant locations in 1993.

Where's the Beef?

It was the year of the Rooster according to the Chinese calendar, but in the United States, 1993 was the year of *E. coli* O157:H7. On January 13, 1993, a doctor reported a cluster of illnesses to the Washington State Department of Health. By the end of February, what later became known as the Jack in the Box outbreak had spread to four western states, causing more than 500 laboratory-confirmed illnesses and four deaths (22). So when a second *E. coli* O157:H7 outbreak surfaced in March 1993 and was linked to a Sizzler steak house, the first reaction was to blame the beef. The cause, however, of the Sizzler outbreak was far more complex than undercooked hamburgers.

The Sizzler saga developed in phases. An initial spate of illnesses in late March 1993 was associated with outlets in Grants Pass and North Bend, Oreg.; a third Sizzler restaurant located in Corvallis, Oreg., was hit in early August 1993, and a fourth one in Seattle, Wash., about a week later. In all, 93 individuals were taken ill; 15 required hospitalization. Fortunately, all 93 survived their bout of illness (57).

When epidemiologists with the state of Oregon and the CDC began their investigation of the initial episode in March 1993, they were surprised to find that there was no apparent link between eating meat and the *E. coli* O157:H7 illness. Instead, their results pointed to consumption of salad bar items—mayonnaise-based foods, salad dressing, carrots, and cantaloupe. The pattern that developed in the two August episodes was similar, though not identical. Suspicion in the August outbreaks focused on cantaloupe, lettuce, tomatoes, and prepackaged shredded cheese.

After reviewing the four outbreaks in their entirety, investigators concluded that cross-contamination was the most likely cause. All four Sizzler locations procured their meat from the same packing plant; some of the raw beef was precut, but much of it was purchased in 5-kg (11-lb) portions. The 5-kg (11-lb) pieces were trimmed, cut, tenderized, and marinated in the restaurant kitchens. Although the designs of the kitchens and storage facilities were not ideal in at least two of the locations—the raw meat was prepared and stored in close proximity to the salad fixings and other foods—investigators uncovered no clear evidence of product mishandling. Sizzler's practice, however, of trimming, tenderizing, and marinating the meat in the restaurant kitchens exposed the work area to the potential for contamination due to spillage or splatter of raw meat juices and marinades. Yet even though the restaurants prepared a wide range of salad bar items

Don't Hold the Mayo

Mayonnaise has been the victim of bad press over the years (74). And, on occasion, it still gets fingered—undeservedly—as a high-risk source of food-borne illness (5, 25). In fact, when prepared correctly, mayonnaise is a very unfriendly environment for the survival and growth of most food-borne pathogens.

Mayonnaise contains only a few ingredients: egg yolk, oil, an acidifying agent (usually citric acid or vinegar), and some seasoning (115). The main key to its safety lies in its acidity. Unlike the situation in Europe, the U.S. version of mayonnaise is very acidic, with a pH range of 3.6 to 4.0 (95, 96). But its pH is not the entire story; organic acids such as citric, lactic, or acetic acid (vinegar) have antimicrobial properties that extend beyond simple acidity. They can interfere with microbial metabolism (59).

The lethal effect of acidified mayonnaise on food-borne pathogens has been well documented in the last 20+ years. *E. coli* O157:H7, *Salmonella*, and *L. monocytogenes* all die off, at least to some extent, in commercial mayonnaise prepared according to U.S. formulation standards (30, 32, 118). This is true for both traditional and reduced-calorie mayonnaises, as well as for mayonnaise-based sauces and salad dressings (30, 44, 46, 109). Contrary to what might be expected, the pathogens die more rapidly at ambient temperature (68 to 77°F [20 to 25°C]) than under refrigerated storage conditions (46, 72, 80, 109).

The antimicrobial effect of mayonnaise also extends a modicum of protection to some mayonnaise-based salads. If proper attention is paid to correct preparation, storage, and handling conditions, salads made using commercial mayonnaise often can be stored safely for several days. *Salmonella* and *L. monocytogenes* both were able to reproduce in temperature-abused (55°F [12.8°C]) home-style chicken salad made with commercial mayonnaise in a 1993 research study (31). *Salmonella* died in macaroni salad—more acidic than the chicken salad—made with the same mayonnaise and held for 2 days at 55°F (12.8°C), whereas *L. monocytogenes* survived but did not grow in the macaroni salad. *L. monocytogenes*, which is usually able to grow slowly at 39°F (4°C), did not multiply in the refrigerated macaroni salad at all over the duration of the 10-day study, but began to grow after a 7-day lag period in the refrigerated chicken salad. In a separate study published in 2005, *L. monocytogenes* grew in ham salad at temperatures ranging between 39 and 54°F (4 and 12°C) but was inactivated in potato salad prepared with the same mayonnaise (56).

in the same kitchens, these activities went on without obvious ill effect—until the meat distributor supplied the four Sizzler locations with *E. coli* O157:H7-contaminated raw beef (61).

As a result of this series of four outbreaks, Sizzler International reviewed and made some significant changes to its internal practices. The company instituted a comprehensive Hazard Analysis and Critical Control Point program and also switched entirely to precut meat rather than cutting meat on location. More than 10 years after those changes were implemented, there have been no further outbreaks associated with Sizzler restaurants (57).

Sizzler, Schwan's, and many other companies and individuals have found themselves in microbiological hot water because they clung to familiar and comfortable old habits. Over the years, this same mindset has greatly facilitated the proliferation of emerging pathogens such as *E. coli* O157:H7 (see chapter 6) and the agent that causes mad cow disease (see chapter 10).

References

1. **Abrishami, S. H., B. D. Tall, T. J. Bruursema, P. S. Epstein, and D. B. Shah.** 1994. Bacterial adherence and viability on cutting board surfaces. *J. Food Safety* **14:**153–172.
2. **Ak, N. O., D. O. Cliver, and C. W. Kaspar.** 1994. Cutting boards of plastic and wood contaminated experimentally with bacteria. *J. Food Prot.* **57:**16–22.
3. **Ak, N. O., D. O. Cliver, and C. W. Kaspar.** 1994. Decontamination of plastic and wooden cutting boards for kitchen use. *J. Food Prot.* **57:**23–30.
4. **Altekruse, S. F., D. A. Street, S. B. Fein, and A. S. Levy.** 1996. Consumer knowledge of foodborne microbial hazards and food-handling practices. *J. Food Prot.* **59:**287–294.
5. **Anonymous.** 16 June 1988. Illness outbreak in Lords traced; *Salmonella* outbreak. *The Times*, London, United Kingdom.
6. **Anonymous.** 27 May 1996. Officials believe 220,000 made ill by tainted ice cream. *Food Inst. Rep.* **69:**10–11.
7. **Bacon, R. T., K. E. Belk, J. N. Sofos, R. P. Clayton, J. O. Reagan, and G. C. Smith.** 2000. Microbial populations on animal hides and beef carcasses at different stages of slaughter in plants employing multiple-sequential interventions for decontamination. *J. Food Prot.* **63:**1080–1086.
8. **Banatvala, N., A. R. Magnano, M. L. Cartter, T. J. Barrett, W. F. Bibb, L. L. Vasile, P. Mshar, M. A. Lambert-Fair, J. H. Green, N. H. Bean, and R. V. Tauxe.** 1996. Meat grinders and molecular epidemiology: two supermarket outbreaks of *Escherichia coli* O157:H7 infection. *J. Infect. Dis.* **173:**480–483.
9. **Barkocy-Gallagher, G. A., T. M. Arthur, G. R. Siragusa, J. E. Keen, R. O. Elder, W. W. Laegreid, and M. Koohmaraie.** 2001. Genotypic analyses of *Escherichia coli* O157:H7 and O157: nonmotile isolates recovered from beef cattle and carcasses at processing plants in the midwestern states of the United States. *Appl. Environ. Microbiol.* **67:**3810–3818.
10. **Bean, N. H., and P. M. Griffin.** 1990. Foodborne disease outbreaks in the United States, 1973–1987: pathogens, vehicles, and trends. *J. Food Prot.* **53:**804–817.
11. **Belœil, P.-A., C. Chauvin, K. Proux, F. Madec, P. Fravalo, and A. Alioum.** 2004. Impact of the *Salmonella* status of market-age pigs and the pre-slaughter process on *Salmonella* caecal contamination at slaughter. *Vet. Res.* **35:**513–530.
12. **Botteldoorn, N., M. Heyndrickx, N. Rijpens, K. Grijspeerdt, and L. Herman.** 2003. *Salmonella* on pig carcasses: positive pigs and cross contamination in the slaughterhouse. *J. Appl. Microbiol.* **95:**891–903.
13. **Bryant, J., D. A. Brereton, and C. O. Gill.** 2003. Implementation of a validated HACCP system for the control of microbiological contamination of pig carcasses at a small abattoir. *Can. Vet. J.* **44:**51–55.
14. **Califano, A. N., G. L. de Antoni, L. Giannuzzi, and R. H. Mascheroni.** 2000. Prevalence of unsafe practices during home preparation of food in Argentina. *Dairy Food Environ. Sanit.* **20:**934–943.
15. **Capita, R., M. Álvarez-Astorga, C. Alonso-Calleja, B. Moreno, and M. del Camino García-Fernández.** 2003. Occurrence of salmonellae in retail chicken carcasses and their products in Spain. *Int. J. Food Microbiol.* **81:**169–173.

16. **Carpentier, B.** 1997. Sanitary quality of meat chopping board surfaces: a bibliographical study. *Food Microbiol.* **14:**31–37.

17. **Castillo, A., L. M. Lucia, G. K. Kemp, and G. R. Acuff.** 1999. Reduction of *Escherichia coli* O157:H7 and *Salmonella typhimurium* on beef carcass surfaces using acidified sodium chlorite. *J. Food Prot.* **62:**580–584.

18. **Centers for Disease Control and Prevention.** 1994. Emerging infectious diseases outbreak of *Salmonella enteritidis* associated with nationally distributed ice cream products—Minnesota, South Dakota, and Wisconsin, 1994. *Morb. Mortal. Wkly. Rep.* **43:**740–741.

19. **Clouser, C. S., S. J. Knabel, M. G. Mast, and S. Doores.** 1995. Effect of type of defeathering system on *Salmonella* cross-contamination during commercial processing. *Poult. Sci.* **74:**732–741.

20. **Cogan, T. A., S. F. Bloomfield, and T. J. Humphrey.** 1999. The effectiveness of hygiene procedures for prevention of cross-contamination from chicken carcases in the domestic kitchen. *Lett. Appl. Microbiol.* **29:**354–358.

21. **Cogan, T. A., J. Slader, S. F. Bloomfield, and T. J. Humphrey.** 2002. Achieving hygiene in the domestic kitchen: the effectiveness of commonly used cleaning procedures. *J. Appl. Microbiol.* **92:**885–892.

22. **Davis, M., C. Osaki, D. Gordon, M. W. Hinds, K. Mottram, C. Winegar, E. D. Avner, P. I. Tarr, D. Jardine, M. Goldoft, B. Bartleson, J. Lewis, J. M. Kobayashi, G. Billman, J. Bradley, S. Hunt, P. Tanner, M. Ginsberg, L. Barrett, S. B. Werner, G. W. Rutherford III, R. W. Jue, H. Root, D. Brothers, R. L. Chehey, R. H. Hudson, F. R. Dixon, D. J. Maxson, L. Empey, O. Ravenholt, V. H. Ueckart, A. DiSalvo, D. S. Kwalick, R. Salcido, and D. Brus.** 1993. Update: multistate outbreak of *Escherichia coli* O157:H7 infections from hamburgers—Western United States, 1992–1993. *Morb. Mortal. Wkly. Rep.* **42:**258–263.

23. **De Wit, J. C., G. Broekhuizen, and E. H. Kampelmacher.** 1979. Cross-contamination during the preparation of frozen chickens in the kitchen. *J. Hyg. Camb.* **83:**27–32.

24. **Dickson, J. S., and M. E. Anderson.** 1992. Microbiological decontamination of food animal carcasses by washing and sanitizing systems: a review. *J. Food Prot.* **55:**133–140.

25. **Donohue, P. G.** 13 December 1999. Faster food poisoning arrives prepackaged. *The Star-Ledger*, Newark, N.J.

26. **Dorsa, W. J., C. N. Cutter, G. R. Siragusa, and M. Koohmaraie.** 1996. Microbial decontamination of beef and sheep carcasses by steam, hot water spray washes, and a steam-vacuum sanitizer. *J. Food Prot.* **59:**127–135.

27. **Duff, S. B., E. A. Scott, M. S. Mafilios, E. C. Todd, L. R. Krilov, A. M. Geddes, and S. J. Ackerman.** 2003. Cost-effectiveness of a targeted disinfection program in household kitchens to prevent foodborne illnesses in the United States, Canada, and the United Kingdom. *J. Food Prot.* **66:**2103–2115.

28. **Elder, R. O., J. E. Keen, G. R. Siragusa, G. A. Barkocy-Gallagher, M. Koohmaraie, and W. W. Laegreid.** 2000. Correlation of enterohemorrhagic *Escherichia coli* O157 prevalence in feces, hides, and carcasses of beef cattle during processing. *Proc. Natl. Acad. Sci. USA* **97:**2999–3003.

29. **Elliot, R. M., J. C. McLay, M. J. Kennedy, and R. S. Simmonds.** 2004. Inhibition of foodborne bacteria by the lactoperoxidase system in a beef cube system. *Int. J. Food Microbiol.* **91:**73–81.

30. **Erickson, J. P., and P. Jenkins.** 1991. Comparative *Salmonella* spp. and *Listeria monocytogenes* inactivation rates in four commercial mayonnaise products. *J. Food Prot.* **54:**913–916.

31. **Erickson, J. P., D. N. McKenna, M. A. Woodruff, and J. S. Bloom.** 1993. Fate of *Salmonella* spp., *Listeria monocytogenes*, and indigenous spoilage microorganisms in home-style salads prepared with commercial real mayonnaise or reduced calorie mayonnaise dressings. *J. Food Prot.* **56:**1015–1021.

32. **Erickson, J. P., J. W. Stamer, M. Hayes, D. N. McKenna, and L. A. van Alstine.** 1995. An assessment of *Escherichia coli* O157:H7 contamination risks in commercial mayonnaise from pasteurized eggs and environmental sources, and behavior in low-pH dressings. *J. Food Prot.* **58:**1059–1064.

33. **Fedorka-Cray, P. J., L. C. Kelley, T. J. Stabel, J. T. Gray, and J. A. Laufer.** 1995. Alternate routes of invasion may affect pathogenesis of *Salmonella typhimurium* in swine. *Infect. Immun.* **63:**2658–2664.

34. **Food and Drug Administration.** 7 November 1994. FDA approves Schwan's corrective steps for ice cream. U.S. Food and Drug Administration. [Online.] http://www.fda.gov/bbs/topics/ANSWERS/ANS00615.html. Accessed 4 June 2006.

35. **Galluzzo, L., and D. O. Cliver.** 1996. Cutting boards and bacteria—oak vs. *Salmonella*. *Dairy Food Environ. Sanit.* **16:**290–293.

36. **Gill, C. O., M. Badoni, and T. Jones.** 1996. Hygienic effects of trimming and washing operations in a beef-carcass-dressing process. *J. Food Prot.* **59:**666–669.

37. **Gill, C. O., and J. Bryant.** 1993. The presence of *Escherichia coli*, *Salmonella* and *Campylobacter* in pig carcass dehairing equipment. *Food Microbiol.* **10:**337–344.

38. **Gill, C. O., and J. Bryant.** 2000. The effects on product of a hot water pasteurizing treatment applied routinely in a commercial beef carcass dressing process. *Food Microbiol.* **17:**495–504.

39. **Gill, C. O., J. Bryant, and D. Bedard.** 1999. The effects of hot water pasteurizing treatments on the appearances and microbiological conditions of beef carcass sides. *Food Microbiol.* **16:**281–289.

40. **Gill, C. O., and L. M. Harris.** 1982. Contamination of red-meat carcasses by *Campylobacter fetus* subsp. *jejuni*. *Appl. Environ. Microbiol.* **43:**977–980.

41. **Gill, C. O., and L. M. Harris.** 1982. Survival and growth of *Campylobacter fetus* subsp. *jejuni* on meat and in cooked foods. *Appl. Environ. Microbiol.* **44:**259–263.

42. **Gill, C. O., and T. Jones.** 1997. Assessment of the hygienic characteristics of a process for dressing pasteurized pig carcasses. *Food Microbiol.* **14:**81–91.

43. **Gill, C. O., J. C. McGinnis, K. Rahn, and A. Houde.** 1996. The hygienic condition of manufacturing beef destined for the manufacture of hamburger patties. *Food Microbiol.* **13:**391–396.

44. **Glass, K. A., and M. P. Doyle.** 1991. Fate of *Salmonella* and *Listeria monocytogenes* in commercial, reduced-calorie mayonnaise. *J. Food Prot.* **54:**691–695.

45. **Graves, T. K., K. K. Bradley, and J. M. Crutcher.** 1998. Outbreak of *Campylobacter* enteritis associated with cross-contamination of food—Oklahoma, 1996. *Morb. Mortal. Wkly. Rep.* **47:**129–131.

46. **Hathcox, A. K., L. R. Beuchat, and M. P. Doyle.** 1995. Death of enterohemorrhagic *Escherichia coli* O157:H7 in real mayonnaise and reduced-calorie

mayonnaise dressing as influenced by initial population and storage temperature. *Appl. Environ. Microbiol.* **61:**4172–4177.

47. **Henkel, J.** 1995. Ice cream firm linked to *Salmonella* outbreak. *FDA Consumer* **29:**30–31.

48. **Hennessy, T. W., C. W. Hedberg, L. Slutsker, K. E. White, J. M. Besser-Wiek, M. E. Moen, J. Feldman, W. W. Coleman, L. M. Edmonson, K. L. Macdonald, M. T. Osterholm, and The Investigation Team.** 1996. A national outbreak of *Salmonella enteritidis* infections from ice cream. *N. Engl. J. Med.* **334:**1281–1286.

49. **Hilts, P. J.** 21 October 1994. *Salmonella* poisoning outbreak may be linked to tanker trucks. *New York Times*, New York, N.Y.

50. **Hinton, A., Jr., J. A. Cason, and K. D. Ingram.** 2004. Tracking spoilage bacteria in commercial poultry processing and refrigerated storage of poultry carcasses. *Int. J. Food Microbiol.* **91:**155–165.

51. **Holah, J. T., and R. H. Thorpe.** 1990. Cleanability in relation to bacterial retention on unused and abraded domestic sink materials. *J. Appl. Bacteriol.* **69:**599–608.

52. **Holtby, I., G. M. Tebbutt, E. Grunert, H. J. Lyle, and M. P. Stenson.** 1997. Outbreak of *Salmonella enteritidis* phage type 6 infection associated with food items provided at a buffet meal. *Commun. Dis. Rep. CDR Rev.* **7:**R87–R90. [Online.] http://www.hpa.org.uk/cdr/archives/CDRreview/1997/cdrr0697.pdf. Accessed 4 June 2006.

53. **Humphrey, T. J., K. W. Martin, and A. Whitehead.** 1994. Contamination of hands and work surfaces with *Salmonella enteritidis* PT4 during the preparation of egg dishes. *Epidemiol. Infect.* **113:**403–409.

54. **Hurd, H. S., J. K. Gailey, J. D. McKean, and M. H. Rostagno.** 2001. Rapid infection in market-weight swine following exposure to a *Salmonella typhimurium*-contaminated environment. *Am. J. Vet. Res.* **62:**1194–1197.

55. **Hurd, H. S., J. D. McKean, R. W. Griffith, I. V. Wesley, and M. H. Rostagno.** 2002. *Salmonella enterica* infections in market swine with and without transport and holding. *Appl. Environ. Microbiol.* **68:**2376–2381.

56. **Hwang, C.-A.** 2005. Effect of mayonnaise pH and storage temperature on the behavior of *Listeria monocytogenes* in ham salad and potato salad. *J. Food Prot.* **68:**1628–1634.

57. **Jackson, L. A., W. E. Keene, J. M. McAnulty, E. R. Alexander, M. Diermayer, M. A. Davis, K. Hedberg, J. Boase, T. J. Barrett, M. Samadpour, and D. W. Fleming.** 2000. Where's the beef? The role of cross-contamination in 4 chain restaurant-associated outbreaks of *Escherichia coli* O157:H7 in the Pacific Northwest. *Arch. Intern. Med.* **160:**2380–2385.

58. **Jacquet, C., B. Catimel, R. Brosch, C. Buchrieser, P. Dehaumont, V. Goulet, A. Lepoutre, P. Veit, and J. Rocourt.** 1995. Investigations related to the epidemic strain involved in the French listeriosis outbreak in 1992. *Appl. Environ. Microbiol.* **61:**2242–2246.

59. **Jay, J. M.** 2000. *Modern Food Microbiology*, 6th ed. Aspen Publishers, Inc., Gaithersburg, Md.

60. **Jay, L. S., D. Comar, and L. D. Govenlock.** 1999. A national Australian food safety telephone survey. *J. Food Prot.* **62:**921–928.

61. **Kight, P.** 21 August 1993. *E. coli* believed transferred to cantaloupe during slicing. *The Oregonian*, Portland, Oreg.

62. **Kistemann, T., F. Dangendorf, L. Krizek, H.-G. Sahl, S. Engelhart, and M. Exner.** 2000. GIS-supported investigation of a nosocomial *Salmonella* outbreak. *Int. J. Hyg. Environ. Health* **203:**117–126.

63. **Klontz, K. C., B. Timbo, S. Fein, and A. Levy.** 1995. Prevalence of selected food consumption and preparation behaviors associated with increased risks of food-borne disease. *J. Food Prot.* **58:**927–930.

64. **Kusumaningrum, H. D., R. Paltinaite, A. J. Koomen, W. C. Hazeleger, F. M. Rombouts, and R. R. Beumer.** 2003. Tolerance of *Salmonella* Enteritidis and *Staphylococcus aureus* to surface cleaning and household bleach. *J. Food Prot.* **66:**2289–2295.

65. **Kusumaningrum, H. D., M. M. van Putten, F. M. Rombouts, and R. R. Beumer.** 2002. Effects of antibacterial dishwashing liquid on foodborne pathogens and competitive microorganisms in kitchen sponges. *J. Food Prot.* **65:**61–65.

66. **Lázaro, N. S., A. Tibana, and E. Hofer.** 1997. *Salmonella* spp. in healthy swine and in abattoir environments in Brazil. *J. Food Prot.* **60:**1029–1033.

67. **L'Ecuyer, P. B., J. Diego, D. Murphy, E. Trovillion, M. Jones, D. F. Sahm, and V. J. Fraser.** 1996. Nosocomial outbreak of gastroenteritis due to *Salmonella senftenberg*. *Clin. Infect. Dis.* **23:**734–742.

68. **McEvoy, J. M., A. M. Doherty, J. J. Sheridan, I. S. Blair, and D. A. McDowell.** 2001. Use of steam condensing at subatmospheric pressures to reduce *Escherichia coli* O157:H7 numbers on bovine hide. *J. Food Prot.* **64:**1655–1660.

69. **Mead, P. S., L. Finelli, M. A. Lambert-Fair, D. Champ, J. Townes, L. Hutwagner, T. Barrett, K. Spitalny, and E. Mintz.** 1997. Risk factors for sporadic infection with *Escherichia coli* O157:H7. *Arch. Intern. Med.* **157:** 204–208.

70. **Meara, P. J., L. N. Melmed, and R. C. Cook.** 1977. Microbiological investigation of meat wholesale premises and beef carcases in Johannesburg. *J. S. Afr. Vet. Assoc.* **48:**255–260.

71. **Meldrum, R. J., D. Tucker, and C. Edwards.** 2004. Baseline rates of *Campylobacter* and *Salmonella* in raw chicken in Wales, United Kingdom, in 2002. *J. Food Prot.* **67:**1226–1228.

72. **Membré, J.-M., V. Majchrzak, and I. Jolly.** 1997. Effects of temperature, pH, glucose, and citric acid on the inactivation of *Salmonella typhimurium* in reduced calorie mayonnaise. *J. Food Prot.* **60:**1497–1501.

73. **Mitakakis, T. Z., M. I. Sinclair, C. K. Fairley, P. K. Lightbody, K. Leder, and M. E. Hellard.** 2004. Food safety in family homes in Melbourne, Australia. *J. Food Prot.* **67:**818–822.

74. **Mitchell, E., M. O'Mahony, D. Lynch, L. R. Ward, B. Rowe, A. Uttley, T. Rogers, D. G. Cunningham, and R. Watson.** 1989. Large outbreak of food poisoning caused by *Salmonella typhimurium* definitive type 49 in mayonnaise. *BMJ* **298:**99–101.

75. **National Advisory Committee on Microbiological Criteria for Foods.** 1994. *Campylobacter jejuni/coli*. *J. Food Prot.* **57:**1101–1121.

76. **Olsen, J. E., D. J. Brown, M. Madsen, and M. Bisgaard.** 2003. Cross-contamination with *Salmonella* on a broiler slaughterhouse line demonstrated by use of epidemiological markers. *J. Appl. Microbiol.* **94:**826–835.

77. **Ono, K., and K. Yamamoto.** 1999. Contamination of meat with *Campylobacter jejuni* in Saitama, Japan. *Int. J. Food Microbiol.* **47:**211–219.

78. **Park, P. K., and D. O. Cliver.** 1996. Disinfection of household cutting boards with a microwave oven. *J. Food Prot.* **59:**1049–1054.

79. **Park, P. K., and D. O. Cliver.** 1997. Disinfection of kitchen sponges and dishcloths by microwave oven. *Dairy Food Environ. Sanit.* **17:**146–149.

80. **Raghubeer, E. V., J. S. Ke, M. L. Campbell, and R. S. Meyer.** 1995. Fate of *Escherichia coli* O157:H7 and other coliforms in commercial mayonnaise and refrigerated salad dressing. *J. Food Prot.* **58:**13–18.

81. **Rajpura, A., K. Lamden, S. Forster, S. Clarke, J. Cheesbrough, S. Gornall, and S. Waterworth.** 2003. Large outbreak of infection with *Escherichia coli* O157 PT21/28 in Eccleston, Lancashire, due to cross contamination at a butcher's counter. *Commun. Dis. Public Health* **6:**279–284.

82. **Redmond, E. C., and C. J. Griffith.** 2003. Consumer food handling in the home: a review of food safety studies. *J. Food Prot.* **66:**130–161.

83. **Richardt-Sylskar, H., L. Seely, and S. Sievers.** 27 May 1995. Schwan's files suit; trucking firm, mix makers named. *Independent*, Marshall, Minn.

84. **Roels, T. H., B. Wickus, H. H. Bostrom, J. J. Kazmierczak, M. A. Nicholson, T. A. Kurzynski, and J. P. Davis.** 1998. A foodborne outbreak of *Campylobacter jejuni* (O:33) infection associated with tuna salad: a rare strain in an unusual vehicle. *Epidemiol. Infect.* **121:**281–287.

85. **Rostagno, M. H., H. S. Hurd, J. D. McKean, C. J. Ziemer, J. K. Gailey, and R. C. Leite.** 2002. Preslaughter holding environment in pork plants is highly contaminated with *Salmonella enterica. Appl. Environ. Microbiol.* **69:**4489–4494.

86. **Rusin, P., P. Orosz-Coughlin, and C. Gerba.** 1998. Reduction of faecal coliform, coliform and heterotrophic plate count bacteria in the household kitchen and bathroom by disinfection with hypochlorite cleaners. *J. Appl. Microbiol.* **85:**819–828.

87. **Ryan, C. A., M. K. Nickels, N. T. Hargrett-Bean, M. E. Potter, T. Endo, L. Mayer, C. W. Langkop, C. Gibson, R. C. McDonald, R. T. Kenney, N. D. Puhr, P. J. McDonnell, R. J. Martin, M. L. Cohen, and P. A. Blake.** 1987. Massive outbreak of antimicrobial-resistant salmonellosis traced to pasteurized milk. *JAMA* **258:**3269–3274.

88. **Salvat, G., M. T. Toquin, Y. Michel, and P. Colin.** 1995. Control of *Listeria monocytogenes* in the delicatessen industries: the lessons of a listeriosis outbreak in France. *Int. J. Food Microbiol.* **25:**75–81.

89. **Sauli, I., J. Danuser, C. Wenk, and K. D. C. Stärk.** 2003. Evaluation of the safety assurance level for *Salmonella* spp. throughout the food production chain in Switzerland. *J. Food Prot.* **66:**1139–1145.

90. **Scott, E., and S. F. Bloomfield.** 1990. The survival and transfer of microbial contamination via cloths, hands and utensils. *J. Appl. Bacteriol.* **68:**271–278.

91. **Simmons, M., D. L. Fletcher, M. E. Berrang, and J. A. Cason.** 2003. Comparison of sampling methods for the detection of *Salmonella* on whole broiler carcasses purchased from retail outlets. *J. Food Prot.* **66:**1768–1770.

92. **Simmons, M., D. L. Fletcher, J. A. Cason, and M. E. Berrang.** 2003. Recovery of *Salmonella* from retail broilers by a whole-carcass enrichment procedure. *J. Food Prot.* **66:**446–450.

93. **Slovut, G.** 16 May 1996. Report says '94 Schwan's outbreak affected 240,000; company official disputes CDC's estimate. *Star Tribune*, Minneapolis, Minn.

94. **Smeltzer, T. I., B. Peel, and G. Collins.** 1979. The role of equipment that has direct contact with the carcase in the spread of *Salmonella* in a beef abattoir. *Aust. Vet. J.* **55:**275–277.

95. **Smittle, R. B.** 1977. Microbiology of mayonnaise and salad dressing: a review. *J. Food Prot.* **40:**415–422.

96. **Smittle, R. B.** 2000. Microbiological safety of mayonnaise, salad dressings, and sauces produced in the United States: a review. *J. Food Prot.* **63:**1144–1153.

97. **Stern, N. J., P. Fedorka-Cray, J. S. Bailey, N. A. Cox, S. E. Craven, K. L. Hiett, M. T. Musgrove, S. Ladely, D. Cosby, and G. C. Mead.** 2001. Distribution of *Campylobacter* spp. in selected U.S. poultry production and processing operations. *J. Food Prot.* **64:**1705–1710.

98. **Stevens, R. A., and J. T. Holah.** 1993. The effect of wiping and spray-wash temperature on bacterial retention on abraded domestic sink surfaces. *J. Appl. Bacteriol.* **75:**91–94.

99. **Stevenson, J., and S. Hanson.** 1996. Outbreak of *Escherichia coli* O157 phage type 2 infection associated with eating precooked meats. *Commun. Dis. Rep. CDR Rev.* **6:**R116–R118. [Online.] http://www.hpa.org.uk/cdr/archives/CDRreview/1996/cdrr0896.pdf. Accessed 4 June 2006.

100. **Swanenburg, M., H. A. P. Urlings, D. A. Keuzenkamp, and J. M. A. Snijders.** 2001. *Salmonella* in the lairage of pig slaughterhouses. *J. Food Prot.* **64:**12–16.

101. **Tauxe, R. V.** 1991. *Salmonella*: a postmodern pathogen. *J. Food Prot.* **54:** 563–568.

102. **Taylor, J. P., B. J. Barnett, L. del Rosario, K. Williams, and S. S. Barth.** 1998. Prospective investigation of cryptic outbreaks of *Salmonella agona* salmonellosis. *J. Clin. Microbiol.* **36:**2861–2864.

103. **Tierney, J., M. Moriarty, and L. Kearney.** 2002. The sink environment as a source of microbial contamination in the domestic kitchen. *Dairy Food Environ. Sanit.* **22:**658–666.

104. **Todd, E. C. D.** 1992. Foodborne disease in Canada—a 10-year summary from 1975 to 1984. *J. Food Prot.* **55:**123–132.

105. **Vanderlinde, P. B., B. Shay, and J. Murray.** 1998. Microbiological quality of Australian beef carcass meat and frozen bulk packed beef. *J. Food Prot.* **61:**437–443.

106. **Wachtel, M. R., and A. O. Charkowski.** 2002. Cross-contamination of lettuce with *Escherichia coli* O157:H7. *J. Food Prot.* **65:**465–470.

107. **Ward, K., R. Hammond, D. Katz, and D. Hallman.** 1997. Outbreak of staphylococcal food poisoning associated with precooked ham—Florida, 1997. *Morb. Mortal. Wkly. Rep.* **46:**1189–1191.

108. **Warriner, K., T. G. Aldsworth, S. Kaur, and C. E. R. Dodd.** 2002. Cross-contamination of carcasses and equipment during pork processing. *J. Appl. Microbiol.* **93:**169–177.

109. **Weagant, S. D., J. L. Bryant, and D. H. Bark.** 1994. Survival of *Escherichia coli* O157:H7 in mayonnaise and mayonnaise-based sauces at room and refrigerated temperatures. *J. Food Prot.* **57:**629–631.

110. **Weklinski, P.** 2001. Oven mitts as a vehicle for cross-contamination in commercial food service establishments. *J. Environ. Health* **64:**27–28.

111. **Wernersson, E. S., E. Johansson, and H. Hakanson.** 2004. Cross-contamination in dishwashers. *J. Hosp. Infect.* **56:**312–317.

112. **Wight, J. P., J. Cornell, P. Rhodes, S. Colley, S. Webster, and A. M. Ridley.** 1996. Four outbreaks of *Salmonella enteritidis* phage type 4 food poisoning linked to a single baker. *Communic. Dis. Rep. CDR Rev.* **6:**R112–R115. [Online.] http://www.hpa.org.uk/cdr/archives/CDRreview/1996/cdrr0896.pdf. Accessed 4 June 2006.

113. **Winquist, A. G., A. Roome, R. Mshar, T. Fiorentino, P. Mshar, and J. Hadler.** 2001. Outbreak of campylobacteriosis at a senior center. *J. Am. Geriatr. Soc.* **49:**304–307.

114. **Worsfold, D., and C. J. Griffith.** 1997. Assessment of the standard of consumer food safety behavior. *J. Food Prot.* **60:**399–406.

115. **Xiong, R., G. Xie, and A. S. Edmondson.** 1999. The fate of *Salmonella enteritidis* PT4 in home-made mayonnaise prepared with citric acid. *Lett. Appl. Microbiol.* **28:**36–40.

116. **Yang, H., Y. Li, and M. G. Johnson.** 2001. Survival and death of *Salmonella typhimurium* and *Campylobacter jejuni* in processing water and on chicken skin during poultry scalding and chilling. *J. Food Prot.* **64:**770–776.

117. **Yang, S., M. G. Leff, D. McTague, K. A. Horvath, J. Jackson-Thompson, T. Murayi, G. K. Boeselager, T. A. Melnik, M. C. Gildemaster, D. L. Ridings, S. F. Altekruse, and F. J. Angulo.** 1998. CDC surveillance summaries. Multistate surveillance for food-handling, preparation, and consumption behaviors associated with foodborne diseases: 1995 and 1996 BRFSS food-safety questions. *Morb. Mortal. Wkly. Rep.* **47**(SS-4)**:**33–57.

118. **Zhao, T., and M. P. Doyle.** 1994. Fate of enterohemorrhagic *Escherichia coli* O157:H7 in commercial mayonnaise. *J. Food Prot.* **57:**780–783.

chapter 6

Birth of a Pathogen

Kevin was just 2½ when his family's nightmare began.

On July 31, 2001, Kevin Kowalcyk developed diarrhea and fever. By August 2, his diarrhea had turned bloody, and he was dehydrated. His doctor ordered Kevin hospitalized. The diagnosis: *Escherichia coli* O157:H7 infection. The next day, the toddler's kidneys began to fail. He had developed hemolytic uremic syndrome (HUS).

Kevin's doctors strove to save him, prescribing repeated blood transfusions and continuous kidney dialysis, but they were powerless to overcome the toxins that were circulating throughout his body and destroying his organs. Despite his doctors' best efforts, Kevin died on August 11, 2001 (124).

Kevin was the victim of a relatively recent addition to the rogues' gallery of disease-causing bacteria. *E. coli* O157:H7 first attracted the attention of food safety experts in 1982, when it caused outbreaks of hemorrhagic colitis (bloody diarrhea) in customers of two McDonald's outlets—one in Michigan and the other in Oregon. It was labeled an emerging pathogen and joined the ranks of other newly recognized pathogens, such as Ebola virus and human immunodeficiency virus. In 2001, the year Kevin died, the U.S. Centers for Disease Control and Prevention (CDC) received reports of 3,287 confirmed cases of *E. coli* O157:H7 infections and 202 cases of HUS (29). *E. coli* O157:H7 had emerged as a full-fledged food-borne pathogen and the primary cause of HUS in North America and Europe.

The Missing Links

It's not often that microbiologists have the chance to witness the birth of a new pathogen. Yet that is what happened with *E. coli* O157:H7.

E. coli O157 was isolated for the first time in 1970 from an Irish piglet that had contracted enteritis (62). It has since been found in a variety of domestic and wild animals on several continents (Table 6.1). But the strains of *E. coli* O157—or even of *E. coli* O157:H7—present in many of these animals are not always the same as those that cause human disease (11, 20, 69, 97).

Table 6.1 Domestic and wild animal reservoirs of *E. coli* O157

Year	Country or region	Reservoir	Reference(s)
1986	United States	Dairy cattle	83
1991	United States	Dairy cattle	137
1995	United States	Range cattle, wild deer	104
1995	United States	Dairy cattle	143
1995	United States	Dairy calves	48
1996	Spain	Cattle	20
1996	Germany	Sheep meat, orangutan	16
1996	United States	Dairy cattle	42
1996	United States	Sheep	72
1997	United Kingdom	Wild birds	134
1997	United States	Feedlot cattle	57, 34
1997	United States	Sheep	73
1998	Canada	Cattle	64
1998	United States	Dairy cattle	117
1998	United States	Dairy cattle	60
1999	United States	Dairy cattle	49
1999	Japan	Swine	94
2000	United States	Sheep	33
2000	Canada	Cattle	115
2000	Czech Republic	Cow	18
2001	United States	Wild deer	45
2001	United States	Cattle	131
2001	United States	Cattle	122
2002	United Kingdom	Cattle, sheep	121
2002	Turkey	Cattle	142
2003	United States	Swine	43
2003	United States	Beef cattle	105
2004	Canada	Cattle	7, 15
2004	Switzerland	Cattle	2
2004	Mexico	Cattle, swine	24

The *E. coli* O157:H7 that has caused such human misery over the past 2 decades is not a direct descendant of the strain that was isolated from the sick piglet in 1970 (138, 139). It is, instead, a close relative to a different serotype—*E. coli* O55:H7—and is believed to have evolved from an ancestor of that serotype (44).

The evolutionary trail of *E. coli* O157:H7 is marked by a series of clues contained in its genetic code. Today's generation of microbiologists and molecular biologists have the tools and the knowledge to follow those clues back in time and deduce the evolutionary path down which *E. coli* O157:H7 has traveled. Not only can the results of this genetic exploration help us understand how the pathogen evolved, but this information eventually might lead regulators and the food industry toward better ways of preventing the explosive development and spread of new food-borne pathogens.

E. coli O157:H7 is one of a group of closely related pathogens, known as a clone complex, that are genetically very similar but differ from each other in a few, relatively minor characteristics. The clone complex includes

Table 6.2 The *E. coli* O157:H7 and *E. coli* O55:H7 clone complex (44)

Variant	Example[a]	Characteristic					
		Serotype	Motility	β-GUD[b] activity	Sorbitol fermentation	Stx1 toxin	Stx2 toxin
A[c]	None	Unknown	Unknown	+	+	−	−
B	DEC 5	O55:H7	+	+	+	−	−
C	USDA 5905	O55:H7	+	+	+	−	+
D[d]	None	O157:H7	+	+	+	−	+
E	493/89	O157:H-	−	+	+	−	+
F	CDC G5101	O157:H7	+	+	−	+	+
G	FDA 413, ATCC 35150	O157:H7	+	−	−	+	+

[a] Example of *E. coli* strain comprising the indicated characteristics.
[b] β-GUD: β-glucuronidase.
[c] Hypothetical ancestor strain.
[d] Hypothetical strain.

strains of both *E. coli* O157:H7 and *E. coli* O55:H7. Typically, *E. coli* O157:H7 is motile, ferments sorbitol slowly (if at all), lacks β-glucuronidase activity, and has the ability to produce two toxins, Stx1 and Stx2. But members of the clone complex differ from each other in several of these individual characteristics (Table 6.2), and these differences are the key to unraveling the evolution of *E. coli* O157:H7.

Researchers at Pennsylvania State University (T. S. Whittam, who was at Pennsylvania State University, is now at Michigan State University), the U.S. Food and Drug Administration (FDA), and the University of Würzburg in Germany have been studying the evolution of *E. coli* O157:H7 for many years. In 1988, Whittam and his colleagues concluded that the isolates of *E. coli* O157:H7 from widely separated North American outbreaks were all very closely related to each other, representing a single pathogenic clone. Conversely, *E. coli* O157:H7 isolates were only distantly related to other serotypes of *E. coli* that produced a similar toxin (138, 139).

In 1993, H. Karch and her colleagues at the University of Würzburg reported on the genetic relationships of a new variant clone of *E. coli* O157 that had first appeared in Germany in 1988 (67). The pathogen was isolated from patients with hemorrhagic colitis and HUS. It differed from the classic profile of *E. coli* O157:H7 in three respects: the newly discovered microbe fermented sorbitol rapidly, was not motile, and only produced one of the two toxins (Stx2) linked to hemorrhagic colitis (55). Was this clone just an oddball variant, or did it represent a missing link in the evolution of *E. coli* O157:H7?

P. Feng and his research colleagues at the FDA proposed an answer to this riddle in 1998 (44). They postulated that *E. coli* O157:H7 evolved from an ancestral strain of enteropathogenic *E. coli* (Table 6.2 and Fig. 6.1). In this evolutionary model, the ancestral strain (variant A) gave rise to *E. coli* O55:H7 (variant B). Gradually, *E. coli* O55:H7 was modified: infection by

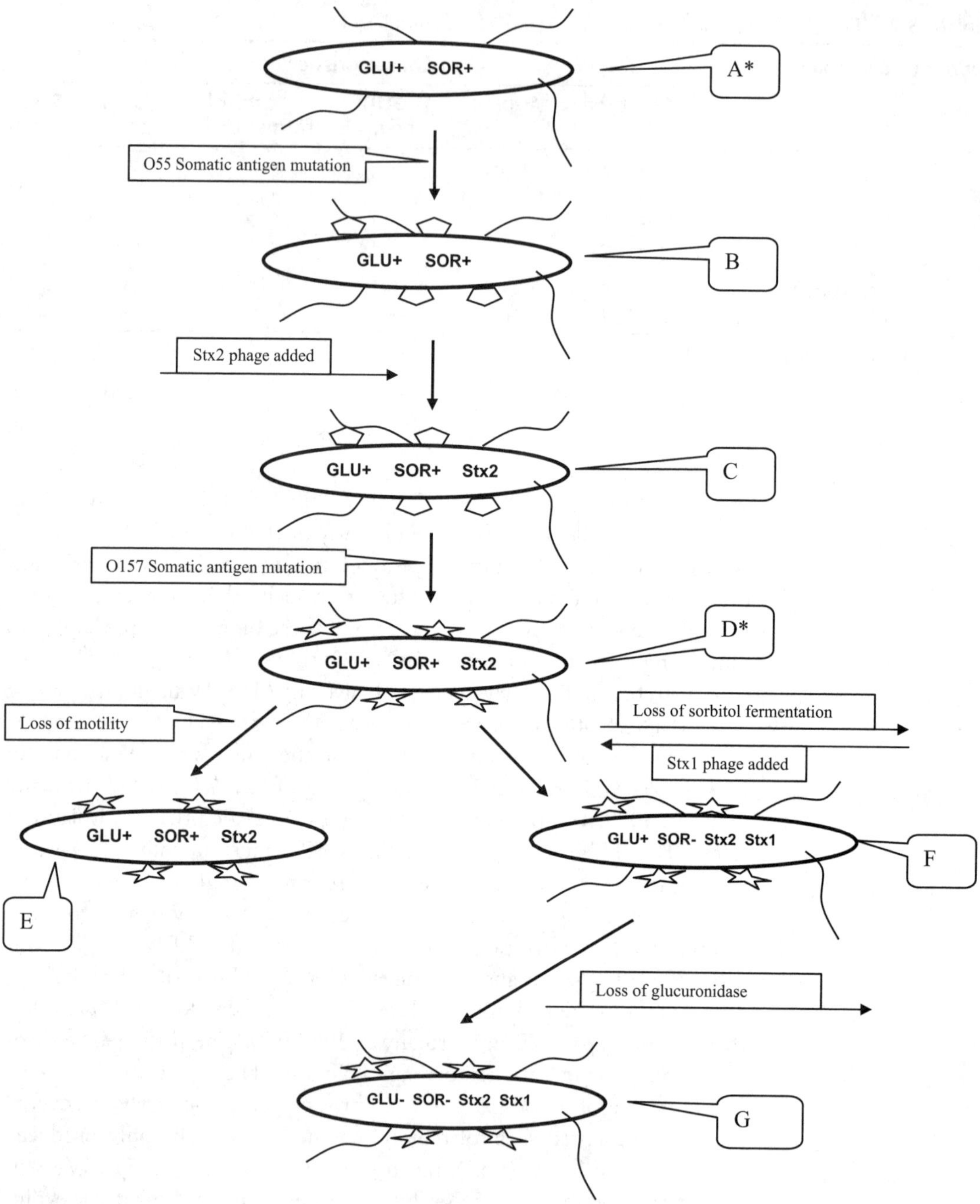

Figure 6.1 Proposed evolution of *E. coli* O157:H7. A* and D* are theoretical strains, not yet isolated in the laboratory. Figure is adapted from Feng et al., 1998 (44). GLU, glucuronidase; SOR, sorbitol fermentation; Stx1 and Stx2, toxins. A, (theoretical) ancestral strain; B, O55:H7 GUD+ SOR+; C, O55:H7 GUD+ SOR+ Stx2+; D, (theoretical) O157:H7 GUD+ SOR+ Stx2+; E, O157:H− GUD+ SOR+ Stx2+; F, O157:H7 GUD+ SOR− Stx2+ Stx1+; G, O157:H7 GUD− SOR− Stx2+ Stx1+.

a bacteriophage conferred on it the ability to produce Stx2 toxin (variant C); a genetic mutation converted the microbe's serotype to *E. coli* O157:H7 (variant D) from *E. coli* O55:H7; it lost the ability to ferment sorbitol and acquired a second bacteriophage that enabled it to produce Stx1 toxin (variant F); and, finally, the microbe lost the ability to produce the β-glucuronidase enzyme (variant G). As of 1998, only two of the evolutionary stages postulated in Feng's model—the original ancestral strain (variant A) and an intermediate strain (variant D)—had not been found in nature (88). According to Feng, the German clone (variant E) that was first identified in 1988 was not an evolutionary missing link; rather, it shared a common ancestor (variant D) with the typical *E. coli* O157:H7 that was first recognized as a pathogen in North America in 1982. Although the existence of the variant D has not yet been confirmed, a recent comparative genomic analysis of 10 strains of pathogenic *E. coli* (three O55:H7, two O157:H−, and five O157:H7) supports this evolutionary hypothesis (140). Although either Stx1 or Stx2 alone can produce symptoms of bloody diarrhea, Stx2 is more closely associated with the development of HUS (135).

HUS, TTP, and *E. coli* O157:H7

In 1955, Gasser and his colleagues described a medical condition that comprised acute kidney failure (sudden, complete shutdown of both kidneys), hemolytic anemia (disintegrating red blood cells), and thrombocytopenia (reduced platelet counts, resulting in poor formation of blood clots). They called the disease hemolytic uremic syndrome, or HUS (96). Within a few years, doctors around the world began to recognize HUS. North and South America, Europe and Japan, Africa and Australia—all reported scattered cases and small outbreaks in the 1960s (10, 23, 54, 59, 61, 86, 100, 118, 119).

In 1968, Rhodesian scientists first suggested a possible connection between HUS and *E. coli* (53). They noticed a similarity between HUS and a disease of pigs and speculated that a strain of *E. coli* infected with a bacteriophage might be the cause of both syndromes. Within 2 years of their published report, in 1970, a sick Irish piglet yielded the world's first confirmed culture of *E. coli* O157 (62). Evidence implicating *E. coli* O157:H7 in HUS came in 1991. British researchers tested stored samples of serum from 16 HUS patients who had been hospitalized in The Netherlands between 1974 and 1977. Six of the 16 serum samples contained antibodies to *E. coli* O157:H7—clear evidence that those six patients had been infected with the pathogen (53).

Although no one realized it at the time, the incidence of HUS quietly increased in parts of North America during the 1970s. Nowhere was this more evident, in retrospect, than in King County, Wash. Between 1971 and 1980, 33 children in the county developed HUS: 14 cases were diagnosed

Hemolytic Uremic Syndrome: Prevention and Treatment

Hemolytic uremic syndrome is a devastating—and sometimes deadly—illness. Patients almost always experience acute kidney failure, and other organs, such as the pancreas, the heart, and even the brain, can also be affected (22). Even when a victim recovers, there are long-term consequences. Survivors are prone to diseases of the intestines, gallbladder, kidneys, and pancreas (21, 22, 120). Some are left with learning disabilities due to brain damage suffered during their struggle to survive (40). Small wonder that medical researchers have been working hard on ways of preventing and treating HUS.

The key to avoiding the medical maelstrom caused by HUS is prevention. For once the symptoms of HUS have appeared, the toxins have already begun to attack the kidneys and other organs, and it is too late for anything but supportive treatments. Dialysis, insulin therapy, and transfusions with packed erythrocytes are some of the more common treatments that are used to counteract the ravages of the syndrome (22). Doctors have also used plasma exchange and plasma infusion therapies, with mixed results (38, 75, 108, 132). Administering an antitoxin or a synthetic toxin-neutralizing agent to HUS patients has also proven futile (102, 127).

To prevent HUS, one must understand its risk factors. Some researchers have determined that antibiotic treatment of children with *E. coli* O157:H7 infection increases the risk of HUS (39, 141). Use of antimotility agents to treat diarrhea is also a risk factor (13, 128). To date, the only proven way to minimize the risk of developing HUS is to avoid becoming infected with *E. coli* O157:H7 or other Shiga toxin-producing microbes (30). But this situation could be changing.

In recent years, researchers have been working on two new approaches to HUS prevention. One relies on immunization of the at-risk population with a toxoid developed from Shiga toxin. The toxoid has been effective in immunizing rabbits but has yet to be proven in humans (76). The second approach uses humanized monoclonal antibodies to neutralize Shiga toxin in a patient's intestine before the toxin enters the bloodstream (91–93). To be effective, this treatment must be carried out very early in the illness. Ideally, any patient complaining of bloody diarrhea would need to be treated immediately—even before a diagnosis of *E. coli* O157:H7 has been confirmed. Monoclonal antibody therapy has been tested in mice, with promising results (116), and a clinical trial has been under way in Canada and Argentina since 2003 (80). If this treatment proves effective, the scourge of HUS might finally be conquered. If not, then avoiding infection with *E. coli* O157:H7 will continue to be the best way—maybe the only way—to prevent HUS.

between 1971 and 1977; 1978 to 1980 saw 19 cases in the county. In 1981 to 1982, an additional 14 cases were reported in just one hospital in King County. And this increase in the number of children with HUS occurred at the same time when the most susceptible population was declining, from more than 300,000 children under the age of 14 in 1970 to fewer than 250,000 in 1980 (129).

HUS was also making inroads in Canada. In the 1970s, doctors at Toronto's Hospital for Sick Children were used to seeing just 5 to 10 cases of HUS each year. Then in 1980, 13 children with HUS were admitted to the hospital in a 10-day period. In all cases, the development of HUS was preceded by a bout of bloody diarrhea. A 14th child, who probably also had developed HUS, died suddenly at home. All of the children had drunk fresh (unpasteurized) apple cider from a local farm 3 to 6 days before their symptoms first developed. The hospital was unable to find any known

Table 6.3 Characteristics of HUS and TTP (8)

Characteristic	HUS[a]	TTP[b]
Kidney failure	Yes	Yes
Low platelet count (<150,000/mm^3)	Yes	Yes
Hemolytic anemia	Yes	Yes
Elevated temperature >100°F (38°C)	No	Yes
Neurological symptoms	No	Yes
Susceptible population	Mainly children	Mainly adults

[a] HUS, hemolytic uremic syndrome.
[b] TTP, thrombotic thrombocytopenic purpura.

pathogen in the children's stools, and the 3-week-old samples of juice yielded no answers, although *E. coli* biotype 1 was found in one juice sample (123). The common food source, the incubation period, the absence of other known pathogens, and the range of symptoms all point to this 1980 Toronto incident as probably having been caused by *E. coli* O157:H7.

Although HUS is the most common complication of *E. coli* O157:H7 infections, it is not the only one. A similar syndrome, known as thrombotic thrombocytopenic purpura, or TTP, was identified in 1924 (68) and was first linked to *E. coli* O157:H7 in the mid-1980s (90, 101). HUS and TTP have many similarities (Table 6.3). Both syndromes affect the kidneys, both damage or destroy red blood cells, and both can be triggered by *E. coli* O157:H7. TTP, which rarely strikes children, is sometimes referred to as the adult form of HUS.

Not all cases of HUS and TTP are due to prior infection with *E. coli* O157:H7. Some arise from infection by other bacteria. *Shigella* has long been known to trigger HUS, as have non-O157 *E. coli* strains, including Shiga toxin-producing strains belonging to serogroups O26, O103, O111, O118, O121, and O145 (4, 17, 78, 125). Viral infections and chemotherapy have also been known to result in the development of HUS, and a few cases even have a genetic origin (74, 112, 113). But in the years since it was first recognized as a human pathogen, *E. coli* O157:H7 has become a major cause of HUS in North America and Europe (8, 31, 82).

E. coli O157:H7—the Early Years

E. coli O157:H7 already had been lurking in the background for several years before doctors and epidemiologists finally became aware of its existence. In 1975, the microbe was isolated from the stool of a California woman who had bloody diarrhea. Three years later, in 1978, Canadian epidemiologists found *E. coli* O157:H7 in an Ottawa patient with similar symptoms and identified five more cases in different parts of the country over the next 5 years (62). Then, in 1982, finding *E. coli* O157:H7 suddenly became more than an annual oddity.

The first recorded outbreak of *E. coli* O157:H7 began in February 1982 in Oregon (106, 136). Twenty-six people between the ages of 8 and 76 fell ill with cramps and bloody diarrhea between February 5 and March 15; 19 were hospitalized. All of the victims had eaten hamburger at a McDonald's—on average, about 4 days before becoming ill. Lab workers couldn't find any of the common gastrointestinal pathogens—*Salmonella*, *Shigella*, *Campylobacter*, or *Yersinia enterocolitica*—in the patients' stools, but *E. coli* O157 was identified in three stool samples. Samples of the raw hamburger patties, however, appeared pathogen free. None of the patients suffered any complications, and the episode appeared to die a natural death.

Two months later, it happened again—this time in Michigan. The second outbreak began on May 28 and lasted for about 1 month. The situation in Michigan was similar to that in Oregon: 21 victims, all complaining of severe cramps and bloody diarrhea; 14 people hospitalized; no deaths; no complications; an average 4-day incubation period. But there was one important difference. This time, in addition to finding *E. coli* O157:H7 in stool samples from four patients, investigators discovered the microbe in a raw hamburger patty from the same production lot that had been served in the Michigan McDonald's outlets. And the lab profiles of the isolates from all of the Oregon and Michigan patients matched perfectly with the *E. coli* O157:H7 found in the hamburger patty (136). The two widely separated episodes were part of a single outbreak.

Just a few months later, in November 1982, 31 residents of a nursing home in Ottawa, Canada, began to complain of symptoms similar to those experienced by victims of the McDonald's outbreak. Four of the patients were hospitalized, and one person died (27). Just as in the U.S. outbreak, *E. coli* O157:H7 was present in the stool of some of the patients. This time, more than half of the stool samples (17 of 31) were positive. Suspicion fell on hamburger prepared in the nursing home's kitchen—a single production lot had been in use through most of November—but the microbe was never found in the raw meat.

In response to the pair of outbreaks traced to their restaurants, McDonald's funded research studies to learn more about the pathogen and how to prevent its recurrence in its product (D. O. Cliver, personal communication; 36, 37). Despite these efforts, during the 10 years that followed the 1982 outbreaks, there were several scattered outbreaks of food-borne illness traced to *E. coli* O157:H7, almost all of them in the United States and Canada (Table 6.4). Together, they accounted for more than 330 illnesses and 17 deaths. Most often, the outbreaks originated in either restaurants or nursing homes. The most deadly episode took place in a Canadian nursing home in 1985. Seventy-three people—most of them elderly—were taken ill, and 12 of them contracted HUS. In all, 17 of the nursing home victims died, including 11 of those with HUS. Yet these outbreaks were no more than a prelude to the events of 1993.

Table 6.4 Meat-borne disease outbreaks due to *E. coli* O157:H7

Year	Country	State or province	Food vehicle	Venue	No. of cases (deaths)	Reference(s)
1982	United States	Michigan, Oregon	Hamburger	Fast-food restaurant	47	106, 136
1982	Canada	Ontario	Hamburger (unconfirmed)	Nursing home	31 (1)	27, 66
1984	United States	Nebraska	Hamburger	Nursing home	34 (4)	111
1985	Canada		Sandwich	Nursing home	73 (17)	26
1986	United States	Washington	Ground beef	Restaurant, nursing home	37 (2)	97
1986	United States	Washington	Ground beef	Nursing home	3	97
1988	United States	Wisconsin	Roast beef	Graduation banquet	61	110
1988	United States	Minnesota	Precooked meat patties	School cafeteria	32	14
1990	United States	North Dakota	Roast beef	Buffet dinner	70	85
1990	Scotland		Unknown	Restaurant	16	81
1993	United States	California	Hamburger	Home	7	133
1993	United States	California, Idaho, Nevada, Washington	Hamburger	Fast-food restaurant	>500 (4)	12, 35, 126
1993	United States	Connecticut	Hamburger	Country club picnic	23	109
1994	United States	California, Washington	Dry-cured salami	N/A[a]	20	1
1994	United States	Virginia	Ground beef	Summer camp	20	47
1994	United States	Connecticut	Ground meat	Supermarket	21	9
1995	United States	Georgia, Tennessee	Hamburger	Fast-food restaurant	10	25
1996	Germany	Bavaria	Meat sausage	Home	28	3
1999	Canada	Manitoba	Hamburger	Family barbecue	7	77
2003	Argentina		Hamburger	Home	1	107
2005	Japan		Beef offal	Restaurant	11	84
2006	Norway		Cured meat sausage	Home	17	114

[a] NA, not available.

Hamburger Becomes a Dirty Word

For every outbreak there is an index case—the first victim in the series. Usually not identified until after the outbreak is well under way, the index case is the one about which second-guessers claim, "If only *They* had paid attention to 'X', *They* could have stopped the outbreak in its tracks."

Some people believe that Lauren Rudolph of San Diego County, Calif., was the index case for what came to be known as the "Jack in the Box outbreak" (46). On December 18, 1992, Lauren ate a cheeseburger at a San Diego area Jack in the Box outlet. Just 10 days later, she was dead—a victim of HUS triggered by *E. coli* O157:H7. By January 18, 1993, at least

45 people from the state of Washington, most of them children, were complaining of cramps and bloody diarrhea (41). Epidemiologists quickly traced the source of the infections to hamburgers from several different Jack in the Box outlets. In all, the outbreak affected more than 500 people in four states (Washington, California, Nevada, and Idaho). Three-quarters of the victims were less than 18 years old; 45 patients developed HUS, and 4 children died—3 in the Seattle, Wash., area and Lauren Rudolph in California (12, 22).

At first, Jack in the Box denied responsibility for the outbreak (89, 103). The restaurant chain had been buying its hamburger patties from a Vons meat processing facility. Jack in the Box purchased the meat for patty production (either directly or through Vons) from three suppliers: Montfort Beef, Service Packing, and Orleans International (103). Investigators traced the contaminated meat to Service Packing, a California-based deboning plant that in turn purchased its meat carcasses from nine slaughterhouses (63). There the trail ended—but not the recriminations (58). Foodmaker, Inc., the parent company of Jack in the Box, promptly sued its suppliers (5).

Lauren Rudolph's death from HUS in San Diego (home city of Jack in the Box) in December 1992 should have caught the attention of California health authorities. But at the time, *E. coli* O157:H7 was not given the same high profile in California as in the state of Washington, and Lauren's case went unnoticed until the outbreak was well under way in Washington.

Washington was the first U.S. state to require that doctors and hospitals notify their county health departments of all *E. coli* O157:H7 infections (98). The state took this measure in response to two apparently unrelated outbreaks that occurred in late 1986 (50). A restaurant in Walla Walla, Wash., was the source of the first outbreak, which began in late October

A Missed Opportunity?

During their investigation of the Jack in the Box outbreak, epidemiologists from the state of Washington and the CDC uncovered another possible index case. On November 22, 1992—almost 1 month before Lauren Rudolph bit into her cheeseburger in San Diego—Kim Weber, a Tacoma, Wash., teenager, indulged in an "ultimate cheeseburger" at a local Jack in the Box. Three days later, she was suffering from severe cramps and, according to her father, was passing "pure blood." Kim was lucky. Although she had to spend 4 days in a Tacoma hospital, she recovered without complications and was back in school on December 3 (99). Was Kim the index case? Should her illness have alerted local health authorities to food safety shortcomings at Jack in the Box restaurants?

Even though the Jack in the Box restaurant at which she had eaten was implicated in the death of one of the outbreak victims, health officials believe that Kim's hamburger probably was not from the same batch of meat patties that was responsible for the multistate outbreak. But had local health authorities investigated the source of Kim's illness, they might have discovered that Jack in the Box was undercooking its hamburgers and, perhaps, alerted the company to its borderline cooking practices several weeks before the outbreak began.

and lasted about 3 weeks. The outbreak went unnoticed until 3 of the 37 victims developed TTP and were transferred to a hospital associated with the University of Washington. The second outbreak was centered in a nursing home in Dayton, a town near Walla Walla. This incident, which began just as the Walla Walla outbreak was ending, would probably have been missed but for the heightened awareness of *E. coli* O157:H7 that had been generated by the Walla Walla cases (97).

In 1987, the year that its surveillance program went into effect, Washington health authorities received reports of 93 cases of *E. coli* O157:H7 infection (98). In May 1992, after consulting with a food safety expert, the state once again led the way in consumer protection by increasing—to 155°F (68°C)—the required internal temperature to which hamburgers had to be cooked (19). The temperature standard in force in California at the time of the outbreak was 145°F (63°C), and the federal standard was only 140°F (60°C) (51, 71). Unfortunately, Jack in the Box hadn't adjusted its procedures to meet the state of Washington's new cooking standards.

Jack in the Box management claimed that they were never notified of Washington's new cooking temperature standard. Later, their president acknowledged that the company had received notice of the change in May 1992, but that top management had never been told of the new requirements (32). On January 18, the same day it recalled the incriminated hamburger meat, the company instructed all of its restaurants to increase their cooking time to 2¼ minutes from 2 minutes (65). Nine days later, a health inspector discovered a Jack in the Box outlet that was still undercooking its hamburgers (70). When asked about this during a congressional hearing, the president of Jack in the Box claimed that the health inspector had prevented his employees from cooking the hamburgers for the allotted time before checking their internal temperature (95).

The Aftermath

One measure of the character of an organization, or an individual, is how well it learns from the past. Many people considered the trio of 1982 outbreaks in the United States and Canada to be an aberration—a one-shot cluster caused by a rare serotype of *E. coli*. As the decade advanced, the list of sporadic illnesses and outbreaks traced to *E. coli* O157:H7 grew, and so did the interest of clinical researchers and epidemiologists. In 1987, Washington became the first U.S. state to institute mandatory reporting of *E. coli* O157:H7 (98). But at the federal level, the FDA and the U.S. Department of Agriculture (USDA) made no substantial changes to their regulatory programs in response to this new threat. Similarly, the CDC did not extend its surveillance program to include *E. coli* O157:H7 until 1994 (28). And while McDonald's took actions to prevent another occurrence, everything else remained "business as usual" in the food industry.

The Jack in the Box outbreak caught the attention of government regulators, consumers, and the food industry in a way that previous *E. coli* O157:H7 outbreaks had failed to do. FDA immediately increased its cooking temperature requirement to 155°F (68°C) from 140°F (60°C). Congress held hearings at which senior management of Jack in the Box testified. USDA began a review of its meat inspection policies and programs. The families of outbreak victims came together to form a new safe food advocacy group, Safe Tables Our Priority, to lobby government for improved food safety legislation. But did anything really change?

The changes at Jack in the Box were dramatic and substantive. The company, having acknowledged its responsibility for the outbreak, made major improvements to its quality assurance program. It hired a new corporate vice president of technical services and introduced a quality program based on Hazard Analysis and Critical Control Points (HACCP) (130). At first, the meat industry and its regulator, the USDA, also acknowledged the need to improve the country's meat inspection system. USDA proposed—and the industry initially accepted—that the old visual inspection system be replaced by a national HACCP program (chapter 7). But meat producers and packers began to drag their feet once the details of the new program were released (6, 52, 56, 79, 87). And, in the last half of the 1990s, the U.S. meat industry paid a fearsome price for its reluctance.

References

1. **Alexander, E. R., J. Boase, M. Davis, L. Kirchner, C. Osaki, T. Tanino, M. Samadpour, P. Tarr, M. Goldoft, S. Lankford, J. Kobayashi, P. Stehr-Green, P. Bradley, B. Hinton, P. Tighe, B. Pearson, G. R. Flores, S. Abbott, R. Bryant, S. B. Werner, and D. J. Vugia.** 1995. *Escherichia coli* O157:H7 outbreak linked to commercially distributed dry-cured salami—Washington and California, 1994. *Morb. Mortal. Wkly. Rep.* **44:**157–160.
2. **Al-Saigh, H., C. Zweifel, J. Blanco, J. E. Blanco, M. Blanco, M. A. Usera, and R. Stephan.** 2004. Fecal shedding of *Escherichia coli* O157, *Salmonella*, and *Campylobacter* in Swiss cattle at slaughter. *J. Food Prot.* **67:**679–684.
3. **Ammon, A., L. R. Petersen, and H. Karch.** 1999. A large outbreak of hemolytic uremic syndrome caused by an unusual sorbitol-fermenting strain of *Escherichia coli* O157:H–. *J. Infect. Dis.* **179:**1274–1277.
4. **Amor, J. S.** 1988. *Shigella dysenteriae* type 1 in tourists to Cancun, Mexico. *Morb. Mortal. Wkly. Rep.* **37:**465.
5. **Anonymous.** 5 February 1993. Jack in the Box sues suppliers over *E. coli*. *The News Tribune*, Tacoma, Wash.
6. **Anonymous.** 2 November 1994. Industry sues to halt meat probe. *Dayton Daily News*, Dayton, Ohio.
7. **Bach, S. J., T. A. McAllister, G. J. Mears, and K. S. Schwartzkopf-Genswein.** 2004. Long-haul transport and lack of preconditioning increases fecal shedding of *Escherichia coli* and *Escherichia coli* O157:H7 by calves. *J. Food Prot.* **67:**672–678.

8. **Banatvala, N., P. M. Griffin, K. D. Greene, T. J. Barrett, W. F. Bibb, J. H. Green, J. G. Wells, and the Hemolytic Uremic Syndrome Study Collaborators.** 2001. The United States national prospective hemolytic uremic syndrome study: microbiologic, serologic, clinical, and epidemiologic findings. *J. Infect. Dis.* **183:**1063–1070.

9. **Banatvala, N., A. R. Magnano, M. L. Cartter, T. J. Barrett, W. F. Bibb, L. L. Vasile, P. Mshar, M. A. Lambert-Fair, J. H. Green, N. H. Bean, and R. V. Tauxe.** 1996. Meat grinders and molecular epidemiology: two supermarket outbreaks of *Escherichia coli* O157:H7 infection. *J. Infect. Dis.* **173:**480–483.

10. **Barnard, P. J., and M. Kibel.** 1965. The haemolytic-uraemic syndrome of infancy and childhood. A report of eleven cases. *Cent. Afr. J. Med.* **11:**4–11.

11. **Barrett, T. J., H. Lior, J. H. Green, R. Khakhria, J. G. Wells, B. P. Bell, K. D. Greene, J. Lewis, and P. M. Griffin.** 1994. Laboratory investigation of a multistate food-borne outbreak of *Escherichia coli* O157:H7 by using pulsed-field gel electrophoresis and phage typing. *J. Clin. Microbiol.* **32:**3013–3017.

12. **Bell, B. P., M. Goldoft, P. M. Griffin, M. A. Davis, D. C. Gordon, P. I. Tarr, C. A. Bartleson, J. H. Lewis, T. J. Barrett, J. G. Wells, R. Baron, and J. Kobayashi.** 1994. A multistate outbreak of *Escherichia coli* O157:H7-associated bloody diarrhea and hemolytic uremic syndrome from hamburgers. The Washington experience. *JAMA* **272:**1349–1353.

13. **Bell, B. P., P. M. Griffin, P. Lozano, D. L. Christie, J. M. Kobayashi, and P. I. Tarr.** 1997. Predictors of hemolytic uremic syndrome in children during a large outbreak of *Escherichia coli* O157:H7 infections. *Pediatrics* **100:**E12. [Online.] http://www.pediatrics.org/cgi/content/full/100/1/e12. Accessed 6 June 2006.

14. **Belongia, E. A., K. L. MacDonald, G. L. Parham, K. E. White, J. A. Korlath, M. N. Lobato, S. M. Strand, K. A. Casale, and M. T. Osterholm.** 1991. An outbreak of *Escherichia coli* O157:H7 colitis associated with consumption of precooked meat patties. *J. Infect. Dis.* **164:**338–343.

15. **Berg, J., T. McAllister, S. Bach, R. Stilborn, D. Hancock, and J. LeJeune.** 2004. *Escherichia coli* O157:H7 excretion by commercial feedlot cattle fed either barley- or corn-based finishing diets. *J. Food Prot.* **67:**666–671.

16. **Beutin, L., G. Knollmann-Schanbacher, W. Rietschel, and H. Seeger.** 1996. Animal reservoirs of *Escherichia coli* O157:[H7]. *Vet. Rec.* **139:**70–71.

17. **Beutin, L., S. Zimmermann, and K. Gleier.** 1998. Human infections with shiga toxin-producing *Escherichia coli* other than serogroup O157 in Germany. *Emerg. Infect. Dis.* **4:**635–639.

18. **Bielaszewska, M., H. Schmidt, A. Liesegang, R. Prager, W. Rabsch, H. Tschäpe, A. Cízek, J. Janda, K. Bláhová, and H. Karch.** 2000. Cattle can be a reservoir of sorbitol-fermenting shiga toxin-producing *Escherichia coli* O157:H– strains and a source of human diseases. *J. Clin. Microbiol.* **38:** 3470–3473.

19. **Blake, J.** 24 January 1993. Jack in the Box poisonings—food safety standards higher here—cooking temperature for hamburgers increased to 155 degrees. *Seattle Times*, Seattle, Wash.

20. **Blanco, M., J. E. Blanco, J. Blanco, E. A. Gonzalez, A. Mora, C. Prado, L. Fernandez, M. Rio, J. Ramos, and M. P. Alonso.** 1996. Prevalence and

characteristics of *Escherichia coli* serotype O157:H7 and other verotoxin-producing *E. coli* in healthy cattle. *Epidemiol. Infect.* **117:**251–257.

21. **Boyce, T. G., D. L. Swerdlow, and P. M. Griffin.** 1995. *Escherichia coli* O157:H7 and the hemolytic-uremic syndrome. *N. Engl. J. Med.* **333:**364–368.
22. **Brandt, J. R., L. S. Fouser, S. L. Watkins, I. Zelikovic, P. I. Tarr, V. Nazar-Stewart, and E. D. Avner.** 1994. *Escherichia coli* O157:H7-associated hemolytic-uremic syndrome after ingestion of contaminated hamburgers. *J. Pediatr.* **125:**519–526.
23. **Caballero, E., E. Cassorla, L. Del Rio, and P. Advis.** 1966. Hemolytic uremic syndrome. (In Spanish.) *Rev. Chil. Pediatr.* **37:**285–292.
24. **Callaway, T. R., R. C. Anderson, G. Tellez, C. Rosario, G. M. Nava, C. Eslava, M. A. Blanco, M. A. Quiroz, A. Olguín, M. Herradora, T. S. Edrington, K. J. Genovese, R. B. Harvey, and D. J. Nisbet.** 2004. Prevalence of *Escherichia coli* O157 in cattle and swine in central Mexico. *J. Food Prot.* **67:**2274–2276.
25. **Cannon, M., H. Thomas, W. Sellers, M. Bates, P. Blake, H. Stetler, K. Toomey, J. Fowler, S. Halford, G. Young, S. Hall, P. Erwin, V. Boaz, and G. Swinger.** 1996. Outbreak of *Escherichia coli* O157:H7 infection—Georgia and Tennessee, June 1995. *Morb. Mortal. Wkly. Rep.* **45:**249–251.
26. **Carter, A. O., A. A. Borczyk, J. A. Carlson, B. Harvey, J. C. Hockin, M. A. Karmali, C. Krishnan, D. A. Korn, and H. Lior.** 1987. A severe outbreak of *Escherichia coli* O157:H7-associated hemorrhagic colitis in a nursing home. *N. Engl. J. Med.* **317:**1496–1500.
27. **Centers for Disease Control.** 1983. International notes. Outbreak of hemorrhagic colitis—Ottawa, Canada. *Morb. Mortal. Wkly. Rep.* **32:**133–134.
28. **Centers for Disease Control and Prevention.** 1995. Summary of notifiable diseases, United States 1994. *Morb. Mortal. Wkly. Rep.* **43:**5.
29. **Centers for Disease Control and Prevention.** 2003. Summary of notifiable diseases—United States, 2001. *Morb. Mortal. Wkly. Rep.* **50:**2.
30. **Chandler, W. L., S. Jelacic, D. R. Boster, M. A. Ciol, G. D. Williams, S. L. Watkins, T. Igarashi, and P. I. Tarr.** 2002. Prothrombotic coagulation abnormalities preceding the hemolytic-uremic syndrome. *N. Engl. J. Med.* **346:**23–32.
31. **Chang, H.-G. H., B. Tserenpuntsag, M. Kacica, P. F. Smith, and D. L. Morse.** 2004. Hemolytic uremic syndrome incidence in New York. *Emerg. Infect. Dis.* **10:**928–931.
32. **Conklin, E. E.** 13 February 1993. Jack in the Box told of cooking rule, boss admits. *Seattle Post-Intelligencer*, Seattle, Wash.
33. **Cornick, N. A., S. L. Booher, T. A. Casey, and H. W. Moon.** 2000. Persistent colonization of sheep by *Escherichia coli* O157:H7 and other *E. coli* pathotypes. *Appl. Environ. Microbiol.* **66:**4926–4934.
34. **Dargatz, D. A., S. J. Wells, L. A. Thomas, D. D. Hancock, and L. P. Garber.** 1997. Factors associated with the presence of *Escherichia coli* O157 in feces of feedlot cattle. *J. Food Prot.* **60:**466–470.
35. **Davis, M., C. Osaki, D. Gordon, M. W. Hinds, K. Mottram, C. Winegar, E. D. Avner, P. I. Tarr, D. Jardine, M. Goldoft, B. Bartleson, J. Lewis, J. M. Kobayashi, G. Billman, J. Bradley, S. Hunt, P. Tanner, M. Ginsberg, L. Barrett, S. B. Werner, G. W. Rutherford III, R. W. Jue, H. Root,**

D. Brothers, R. L. Chehey, R. H. Hudson, F. R. Dixon, D. J. Maxson, L. Empey, O. Ravenholt, V. H. Ueckart, A. DiSalvo, D. S. Kwalick, R. Salcido, and D. Brus. 1993. Update: multistate outbreak of *Escherichia coli* O157:H7 infections from hamburgers—Western United States, 1992–1993. *Morb. Mortal. Wkly. Rep.* **42:**258–263.

36. **Doyle, M. P., and J. L. Schoeni.** 1984. Survival and growth characteristics of *Escherichia coli* associated with hemorrhagic colitis. *Appl. Environ. Microbiol.* **48:**855–856.

37. **Doyle, M. P., and J. L. Schoeni.** 1987. Isolation of *Escherichia coli* O157:H7 from retail fresh meats and poultry. *Appl. Environ. Microbiol.* **53:**2394–2396.

38. **Dundas, S., J. Murphy, R. L. Soutar, G. A. Jones, S. J. Hutchinson, and W. T. Todd.** 1999. Effectiveness of therapeutic plasma exchange in the 1996 Lanarkshire *Escherichia coli* O157:H7 outbreak. *Lancet* **354:**1327–1330.

39. **Dundas, S., W. T. A. Todd, A. I. Stewart, P. S. Murdoch, A. K. R. Chaudhuri, and S. J. Hutchinson.** 2001. The Central Scotland *Escherichia coli* O157:H7 outbreak: risk factors for the hemolytic uremic syndrome and death among hospitalized patients. *Clin. Infect. Dis.* **33:**923–931.

40. **Durbin, B.** 23 September 1997. *E. coli* one family's nightmare. Series: Risky business (3rd of 3 parts). *The Oregonian*, Portland, Oreg.

41. **Eng, J. L.** 18 January 1993. Undercooked beef linked to illness affecting 45. *Seattle Times*, Seattle, Wash.

42. **Faith, N. G., J. A. Shere, R. Brosch, K. W. Arnold, S. E. Ansay, M.-S. Lee, J. B. Luchansky, and C. W. Kaspar.** 1996. Prevalence and clonal nature of *Escherichia coli* O157:H7 on dairy farms in Wisconsin. *Appl. Environ. Microbiol.* **62:**1519–1525.

43. **Feder, I., F. M. Wallace, J. T. Gray, P. Fratamico, P. J. Fedorka-Cray, R. A. Pearce, J. E. Call, R. Perrine, and J. B. Luchansky.** 2003. Isolation of *Escherichia coli* O157:H7 from intact colon fecal samples of swine. *Emerg. Infect. Dis.* **9:**380–383.

44. **Feng, P., K. A. Lampel, H. Karch, and T. S. Whittam.** 1998. Genotypic and phenotypic changes in the emergence of *Escherichia coli* O157:H7. *J. Infect. Dis.* **177:**1750–1753.

45. **Fischer, J. R., T. Zhao, M. P. Doyle, M. R. Goldberg, C. A. Brown, C. T. Sewell, D. M. Kavanaugh, and C. D. Bauman.** 2001. Experimental and field studies of *Escherichia coli* O157:H7 in white-tailed deer. *Appl. Environ. Microbiol.* **67:**1218–1224.

46. **Fox, N.** 1997. *Spoiled. The Dangerous Truth About a Food Chain Gone Haywire*, p. 3–9. BasicBooks, A Division of HarperCollins Publishers, Inc., New York, N.Y.

47. **Frost, B., C. Chaos, L. Ladaga, W. Day, M. Tenney, D. McWilliams, E. Barrett, L. Branch, S. Jenkins, M. Linn, E. Turf, D. Woolard, G. B. Miller, Jr., S. Henderson, B. Campbell, M. Mismas, J. Dvorak, D. Patel, D. Peery, J. Morano, and K. Campbell.** 1995. *Escherichia coli* O157:H7 outbreak at a summer camp—Virginia, 1994. *Morb. Mortal. Wkly. Rep.* **44:**419–421.

48. **Garber, L. P., S. J. Wells, D. D. Hancock, M. P. Doyle, J. Tuttle, J. A. Shere, and T. Zhao.** 1995. Risk factors for fecal shedding of *Escherichia coli* O157:H7 in dairy calves. *J. Am. Vet. Med. Assoc.* **207:**46–49.

49. **Garber, L., S. Wells, L. Schroeder-Tucker, and K. Ferris.** 1999. Factors associated with fecal shedding of verotoxin-producing *Escherichia coli* O157 on dairy farms. *J. Food Prot.* **62:**307–312.

50. **Gilmore, S.** 31 January 1993. Walla Walla cases prepared the state for this outbreak—Washington became a pioneer in reporting, cooking standards. *Seattle Times*, Seattle, Wash.

51. **Gorman, T., and D. Conner.** 23 January 1993. Restaurant not to blame for tainted meat, health officials say: bacteria that caused one death are believed to be from slaughterhouse, not Jack in the Box or Vons. *Los Angeles Times*, Los Angeles, Calif.

52. **Greene, R.** 25 May 1994. Beef packers say health rule boomerangs. Industry claims stringent inspection has increased germ problem. *Buffalo News*, Buffalo, N.Y.

53. **Griffin, P. M., and R. V. Tauxe.** 1991. The epidemiology of infections caused by *Escherichia coli* O157:H7, other enterohemorrhagic *E. coli*, and the associated hemolytic uremic syndrome. *Epidemiol. Rev.* **13:**60–98.

54. **Griffiths, J., and K. G. Irving.** 1961. A haemolytic-uraemic syndrome in infancy. *Arch. Dis. Child.* **36:**500–506.

55. **Gunzer, F., H. Böhm, H. Rüssmann, M. Bitzan, S. Aleksic, and H. Karch.** 1992. Molecular detection of sorbitol-fermenting *Escherichia coli* O157 in patients with hemolytic-uremic syndrome. *J. Clin. Microbiol.* **30:**1807–1810.

56. **Hammel, P., and J. Thompson.** 15 December 1997. Europeans can track an *E. coli* outbreak right to the farmer's doorstep. *Omaha World-Herald*, Omaha, Nebr.

57. **Hancock, D. D., D. H. Rice, L. A. Thomas, D. A. Dargatz, and T. E. Besser.** 1997. Epidemiology of *Escherichia coli* O157 in feedlot cattle. *J. Food Prot.* **60:**462–465.

58. **Hanson, C.** 12 March 1993. *E. coli* source not likely to be found. *Seattle Post-Intelligencer*, Seattle, Wash.

59. **Havnen, J.** 1965. Acute renal failure in children. A case of hemolytic-uremic syndrome treated with artificial kidney dialysis. (In Swedish, English abstract.) *Tidsskr. Nor. Laegeforen.* **85:**459–460.

60. **Herriott, D. E., D. D. Hancock, E. D. Ebel, L. V. Carpenter, D. H. Rice, and T. E. Besser.** 1998. Association of herd management factors with colonization of dairy cattle by shiga toxin-positive *Escherichia coli* O157. *J. Food Prot.* **61:**802–807.

61. **Hinton, P., and B. Koch.** 1968. Hemolytic uremic syndrome: report of two cases. *Can. Med. Assoc. J.* **98:**819–826.

62. **Hockin, J. C., and H. Lior.** 1986. Hemorrhagic colitis due to *Escherichia coli* O157:H7. A rare disease? *Can. Med. Assoc. J.* **134:**25–26.

63. **Ingram, E.** 23 February 1993. Investigators tracking tainted meat. Probe traces hamburger from northwest to LA packing plant. *San Francisco Chronicle*, San Francisco, Calif.

64. **Jackson, S. G., R. B. Goodbrand, R. P. Johnson, V. G. Odorico, D. Alves, K. Rahn, J. B. Wilson, M. K. Welch, and R. Khakhria.** 1998. *Escherichia coli* O157:H7 diarrhoea associated with well water and infected cattle on an Ontario farm. *Epidemiol. Infect.* **120:**17–20.

65. **Jamieson, R. L., Jr., and E. Houston.** 26 January 1993. Lawsuits filed over burgers; cases increase. *Seattle Post-Intelligencer*, Seattle, Wash.

66. **Johnson, W. M., H. Lior, and G. S. Bezanson.** 1983. Cytotoxic *Escherichia coli* O157:H7 associated with haemorrhagic colitis in Canada. *Lancet* **i:**76.

67. **Karch, H., H. Böhm, H. Schmidt, F. Gunzer, S. Aleksic, and J. Heesemann.** 1993. Clonal structure and pathogenicity of shiga-like toxin-producing, sorbitol-fermenting *Escherichia coli* O157:H–. *J. Clin. Microbiol.* **31:**1200–1205.

68. **Karmali, M. A.** 1989. Infection by verocytotoxin-producing *Escherichia coli*. *Clin. Microbiol. Rev.* **2:**15–38.

69. **Kim, J., J. Nietfeldt, and A. K. Benson.** 1999. Octamer-based genome scanning distinguishes a unique subpopulation of *Escherichia coli* O157:H7 strains in cattle. *Proc. Natl. Acad. Sci. USA* **96:**13288–13293.

70. **King, W.** 28 January 1993. Officials close Jack in the Box store; burgers undercooked. *Seattle Times*, Seattle, Wash.

71. **King, W., and J. Blake.** 22 January 1993. Tests confirm meat as source of illness. *Seattle Times*, Seattle, Wash.

72. **Kudva, I. T., P. G. Hatfield, and C. J. Hovde.** 1996. *Escherichia coli* O157:H7 in microbial flora of sheep. *J. Clin. Microbiol.* **34:**431–433.

73. **Kudva, I. T., P. G. Hatfield, and C. J. Hovde.** 1997. Characterization of *Escherichia coli* O157:H7 and other shiga toxin-producing *E. coli* serotypes isolated from sheep. *J. Clin. Microbiol.* **35:**892–899.

74. **Lee, M. H., K. S. Cho, K. W. Kahng, and C. M. Kang.** 1998. A case of hemolytic uremic syndrome associated with Epstein-Barr virus infection. *Korean J. Intern. Med.* **13:**131–135.

75. **Loirat, C., E. Sonsino, N. Hinglais, J. P. Jais, P. Landais, and J. Fermanian.** 1988. Treatment of the childhood haemolytic uraemic syndrome with plasma. A multicentre randomized controlled trial. *Pediatr Nephrol.* **2:**279–285.

76. **Ludwig, K., M. A. Karmali, C. R. Smith, and M. Petric.** 2002. Cross-protection against challenge by intravenous *Escherichia coli* verocytotoxin 1 (VT1) in rabbits immunized with VT2 toxoid. *Can. J. Microbiol.* **48:**99–103.

77. **Macdonald, C., J. Drew, R. Carlson, S. Dzogan, S. Tataryn, A. Macdonald, A. Ali, R. Amhed, R. Easy, C. Clark, and F. Rodgers.** 2000. Outbreak of *Escherichia coli* O157:H7 leading to the recall of retail ground beef—Winnipeg, Manitoba, May 1999. *Can. Comm. Dis. Rep.* **26:**109–111. [Online.] http://www.phac-aspc.gc.ca/publicat/ccdr-rmtc/00vol26/dr2613ea.html. Accessed 6 June 2006.

78. **Maidhof, H., B. Guerra, S. Abbas, H. M. Elsheikha, T. S. Whittam, and L. Beutin.** 2002. A multiresistant clone of shiga toxin-producing *Escherichia coli* O118:[H16] is spread in cattle and humans over different European countries. *Appl. Environ. Microbiol.* **68:**5834–5842.

79. **Manning, A.** 29 December 1999. Texas beef plant fights *Salmonella* standards; *E. coli* prompted processor to recall 90 tons. *USA Today*, McLean, Va.

80. **Marowits, R.** 30 July 2003. Burger disease drug put to test. Study targets *E. coli* bacteria. Illness proves deadly for kids. *Toronto Star*, Toronto, Ontario, Canada.

81. **Marsh, J., A. F. MacLeod, M. F. Hanson, F. X. S. Emmanuel, J. A. Frost, and A. Thomas.** 1992. A restaurant-associated outbreak of *E. coli* O157 infection. *J. Public Health Med.* **14:**78–83.

82. **Martin, D. L., K. L. MacDonald, K. E. White, J. T. Soler, and M. T. Osterholm.** 1990. The epidemiology and clinical aspects of the hemolytic uremic syndrome in Minnesota. *N. Engl. J. Med.* **323:**1161–1167.

83. **Martin, M. L., L. D. Shipman, J. G. Wells, M. E. Potter, K. Hedberg, I. K. Wachsmuth, R. V. Tauxe, J. P. Davis, J. Arnoldi, and J. Tilleli.** 1986. Isolation of *Escherichia coli* O157:H7 from dairy cattle associated with two cases of haemolytic uraemic syndrome. *Lancet* **ii:**1043.

84. **Maruzumi, M., M. Morita, Y. Matsuoka, A. Uekawa, T. Nakamura, and K. Fuji.** 2005. Mass food poisoning caused by beef offal contaminated by *Escherichia coli* O157. *Jpn. J. Infect. Dis.* **58:**397.

85. **McDonough, S., F. Heer, and L. Shireley.** 1991. Epidemiologic notes and reports. Foodborne outbreak of gastroenteritis caused by *Escherichia coli* O157: H7—North Dakota, 1990. *Morb. Mortal. Wkly. Rep.* **40:**265–267.

86. **McLean, M. M., C. H. Jones, and D. A. Sutherland.** 1966. Haemolytic-uraemic syndrome. A report of an outbreak. *Arch. Dis. Child.* **41:**76–81.

87. **Means, M.** 2 June 1995. Delays in reform could serve up another helping of *E. coli*. *Seattle Post–Intelligencer*, Seattle, Wash.

88. **Monday, S. R., T. S. Whittam, and P. C. H. Feng.** 2001. Genetic and evolutionary analysis of mutations in the *gusA* gene that cause the absence of β-glucuronidase activity in *Escherichia coli* O157:H7. *J. Infect. Dis.* **184:** 918–921.

89. **Moriwaki, L., and K. Kusumoto.** 1 February 1993. Burger chain offers to pay medical bills. *Seattle Times*, Seattle, Wash.

90. **Morrison, D. M., D. L. Tyrrell, and L. D. Jewell.** 1986. Colonic biopsy in verotoxin-induced hemorrhagic colitis and thrombotic thrombocytopenic purpura (TTP). *Am. J. Clin. Pathol.* **86:**108–112.

91. **Mukherjee, J., K. Chios, D. Fishwild, D. Hudson, S. O'Donnell, S. M. Rich, A. Donohue-Rolfe, and S. Tzipori.** 2002. Human stx2-specific monoclonal antibodies prevent systemic complications of *Escherichia coli* O157: H7 infection. *Infect. Immun.* **70:**612–619.

92. **Mukherjee, J., K. Chios, D. Fishwild, D. Hudson, S. O'Donnell, S. M. Rich, A. Donohue-Rolfe, and S. Tzipori.** 2002. Production and characterization of protective human antibodies against shiga toxin 1. *Infect. Immun.* **70:**5896–5899.

93. **Nakao, H., N. Kiyokawa, J. Fujimoto, S. Yamasaki, and T. Takeda.** 1999. Monoclonal antibody to shiga toxin 2 which blocks receptor binding and neutralizes cytotoxicity. *Infect. Immun.* **67:**5717–5722.

94. **Nakazawa, M., M. Akiba, and T. Sameshima.** 1999. Swine as a potential reservoir of shiga toxin-producing *Escherichia coli* O157:H7 in Japan. *Emerg. Infect. Dis.* **5:**833–834.

95. **Nelson, R. T.** 6 February 1993. Jack in Box blames inspectors for failed temperature test. *Seattle Times*, Seattle, Wash.

96. **O'Brien, A. D., and R. K. Holmes.** 1987. Shiga and shiga-like toxins. *Microbiol. Rev.* **51:**206–220.

97. **Ostroff, S. M., P. M. Griffin, R. V. Tauxe, L. D. Shipman, K. D. Greene, J. G. Wells, J. H. Lewis, P. A. Blake, and J. M. Kobayashi.** 1990. A statewide outbreak of *Escherichia coli* O157:H7 infections in Washington state. *Am. J. Epidemiol.* **132:**239–247.

98. **Ostroff, S. M., J. M. Kobayashi, and J. H. Lewis.** 1989. Infections with *Escherichia coli* O157:H7 in Washington state. The first year of statewide disease surveillance. *JAMA* **262:**355–359.

99. **Porterfield, E., and S. Gordon.** 30 January 1993. An early—and lucky—*E. coli* victim. Teen hit in November; case wasn't pursued. *The News Tribune*, Tacoma, Wash.

100. **Quoyle, J., E. Hirst, and D. Painter.** 1965. The haemolytic uraemic syndrome. *Med. J. Aust.* **2:**386–391.

101. **Ramsey, P. G., and M. A. Neill.** 1986. Epidemiologic notes and reports. Thrombotic thrombocytopenic purpura associated with *Escherichia coli* O157:H7—Washington. *Morb. Mortal. Wkly. Rep.* **35:**549–551.

102. **Ray, P., D. Acheson, R. Chitrakar, A. Cnaan, K. Gibbs, G. H. Hirschman, E. Christen, H. Trachtman, and The Investigators of the Hemolytic Uremic Syndrome-Synsorb PK Multicenter Clinical Trial.** 2002. Basic fibroblast growth factor among children with diarrhea-associated hemolytic uremic syndrome. *J. Am. Soc. Nephrol.* **13:**699–707.

103. **Reza, H. G.** 22 January 1993. Firms reject blame in food poisoning. *Los Angeles Times*, Los Angeles, Calif.

104. **Rice, D. H., D. D. Hancock, and T. E. Besser.** 1995. Verotoxigenic *E. coli* O157 colonisation of wild deer and range cattle. *Vet. Rec.* **137:**524.

105. **Riley, D. G., J. T. Gray, G. H. Loneragan, K. S. Barling, and C. C. Chase, Jr.** 2003. *Escherichia coli* O157:H7 prevalence in fecal samples of cattle from a southeastern beef cow-calf herd. *J. Food Prot.* **66:**1778–1782.

106. **Riley, L. W., R. S. Remis, S. D. Helgerson, H. B. McGee, J. G. Wells, B. R. Davis, R. J. Hebert, E. S. Olcott, L. M. Johnson, N. T. Hargrett, P. A. Blake, and M. L. Cohen.** 1983. Hemorrhagic colitis associated with a rare *Escherichia coli* serotype. *New Engl. J. Med.* **308:**681–685.

107. **Rivas, M., M. G. Caletti, I. Chinen, S. M. Refi, C. D. Roldán, G. Chillemi, G. Fiorilli, A. Bertolotti, L. Aguerre, and S. S. Estani.** 2003. Home-prepared hamburger and sporadic hemolytic uremic syndrome, Argentina. *Emerg. Infect. Dis.* **9:**1184–1186.

108. **Rizzoni, G., A. Claris-Appiani, A. Edefonti, P. Facchin, F. Franchini, R. Gusmano, E. Imbasciati, L. Pavanello, F. Perfumo, and G. Remuzzi.** 1988. Plasma infusion for hemolytic-uremic syndrome in children: results of a multicenter controlled trial. *J. Pediatr.* **112:**284–290.

109. **Roberts, C. L., P. A. Mshar, M. L. Cartter, J. L. Hadler, D. M. Sosin, P. S. Hayes, and T. J. Barrett.** 1995. The role of heightened surveillance in an outbreak of *Escherichia coli* O157:H7. *Epidemiol. Infect.* **115:**447–454.

110. **Rodrigue, D. C., E. E. Mast, K. D. Greene, J. P. Davis, M. A. Hutchinson, J. G. Wells, T. J. Barrett, and P. M. Griffin.** 1995. A university outbreak of *Escherichia coli* O157:H7 infections associated with roast beef and an unusually benign clinical course. *J. Infect. Dis.* **172:**1122–1125.

111. **Ryan, C. A., R. V. Tauxe, G. W. Hosek, J. G. Wells, P. A. Stoesz, H. W. McFadden, Jr., P. W. Smith, G. F. Wright, and P. A. Blake.** 1986.

Escherichia coli O157:H7 diarrhea in a nursing home: clinical, epidemiological, and pathological findings. *J. Infect. Dis.* **154:**631–638.

112. **Saif, M. W., and P. J. McGee.** 2005. Hemolytic-uremic syndrome associated with gemcitabine: a case report and review of literature. *J. Pancreas* **6:** 369–374.

113. **Saunders, R. E., T. H. Goodship, P. F. Zipfel, and S. J. Perkins.** 2006. An interactive web database of factor H-associated hemolytic uremic syndrome mutations: insights into the structural consequences of disease-associated mutations. *Hum. Mutat.* **27:**21–30.

114. **Schimmer, B.** 2006. Outbreak of haemolytic uraemic syndrome in Norway: update. *Eur. Surveill. Wkly.* **11**(4). [Online.] http://www.eurosurveillance.org/ew/2006/060406.asp#2. Accessed 6 June 2006.

115. **Schurman, R. D., H. Hariharan, S. B. Heaney, and K. Rahn.** 2000. Prevalence and characteristics of shiga toxin-producing *Escherichia coli* in beef cattle slaughtered on Prince Edward Island. *J. Food Prot.* **63:**1583–1586.

116. **Sheoran, A. S., S. Chapman, P. Singh, A. Donohue-Rolfe, and S. Tzipori.** 2003. Stx2-specific human monoclonal antibodies protect mice against lethal infection with *Escherichia coli* expressing Stx2 variants. *Infect. Immun.* **71:**3125–3130.

117. **Shere, J. A., K. J. Bartlett, and C. W. Kaspar.** 1998. Longitudinal study of *Escherichia coli* O157:H7 dissemination on four dairy farms in Wisconsin. *Appl. Environ. Microbiol.* **64:**1390–1399.

118. **Shibagaki, M., T. Yamamoto, H. Ikuta, P. Y. Kim, and A. Osawa.** 1968. The hemolytic uremic syndrome with pulmonary involvement. An autopsied case treated with hemodialysis for three months. *Acta Paediatr. Jpn.* **10:**6–11.

119. **Shumway, C. N., and K. L. Terplan.** 1964. Hemolytic anemia, thrombocytopenia and renal disease in childhood. The hemolytic-uremic syndrome. *Pediatr. Clin. North Am.* **11:**577–591.

120. **Siegler, R. L., A. T. Pavia, R. D. Christofferson, and M. K. Milligan.** 1994. A 20-year population-based study of postdiarrheal hemolytic uremic syndrome in Utah. *Pediatrics* **94:**35–40.

121. **Small, A., C.-A. Reid, S. M. Avery, N. Karabasil, C. Crowley, and S. Buncic.** 2002. Potential for the spread of *Escherichia coli* O157, *Salmonella*, and *Campylobacter* in the lairage environment at abattoirs. *J. Food Prot.* **65:** 931–936.

122. **Smith, D., M. Blackford, S. Younts, R. Moxley, J. Gray, L. Hungerford, T. Milton, and T. Klopfenstein.** 2001. Ecological relationships between the prevalence of cattle shedding *Escherichia coli* O157:H7 and characteristics of the cattle or conditions of the feedlot pen. *J. Food Prot.* **64:**1899–1903.

123. **Steele, B. T., N. Murphy, G. S. Arbus, and C. P. Rance.** 1982. An outbreak of hemolytic uremic syndrome associated with ingestion of fresh apple juice. *J. Pediatr.* **101:**963–965.

124. **Surendran, A.** 4 June 2003. Unsafe meat ended or tragically changed their lives. *The Philadelphia Inquirer*, Philadelphia, Pa.

125. **Tarr, C. L., T. M. Large, C. L. Moeller, D. W. Lacher, P. I. Tarr, D. W. Acheson, and T. S. Whittam.** 2002. Molecular characterization of a serotype O121:H19 clone, a distinct shiga toxin-producing clone of pathogenic *Escherichia coli*. *Infect. Immun.* **70:**6853–6859.

126. **Tarr, P. I.** 1994. Review of 1993 *Escherichia coli* O157:H7 outbreak: western United States. *Dairy Food Environ. Sanit.* **14:**372–373.

127. **Tarr, P. I.** 2002. Basic fibroblast growth factor and shiga toxin-O157: H7-associated hemolytic uremic syndrome. *J. Am. Soc. Nephrol.* **13:** 817–820.

128. **Tarr, P. I., and D. L. Christie.** 1999. Antimotility agents and *E. coli* infection. *Can. Med. Assoc. J.* **160:**984, 986.

129. **Tarr, P. I., and R. O. Hickman.** 1987. Hemolytic uremic syndrome epidemiology: a population-based study in King County, Washington, 1971 to 1980. *Pediatrics* **80:**41–45.

130. **Theno, D., and S. Bjerklie.** 1995. HACCP's role in the Jack in the Box recovery. *Meat and Poultry* **41**(3)**:**42, 44, 46, 53–54.

131. **Thran, B. H., H. S. Hussein, M. R. Hall, and S. F. Khaiboullina.** 2001. Shiga toxin-producing *Escherichia coli* in beef heifers grazing an irrigated pasture. *J. Food Prot.* **64:**1613–1616.

132. **Tsai, H.-M., W. L. Chandler, R. Sarode, R. Hoffman, S. Jelacic, R. L. Habeeb, S. L. Watkins, C. S. Wong, G. D. Williams, and P. I. Tarr.** 2001. Von Willebrand factor and von Willebrand factor-cleaving metalloprotease activity in *Escherichia coli* O157:H7–associated hemolytic uremic syndrome. *Pediatr. Res.* **49:**653–659.

133. **Turney, C., M. Green-Smith, M. Shipp, C. Mordhorst, C. Whittingslow, L. Brawley, D. Koppel, E. Bridges, G. Davis, J. Voss, R. Lee, M. Jay, S. Abbott, R. Bryant, K. Reilly, S. B. Werner, L. Barrett, R. J. Jackson, G. W. Rutherford III, and H. Lior.** 1994. *Escherichia coli* O157:H7 outbreak linked to home-cooked hamburger—California, July 1993. *Morb. Mortal. Wkly. Rep.* **43:**213–216.

134. **Wallace, J. S., T. Cheasty, and K. Jones.** 1997. Isolation of vero cytotoxin-producing *Escherichia coli* O157 from wild birds. *J. Appl. Microbiol.* **82:** 399–404.

135. **Wang, G., C. G. Clark, and F. G. Rodgers.** 2002. Detection in *Escherichia coli* of the genes encoding the major virulence factors, the genes defining the O157:H7 serotype, and components of the type 2 shiga toxin family by multiplex PCR. *J. Clin. Microbiol.* **40:**3613–3619.

136. **Wells, J. G., B. R. Davis, I. K. Wachsmuth, L. W. Riley, R. S. Remis, R. Sokolow, and G. K. Morris.** 1983. Laboratory investigation of hemorrhagic colitis outbreaks associated with a rare *Escherichia coli* serotype. *J. Clin. Microbiol.* **18:**512–520.

137. **Wells, J. G., L. D. Shipman, K. D. Greene, E. G. Sowers, J. H. Green, D. N. Cameron, F. P. Downes, M. L. Martin, P. M. Griffin, S. M. Ostroff, M. E. Potter, R. V. Tauxe, and I. K. Wachsmuth.** 1991. Isolation of *Escherichia coli* serotype O157:H7 and other shiga-like-toxin-producing *E. coli* from dairy cattle. *J. Clin. Microbiol.* **29:**985–989.

138. **Whittam, T. S., I. K. Wachsmuth, and R. A. Wilson.** 1988. Genetic evidence of clonal descent of *Escherichia coli* O157:H7 associated with hemorrhagic colitis and hemolytic uremic syndrome. *J. Infect. Dis.* **157:** 1124–1133.

139. **Whittam, T. S., and R. A. Wilson.** 1988. Genetic relationships among pathogenic *Escherichia coli* of serogroup O157. *Infect. Immun.* **56:**2467–2473.

140. **Wick, L. M., W. Qi, D. W. Lacher, and T. S. Whittam.** 2005. Evolution of genomic content in the stepwise emergence of *Escherichia coli* O157:H7. *J. Bacteriol.* **187:**1783–1791.

141. **Wong, C. S., S. Jelacic, R. L. Habeeb, S. L. Watkins, and P. I. Tarr.** 2000. The risk of the hemolytic-uremic syndrome after antibiotic treatment of *Escherichia coli* O157:H7 infections. *N. Engl. J. Med.* **342:**1930–1936.

142. **Yilmaz, A., H. Gun, and H. Yilmaz.** 2002. Frequency of *Escherichia coli* O157:H7 in Turkish cattle. *J. Food Prot.* **65:**1637–1640.

143. **Zhao, T., M. P. Doyle, J. Shere, and L. Garber.** 1995. Prevalence of enterohemorrhagic *Escherichia coli* O157:H7 in a survey of dairy herds. *Appl. Environ. Microbiol.* **61:**1290–1293.

chapter 7

USDA, HACCP, and *E. coli* O157:H7

THE OUTBREAK BEGAN IN COLORADO IN MID-JUNE 1997, with a small cluster of unusual illnesses. By July 14, health officials had confirmed that 15 patients were infected with *Escherichia coli* O157:H7. The abnormal spike in *E. coli* O157:H7 cases worried Pam Shillam, an epidemiologist with Colorado's Department of Health, but she was unable to find a common source for the illnesses (55).

Epidemiologists are disease detectives. Just like Columbo, Hercule Poirot, and Sgt. Friday, Pam Shillam and others like her rely on a combination of examining physical evidence, interviewing victims and witnesses, and piecing together the jigsaw puzzle of evidence to find the cause of a disease outbreak. And sometimes a dollop of good luck provides the missing link that allows a case to be solved.

This time, Sandra Gallegos, a nurse with the Pueblo, Colo., health department, furnished that link. She received a report about Lee Harding, a supermarket worker who had come down with *E. coli* O157:H7 infection. To find out what might have caused Mr. Harding's illness, Sandra interviewed both Lee and his wife. When she heard that Mr. Harding had grilled and eaten packaged frozen hamburgers on July 9, she jumped on the clue, even though Mrs. Harding had eaten the same meal and had not become ill. Fortunately, some of the patties—in their original Hudson Foods packaging and showing the production batch number—were still in the Hardings' freezer. Acting on a hunch, Sandra arranged to have the remaining patties tested by the U.S. Department of Agriculture (USDA) lab in Athens, Ga. The lab found *E. coli* O157:H7 in Lee Harding's frozen patties. And on August 7, when the state lab confirmed that the *E. coli* O157:H7 detected in the patties was identical to the strain that had sickened Lee Harding and 15 other people, the picture was complete. Frozen meat patties produced by Hudson Foods were responsible for the outbreak (64).

Colorado advised USDA of the link between its *E. coli* O157:H7 outbreak and the Hudson Foods frozen patties. USDA sent a team of investigators to Hudson's meat grinding plant to figure out how the patties had become contaminated and to establish how much meat should be recalled. Based on the initial information USDA received from Hudson, the government requested—and Hudson agreed to—a recall of 20,000 pounds of frozen beef patties on August 12 (20). A second 20,000-pound recall was announced the next day. Then, on August 15, USDA advised the public that the company had expanded its recall to include additional production batches totaling 1.2 million pounds of meat (21). Finally, on August 21, Hudson agreed to USDA's recommendation that all of its outstanding production—25 million pounds of meat—be recalled and that the plant be closed (22).

What Happened at Hudson?

How did Hudson get in so deep? The company followed a standard industry practice known as rework—mixing leftover scraps of meat from one day's production into another production batch. Hudson had some 3,400 pounds of meat scraps left over at the end of its production day on June 5, 1997—the batch that was later implicated in the *E. coli* O157:H7 outbreak (40).

The 3,400 pounds of meat scraps were blended into several subsequent production lots. But, due to poor record-keeping, the company wasn't sure which of those lots contained the scraps from June 5. At first, Hudson told USDA investigators that all 3,400 pounds had been used on June 6. This triggered the two successive recalls of 20,000 pounds on August 12 and 13. The company soon determined that the scrap meat had also found its way into additional production lots, resulting in the August 15 recall of 1.2 million pounds. When it became clear that there was no way to be certain into which lots the June 5 meat scraps had been blended, Hudson reluctantly agreed to recall all of its outstanding meat inventories (40).

The History of HACCP

Before 1993, Americans were complacent about the safety of their food supply. Notwithstanding the Jewel Dairy *Salmonella* outbreak of 1985 (chapter 2), a rising incidence of *Salmonella enterica* serotype Enteritidis in shell eggs (chapter 3), and a chronically high level of *Salmonella* and *Campylobacter* in raw poultry (chapter 5), former President George H.W. Bush proclaimed in 1992 that the U.S. food supply was "the safest in the world" (7). But after the 1993 Jack in the Box outbreak (chapter 6) and a second *E. coli* O157:H7 outbreak that was traced to four Sizzler restaurant outlets that same year (chapter 5), consumers demanded an overhaul of the country's food safety regulatory system. And, although government officials continued to extol the safety of the U.S. food supply, regulators sought a new approach to managing food safety and rediscovered Hazard Analysis and Critical Control Points (HACCP) (61, 63).

In a way, the food industry has President John F. Kennedy to thank for HACCP. It was his vision to land a man on the moon before the end of the 1960s and return him safely to Earth. The engineering challenges were huge. Each component, from rocket booster to toggle switch, had to work right every time. But sending a team of astronauts to the moon required much more than just a functioning space capsule. The project demanded foolproof life support systems—breathable air, potable water, a means of dealing with bodily waste, and a supply of safe, nutritious food. The Pillsbury Company, which received a contract from the National Aeronautics and Space Administration (NASA) to manufacture food for the space program, was faced with a daunting challenge—to produce food with an extraordinarily high degree of microbiological safety. Since detecting low-level contamination of a processed food by a pathogen is at least as difficult as finding the proverbial needle in a haystack, finished product testing alone would not provide the guarantee that NASA needed (68). Pillsbury, together with NASA and the U.S. Army Natick Laboratories, decided instead on a new approach—to engineer safety into the food from the very beginning of the production process. The Hazard Analysis and Critical Control Points system was born. And gradually, over the subsequent 20+ years, HACCP was adopted by many food processors and government agencies worldwide (Table 7.1).

Table 7.1 Milestones in the development and adoption of HACCP as a food safety system

Year	Event
1971	HACCP introduced at National Conference on Food Protection (10)
1973	Pillsbury Company publishes "Food Safety Through the Hazard Analysis and Critical Control Point System" (10)
1973	FDA mandates HACCP for canned acidified and low-acid canned foods, the first time a food safety regulation is based on HACCP principles (10)
1978	HACCP applied to a food service environment (3–5)
1985	National Academy of Sciences recommends the widespread adoption of HACCP for food safety regulation (67)
1989	National Advisory Committee on Microbiological Criteria for Foods (NACMCF) develops its first comprehensive HACCP guide (10)
1992	Canada's Department of Fisheries and Oceans adopts mandatory HACCP-based regulatory program for the Canadian fish processing industry (65)
1992	NACMCF revises and updates its HACCP guide (10)
1995	FDA mandates HACCP for fish and fishery products industry (15)
1996	USDA mandates HACCP for meat and poultry industry, but does not require companies to test for pathogens (16)
1997	NACMCF issues a further revision and update of its HACCP guide (10)
1997	The United Nations Food and Agricultural Organization incorporates HACCP into the Codex Alimentarius (10)
2001	FDA mandates HACCP for juice industry (17)
2002	USDA amends HACCP regulation to address the risks posed by *E. coli* O157:H7 in raw meat (18)
2003	USDA amends HACCP regulation to address the risks posed by *Listeria monocytogenes* in ready-to-eat meat and poultry (19)
2004	European Union announces that the Codex Alimentarius HACCP protocol will become mandatory for all food businesses except primary producers in 2006 (13)

HACCP is a detailed step-by-step approach to producing safe food. The first step in developing a HACCP plan is hazard analysis. Food safety experts identify all of the potential microbiological, chemical, and physical hazards (e.g., *Salmonella*, *Staphylococcus aureus*, *Clostridium botulinum*, pesticide residues, and splinters of broken glass) that might be associated with each component of every product. Next, they establish what ingredient specifications, processing treatments, or handling procedures are needed to overcome each potential hazard. Then, they identify the critical control points—one or more stages in the food production process at which each hazard must be controlled. Finally, they develop procedures to monitor and document the correct performance of each critical control point (62).

The food service industry soon recognized the benefits of HACCP-based safety programs. Hospital dietitians and administrators welcomed an approach that would reduce the risk of food-borne illness in hospital food service operations. In 1978, Bobeng and David reported having developed and validated several model HACCP systems for hospital kitchens engaged in conventional cook-and-serve as well as cook/chill and cook/freeze operations (4, 5).

It also wasn't long before the U.S. Food and Drug Administration (FDA) realized that HACCP made good sense, too. In 1973, the same year that Pillsbury published its HACCP manual, FDA incorporated this concept into its regulations for the manufacture of acidified and low-acid canned foods (67). The Pillsbury manual formed the basis for FDA's internal HACCP training program. But then HACCP went into hibernation—as far as government regulators were concerned.

It took several more years before the U.S. government reawakened to the potential benefits of HACCP. In September 1980, four federal agencies with responsibility for food safety regulation (FDA, USDA, National Marine Fisheries Service, and the U.S. Army Natick Research and Development Center) asked the U.S. National Research Council (NRC) to undertake a study of the country's food safety regulatory system. The NRC report, published in 1985, strongly urged that the nation's food safety regulations be based on the concepts inherent in HACCP. NRC's proposal that a committee be set up to develop microbiological criteria for food safety resulted in the formation of a National Advisory Committee on Microbiological Criteria for Foods (NACMCF). But the report's recommendation that government agencies adopt HACCP as the foundation for the country's food safety programs appeared to fall on deaf ears—at least in the United States.

Canada, however, acted on recommendations contained in the NRC report. In 1992, the Department of Fisheries and Oceans introduced the first regulatory framework in the world (except for the U.S. canned food regulations) to be based on a HACCP program (65). But Canada did not extend its HACCP approach beyond fish and seafood to other sectors of the food industry for many years.

Even while government agencies largely ignored HACCP as a regulatory tool, experts in academia and industry pursued the concept as a food safety management strategy (Table 7.1). In 1989, NACMCF unveiled a comprehensive guide describing and explaining the seven principles of HACCP. The committee continued to refine its approach to HACCP and published updates to its original program in 1992 and 1997 (10). And in 1992 and 1993, the National Food Processors Association (NFPA) added its support to the HACCP system (48, 49).

Despite the behind-the-scenes work of NACMCF and NFPA, the 1985 NRC report appeared to have achieved very little impact in the first 7 years after its publication. Then, in January 1993, *E. coli* O157:H7 met Jack in the Box, and the world of food safety changed. The magnitude of the 1993 outbreak and the death of four children guaranteed that its ramifications would be many and long lasting. Surviving family members of the dead children joined together to form a food safety public interest group that evolved into the organization Safe Tables Our Priority, or S.T.O.P. This organization lobbied strenuously for major improvements to the way in which USDA and FDA regulated the food industry. USDA's adoption of HACCP-based regulation of the meat and poultry industry in 1996 was a direct result of the public testimony and lobbying efforts of S.T.O.P. and other consumer advocacy groups (16).

The USDA based its regulation on the 1992 version of NACMCF's HACCP system. The Final Rule, issued on July 25, 1996, set out a three-stage implementation schedule. Large meat and poultry processors were required to comply within 18 months (January 1998), smaller processors within 2½ years (January 1999), and the smallest operators within 3½ years (January 2000). The FDA also embraced HACCP in the 1990s, adopting a HACCP-based program for regulating the fish and fishery products industry in 1995 (15) and juice processors in 2001 (17). Both FDA and USDA incorporated into their regulations the seven HACCP principles enunciated by NACMCF.

USDA, HACCP, and the Meat Industry

For a HACCP program to be effective, it must be well conceived, well executed, and well maintained. This requires commitment on the part of all levels within a company, from the chief executive to the assistant floor sweeper. HACCP is a philosophy that encompasses every aspect of a company's food safety program; it cannot succeed if company management does not support it actively (36, 45). Jack in the Box was an early convert to HACCP, having learned from the 1993 outbreak the critical importance of critical control points (69). Unfortunately, many meat and poultry industry members were not solidly behind USDA's 1996 foray into HACCP.

HACCP—The Seven-Step Program (10)

The HACCP program developed in the 1960s for the space program comprised just three basic principles: assessing hazards, determining critical control points, and establishing monitoring procedures. By 1989, the philosophy of HACCP had evolved further. The most recent version of HACCP, as described in the 1997 revision to the NACMCF system, comprises seven principles (58).

1. Conduct a Hazard Analysis

NACMCF defines a hazard as a "… biological, chemical, or physical property that may cause a food to be unsafe for consumption." Hazard analysis consists of (i) identifying all possible microbiological, chemical, or physical agents associated with a food that could cause food-borne illness, (ii) establishing the likelihood that those agents might be present in the food, in any of its ingredients, in the packaging material, or in the plant environment, and (iii) determining the seriousness of the potential health risk posed by each identified hazard. Not all hazards are equally likely to occur, and not all hazards carry the same degree of risk to life and health. The 1989 NACMCF HACCP system identified six criteria that could be used to determine the seriousness of microbiological, chemical, and physical hazards (10).

2. Determine the Critical Control Points

Critical control points are those steps in food production at which hazards can be eliminated, reduced, or controlled (58). Cooking steps can be critical control points (47). Raw material microbial load, line speeds, antimicrobial carcass rinses, speed of chilling or freezing, pH, and water activity (a_w) are other likely candidates for microbiological critical control points (47, 49, 57). Depending on the nature of a processing operation and finished product, a HACCP program may include one or more critical control points (or even none at all) for each identified hazard (6, 49, 58, 66, 70).

3. Establish Critical Limits

Examples of critical limits would be a minimum cooking temperature, a target pH range, maximum exposure time to temperatures within microbial growth range, or a maximum a_w value for finished product. To be effective, a critical limit must be clearly defined, feasible, and make scientific sense. Ideally, it should also be quantitative to allow easy monitoring. Separate critical limits must be established for each critical control point (58).

4. Establish Monitoring Procedures

Well-chosen critical limits should be easy to monitor. Temperature, for example, can be monitored continuously and the monitoring equipment set to send an alarm if the temperature drifts outside a specified range. Processing line speeds are also easy to track, as are parameters such as ingredient concentration, a_w, and pH.

5. Establish Corrective Actions

A well-designed HACCP plan should leave as little to the imagination as possible. For every critical control point limit, there must be a defined corrective action to be taken as soon as a problem is detected. The correction might be a simple adjustment while a production line continues to run, or a line shutdown might be needed. Personnel should never be left in the dark as to the appropriate corrective action to take for any eventuality.

6. Establish Verification Procedures

HACCP is not a static system. It cannot be designed and then allowed to run unsupervised. The ultimate test of how well a HACCP system has been designed is the safety of the finished food product. Verification of an establishment's HACCP system can comprise multiple elements, including periodic outside audit of HACCP records and the use of appropriately chosen quantitative microbial indicator tests for in-process samples and finished products (32, 43, 44).

7. Establish Record-Keeping and Documentation Procedures

If work has not been documented, it might just as well have not been done. Without proper record-keeping, even a well-designed program quickly becomes meaningless.

The 1993 *E. coli* O157:H7 outbreaks exposed several weaknesses in the U.S. meat regulatory system: tracing contaminated meat back to its source was nearly impossible; USDA had no data on the incidence of *E. coli* O157:H7 in meat; no one—neither the government nor the industry—was testing meat for pathogens on a routine basis; USDA's inspection procedures were still entirely visual, though an existing "zero tolerance" for visual fecal contamination on meat carcasses was being ignored; and the USDA had no authority to mandate a food recall (53). The government moved relatively quickly to address two of these areas. In March 1993, the USDA began to enforce the fecal "zero tolerance" rule rigorously (38). The following year, the agency announced that raw meat found to contain *E. coli* O157:H7 would be considered adulterated and initiated a testing program to determine the incidence of the pathogen in ground meat (1).

The meat industry's reaction to these new initiatives was decisive—and negative. Industry representatives denounced the "zero tolerance" rule as counterproductive. Meat industry spokesmen claimed that the additional handling required to remove areas of fecal contamination from the surfaces of carcasses would result in higher overall contamination and reduced consumer safety (38). In fact, a Canadian government study, published in 1996, showed that, while knife trimming did not increase overall contamination, the benefits obtained by trimming visual fecal contamination from beef carcasses were only cosmetic (34). Bacterial counts were essentially unaffected by trimming visible contamination, although routine trimming of areas known to be subject to heavy contamination—whether or not that contamination was visible—proved to be beneficial (35).

Industry reaction to USDA's identification of *E. coli* O157:H7 as an adulterant and to its random testing program to determine the incidence level of *E. coli* O157:H7 in the U.S. ground beef supply was fierce. The American Meat Institute asked the federal court to prevent USDA from implementing this new program, claiming (among other arguments) that the agency had exceeded its regulatory authority and—since USDA methods in effect at the time for processing ground beef were unable to prevent meat from become contaminated—that the declaration of *E. coli* O157:H7 as an adulterant was inappropriate (41). But the declaration and the random testing program went ahead despite industry opposition.

In February 1995, USDA announced its intention to revamp its entire approach to meat and poultry regulation and to mandate that the industry develop and adopt HACCP food safety programs. The Proposed Rule also required that processors use at least one antimicrobial treatment during slaughter, clarified industry management's responsibility for ensuring that sanitation requirements were complied with, required prompt chilling of carcasses, mandated daily microbiological testing, and established interim targets for pathogen reduction (14). USDA received more than 6,800 comments on its proposal from industry and consumer representatives, other food safety experts, and individual consumers.

Although most of the comments supported USDA's HACCP concept, some of the other ideas contained in the 1995 proposal encountered serious opposition. The meat industry was unhappy about the mandated carcass chilling provisions, the antimicrobial treatment requirements, and the *Salmonella* pathogen reduction targets—including the need to test daily for the presence of *Salmonella*. The mandatory antimicrobial treatment and chilling provisions were both eliminated from the final version of the regulation. And although the agency maintained the *Salmonella* pathogen reduction target, it replaced *Salmonella* testing with *E. coli* biotype 1 (the so-called generic *E. coli*) and changed from mandatory daily testing to a testing frequency based on the production volume at each plant. Finally, USDA extended its HACCP implementation deadlines by 6 months, to January 1998, 1999, and 2000 for large, smaller, and smallest facilities, respectively (16). USDA's HACCP initiative was finally a reality, but its implementation was too late to prevent the largest meat recall in U.S. history.

Embracing HACCP

Could a HACCP program have prevented the demise of Hudson Foods? The experience of Jack in the Box provides some answers. Soon after its 1993 *E. coli* O157:H7 outbreak, the company hired a new vice president of quality assurance and product safety, Dr. David Theno, and committed itself to a comprehensive program of food safety. Dr. Theno orchestrated the development and implementation of a HACCP program for Jack in the Box that began at the farm and ended at the restaurant table. The company established very strict specifications for all of its suppliers to meet, demanding that their suppliers be able to trace meat all the way back to the ranch. Each supplier also was required to develop and implement its own HACCP program, including critical control points for *E. coli* O157:H7. Two years after the introduction of its new food safety program, Jack in the Box noticed a dramatic improvement in the overall microbiological quality of its raw hamburger patties. The new, more demanding supplier specifications were working (69). But supplier specifications were only one element of the program.

Jack in the Box also instituted tight controls over its own operations, from the moment the raw patties and other supplies arrived at the warehouse or restaurant door until the meal was in the customer's hands. Storage temperature, delivery procedures, cooking conditions, and sanitation were all part of the company's HACCP umbrella. Though impressive, this program would have been useless without documentation. Paperwork is the essential tool for tracking the effectiveness of a HACCP protocol and for picking up the pieces when something goes wrong. The restaurant chain's contributions to the advancement of food safety were acknowledged in 2004, when it received the Black Pearl Award from the International Association for Food Protection (42).

Even a properly designed and executed internal HACCP program would not have prevented the contamination of Hudson's patties by *E. coli* O157:H7. The company purchased its meat from as many as 15 slaughterhouses. The contamination might have come from any of them—the ultimate source of the *E. coli* O157:H7 was never confirmed. But the documentation trail that is part of a well-designed HACCP program would have enabled Hudson to pinpoint exactly which batches of frozen patties had been exposed to the contaminated meat. The company would have been able to limit its recall to the batches it knew to be at risk, and could have avoided the largest meat recall in U.S. history. Instead, the lack of documentation caused Hudson to provide the USDA with incomplete and faulty information.

At first, Hudson had advised USDA that all of the contaminated meat was confined to a single lot, about 3 days of production. That lot was recalled. But the company later determined that the lot it had singled out for recall was never distributed inside Colorado, the state where the outbreak had taken place (2). As Hudson's employees continued to change their estimates of the amount of product implicated, USDA finally lost confidence in the reliability of the company's information and insisted on the recall of the entire 25 million pounds.

As USDA began to implement its HACCP-based regulatory program, consumers waited and hoped that the agency's new approach would prevent another Hudson-style episode. And for a few years, their prayers appeared to have been answered. While the number of recalls due to *E. coli* O157:H7 continued to increase (Table 7.2), there were no new catastrophic outbreaks—until mid-2002.

Table 7.2 *E. coli* O157:H7 U.S. meat recalls, 1994–2004[a]

Year	No. of recalls	Pounds of meat recalled (thousands)
1994	3	873
1995	5	949
1996	2	164
1997	6	25,614
1998	13	2,058
1999	10	731
2000	21	2,707
2001	26	2,177
2002	35	24,255
2003	13	2,205
2004	6	1,199
2005	5	1,248

[a] Summary of data on recalls under jurisdiction of the Food Safety and Inspection Service, USDA (31).

Déjà Vu?

On June 30, 2002, USDA announced that it had detected *E. coli* O157:H7 in a sample of ground meat produced by ConAgra's Greeley, Colo., plant one month earlier (May 31). The company was recalling approximately 354,000 pounds of meat—a single day's production. At the time of the press release announcing the recall, no illnesses had yet been associated with the contaminated meat (24). But that would soon change.

On July 8, a little more than 1 week after the recall was announced, the state of Colorado advised the Centers for Disease Control and Prevention (CDC) of a possible *E. coli* O157:H7 outbreak—a cluster of 16 cases that had been reported in the latter half of June (8). Using molecular fingerprinting, the state lab was able to link the outbreak to the production lot of ground meat that had been recalled by ConAgra the week before. On July 19, ConAgra expanded its recall to encompass nearly 19 million pounds of ground beef and beef trimmings (26). By August 13, CDC had identified a total of 38 victims: 22 in Colorado and 16 scattered throughout nine other states. Six people were stricken with hemolytic uremic syndrome (HUS), and one of them died (9). The final tally of victims was 46, spread over 16 U.S. states and traced to meat from both phases of the recall (50, 60).

At first, the ConAgra outbreak and recall—the second-largest meat recall in U.S. history—appeared to be a major blow to supporters of HACCP-based regulatory reform. Was HACCP not designed to prevent just such a food safety disaster? ConAgra's Greeley, Colo., plant, as a large producer of ground meat, had a HACCP program in place since January 1998. In addition to meeting USDA's required testing for *E. coli* biotype 1 as an indicator of possible fecal contamination, the company conducted regular tests for *E. coli* O157:H7. The contamination of ConAgra's meat by *E. coli* O157:H7 should not have taken either the company or USDA by surprise. Both ConAgra and USDA officials had known of or suspected problems at the Greeley plant for several months prior to the outbreak.

In January 2002, USDA inspectors found *E. coli* O157:H7 in ground meat produced by Montana Quality Foods, a ConAgra customer (60). Although Montana's management said that they had purchased the contaminated meat from ConAgra, agency officials decided not to alert ConAgra to their finding, as they could not be 100% certain of the origin of the contamination (54). Then USDA found *E. coli* O157:H7 in ground meat produced on May 9 by another ConAgra customer, Galligan's Wholesale Meat Co. But the contaminated batch contained meat from more than one supplier, so USDA conducted additional tests. The agency found *E. coli* O157:H7 in ground meat produced by Galligan's on June 17 and June 19—both batches made exclusively using meat supplied by ConAgra. There could be no further doubt. ConAgra, however, was not informed of the Galligan finding by USDA until June 27 (51).

Even in the absence of official notification from USDA, the company was far from ignorant of the developing problem. USDA cited ConAgra 31 times between July 1, 2001, and August 27, 2002, for visible fecal contamination on its beef carcasses (53). In fact, the number of citations understated the number of occasions on which fecal contamination occurred since, in many cases, inspectors simply advised ConAgra management of the infraction verbally (60). Not only had it received a citation from USDA for fecal contamination as recently as April 11, 2002, but also ConAgra's own testing of its beef trimmings produced 34 positive results for *E. coli* O157:H7 between April 12 and July 11 (53). Each time, the positive result was recorded as an "unforeseen occurrence." These results were not reported to USDA, and the company made no adjustments to its procedures or to its HACCP program to prevent future occurrences (12). ConAgra withheld its internal *E. coli* O157:H7 test results from USDA even after it was advised of the government's findings. And company management delayed announcing the initial recall of approximately 354,000 pounds of meat until June 30, on the pretext that it was awaiting confirmation of the government test results—even though its own tests had already corroborated the government's data (52).

What Happened to HACCP?

Unlike the seafood industry, which supported strongly FDA's introduction of HACCP-based regulations in 1995 (15), members of the meat industry, with a few exceptions, never fully embraced USDA's HACCP program. Nor did they welcome other changes, including the agency's enhanced scrutiny of their products for microbiological contaminants such as *Salmonella* and *E. coli* O157:H7 and the requirement for safe handling labels on raw meats (1, 37, 46). In the absence of a full, honest, and continuing commitment to HACCP at all levels of an organization, eventually the program will fail, as it did at ConAgra.

ConAgra's original HACCP plans were based on the faulty assumption that *E. coli* O157:H7 was not a hazard that was likely to occur in its plant operations (60). Nevertheless, as specified under its HACCP program, the company tested beef carcasses on a regular basis for *E. coli* O157:H7. Every one of the test results in 2001 and 2002 was negative. Since this testing program was part of the company's HACCP plan, these results were available to USDA inspectors. On the other hand, ConAgra's HACCP program did not call for testing of its beef trim—a product that is frequently used to make ground beef—for *E. coli* O157:H7. But it performed the tests anyway, with at least 63 positive results in 2002. Because this testing was not part of the company's HACCP program, ConAgra was not obliged to share this information with USDA. ConAgra did not test its ground beef for *E. coli* O157:H7. Neither did the USDA.

USDA mandated, as part of its HACCP regulatory program, that companies monitor their process microbiologically by testing carcasses for *E. coli* biotype 1, the so-called generic *E. coli*, according to a sampling protocol defined in the HACCP Final Rule (16). As early as April 2002, ConAgra's records showed increased levels of *E. coli* biotype 1 on their beef carcasses. The company reacted by increasing its acceptable limit for this indicator of fecal contamination, thus defining the problem out of existence. And it went even further, establishing a more liberal limit—almost double—for the microbe on one shift than on the other (60).

ConAgra performed an annual review and reassessment of its HACCP plans in April 2002. Faced with a rising incidence of fecal contamination as measured by the *E. coli* biotype 1 test results, and the increased number of beef trim samples containing *E. coli* O157:H7, company officials decided, nevertheless, that the HACCP plans needed no change whatsoever. They concluded that *E. coli* O157:H7 contamination of their products was still a hazard that was unlikely to occur. Only a corporate commitment to HACCP as window dressing could have justified such a conclusion. And only USDA's commitment to giving the meat industry the benefit of every doubt could have explained the agency's actions in the ConAgra affair.

In designing its HACCP-based regulatory program, USDA's Food Safety and Inspection Service (FSIS) made several policy decisions that placed control of meat safety squarely in the hands of the industry. Many of these decisions played a significant role in allowing the ConAgra situation to get out of control.

1. **FSIS did not sample ground beef for *E. coli* O157:H7 from facilities that carried out their own tests for the pathogen.**
   FSIS Directive 10,010.1, issued in 1998, specifically enjoined meat inspectors from collecting ground beef samples for *E. coli* O157:H7 testing from establishments that were conducting "… routine daily testing of their raw ground beef products or boneless beef to be used in raw ground products for *E. coli* O157:H7" (23). Since ConAgra was conducting routine tests of its boneless beef for *E. coli* O157:H7 and sharing those results, all of which were negative, with USDA, the government carried out no independent tests on the company's ground beef (60). FSIS changed its policy in the wake of the 2002 *E. coli* O157: H7 outbreak and now tests ground beef from all meat producers, whether or not the companies do their own testing (28, 60).
2. **FSIS policy barred access by meat inspectors to test results that were obtained by an establishment outside that establishment's HACCP plans.**
   ConAgra did not include beef trim as a critical control point in its HACCP plans. Therefore, the company had no obligation to share the test results from its beef trim sampling program with FSIS. In

consequence, the agency remained ignorant of the rising number of positive *E. coli* O157:H7 results that ConAgra was finding in its beef trim (60).

3. **FSIS policy barred meat inspectors from taking samples at ConAgra after finding *E. coli* O157:H7 in ground beef from a ConAgra customer.**
   When FSIS detected *E. coli* O157:H7 in a ground beef sample from Montana Quality Foods (a customer of ConAgra) in January 2002, the meat inspector requested permission to sample meat from ConAgra. USDA's Technical Service Center, whose approval was needed for the action, denied the inspector's request. It took until May 2002—when USDA found *E. coli* O157:H7 in ground meat produced by a second ConAgra customer—before the agency traced the contamination back to ConAgra (60).

In addition to these industry-favoring policies, USDA also chose not to confront ConAgra over sanitation problems and questionable food safety decisions. FSIS cited the Greeley plant repeatedly for visible fecal contamination on beef carcasses, yet took no other enforcement actions, and closed its eyes when ConAgra increased its acceptable limit on *E. coli* biotype 1 levels (a key indicator of fecal contamination) on those carcasses. The agency never questioned ConAgra's assumption that *E. coli* O157:H7 was a hazard that was unlikely to occur in its products, even in the face of other outbreaks and meat recalls traced to the presence of that pathogen in ground beef. And FSIS allowed ConAgra to resell some of its contaminated beef without verifying that the meat would not be resold as raw ground beef (60).

USDA had been alerted to many of the shortcomings in its policies and practices more than 2 years before the ConAgra incident. The General Accounting Office (now the Government Accountability Office) reviewed USDA's HACCP implementation progress in 1999, and the agency's Office of Inspector General audited FSIS implementation of HACCP regulations in 2000 (33, 59). These reviews highlighted several glaring problems.

1. FSIS allowed establishments arbitrarily to reduce or limit the number of critical control points in their HACCP plans, limiting government oversight of their operations.
2. FSIS did not always review an establishment's testing plans.
3. FSIS lacked access to plant records.
4. Auditors found two plants whose internal testing showed a potential for pathogen contamination in their products, but those results were not available to FSIS.
5. Training of inspectors in the details of the HACCP regulatory program was weak in some areas.

6. FSIS did not have the authority to either approve or disapprove a company's HACCP plan, nor could inspectors verify aspects of a company's food safety program that fell outside the HACCP plan.

These shortcomings, and others, increased the likelihood of an incident the size and scope of the ConAgra contamination. When the need for a massive recall became evident, ConAgra did not have all of the information it required to trace its meat shipments. And USDA was unprepared to supervise the recall adequately.

Total Recall

Mandatory recall authority for USDA and FDA has been a goal of consumer groups and a boogeyman for the food industry for more than a decade. Senator Tom Daschle, acting in the wake of the 1993 Jack in the Box outbreak, first proposed to Congress that USDA be granted the ability to mandate meat recalls. The Senator's bill also provided for enhanced animal quarantine laws and would have required that meat companies be able to trace contaminated animals from the slaughterhouse back to the farm (39). The bill was never passed.

The 1997 Hudson Foods recall shined an embarrassing spotlight on the shortcomings of the voluntary recall process. The recovery of just 10 to 11 million pounds, less than half of the 25 million pounds of recalled meat, convinced then-Secretary of Agriculture Dan Glickman to ask Congress once more for mandatory recall authority. Again, Congress declined to act, and the stage was set for the ConAgra recall (11, 40).

Recall logistics are necessarily complex, especially when the recalled product has passed through several levels of reprocessing and distribution. ConAgra produced and sold both beef trim and coarse-ground beef to companies that, in turn, reground the beef and sold it to supermarkets and other retail outlets. Two of these regrinding operations were Montana Quality Foods and Galligan's—the ConAgra customers in whose meat USDA had found *E. coli* O157:H7 in January 2002 and May 2002, respectively. Neither Montana nor Galligan's had a recall plan in place, and Galligan's production records did not contain enough information to enable its ground beef products to be traced to their ultimate destinations, making an effective recall nearly impossible. As a result, some distributors of the meat were never notified that a recall had been initiated. And USDA's actions didn't help (60).

FSIS policy limiting traceback of contamination to the ultimate producer—in this case, ConAgra—delayed the initiation of the ConAgra recall by a week. At first, the local inspector was denied permission to sample meat from ConAgra for *E. coli* O157:H7. It was only after the inspector obtained permission from headquarters that meat produced by ConAgra on May 31—the production lot that was implicated in Galligan's contamination—was sent to the FSIS lab for analysis. During that week, consumers were exposed unnecessarily to potentially deadly contamination. In addition, FSIS was both dilatory and derelict in carrying out its responsibilities. Government policy requires that the agency carry out timely checks of the effectiveness of a recall, yet some of those checks were not carried out until 4 months after the recall began, and a number of them were never completed (60).

The ConAgra meat recall encompassed more than 18 million pounds of beef trim and ground beef produced between April 12 and June 29, 2002 (26). The company only recovered 3 million pounds—less than 20% of the recalled meat. Nevertheless, the FSIS District Office responsible for monitoring the recall described it as effective, on the sole basis that no additional illnesses had been attributed to the recalled meat (60).

The feeble performance of ConAgra, its distributors, and FSIS in this recall was, in part, a consequence of USDA's lack of mandatory recall authority. In the absence of this authority, FSIS could not insist that slaughterhouses, meat processors, and grinding operations have a recall procedure in place and available for the agency's review. ConAgra had developed a recall plan, but it was kept separate from the company's HACCP plans and was not subject to FSIS scrutiny. Montana Quality Foods didn't consider that it needed a recall plan because it was a small plant, and Galligan's claimed a recall plan was unnecessary, since it held its meat until receiving an "all clear" from the government-run tests (60).

In responding to the findings of the Office of Inspector General in its post-recall audit, FSIS claimed that it was powerless to require meat plants to include recall procedures in their HACCP programs. Yet its parent agency, USDA, has rejected repeated congressional attempts to legislate mandatory recall authority (56). In this, as in so many of its policy positions that led to the ConAgra recall, USDA has favored the interests of the meat industry over the health concerns of consumers. And by its laissez-faire approach to meat safety regulation, the agency bears a portion of the responsibility for the toll of illness and death caused by ConAgra's *E. coli* O157:H7-contaminated ground meat. But some things have changed, and the changes have borne fruit.

The Aftermath

The ConAgra affair caused USDA to rethink certain of its policies and procedures. On July 15, 2002, Linda Swacina, the acting administrator of FSIS, announced that the agency was implementing a new plan to improve its ability to trace back contaminated meat to its source and was also changing its previous policy on notifying suppliers of contaminated meat (25). In October 2002, FSIS ordered all meat plants producing raw beef products to review their HACCP programs and data with the purpose of determining whether or not *E. coli* O157:H7 was a "hazard likely to occur" in their plant operations, and to amend their HACCP programs accordingly (18). And USDA went even further.

The agency reviewed and revised its directives to FSIS inspectors in May 2003, giving them more detailed guidance on evaluating food safety and HACCP issues in all FSIS-inspected plants (27). In March 2004, the agency revised its microbiological testing program for *E. coli* O157:H7 in raw beef (28). And in April 2004, FSIS issued a comprehensive set of compliance guidelines for *E. coli* O157:H7 sampling and verification to meat plants operating under its regulatory authority (29).

The results of these initiatives are encouraging. Recalls due to *E. coli* O157:H7 contamination of meat products dropped from a peak of 35 (comprising more than 24 million pounds) in 2002 to just five recalls (totaling 1.25 million pounds) in 2005 (Table 7.2). The percentage of *E. coli* O157:H7-positive samples detected by FSIS also decreased—from 0.78% in 2002 to 0.30% in 2003, and to just 0.17% in 2004 (30). And consumers reaped the benefit of these improvements. The number of illnesses due to *E. coli* O157:H7 reported in the United States declined by more than 40% between 2002 and 2004, although the incidence of reported illness due to this pathogen increased slightly in 2005 to 1.06 per 100,000 people, compared to 0.90 per 100,000 people in 2004 (71, 72).

References

1. **Anonymous.** 2 November 1994. Suit on meat testing. *The Washington Post*, Washington, D.C.
2. **Belluck, P.** 17 December 1998. U.S. indicts producer of contaminated beef. *New York Times*, New York, N.Y.
3. **Bobeng, B. J., and B. D. David.** 1977. HACCP models for quality control of entree production in foodservice systems. *J. Food Prot.* **40:**632–638.
4. **Bobeng, B. J., and B. D. David.** 1978. HACCP models for quality control of entrée production in hospital foodservice systems. I. Development of hazard analysis critical control point models. *J. Am. Diet. Assoc.* **73:**524–529.
5. **Bobeng, B. J., and B. D. David.** 1978. HACCP models for quality control of entrée production in hospital foodservice systems. II. Quality assessment of beef loaves utilizing HACCP models. *J. Am. Diet. Assoc.* **73:**530–535.
6. **Bryan, F. L.** 1996. Another decision-tree approach for identification of critical control points. *J. Food Prot.* **59:**1242–1247.
7. **Bush, G. H. W.** 2 September 1992. Remarks to Shallowater Co-op Gin Company employees in Shallowater, Texas. The Museum at the George Bush Presidential Library, College Station, Tex. [Online.] http://bushlibrary.tamu.edu/research/papers/1992/92090201.html. Accessed 7 June 2006.
8. **Centers for Disease Control and Prevention.** 19 July 2002. Backgrounder. Division of Media Relations, Centers for Disease Control and Prevention. Atlanta, Ga. [Online.] http://www.cdc.gov/od/oc/media/pressrel/b020719.htm. Accessed 9 June 2006.

9. **Centers for Disease Control and Prevention.** 13 August 2002. Backgrounder. Division of Media Relations, Centers for Disease Control and Prevention. Atlanta, Ga. [Online.] http://www.cdc.gov/od/oc/media/pressrel/b020813.htm. Accessed 9 June 2006.
10. **Corlett, D. A., Jr.** 1998. *HACCP User's Manual.* Aspen Publishers, Inc., Gaithersburg, Md.
11. **Cuadros, P., P. J. Kiger, and B. Allison.** 1998. Recall and recovery, p. 51–52. In W. O'Sullivan (ed.), *Safety Last. The Politics of E. coli and Other Food-borne Killers.* The Center for Public Integrity, Washington, D.C.
12. **Day, S.** 13 March 2003. U.S. is urged to investigate beef company over tainting. *New York Times*, New York, N.Y.
13. **Elson, R.** 8 December 2004. Overview of incoming changes to European food safety and hygiene legislation. *Eur. Surveill. Wkly.* **8.** [Online.] http://www.eurosurveillance.org/ew/2004/041208.asp#1. Accessed 8 June 2006.
14. **Federal Register.** 1995. Pathogen reduction; hazard analysis and critical control point (HACCP) systems; Proposed Rule. *Fed. Regist.* **60:**6773–6889.
15. **Federal Register.** 1995. Procedures for the safe and sanitary processing and importing of fish and fishery products; Final Rule. *Fed. Regist.* **60:** 65096–65202.
16. **Federal Register.** 1996. Pathogen reduction; hazard analysis and critical control point (HACCP) systems; Final Rule. *Fed. Regist.* **61:**38806–38855.
17. **Federal Register.** 2001. Hazard analysis and critical control point (HAACP); procedures for the safe and sanitary processing and importing of juice; Final Rule. *Fed. Regist.* **66:**6138–6202.
18. **Federal Register.** 2002. *E. coli* O157:H7 contamination of beef products. *Fed. Regist.* **67:**62325–62334.
19. **Federal Register.** 2003. Control of *Listeria monocytogenes* in ready-to-eat meat and poultry products; Interim Final Rule. *Fed. Regist.* **68:**34208–34254.
20. **Food Safety and Inspection Service.** 12 August 1997. Hudson Foods recalls beef burgers nationwide for *E. coli* O157:H7. Release No. 0272.97. Food Safety and Inspection Service, U.S. Department of Agriculture, Washington, D.C. [Online.] http://www.fsis.usda.gov/OA/recalls/prelease/pr015-97.htm. Accessed 9 June 2006.
21. **Food Safety and Inspection Service.** 15 August 1997. USDA announces recall of additional Hudson frozen ground beef. Release No. 0276.97. Food Safety and Inspection Service, U.S. Department of Agriculture, Washington, D.C. [Online.] http://www.fsis.usda.gov/oa/recalls/prelease/pr015-97a.htm. Accessed 9 June 2006.
22. **Food Safety and Inspection Service.** 21 August 1997. Glickman announces Hudson to act on USDA recommendation to close Nebraska plant, recall all Hudson beef products. Release No. 0283.97. U.S. Department of Agriculture, Washington, D.C. [Online.] http://www.usda.gov/news/releases/1997/08/0283. Accessed 9 June 2006.
23. **Food Safety and Inspection Service.** 1 February 1998. FSIS Directive 10,010.1. Microbiological testing program for *Escherichia coli* O157:H7 in raw ground beef. Food Safety and Inspection Service, U.S. Department of Agriculture, Washington, D.C.

24. **Food Safety and Inspection Service.** 30 June 2002. Colorado firm recalls ground beef products for possible *E. coli* O157:H7. Recall Release FSIS-RC-055-2002. Food Safety and Inspection Service, U.S. Department of Agriculture, Washington, D.C. [Online.] http://www.fsis.usda.gov/oa/recalls/prelease/pr055-2002a.htm. Accessed 9 June 2006.

25. **Food Safety and Inspection Service.** 15 July 2002. Statement—ConAgra beef recall. Food Safety and Inspection Service, U.S. Department of Agriculture, Washington, D.C. [Online.] http://www.fsis.usda.gov/oa/news/2002/swacina071502.htm. Accessed 9 June 2006.

26. **Food Safety and Inspection Service.** 19 July 2002. Colorado firm recalls beef trim and ground beef products for possible *E. coli* O157:H7. Recall Release FSIS-RC-055-2002. Food Safety and Inspection Service, U.S. Department of Agriculture, Washington, D.C. [Online.] http://www.fsis.usda.gov/oa/recalls/prelease/pr055-2002.htm. Accessed 9 June 2006.

27. **Food Safety and Inspection Service.** 21 May 2003. FSIS Directive 5000.1, Revision 1. Verifying an establishment's food safety system. Food Safety and Inspection Service, U.S. Department of Agriculture, Washington, D.C.

28. **Food Safety and Inspection Service.** 31 March 2004. FSIS Directive 10,010.1, Revision 1. Microbiological testing program and other verification activities for *Escherichia coli* O157:H7 in raw ground beef products and raw ground beef components and beef patty components. Food Safety and Inspection Service, U.S. Department of Agriculture, Washington, D.C. [Online.] http://www.fsis.usda.gov/OPPDE/rdad/FSISDirectives/10.010.1.pdf. Accessed 9 June 2006.

29. **Food Safety and Inspection Service.** 13 April 2004. Compliance guidelines for establishments on the FSIS microbiological testing program and other verification activities for *Escherichia coli* O157:H7. Food Safety and Inspection Service, U.S. Department of Agriculture, Washington, D.C. [Online.] http://www.fsis.usda.gov/OPPDE/rdad/fsisdirectives/10010_1/ecolio157h7dirguid4-13-04.pdf. Accessed 9 June 2006.

30. **Food Safety and Inspection Service.** 28 February 2005. FSIS ground beef sampling shows substantial *E. coli* O157:H7 decline in 2004. Food Safety and Inspection Service, U.S. Department of Agriculture, Washington, D.C. [Online.] http://www.fsis.usda.gov/News_&_Events/NR_022805_01/index.asp. Accessed 9 June 2006.

31. **Food Safety and Inspection Service.** 2005. FSIS recalls. Food Safety and Inspection Service, U.S. Department of Agriculture, Washington, D.C. [Online.] http://www.fsis.usda.gov/Fsis_Recalls/index.asp. Accessed 9 June 2006.

32. **Friedhoff, R. A., A. P. M. Houben, J. M. J. Leblanc, J. M. W. M. Beelen, J. T. Jansen, and D. A. A. Mossel.** 2005. Elaboration of microbiological guidelines as an element of codes of hygienic practices for small and/or less developed businesses to verify compliance with hazard analysis critical control point. *J. Food Prot.* **68:**139–145.

33. **General Accounting Office.** 1999. Meat and poultry. Improved oversight and training will strengthen new food safety system. Report No. GAO/RCED-00-16. U.S. General Accounting Office, Washington, D.C. [Online.] http://www.gao.gov/new.items/rc00016.pdf. Accessed 8 June 2006.

34. **Gill, C. O., M. Badoni, and T. Jones.** 1996. Hygienic effects of trimming and washing operations in a beef-carcass-dressing process. *J. Food Prot.* **59:**666–669.

35. **Gill, C. O., and T. Jones.** 1999. The microbiological effects of breaking operations on hanging beef carcass sides. *Food Res. Int.* **32:**453–459.

36. **Gilling, S. J., E. A. Taylor, K. Kane, and J. Z. Taylor.** 2001. Successful hazard analysis critical control point implementation in the United Kingdom: understanding the barriers through the use of a behavioral adherence model. *J. Food Prot.* **64:**710–715.

37. **Greene, R.** 9 October 1993. U.S. delays meat labeling rules. *Chicago Sun-Times*, Chicago, Ill.

38. **Greene, R.** 25 May 1994. Meat industry says federal regulation may endanger consumers. *Houston Chronicle*, Houston, Tex.

39. **Greene, R.** 20 September 1994. Administration unveils new food-safety bill. *The Oregonian*, Portland, Oreg.

40. **Guthrie, P.** 22 August 1998. *E. coli* beef scare: one year later. The issue of oversight: progress made, but USDA's lack of authority to recall meat remains a problem. *The Atlanta Journal-Constitution*, Atlanta, Ga.

41. **Hall, J. M.** 1995. R-Tech Review. USDA HACCP/Meat safety proposal. *Food Quality* **1:**13–16.

42. **International Association for Food Protection.** 2006. IAFP 2004 Black Pearl Award winner. Jack in the Box, Inc., San Diego, California. International Association for Food Protection, Des Moines, Iowa. [Online.] http://www.foodprotection.org/meetingsEducation/JackintheBoxBlackPearl.asp. Accessed 8 June 2006.

43. **Jericho, K. W. F., G. C. Kozub, V. P. J. Gannon, and C. M. Taylor.** 2000. Microbiological testing of raw, boxed beef in the context of hazard analysis critical control point at a high-line-speed abattoir. *J. Food Prot.* **63:**1681–1686.

44. **Kvenberg, J. E., and D. J. Schwalm.** 2000. Use of microbial data for hazard analysis and critical control point verification—Food and Drug Administration perspective. *J. Food Prot.* **63:**810–814.

45. **Long, S.** 2003. Evaluation of HACCP program in Plano, Texas resulted in new approach to guidelines and inspections. *Food Prot. Trends* **23:**100, 95.

46. **Manning, A.** 29 December 1999. Texas beef plant fights *Salmonella* standards. *E. coli* prompted processor to recall 90 tons. *USA Today*, McLean, Va.

47. **Michel, M. E., J. T. Keeton, and G. R. Acuff.** 1991. Pathogen survival in precooked beef products and determination of critical control points in processing. *J. Food Prot.* **54:**767–772.

48. **Microbiology and Food Safety Committee of the National Food Processors Association.** 1992. HACCP and total quality management—winning concepts for the 90's: a review. *J. Food Prot.* **55:**459–462.

49. **Microbiology and Food Safety Committee of the National Food Processors Association.** 1993. Implementation of HACCP in a food processing plant. *J. Food Prot.* **56:**548–554.

50. **Migoya, D.** 8 August 2002. Two more Colo. kids sickened. Same *E. coli* strain also killed woman in Ohio last month. *Denver Post*, Denver, Colo.

51. **Migoya, D.** 23 August 2002. Meat recall timing probed. Inspectors waited to tell ConAgra of *E. coli* finding. *Denver Post*, Denver, Colo.

52. **Migoya, D.** 6 September 2002. ConAgra delayed beef recall for 2 days. Firm waited for USDA to recheck *E. coli* tests. *Denver Post*, Denver, Colo.

53. **Migoya, D.** 19 September 2002. ConAgra had list of violations. Tainted meat found dozens of times at Greeley facility. *Denver Post*, Denver, Colo.

54. **Migoya, D., and A. Sherry.** 21 July 2002. Critics urge reform of beef-recall rules. Orders often too late to protect consumers. *Denver Post*, Denver, Colo.

55. **Morganthau, T.** 31 August 1997. Health pros' detective work helps arrest villain *E. coli*. *The Oregonian*, Portland, Oreg.

56. **Mulkern, A. C.** 12 February 2003. U.S. food safety at high risk, group warns. *Denver Post*, Denver, Colo.

57. **National Advisory Committee on Microbiological Criteria for Foods.** 1997. Generic HACCP application in broiler slaughter and processing. *J. Food Prot.* **60:**579–604.

58. **National Advisory Committee on Microbiological Criteria for Foods.** 1998. Hazard analysis and critical control point principles and application guidelines. *J. Food Prot.* **61:**1246–1259.

59. **Office of Inspector General.** 2000. Food Safety and Inspection Service implementation of the hazard analysis and critical control point system. Report No. 24001-3-AT. U.S. Department of Agriculture, Office of Inspector General, Washington, D.C. [Online.] http://www.usda.gov/oig/webdocs/haccp.pdf. Accessed 9 June 2006.

60. **Office of Inspector General.** 2003. Audit Report. Food Safety and Inspection Service oversight of production process and recall at ConAgra plant (Establishment 969). Report No. 24601-2-KC. U.S. Department of Agriculture, Office of Inspector General, Great Plains Region, Washington, D.C. [Online.] http://www.usda.gov/oig/webdocs/24601-2-KC%20conagra%20091603.pdf. Accessed 8 June 2006.

61. **Office of Management and Budget.** 23 June 1998. H.R. 4101—Agriculture, Rural Development, Food and Drug Administration, and related agencies appropriations bill, FY 1999. Office of Management and Budget, The Executive Office of the President, Washington, D.C. [Online.] http://www.whitehouse.gov/OMB/legislative/sap/105-2/HR4101-h.html. Accessed 7 June 2006.

62. **Ray, B.** 1996. *Fundamental Food Microbiology*. CRC Press, Inc., Boca Raton, Fla.

63. **Shalala, D.** 15 July 1999. Food safety remarks by HHS Secretary Shalala. Department of Health and Human Services, Washington, D.C. [Online.] http://vm.cfsan.fda.gov/~dms/fs-hhs01.html. Accessed 7 June 2006.

64. **Shillam, P., D. Heltzel, J. Beebe, and R. Hoffman.** 1997. *Escherichia coli* O157:H7 infections associated with eating a nationally distributed commercial brand of frozen ground beef patties and burgers—Colorado, 1997. *Morb. Mortal. Wkly. Rep.* **46:**777–778.

65. **Spencer, H.** 1992. The role of government in a mandatory HACCP based program. *Dairy Food Environ. Sanit.* **12:**501–505.

66. **Stier, R. F.** 2003. HACCP myths and misunderstandings. *Food Prot. Trends* **23:**712, 704.

67. **Subcommittee on Microbiological Criteria, Committee on Food Protection, Food and Nutrition Board, National Research Council.** 1985. *An Evaluation of the Role of Microbiological Criteria for Foods and Food Ingredients.* The National Academies Press, Washington, D.C. [Online.] http://books.nap.edu/books/0309034973/html/index.html. Accessed 9 June 2006.

68. **Swanson, K. M. J., and J. E. Anderson.** 2000. Industry perspectives on the use of microbial data for hazard analysis and critical control point validation and verification. *J. Food Prot.* **63:**815–818.

69. **Theno, D., and S. Bjerklie.** 1995. HACCP's role in the Jack in the Box recovery. *Meat and Poultry* **41:**42, 44, 46, 53–54.

70. **Vadhanasin, S., A. Bangtrakulnonth, and T. Chidkrau.** 2004. Critical control points for monitoring salmonellae reduction in Thai commercial frozen broiler processing. *J. Food Prot.* **67:**1480–1483.

71. **Vugia, D., A. Cronquist, J. Hadler, M. Tobin-D'Angelo, D. Blythe, K. Smith, K. Thornton, D. Morse, P. Cieslak, T. Jones, K. Holt, J. Guzewich, O. Henao, E. Scallan, F. Angulo, P. Griffin, R. Tauxe, and E. Barzilay.** 2006. Preliminary FoodNet data on the incidence of infection with pathogens transmitted commonly through food—10 states, United States, 2005. *Morb. Mortal. Wkly. Rep.* **55:**392–395.

72. **Vugia, D., A. Cronquist, J. Hadler, M. Tobin-D'Angelo, D. Blythe, K. Smith, K. Thornton, D. Morse, P. Cieslak, T. Jones, R. Varghese, J. Guzewich, F. Angulo, P. Griffin, R. Tauxe, and J. Dunn.** 2005. Preliminary FoodNet data on the incidence of infection with pathogens transmitted commonly through food—10 sites, United States, 2004. *Morb. Mortal. Wkly. Rep.* **54:**352–356.

chapter 8

Crossing Over

NOT EVERY *ESCHERICHIA COLI* STRAIN is a pathogen (most of them are not), and not every pathogenic *E. coli* strain belongs to the O157:H7 serotype. Pathogenic strains of *E. coli* fall into five distinct categories: enteropathogenic, enteroaggregative, enteroinvasive, enterotoxigenic, and enterohemorrhagic, also known as verotoxigenic or Shiga toxin-producing *E. coli* (53). The enterohemorrhagic *E. coli* strains comprise numerous serotypes, including, for example, O103, O111, O26:H11, O121:H19, O104:H21, and O157:H7 (23, 53, 74, 97, 105). Although enterohemorrhagic strains of *E. coli* are a worldwide problem, for many years *E. coli* O157:H7 was thought to be confined mainly to the United States, Canada, and the United Kingdom. This changed abruptly in 1996.

Sprouts for Lunch

School lunch programs in Japan are different from those in the United States and Canada. In Japanese elementary schools, every child in a school (or sometimes in an entire school district) receives the same lunch, and every child is encouraged to clean his or her plate. Therefore, when school children in Sakai City began to complain of diarrhea in July 1996, epidemiologists were faced with a daunting task. Lab tests quickly established that *E. coli* O157:H7 was the cause of the illnesses (4). But how were investigators to determine the probable source of the food poisoning outbreak when nearly all of the children had eaten every single item on the menu?

Unfortunately for the victims—but fortunately for the epidemiologists—Sakai City was not the only outbreak location. There were 16 outbreaks of *E. coli* O157:H7 in Japan between May and December 1996, encompassing more than 11,800 cases and 12 deaths (69). Three of the incidents were connected with meals served in nursing homes; one was linked to a factory cafeteria; two could not be traced to a specific origin; and nine outbreaks were tied to school lunches (68, 104). News of the first outbreak reached the public at the beginning of June.

The Pathogenic and Toxigenic *E. coli*

Although it has long been known to cause urinary tract infections, *E. coli* was once thought to be a harmless inhabitant of the intestinal tract (72). That is no longer the case. Whereas many *E. coli* strains are innocuous, or even helpful, components of the intestinal flora, we now recognize a significant number of pathogenic and toxigenic strains of *E. coli*. These strains fall into five categories, based on their mode of action (2). Enterohemorrhagic *E. coli* strains, including *E. coli* O157:H7, are zoonotic pathogens—resident in the intestines of several ruminant species (77). The other pathogenic and toxigenic *E. coli* strains either are human specific or their host range has not been determined (58, 78, 89).

1. Enterotoxigenic *E. coli*

One of the leading causes of traveler's diarrhea, enterotoxigenic *E. coli* (ETEC) causes a nonbloody watery diarrhea that begins with little or no warning—symptoms often reminiscent of *Vibrio cholerae*. ETEC can be either food borne or waterborne (29). It passes through the stomach and colonizes the small intestine, attaching to the intestine wall by means of fimbriae. Once there, the microbes produce two enterotoxins, a heat-stable toxin and a heat-labile toxin. These toxins activate an intestinal enzyme (adenylate cyclase), triggering diarrhea. The incubation period for ETEC gastroenteritis is typically 1 to 2 days (although it can be longer), and symptoms usually last for at least 4 days (17, 53, 89).

2. Enteropathogenic *E. coli*

Enteropathogenic strains of *E. coli* (EPEC), like their enterotoxigenic relatives, colonize the small intestine. But EPEC strains don't produce enterotoxins. Instead, they bind to the intestinal mucosa and, in a three-stage process, destroy the brush border microvilli, forming attachment-effacement lesions and triggering a profuse watery diarrhea that can last for more than 2 weeks (2, 30, 61). The incidence of illness due to EPEC has declined since the 1960s, but these strains are still a leading cause of infantile diarrhea in underdeveloped countries (61, 88). The incubation period in infants is not known. Adult volunteers, when fed a high dose (10^9 to 10^{10} cells) of an EPEC strain, develop watery diarrhea, usually accompanied by fever and vomiting, within 12 to 24 hours. In younger children, the diarrhea can last as long as 14 days or more (2).

3. Enteroinvasive *E. coli*

As their name implies, these strains of *E. coli* are capable of invading the epithelial cells of the colon, multiplying, and spreading from cell to cell. Their destructive spread results in a dysentery-like illness, the symptoms of which usually include mucoid diarrhea that is often bloody, fever, chills, headache, vomiting, muscle pain, and abdominal cramps (2). The typical incubation period is between 2 and 48 hours (average of 18 hours).

4. Enteroaggregative *E. coli*

The pathogenicity of the enteroaggregative *E. coli* (EAEC) was recognized relatively recently. EAEC strains earn their name from the characteristic bricklike stacking arrangement of the bacterial cells adhering to the surfaces of tissue culture monolayers. EAEC is related to EPEC, and both groups of pathogenic strains share many characteristics. They both colonize the small intestine and produce watery diarrhea, low-grade fever, and vomiting. The incubation period for EAEC (based on volunteer feeding studies) is 8 to 18 hours. The length of the illness caused by EAEC varies, but the diarrhea can become chronic, lasting longer than 14 days in some cases (43, 78).

5. Enterohemorrhagic *E. coli*

Also known as verotoxigenic or verocytotoxigenic *E. coli*, enterohemorrhagic *E. coli* (EHEC) first made its presence known in the 1982 McDonald's outbreak described in chapter 6. These strains of *E. coli* produce one or both of two cytotoxins, Stx1 and Stx2, which are similar to toxins produced by *Shigella dysenteriae*. Like *Shigella*, infection with EHEC results in bloody diarrhea. The incubation period is typically 3 to 8 days; symptoms usually last at least 3 days, and often more than a week. Five to 10 percent of children who are infected with EHEC develop hemolytic uremic syndrome (53).

On June 1, 1996, a child died in Okayama, Japan—one of approximately 340 children from that area who were stricken with *E. coli* O157:H7 (13). By June 8, the number of victims had grown to 416, and a second student had died. Health officials, suspecting that school lunches were the source of the outbreak, closed four schools and suspended the lunch program at other schools in the affected area (3). A few weeks later, Sakai City was hit.

Sakai City health authorities received word on July 12 that school children were complaining of diarrhea, some of it bloody (68). By July 17, the number of victims was approaching 6,000. To stop the outbreak from spreading further, city officials closed all of Sakai City's 92 public elementary schools (4). And they began to search for the culprit.

Food inspectors and public health nurses first interviewed all of the more than 200 children hospitalized during the Sakai City outbreak. Next, they contacted other students (who had not been hospitalized) from the same district in order to identify additional victims and determine their symptoms. The children were asked to list all of the individual components of the school lunches they had eaten between July 1 and July 10. Meanwhile, investigators also interviewed the people who had prepared the lunches to establish what cooking procedures had been used, and they tested the water that was being supplied to the individual schools.

While the interviews were taking place, the Sakai City Health Institute lab began to analyze individual components of the lunches that had been served in the city's schools between July 8 and July 12. Institute scientists also obtained and tested environmental and ingredient samples from the facilities that had prepared the lunches. After examining a total of 1,626 samples, the Institute reported that it had failed to detect *E. coli* O157:H7 in even a single sample (68).

Epidemiologists working on the investigation had better luck. The results of their interviews pointed to uncooked white radish sprouts. The sprouts—all produced on the same farm—were an ingredient in the sweet and sour chicken dish served on July 8 in one of the affected school districts, in the chilled Japanese noodles served on July 9 in a second district, and in the July 10 chicken and lettuce meal served in a third district (68). On August 7, Japan's Minister of Health and Welfare announced that white radish sprouts from a single supplier were suspected of causing the Sakai City outbreak. The minister added that he was ordering an inspection of the sprout supplier's facilities (5).

Inspectors visited the radish sprout farm and obtained samples of well water, sewer water, sprouting water, radish seeds, and radish sprouts. None of the samples yielded *E. coli* O157:H7. Fecal samples from the farm's employees were also negative. *E. coli* O157:H7 could not be detected in environmental samples, nearby groundwater, fecal material from area cattle, or sewage from animal quarters. Nor was the pathogen found in samples of imported radish seeds that had been used to produce sprouts on that farm in early July 1996. Radish seeds from other seed producers also were negative for *E. coli* O157:H7, although a different *E. coli* serotype was found in one batch of seeds (68). On September 26, the Minister of Health and Welfare issued his government's final report. He stated that, while there was no definitive proof, epidemiological evidence pointed to the white radish sprouts from the suspect farm as having been the most likely source of the outbreak (6).

Determining the source of a food poisoning outbreak based solely on epidemiological evidence is a bit like convicting a criminal without an eyewitness. In both situations, investigators rely on circumstantial evidence. In the 1996 outbreak, all leads pointed to a single producer of white radish sprouts. The only missing element was the smoking gun—direct evidence that the radish seeds or sprouts were contaminated with *E. coli* O157:H7. As luck would have it, that evidence turned up the following year.

In March and April 1997, Japan once again experienced an increase in illnesses caused by *E. coli* O157:H7 (8, 9). By April 26, more than 190 cases had been reported, including one death. This time, a container of sprouts found at the home of one of the affected families yielded *E. coli* O157:H7 (9). The sprouts implicated in the 1997 outbreak were produced at a different facility than the 1996 sprouts, but both sprouting farms had purchased their supply of seeds from the same distributor—a U.S. company—in 1995. And the DNA profile of the strain that investigators found in the package of sprouts matched the profiles of both the *E. coli* O157:H7 strain isolated from patients in the 1996 school lunch outbreak and the 1997 outbreak strain (67).

Further evidence that *E. coli* O157:H7 could contaminate seed sprouts soon followed. In mid-1997, a single strain of the pathogen felled 85 individuals living in Virginia and Michigan. All of the victims had eaten alfalfa sprouts produced by sprouting facilities in those two states. While investigators never succeeded in isolating *E. coli* O157:H7 from either the sprouts or the seeds, the epidemiological evidence was compelling. The outbreak was limited to the two states in which the implicated sprouts had been distributed, both sprouting operations had used the same batch of alfalfa seeds (from an Idaho grower), and the outbreak ended once the sprouts were removed from the market (20).

One of the difficulties faced by investigators of food-borne outbreaks is the effectiveness of available lab methods to detect pathogens like *E. coli* O157:H7. As those methods have improved, microbiologists have enjoyed more success in finding the microbe in seed sprouts (48, 80). There have been at least two recalls of sprouts due to *E. coli* O157:H7 in the United States since the 1997 outbreak. The first, in 2000, was limited to California and was the result of *E. coli* O157:H7 having been detected in samples of sprouts at the processing plant (22). The second recall, which took place in 2004, resulted from the pathogen having been found in irrigation

Table 8.1 Selected food-borne disease outbreaks associated with consumption of raw sprouts

Type of sprout	Pathogen[a]	Location of outbreak	No. of victims (deaths)	Year	Reference(s)
Soybean?	*B. cereus*	United States	4	1973	83
Alfalfa	Bovismorbificans	Finland	210	1994	82, 84
Alfalfa	Stanley	United States, Finland	≥242	1995	64, 84
Alfalfa	Newport	United States, Canada	133	1995	99
Radish	*E. coli* O157:H7	Japan	11,826 (12)	1996	69
Alfalfa	Montevideo	United States	417 (1)	1996	73
Radish	*E. coli* O157:H7	Japan	>200	1997	10
Alfalfa	*E. coli* O157:H7	United States	85	1997	20, 27
Alfalfa/clover	*E. coli* O157:NM	United States	8	1998	73
Alfalfa	Mbandaka	United States	87	1999	44
Beans	Enteritidis	Netherlands	27	2000	100
Mung bean	Enteritidis	United States	45	2000	21
Alfalfa	Kottbus	United States	31	2001	106

[a] Bacterial species or *S. enterica* serotype.

water used to produce the sprouts (39). In both instances, the contaminant was discovered and the sprouts were recalled in time to prevent a disease outbreak.

E. coli O157:H7 is not the only pathogen that has been associated with sprouts (Table 8.1). In 1973, four people fell ill after they ate sprouts produced using a do-it-yourself kit purchased from a health food store. The soy seeds that they had sprouted were later found to contain almost a pure culture of *Bacillus cereus* (83). There have also been several outbreaks caused by *Salmonella*-contaminated sprouts.

Seeds of Doubt

A seed is a seed, whether it is used to produce a field crop or a sprout. Sprout producers purchase their seeds—whether soybean or mung bean, alfalfa, or radish—from the same people who sell seeds to farmers for planting in their fields. If some of the seeds happen to be contaminated with *E. coli* O157:H7, it makes no difference to a farmer, but a few bad seeds can mean disaster for a sprout producer.

Seeds used for sprouting are grown as field crops and are susceptible to contamination. Fertilizers (especially homemade fertilizers that contain fresh manure), contaminated irrigation water, livestock, and native wild animals are all potential sources of pathogens. Harvesting equipment, warehousing, and handling practices also contribute their share to the microbial load. It's not surprising that a portion of the seeds may become contaminated. Unfortunately, washing and surface disinfection cannot be relied on to produce clean, pathogen-free seeds for sprouting (73).

B. cereus—Two Pathogens in One

B. cereus is a gram-positive, spore-forming, rod-shaped bacterium that can grow either aerobically or anaerobically (86). It is ubiquitous—a common inhabitant of soil worldwide—and has been found in many different types of food, including meats, grains, dairy products, spices, and soybeans (12, 32, 46, 55, 79, 81, 86, 91). The microbe's ability to produce spores enables it to survive mild heat treatments such as pasteurization, as well as typical conditions of home cooking (66). Depending on the strain of *B. cereus*, its growth in food can result in two different forms of food poisoning.

It once was common (and might still be so) for Chinese restaurants to prepare a large quantity of boiled rice well ahead of time, hold it at room temperature for several hours (sometimes even overnight), and stir-fry individual portions of rice with other ingredients just before serving. When *B. cereus* spores were present in the uncooked rice, they would often survive cooking, then germinate and grow in the cooked rice as it cooled. And while they grew, some strains of *B. cereus* produced a heat-stable toxin capable of surviving stir-frying.

The heat-stable toxin manufactured by *B. cereus* produces an acute illness, an emetic syndrome that resembles *Staphylococcus aureus* food poisoning. After a short incubation period of 1 to 6 hours, victims suffer a relatively short, severe bout of nausea and vomiting (lasting just a few hours), sometimes accompanied by abdominal cramps and diarrhea (59).

Not all strains of *B. cereus* produce a toxin that triggers the emetic syndrome. Some strains more closely resemble *Clostridium perfringens* in their pathogenic activity. Instead of manufacturing their toxin outside their host, they elaborate an enterotoxin while growing inside the intestines. The diarrheal syndrome produced by this form of *B. cereus* illness is slower to develop (incubation period of 8 to 16 hours) than the emetic syndrome and is a milder ailment. The common symptoms are nausea, cramps, and watery diarrhea, usually lasting 6 to 12 hours; vomiting is rare (53).

Alfalfa, bean, and other types of sprouts are usually produced using hydroponic methods and under conditions that are almost ideal for bacterial growth. Seeds are first soaked in water to induce germination, then incubated at room temperature for several days in a moist environment. If either the water or the seeds are contaminated with *E. coli* O157:H7, the high moisture level and warm incubation temperature almost guarantee that the pathogen will spread through a significant portion of a production batch (47). And once contaminated, the sprouts cannot be disinfected easily or completely. In 1998, Japanese researchers demonstrated that the roots of radish sprouts were able to take up *E. coli* O157:H7 from contaminated water. The bacteria traveled up the vascular system and were detected inside the edible portions (hypocotyls and cotyledons) of the sprouts (51). Similarly, in a 2001 study, *Salmonella enterica* serotype Stanley was able to penetrate to a depth of 12 μm into the tissue of alfalfa sprouts (42). The presence of pathogens within sprout tissue, rather than just on the surface, renders surface decontamination with agents such as chlorine ineffective.

The 1996 Japanese radish sprout outbreak marked the first time that *E. coli* O157:H7 was linked to seed sprouts. And it still holds the record as the largest and longest sprout-related outbreak of food-borne illness, whether caused by *E. coli* O157:H7, *Salmonella*, or any other food-borne disease agent. But sprouts represent just one type of produce that has been identified as a vehicle of food-borne *E. coli* O157:H7 in the last couple of decades. Any field or orchard crop has the potential to become a carrier of disease.

Turning Over a New Leaf

Take a drive on the Interstate 5 through California's San Joaquin Valley and you will pass fields of produce, open irrigation canals, and cattle feedlots. Irrigation is a way of life in this semiarid agricultural area; rain is a rare visitor, except during the winter months. But when it rains, it pours. The soil becomes saturated, and excess water drains off the fields and feedlots. Some of the runoff finds its way into the irrigation canals, bringing with it pathogens such as *E. coli* O157:H7 (52, 98).

Irrigating with contaminated water introduces pathogens into the fields and onto crops (93, 101). Drip irrigation carries the microbes into the soil; once there, they can find their way to the root systems of the plants. Spray irrigation contaminates plants with even greater efficiency; it inoculates the soil, the leaves, and the stems simultaneously (92, 95). Fertilizing with untreated manure also can introduce pathogens into soil (41). And *E. coli* O157:H7 can survive in manure, water, and soil for a long time (25, 60, 102, 103).

The presence of *E. coli* O157:H7 in the soil or on the surface of crops wouldn't be a hazard except that, once in contact with the plants, the microbes are extremely difficult to eliminate. *E. coli* O157:H7 attaches very firmly to plant tissue. On lettuce, for example, the microbes stick especially well to cut or torn leaf surfaces, penetrating as far as 100 μm into the tissue—out of reach of most surface sanitizing treatments (90, 96).

The risks associated with contaminated fields and irrigation water are more than just theoretical. In July 1995, 70 people were infected with *E. coli* O157:H7 in Montana. Although investigators were unable to confirm the source of the outbreak in the laboratory, circumstantial evidence pointed to locally grown leaf lettuce—the first time that lettuce had been associated with an outbreak of *E. coli* O157:H7. The implicated farm fertilized its crops with a nutrient mixture that included cow manure from a local dairy. In addition, cattle had access to streams above the pond that supplied irrigation water to the field, adding another possible source of contamination (1).

The 1995 incident was not a fluke. Less than one year later, in May and June 1996, mesclun lettuce was the source of a three-state *E. coli* O157:H7 outbreak (40, 109). It began, so investigators believe, at Fancy Cutt Farms, a small California producer of washed, mixed organic salad greens. The company started up business earlier that year but had never registered with the state—as it should have done—and had never been inspected. Yet the lettuce that it produced found its way through the U.S. food distribution chain to at least 11 restaurants and stores in New York, Illinois, and Connecticut (15).

The first case turned up in Illinois on May 28, 1996. One day later, the first Connecticut victim fell ill. In all, 61 people in Connecticut, New York, and Illinois were affected, including 3-year-old Haylee Bernstein, who spent 11 weeks in intensive care—a victim of hemolytic uremic syndrome. Haylee, who managed to survive severe bleeding in her brain, was temporarily blinded and suffered major damage to her kidneys and pancreas. Eight years after her return from the brink of death, Haylee's vision was still impaired, her kidneys had not grown for 2 years, and she was an insulin-dependent diabetic (40).

State and federal epidemiologists carried out a case-control study of the outbreak by interviewing victims and healthy individuals in Connecticut and Illinois. The results of their investigation pointed to mesclun lettuce. The Connecticut victims identified a total of five restaurants at which they had eaten the lettuce. All five restaurants purchased their lettuce from the same distributor, who had obtained his mesclun lettuce from a single supplier, Fancy Cutt Farms. And Fancy Cutt was also one of the suppliers of mesclun to the Illinois outlets that were associated with the outbreak investigation (15, 49).

When investigators visited Fancy Cutt's farm and processing facility, they found several possible avenues through which *E. coli* O157:H7 might have contaminated the lettuce in the field. Cattle pens from a beef cattle operation were located across a dirt track from the lettuce fields, and chickens roamed freely between the cattle and the fields. And some of the lettuce had been grown in a field in which cattle had been allowed to graze the previous winter. In addition, a common irrigation system was used both to flood the cattle pasture and to spray-irrigate the lettuce. The spigots in the cattle pasture were sometimes submerged in manure-contaminated pools of water. Without backflow protection, contaminated water could be drawn through these spigots back into the irrigation system and sprayed onto the lettuce. Nor was the lettuce safe from contamination once it was harvested.

Fancy Cutt rinsed and packaged its mesclun lettuce in a three-sided shed. Dust and dirt from the cattle pens could be blown in by gusts of wind or tracked in by farm workers, birds, or animals. The workers washed the lettuce with unchlorinated well water that was stored in a holding tank and circulated through filters. The filtration system failed on May 30—ironically, just after the start of the outbreak—and from that date until the day inspectors arrived at the facility in July 1996, the tank water was simply changed three times a day.

During their July 1996 visit, the inspection team sampled the wash tank filters, wash water, finished lettuce, cutting board and knife, well water, water from a cattle trough, water from the cattle pasture, and cow and chicken manure. They found no *E. coli* O157:H7, but *E. coli* biotype 1, a recognized indicator of fecal contamination, was present in the wash tank water, the recirculating water system, and some samples of mesclun lettuce (49).

The results of the epidemiological studies, the numerous flaws in Fancy Cutt's sanitation and processing practices, and the presence of *E. coli* biotype 1 were enough—even in the absence of a smoking gun—to convince federal and California investigators that the company's washed lettuce was the source of this outbreak. But California allowed the company to continue operating while it corrected its deficiencies. And although a follow-up inspection carried out in 1997 revealed that several serious flaws in Fancy Cutt's operations still existed, the company president, Robert Chavez, told the *New York Times* that he believed his company had been cleared of responsibility for the 1996 outbreak. After all, he reasoned, the state had never charged him with any infractions and never closed him down, not even temporarily (15).

On January 8, 1998, 1½ years after the *E. coli* O157:H7 outbreak was traced to Fancy Cutt's lettuce, the state of California finally filed charges against the company for having violated food safety laws (14). Fancy Cutt management settled the suit out of court by agreeing to shut down the operation—even while continuing to insist that the company's product had not been the source of the 1996 outbreak (40).

Being unable to detect the outbreak strain of *E. coli* O157:H7 in suspect lettuce or its production environment has been the rule, rather than the exception, in lettuce-related outbreaks (54, 65). Although many research studies have probed the potential for the pathogen to infect lettuce leaves in the field and to survive for extended periods in the lettuce, no one has yet been able to recover an outbreak strain of *E. coli* O157:H7 directly from an implicated batch of lettuce, from the grower's facility, or from the field in which the lettuce was grown. Scandinavian investigators ran into a similar problem when trying to trace the source of an international outbreak of *Shigella sonnei* linked to iceberg lettuce (56). This should not be surprising. Lettuce is a very perishable commodity and has usually been either consumed or discarded by the time an outbreak investigation gets under way. But until investigators get lucky and manage to find the contaminant in an implicated batch of lettuce, growers such as Robert Chavez of Fancy Cutt will continue to disbelieve that their product could be the source of an *E. coli* O157:H7 outbreak.

Apples to Apples—to Cider

Certain numbers can take on a mystical, almost magical, significance. Three, representing the Trinity, is an important number for most Christians. In many Western societies, 7 is considered to be lucky, whereas 13 is an unlucky number. The Chinese believe 8 to be an auspicious number. Four symbolizes death to Japanese; 18 means life to Jews. For food scientists, 4.6 is a magic number.

The aura surrounding 4.6 originates from the canned food industry and its early struggles with *Clostridium botulinum*. This pathogen does not grow or produce its deadly toxins in an acid environment below pH 4.6 (53). The U.S. Food and Drug Administration (FDA) recognized the significance of 4.6 in developing and promulgating its regulations for the safe processing of low-acid canned foods by defining low-acid foods as "... any foods, other than alcoholic beverages, with a finished equilibrium pH greater than 4.6 and a water activity (a_w) greater than 0.85 . . ." (38). Also, according to the FDA Food Code, "... food with a pH level of 4.6 or below when measured at 24°C (75°F) . . ." is not a potentially hazardous food (37).

Under the circumstances, it should surprise no one that apple cider producers believed their product, with a pH typically near or below 4.0, to be safe from pathogen contamination (31, 53). But FDA's safety definitions were based on the inability of *C. botulinum* to grow under highly acidic conditions (pH below 4.6). Survival (as opposed to growth) of pathogens such as *E. coli* O157:H7, *Salmonella*, and *Listeria monocytogenes* in acidic environments is an entirely different matter.

The earliest hint that unpasteurized apple cider might not be entirely safe came on October 18, 1974. A New Jersey hospital lab found *S. enterica* serotype Typhimurium in samples from five patients who were suffering from gastroenteritis and fever. By October 23, 13 patients were confirmed to be suffering from serotype Typhimurium infection. All of them had consumed apple cider from the same producer within 48 hours before their symptoms appeared. Health authorities acted quickly to stop production and sale of the cider, but the damage had been done. Before the outbreak ended, 296 people had been affected.

There was no doubt as to the source of the *Salmonella*. The microbe was isolated from 6 of the 30 cider production employees and from 2 bottles of cider taken from the homes of ill patients. But no one ever established how the *Salmonella* got into the cider. It could have come from fallen apples contaminated with manure that had been used to fertilize the apple trees, or manure from wild animals—perhaps deer—grazing in the orchard. Or it might have been lurking in the equipment, which was rinsed with cold water but not sanitized. No matter its point of entry, the outbreak strain was able to survive in an acid environment. The pH of the apple cider, including samples that had been produced both before and after the outbreak, was between 3.4 and 3.9 (24).

The 1974 outbreak did not appear to set off any alarms in the industry or among food safety regulators. A 1980 Canadian outbreak of hemolytic uremic syndrome traced to apple cider also went largely unnoticed at the time. Investigators failed to detect any known pathogens in specimens from even 1 of the 14 patients or from the implicated lot of cider (94). *E. coli* O157:H7 was still 2 years away from being recognized as a human pathogen, but in retrospect, it was the most likely cause of this outbreak. The symptoms, incubation period, and probable vehicle of contamination all fit the pattern (108). It wasn't until 1991, however, that investigators first determined that *E. coli* O157:H7 could contaminate unpasteurized fresh-pressed apple cider and survive in the cider long enough to cause disease.

Failing the acid test

Children's Hospital in Boston must have admitted many ill children between October 24 and November 20, 1991, but four of those kids had at least two things in common. All four came from the Fall River area of southeastern Massachusetts, and all four were suffering from hemolytic uremic syndrome. Health officials quickly recognized the unusual cluster of illnesses and reviewed emergency department and hospital admission records from the Fall River area. They identified an additional 19 victims, ranging in age between 2 and 70 years. Six of the victims—including the four children in Boston—were hospitalized (18).

Interviews with patients and unaffected people from the same area (case-control study) quickly pointed investigators to apple cider as a common link. Seventy-two percent of the patients and only 33% of controls remembered drinking apple cider immediately prior to or during the outbreak. And 92% of those who drank cider had obtained it from the same source, identified in the investigation report simply as "cider mill A." Only 13% of control individuals had purchased cider produced by that mill.

Investigators visited the mill at the end of December 1991, 7 weeks after the onset of the last cider-related case and 1 week after the mill had closed for the season. They also inspected the two orchards that supplied apples to the mill and collected environmental samples from the mill and from both orchards, including rotten apples and animal droppings. None of the environmental samples yielded *E. coli* O157:H7, nor was the microbe recovered from any samples of cider from the mill. However, the physical inspections provided valuable clues as to how the apple cider might have become contaminated.

Ninety percent of the apples used for cider production at cider mill A were dropped apples—apples that had fallen off the trees and were harvested from the ground. Those apples might have become contaminated by cow manure, either from grazing cattle or from environmental runoff. The mill operator pressed all of the apples—picked or dropped—without first washing or brushing them to remove debris. The practices followed by this mill operator were fairly typical for the industry at that time. A survey, carried out among cider mill representatives at a New England trade show after the outbreak, established that every one of the mills used dropped apples. And only one-third of the operators, most of them large, claimed to wash or brush their fruit before pressing. Just one out of four small cider mills did so. Most of the producers never used a preservative, either. Instead, they relied on the natural acidity of the cider to keep it safe. But, as researchers have since confirmed, natural acidity is not enough.

The strain of *E. coli* O157:H7 that caused the 1991 outbreak proved to be quite acid-tolerant. When inoculated at a concentration of only 100 cells per ml into apple cider samples having pH values of 3.6 to 4.0, the pathogen could still be detected after more than 10 days of storage at 46°F (8°C). It survived even longer—at least 1 month—when a higher concentration (10^5 cells per ml) was used. The microbe died off more quickly, though, when the cider was stored at room temperature. Even with an initial inoculum of 10^5 cells per ml, *E. coli* O157:H7 could not be detected after 6 days at 77°F (25°C). Sodium benzoate at a concentration of 0.1% also reduced the survival of the pathogen in cider held at 46°F (8°C) to less than 1 week (108). Between 1994 and 1996, several other groups of researchers, including FDA scientists, corroborated these findings (11, 16, 62, 63, 71). Clearly, 4.6 could no longer be considered a magic number. But these results were not enough to change either the industry's standard operating procedures or FDA's regulatory approach, although the researchers from the Centers for Disease Control and Prevention (CDC) who investigated the outbreak recommended that processors should wash and brush the apples before pressing them and should add sodium benzoate to the juice (23).

The wake-up call

There is nothing like a major food-borne disease outbreak to make industry, government, and consumers sit up and take notice. Just as the 1993 Jack in the Box outbreak provoked a drastic reevaluation of the health risks associated with *E. coli* O157:H7 in ground beef, the 1996 Odwalla outbreak, coupled with two nearly simultaneous cider-related outbreaks in the U.S. Northeast, produced a quantum change in the perceived safety of fresh, unpasteurized apple cider.

New Haven County, Conn., was the first area to be hit. Alerted on October 11, 1996, to a cluster of four cases of *E. coli* O157:H7 infection, the state's Department of Public Health began a search for additional victims. They identified a total of 14 individuals, 5 of whom were hospitalized as a result of their infection. Epidemiologists determined that apple cider produced by a local mill was the probable source (50).

The mill purchased its apples from several different orchards. As usual, some of the apples were harvested from the ground (dropped apples), rather than handpicked directly from the trees. But, unlike the situation in the 1991 outbreak, the mill had brushed and washed all of its apples in a flow-through system using potable municipal water. And, although the cider was unpasteurized, the producer had added 0.1% potassium sorbate, a preservative later found to be effective against *E. coli* O157:H7 in apple cider, to protect against pathogens (108).

The second outbreak struck New York State. A local hospital detected *Cryptosporidium* in specimens from 10 residents of Cortland County in less than 1 week. When state agencies investigated further, they were able to find 20 confirmed and 11 suspected victims, all of whom became ill between September 28 and October 19, 1996. All 31 individuals complained of diarrhea; many also reported abdominal cramps, vomiting, or fever. A few had bloody diarrhea. Their symptoms lasted for 1 to 21 days (median of 6 days). Nearly all of the victims remembered drinking unpasteurized cider from the same mill (76).

Unlike the situation in Connecticut, the New York cider mill purchased all of its apples from a single orchard. The owner of the orchard claimed

Cryptosporidium parvum and Cryptosporidiosis

The first report of what was probably a *Cryptosporidium* was published in 1895, but the microbe wasn't described in detail or named until 1910. *Cryptosporidium* is a protozoan that is parasitic in animals and humans. It reproduces within the epithelial cells of the respiratory and digestive organs of vertebrate animals. The genus comprises up to 10 species, each one named for the host in which it was first found. Most *Cryptosporidium* spp. once were believed to be host-specific, but cross-transmission between species occurs. Avian *Cryptosporidium* can infect a range of birds; *Cryptosporidium* spp. that infect one species of mammal can be transmitted to other species. *C. parvum*, the most common *Cryptosporidium* species to infect humans, is zoonotic, and harbored by a number of mammals (33, 75).

Cryptosporidium reproduces sexually, resulting in the formation of oocysts. Oocysts of *C. parvum* are spherical to ovoid, measuring 5.0 μm by 4.5 μm. Each oocyst contains four sporozoites that, if released, can infect a host's epithelial cells. Some intact oocysts are shed in the feces of an infected individual, contributing to the continuation of the outbreak. While it is possible to grow *C. parvum* in cell culture monolayers, *Cryptosporidium* contamination is usually detected by examination of stool, water, or environmental samples for the presence of oocytes—either directly or after sample concentration—using direct microscopy, fluorescent antibody staining, or nucleic acid amplification methods (19, 45, 107).

The first evidence that *Cryptosporidium* could cause illness in humans was documented in 1976. Until 1980, it was thought to be no more than an opportunistic pathogen of immunocompromised individuals, but by 1982, *C. parvum* was recognized as an important cause of diarrhea in humans (28, 33). The symptoms of *C. parvum* infection vary, depending on the health of a victim's immune system. In otherwise healthy individuals, the disease causes watery diarrhea, sometimes accompanied by abdominal cramps, nausea, vomiting, and fever. The symptoms can disappear spontaneously in as little as 3 days or might last as long as 3 weeks. The infection is rarely fatal, except for people with weakened immune systems.

C. parvum behaves differently when it infects an immunocompromised individual. The infection begins in the small intestine (the ileum and jejunum), causing a severe cholera-like illness, and can spread to other organs and systems, including the respiratory tract, the gallbladder, and the pancreas. It is not easily treated. Chemotherapy patients, people with human immunodeficiency virus infection, and those on immune-suppressive therapy are all at risk of severe illness, even death, as a result of *C. parvum* infection (28).

that he had not sold the mill any dropped apples; he only supplied apples that had been picked from trees. The mill brushed and washed the apples with water from a 45-foot drilled well. The cider was not pasteurized, nor were any preservatives added to extend its shelf life.

Lab tests came up empty. *Cryptosporidium* could not be found in the remaining cider or on the mill's production equipment. Investigators also tested several samples of the mill's well water, to no avail. They found *E. coli* biotype 1 in one of the samples, though, which suggested that the well might have been contaminated with fecal material. There was a dairy farm across the road from the mill, and surface water from the farm could have found its way into the well.

While public health officials in Connecticut and New York were wrestling with their respective outbreaks, a major problem was developing for a West Coast juice producer, Odwalla, Inc., of Half Moon Bay, Calif. Odwalla boasted of the fresh taste and healthful benefits of its unpasteurized juice products. And it had developed a large following in the health- and quality-conscious western regions of the United States and Canada. By the summer of 1996, the company was struggling to keep up with the demand for its products. In some ways, Odwalla became a victim of its own success (31).

Odwalla's stated policy was to accept only handpicked apples from its growers, but it had no way to ensure compliance with that standard. Each shipment of apples that it received was required to contain less than 10% of decay by weight. In the weeks prior to the outbreak, however, management decided to relax its standards and began to accept shipments containing a higher percentage of blemished fruit. Anita Velasquez, a quality manager for Odwalla who was interviewed by the *New York Times* in 1998, claimed that some shipments received between mid-September and October 7, 1996—a critical day in the life of the company—contained 25 to 30% defective fruit (31).

The first indication of a problem with Odwalla's juice surfaced in the state of Washington. Ten patients in the Seattle-King County area were diagnosed with *E. coli* O157:H7 infection. All 10 had consumed Odwalla apple juice. By November 6, the number of cases had risen to 45 people in British Columbia (Canada's westernmost province), California, Washington, and Colorado (7). The final tally was 70 victims, 25 of whom were hospitalized. Fourteen patients developed hemolytic uremic syndrome, and one person died (26). As soon as the company learned of the connection between the outbreak and its product, it recalled all of its apple-based juices from the market.

All but two of the outbreak victims drank Odwalla apple juice that had been produced on October 7, 1996. The FDA, the CDC, and California state labs tested 184 bottles of juices produced between September 23 and October 29. They found *E. coli* O157:H7 in a single unopened container of apple juice produced on October 7. Two packinghouses and six farms supplied the three suspect lots of apple used that day (26).

One of the two packinghouses—the supplier of one of the three suspect lots—changed its wash water only once a day. This facility also waxed its apples between washing them and sorting them by grade, a practice that could have sealed any pathogens onto the skin of the apples and would also have reduced the effectiveness of one of the washing solutions used by Odwalla. Water sampled from the wash-water reservoir drain on November 5 contained *E. coli* biotype 1 (11.1 cells per 100 ml), but no *E. coli* O157:H7. Two of the five farms that supplied apples for this lot admitted to including some dropped apples in their shipment despite Odwalla's policy requiring that only handpicked apples be supplied.

All of the apples in the other two lots came from one farm. The apple pickers on those farms were directed by the owner not to harvest dropped apples, but this policy could not be easily enforced. The orchard was fertilized with composted turkey manure, samples of which were negative for *E. coli* O157:H7. The farm was adjacent to a wildlife refuge area, and a large deer population grazed among the apple trees. Since deer can carry *E. coli* O157:H7, investigators sampled fresh deer droppings from the wildlife refuge on December 18 (57, 87). One of the samples yielded *E. coli* O157:H7, but its genomic fingerprint pattern did not match the outbreak strain (26).

The Odwalla plant had undergone two inspections in the 4 months prior to the outbreak, by the U.S. Army in June 1996 and by FDA in July. The U.S. Army inspector concluded that Odwalla's unpasteurized juices contained too high a level of bacteria and that the company did not have control mechanisms in place to minimize the risk of food-borne disease (85). The FDA inspector, in contrast, identified only minor deficiencies (31). The outbreak investigators agreed that the plant was clean and well run; there were no apparent internal sources of contamination. The company, however, had made one significant change to its procedures in the summer of 1996. It had switched from using a chlorine-based washing solution for cleaning the fruit to using a phosphoric acid-based solution. The outbreak investigation team found that neither of the two brands of phosphoric acid washes was being used correctly—one was not designed for use on waxed produce, and the other one was not meant to be used on any kind of produce, waxed or unwaxed. In addition, the second brand of phosphoric acid wash was not always used at the correct concentration (26).

The Odwalla outbreak, coming on the heels of the two smaller East Coast cider-related episodes, caused the FDA and the fresh juice industry to realize that existing methods for screening and cleaning fruit could not guarantee the safety of unpasteurized juices. Following a series of meetings between regulators, producers, and fruit growers, FDA announced two new regulations in 1998. Effective November 5, 1998, all unpasteurized fruit and vegetable juices were required to carry a warning label (35), and the federal government published a Proposed Rule requiring juice processors to develop and adhere to a Hazard Analysis and Critical Control Points (HACCP) program (34). The FDA's HACCP juice regulations were finalized in 2001 (36).

Action or Reaction?

Regulators, researchers, and juice producers have known since 1975 that *Salmonella* can survive in apple cider and still be able to cause gastroenteritis (24). Unpasteurized cider was blamed for a 1980 outbreak of hemolytic uremic syndrome (94). Fresh cider also was the source of an outbreak of *E. coli* O157:H7 in 1991 and of cryptosporidiosis in 1993 (18, 70). By 1995, researchers had established beyond any doubt that *E. coli* O157:H7 could survive for as long as 1 month in fresh-pressed apple cider (11, 16, 62). Yet the FDA took no action to prevent new outbreaks or improve the safety of fruit and vegetable juices—until after the Odwalla outbreak.

This pattern of inertia in the face of a gradually growing body of knowledge has shown itself over and over again. The U.S. Department of Agriculture's slow response to the hazards associated with *E. coli* O157:H7 in ground beef and *L. monocytogenes* in processed meats, and the British government's reaction to the threat posed by bovine spongiform encephalopathy (chapter 10), are examples of the tendency of government agencies to wait for a crisis to develop before they act. Why is this?

Perhaps the answer lies in the tendency of government agencies in many countries to lead by consensus rather than by fiat. New regulations are expensive, both for industries that must comply with them and for governments that must enforce them. Unless there is a clearly recognized need for a new safety law (such as developed after the Jewel Dairy *Salmonella*, the 1993 *E. coli* O157:H7 ground beef, or the Odwalla juice outbreak), introducing new and costly demands on industry can be a tough sell. Unfortunately, sometimes it takes the shock of a large disease outbreak—and even one or more deaths—to catalyze the introduction of improvements to our food safety system.

References

1. **Ackers, M.-L., B. E. Mahon, E. Leahy, B. Goode, T. Damrow, P. S. Hayes, W. F. Bibb, D. H. Rice, T. J. Barrett, L. Hutwagner, P. M. Griffin, and L. Slutsker.** 1998. An outbreak of *Escherichia coli* O157:H7 infections associated with leaf lettuce consumption. *J. Infect. Dis.* **177:**1588–1593.
2. **Andrews, W. H.** 1997. *Escherichia coli* organisms turn deadly. *Inside Lab. Management* **1:**27–29.
3. **Anonymous.** 8 June 1996. Food poison outbreak spreads. *Mainichi Daily News* (English edition), Tokyo, Japan.

4. **Anonymous.** 17 July 1996. Tainted food sickens 6,000 kids in Japan. Health officials suspect the outbreak of *E. coli* was caused by school lunches of eel sushi. *Orlando Sentinel*, Orlando, Fla.

5. **Anonymous.** 8 August 1996. Kan says radish sprouts may be source of Sakai outbreak. *Mainichi Daily News* (English edition), Tokyo, Japan.

6. **Anonymous.** 27 September 1996. Ministry report fails to pinpoint source of food poisoning. *Mainichi Daily News* (English edition), Tokyo, Japan.

7. **Anonymous.** 1996. Outbreak of *Escherichia coli* O157:H7 infections associated with drinking unpasteurized commercial apple juice—British Columbia, California, Colorado, and Washington, October 1996. *Morb. Mortal. Wkly. Rep.* **45:**975.

8. **Anonymous.** 5 April 1997. Markets pull sprouts amid O157 fears. *Mainichi Daily News* (English edition), Tokyo, Japan.

9. **Anonymous.** 26 April 1997. O157 detected in radish sprouts in Yokohama. *Mainichi Daily News* (English edition), Tokyo, Japan.

10. **Anonymous.** 16 May 1997. Radish sprouts disappearing from stores: action linked to gov't report. *Mainichi Daily News* (English edition), Tokyo, Japan.

11. **Arnold, K. W., and C. W. Kaspar.** 1995. Starvation and stationary-phase-induced acid tolerance in *Escherichia coli* O157:H7. *Appl. Env. Microbiol.* **61:**2037–2039.

12. **Becker, H., G. Schaller, W. von Wiese, and G. Terplan.** 1994. *Bacillus cereus* in infant foods and dried milk products. *Int. J. Food Microbiol.* **23:**1–15.

13. **Bellamy, P.** 4 June 1996. Student dies in bacteria outbreak. *South China Morning Post*, Hong Kong.

14. **Belluck, P.** 10 January 1998. Criminal case being brought over lettuce in California. *New York Times*, New York, N.Y.

15. **Belluck, P., and C. Drew.** 5 January 1998. Tracing bout of illness to small lettuce farm; fresh hazards. *New York Times*, New York, N.Y.

16. **Benjamin, M. M., and A. R. Datta.** 1995. Acid tolerance of enterohemorrhagic *Escherichia coli*. *Appl. Env. Microbiol.* **61:**1669–1672.

17. **Benoit, V., P. Raiche, M. G. Smith, J. Guthrie, E. F. Donnelly, E. M. Julian, R. Lee, S. DiMaio, M. Rittmann, and B. T. Matyas.** 1994. Foodborne outbreaks of enterotoxigenic *Escherichia coli*—Rhode Island and New Hampshire, 1993. *Morb. Mortal. Wkly. Rep.* **43:**81, 87–89.

18. **Besser, R. E., S. M. Lett, J. T. Weber, M. P. Doyle, T. J. Barrett, J. G. Wells, and P. M. Griffin.** 1993. An outbreak of diarrhea and hemolytic uremic syndrome from *Escherichia coli* O157:H7 in fresh-pressed apple cider. *JAMA* **269:**2217–2220.

19. **Brasseur, P., C. Uguen, A. Moreno-Sabater, L. Favennec, and J. J. Ballet.** 1998. Viability of *Cryptosporidium parvum* oocysts in natural waters. *Folia Parasitol.* **45:**113–116.

20. **Breuer, T., D. H. Benkel, R. L. Shapiro, W. N. Hall, M. M. Winnett, M. J. Linn, J. Neimann, T. J. Barrett, S. Dietrich, F. P. Downes, D. M. Toney, J. L. Pearson, H. Rolka, L. Slutsker, P. M. Griffin, and the Investigation Team.** 2001. A multistate outbreak of *Escherichia coli* O157:H7 infections linked to alfalfa sprouts grown from contaminated seeds. *Emerg. Infect. Dis.* **7:**977–982.

21. **California Department of Health Services.** 19 April 2000. Press Release No. 22-00. Salmonellosis outbreak associated with raw mung bean sprouts. California Department of Health Services, Sacramento, Calif. [Online.] http://www.applications.dhs.ca.gov/pressreleases/store/PressReleases/22-00.html. Accessed 12 June 2006.

22. **California Department of Health Services.** 7 July 2000. Press Release No. 35-00. State health department warns consumers about sprouts. California Department of Health Services, Sacramento, Calif. [Online.] http://www.applications.dhs.ca.gov/pressreleases/store/PressReleases/35-00.html. Accessed 12 June 2006.

23. **Caprioli, A., A. E. Tozzi, G. Rizzoni, and H. Karch.** 1997. Non-O157 shiga toxin-producing *Escherichia coli* infections in Europe. *Emerg. Infect. Dis.* **3:**578.

24. **Center for Disease Control.** 1975. *Salmonella typhimurium* outbreak traced to a commercial apple cider—New Jersey. *Morb. Mortal. Wkly. Rep.* **24:** 87–88.

25. **Cieslak, P. R., T. J. Barrett, P. M. Griffin, K. F. Gensheimer, G. Beckett, J. Buffington, and M. G. Smith.** 1993. *Escherichia coli* O157:H7 infection from a manured garden. *Lancet* **342:**367.

26. **Cody, S. H., M. K. Glynn, J. A. Farrar, K. L. Cairns, P. M. Griffin, J. Kobayashi, M. Fyfe, R. Hoffman, A. S. King, J. H. Lewis, B. Swaminathan, R. G. Bryant, and D. J. Vugia.** 1999. An outbreak of *Escherichia coli* O157:H7 infection from unpasteurized commercial apple juice. *Ann. Intern. Med.* **130:**202–209.

27. **Como-Sabetti, K., S. Reagan, S. Allaire, K. Parrott, C.M. Simonds, S. Hrabowy, B. Ritter, W. Hall, J. Altamirano, R. Martin, F. Downes, G. Jennings, R. Barrie, M. F. Dorman, N. Keon, M. Kucab, A. Al Shab, B. Robinson-Dunn, S. Dietrich, L. Moshur, L. Reese, J. Smith, K. Wilcox, J. Tilden, G. Wojtala, J. D. Park, M. Winnett, L. Petrilack, L. Vasquez, S. Jenkins, E. Barrett, M. Linn, D. Woolard, R. Hackler, H. Martin, D. McWilliams, B. Rouse, S. Willis, J. Rullan, G. Miller, Jr., S. Henderson, J. Pearson, J. Beers, R. Davis, and D. Saunders.** 1997. Outbreaks of *Escherichia coli* O157:H7 infection associated with eating alfalfa sprouts—Michigan and Virginia, June–July 1997. *Morb. Mortal. Wkly. Rep.* **46:**741–744.

28. **Current, W. L., and L. S. Garcia.** 1991. Cryptosporidiosis. *Clin. Microbiol. Rev.* **4:**325–358.

29. **Danielsson, M.-L., R. Möllby, H. Brag, N. Hansson, P. Jonsson, E. Olsson, and T. Wadström.** 1979. Enterotoxigenic enteric bacteria in foods and outbreaks of food-borne diseases in Sweden. *J. Hyg. Camb.* **83:**33–40.

30. **Donnenberg, M. S., and J. B. Kaper.** 1992. Enteropathogenic *Escherichia coli*. *Infect. Immun.* **60:**3953–3961.

31. **Drew, C., and P. Belluck.** 4 January 1998. Deadly bacteria a new threat to fruit and produce in U.S.; fresh hazards. *New York Times*, New York, N.Y.

32. **Fang, T. J., C.-Y. Chen, and W.-Y. Kuo.** 1999. Microbiological quality and incidence of *Staphylococcus aureus* and *Bacillus cereus* in vegetarian food products. *Food Microbiol.* **16:**385–391.

33. **Fayer, R., and B. L. P. Ungar.** 1986. *Cryptosporidium* spp. and cryptosporidiosis. *Microbiol. Rev.* **50:**458–483.

34. **Federal Register.** 1998. Hazard analysis and critical control point (HACCP); procedures for the safe and sanitary processing and importing of juice. *Fed. Regist.* **63:**20449–20486.

35. **Federal Register.** 1998. Food labeling: warning and notice statements; labeling of juice products. Proposed Rule. *Fed. Regist.* **63:**20486–20493.

36. **Federal Register.** 2001. Hazard analysis and critical control point (HAACP); procedures for the safe and sanitary processing and importing of juice; Final Rule. *Fed. Regist.* **66:**6138–6202.

37. **Food and Drug Administration.** 2001. Food Code. U.S. Department of Health and Human Services, Public Health Service, Food and Drug Administration, Washington, D.C. [Online.] http://www.cfsan.fda.gov/~dms/fc01-toc.html. Accessed 12 June 2006.

38. **Food and Drug Administration.** 2001. Thermally processed low-acid foods packaged in hermetically sealed containers. Definitions. *Code of Federal Regulations* **21**(113.3).

39. **Food and Drug Administration.** 22 July 2004. Marjon Specialty Foods, Inc. recalls sprout products because of possible health risk. U.S. Food and Drug Administration, Washington, D.C. [Online.] http://www.fda.gov/oc/po/firmrecalls/marjon07_04.html. Accessed 12 June 2006.

40. **Fried, J. P.** 4 July 2004. Legal battle goes on over tainted lettuce. *New York Times*, New York, N.Y.

41. **Gagliardi, J. V., and J. S. Karns.** 2000. Leaching of *Escherichia coli* O157:H7 in diverse soils under various agricultural management practices. *Appl. Env. Microbiol.* **66:**877–883.

42. **Gandhi, M., S. Golding, S. Yaron, and K. R. Matthews.** 2001. Use of green fluorescent protein expressing *Salmonella* Stanley to investigate survival, spatial location, and control on alfalfa sprouts. *J. Food Prot.* **64:**1891–1898.

43. **Gascón, J., M. Vargas, L. Quintó, M. Corachán, M. T. Jimenez de Anta, and J. Vila.** 1998. Enteroaggregative *Escherichia coli* strains as a cause of traveler's diarrhea: a case-control study. *J. Infect. Dis.* **177:**1409–1412.

44. **Gill, C. J., W. E. Keene, J. C. Mohle-Boetani, J. A. Farrar, P. L. Waller, C. G. Hahn, and P. R. Cieslak.** 2003. Alfalfa seed decontamination in a *Salmonella* outbreak. *Emerg. Infect. Dis.* **9:**474–479.

45. **Gomez-Bautista, M., L. M. Ortega-Mora, E. Tabares, V. Lopez-Rodas, and E. Costas.** 2000. Detection of infectious *Cryptosporidium parvum* oocysts in mussels (*Mytilus galloprovincialis*) and cockles (*Cerastoderma edule*). *Appl. Environ. Microbiol.* **66:**1866–1870.

46. **Granum, P. E., S. Brynestad, and J. M. Kramer.** 1993. Analysis of enterotoxin production by *Bacillus cereus* from dairy products, food poisoning incidents and non-gastrointestinal infections. *Int. J. Food Microbiol.* **17:**269–279.

47. **Hara-Kudo, Y., H. Konuma, M. Iwaki, F. Kasuga, Y. Sugita-Konishi, Y. Ito, and S. Kumagai.** 1997. Potential hazard of radish sprouts as a vehicle of *Escherichia coli* O157:H7. *J. Food Prot.* **60:**1125–1127.

48. **Hara-Kudo, Y., Y. Onoue, H. Konuma, H. Nakagawa, and S. Kumagai.** 1999. Comparison of enrichment procedures for isolation of *Escherichia coli* O157:H7 from ground beef and radish sprouts. *Int. J. Food Microbiol.* **50:** 211–214.

49. **Hilborn, E. D., J. H. Mermin, P. A. Mshar, J. L. Hadler, A. Voetsch, C. Wojtkunski, M. Swartz, R. Mshar, M.-A. Lambert-Fair, J.A. Farrar, M. K. Glynn, and L. Slutsker.** 1999. A multistate outbreak of *Escherichia coli* O157:H7 infections associated with consumption of mesclun lettuce. *Arch. Intern. Med.* **159:**1758–1764.

50. **Hilborn, E. D., P. A. Mshar, T. R. Fiorentino, Z. F. Dembek, T. J. Barrett, R. T. Howard, and M. L. Cartter.** 2000. An outbreak of *Escherichia coli* O157:H7 infections and haemolytic uraemic syndrome associated with consumption of unpasteurized apple cider. *Epidemiol. Infect.* **124:** 31–36.

51. **Itoh, Y., Y. Sugita-Konishi, F. Kasuga, M. Iwaki, Y. Hara-Kudo, N. Saito, Y. Noguchi, H. Konuma, and S. Kumagai.** 1998. Enterohemorrhagic *Escherichia coli* O157:H7 present in radish sprouts. *Appl. Environ. Microbiol.* **64:**1532–1535.

52. **Jackson, S. G., R. B. Goodbrand, R. P. Johnson, V. G. Odorico, D. Alves, K. Rahn, J. B. Wilson, M. K. Welch, and R. Khakhria.** 1998. *Escherichia coli* O157:H7 diarrhoea associated with well water and infected cattle on an Ontario farm. *Epidemiol. Infect.* **120:**17–20.

53. **Jay, J. M.** 2000. *Modern Food Microbiology*, 6th ed. Aspen Publishers, Inc., Gaithersburg, Md.

54. **Johnson, C. K.** 21 October 2003. Plaintiff seeks to add eight to *E. coli* suit; more dance camp participants, others may be included in lawsuit against Spokane Produce. *Spokesman Review*, Spokane, Wash.

55. **Kamat, A. S., D. P. Nerkar, and P. M. Nair.** 1989. *Bacillus cereus* in some Indian foods. Incidence and antibiotic, heat and radiation resistance. *J. Food Safety* **10:**31–41.

56. **Kapperud, G., L. M. Rørvik, V. Hasseltvedt, E. A. Høiby, B. G. Iversen, K. Staveland, G. Johnsen, J. Leitao, H. Herikstad, Y. Andersson, G. Langeland, B. Gondrosen, and J. Lassen.** 1995. Outbreak of *Shigella sonnei* infection traced to imported iceberg lettuce. *J. Clin. Microbiol.* **33:** 609–614.

57. **Keene, W. E., E. Sazie, J. Kok, D. H. Rice, D. D. Hancock, V. K. Balan, T. Zhao, and M. P. Doyle.** 1997. An outbreak of *Escherichia coli* O157:H7 infections traced to jerky made from deer meat. *JAMA* **277:**1229–1231.

58. **Ketyi, I.** 1989. Epidemiology of the enteroinvasive *Escherichia coli*. Observations in Hungary. *J. Hyg. Epidemiol. Microbiol. Immunol.* **33:**261–267.

59. **Khodr, M., S. Hill, L. Perkins, S. Stiefel, C. Comer-Morrison, S. Lee, D. R. Patel, D. Peery, C. W. Armstrong, and G. B. Miller, Jr.** 1994. *Bacillus cereus* food poisoning associated with fried rice at two child day care centers—Virginia, 1993. *Morb. Mortal. Wkly. Rep.* **43:**177–178.

60. **Kudva, I. T., K. Blanch, and C. J. Hovde.** 1998. Analysis of *Escherichia coli* O157:H7 survival in ovine or bovine manure and manure slurry. *Appl. Environ. Microbiol.* **64:**3166–3174.

61. **Law, D.** 1994. Adhesion and its role in the virulence of enteropathogenic *Escherichia coli*. *Clin. Microbiol. Rev.* **7:**152–173.

62. **Leyer, G. J., L.-L. Wang, and E. A. Johnson.** 1995. Acid adaptation of *Escherichia coli* O157:H7 increases survival in acidic foods. *Appl. Environ. Microbiol.* **61:**3752–3755.

63. **Lin, J., M. P. Smith, K. C. Chapin, H. S. Baik, G. N. Bennett, and J. W. Foster.** 1996. Mechanisms of acid resistance in enterohemorrhagic *Escherichia coli*. *Appl. Environ. Microbiol.* **62:**3094–3100.

64. **Mahon, B. E., A. Pönkä, W. N. Hall, K. Komatsu, S. E. Dietrich, A. Siitonen, G. Cage, P. S. Hayes, M. A. Lambert-Fair, N. H. Bean, P. M. Griffin, and L. Slutsker.** 1997. An international outbreak of *Salmonella* infections caused by alfalfa sprouts grown from contaminated seeds. *J. Infect. Dis.* **175:**876–882.

65. **McGrath, M.** 6 May 1999. *E. coli*-lettuce link studied. *Omaha World-Herald*, Omaha, Nebr.

66. **McKillip, J. L.** 2000. Prevalence and expression of enterotoxins in *Bacillus cereus* and other *Bacillus* spp., a literature review. *Antonie van Leeuwenhoek* **77:**393–399.

67. **Mermin, J. H., and P. M. Griffin.** 1999. Invited commentary: public health in crisis: outbreaks of *Escherichia coli* O157:H7 infections in Japan. *Am. J. Epidemiol.* **150:**797–803.

68. **Michino, H., K. Araki, S. Minami, S. Takaya, N. Sakai, M. Miyazaki, A. Ono, and H. Yanagawa.** 1999. Massive outbreak of *Escherichia coli* O157:H7 infection in schoolchildren in Sakai City, Japan, associated with consumption of white radish sprouts. *Am. J. Epidemiol.* **150:**787–796.

69. **Michino, H., and K. Otsuki.** 2000. Risk factors in causing outbreaks of food-borne illness originating in school lunch facilities in Japan. *J. Vet. Med. Sci.* **62:**557–560.

70. **Millard, P. S., K. F. Gensheimer, D. G. Addiss, D. M. Sosin, G. A. Beckett, A. Houck-Jankoski, and A. Hudson.** 1994. An outbreak of cryptosporidiosis from fresh-pressed apple cider. *JAMA* **272:**1592–1596.

71. **Miller, L. G., and C. W. Kaspar.** 1994. *Escherichia coli* O157:H7 acid tolerance and survival in apple cider. *J. Food Prot.* **57:**460–464.

72. **Mobley, H. L. T.** 2000. Virulence of the two primary uropathogens. *ASM News* **66:**403–410.

73. **Mohle-Boetani, J. C., J. A. Farrar, S. B. Werner, D. Minassian, R. Bryant, S. Abbott, L. Slutsker, and D. J. Vugia, for the Investigation Team.** 2001. *Escherichia coli* O157 and *Salmonella* infections associated with sprouts in California, 1996–1998. *Ann. Intern. Med.* **135:**239–247.

74. **Moore, K., T. Damrow, D. O. Abbott, and S. Jankowski.** 1995. Outbreak of acute gastroenteritis attributable to *Escherichia coli* serotype O104:H21—Helena, Montana, 1994. *Morb. Mortal. Wkly. Rep.* **44:**501–503.

75. **Morgan, U. M., L. Xiao, P. Monis, A. Fall, P. J. Irwin, R. Fayer, K. M. Denholm, J. Limor, A. Lal, and R. C. A. Thompson.** 2000. *Cryptosporidium* spp. in domestic dogs: the "dog" genotype. *Appl. Environ. Microbiol.* **66:**2220–2223.

76. **Mshar, P. A., Z. F. Dembek, M. L. Cartter, J. L. Hadler, T. R. Fiorentino, R. A. Marcus, J. McGuire, M. A. Shiffrin, A. Lewis, J. Feuss, J. Van Dyke, M. Toly, M. Cambridge, J. Guzewich, J. Keithly, D. Dziewulski, E. Braun-Howland, D. Ackman, P. Smith, J. Coates, and J. Ferrara.** 1997. Outbreaks of *Escherichia coli* O157:H7 infection and cryptosporidiosis associated with drinking unpasteurized apple cider—Connecticut and New York, October, 1996. *Morb. Mortal. Wkly. Rep.* **46:**4–8.

77. **Nataro, J. P., and J. B. Kaper.** 1998. Diarrheagenic *Escherichia coli. Clin. Microbiol. Rev.* **11:**142–201.

78. **Nataro, J. P., T. Steiner, and R. L. Guerrant.** 1998. Enteroaggregative *Escherichia coli. Emerg. Infect. Dis.* **4:**251–261.

79. **Nortjé, G. L., S. M. Vorster, R. P. Greebe, and P. L. Steyn.** 1999. Occurrence of *Bacillus cereus* and *Yersinia enterocolitica* in South African retail meats. *Food Microbiol.* **16:**213–217.

80. **Onoue, Y., H. Konuma, H. Nakagawa, Y. Hara-Kudo, T. Fujita, and S. Kumagai.** 1999. Collaborative evaluation of detection methods for *Escherichia coli* O157:H7 from radish sprouts and ground beef. *Int. J. Food Microbiol.* **46:**27–36.

81. **Pirttijärvi, T. S. M., L. M. Ahonen, L. M. Maunuksela, and M. S. Salkinoja-Salonen.** 1998. *Bacillus cereus* in a whey process. *Int. J. Food Microbiol.* **44:**31–41.

82. **Pönkä, A., Y. Andersson, A. Siitonen, B. de Jong, M. Jahkola, O. Haikala, A. Kuhmonen, and P. Pakkala.** 1995. *Salmonella* in alfalfa sprouts. *Lancet* **345:**462–463.

83. **Portnoy, B. L., J. M. Goepfert, and S. M. Harmon.** 1976. An outbreak of *Bacillus cereus* food poisoning resulting from contaminated vegetable sprouts. *Am. J. Epidemiol.* **103:**589–594.

84. **Puohiniemi, R., T. Heiskanen, and A. Siitonen.** 1997. Molecular epidemiology of two international sprout-borne *Salmonella* outbreaks. *J. Clin. Microbiol.* **35:**2487–2491.

85. **Raine, G.** 8 April 1998. U.S. Army declined to carry Odwalla; inspector felt unpasteurized juice posed health hazard to military consumers. *The Fresno Bee*, Fresno, Calif.

86. **Reed, G. H.** 1994. Foodborne illness (part 4). *Bacillus cereus* gastroenteritis. *Dairy Food Environ. Sanit.* **14:**87.

87. **Rice, D. H., D. D. Hancock, and T. E. Besser.** 1995. Verotoxigenic *E. coli* O157 colonisation of wild deer and range cattle. *Vet. Rec.* **137:**524.

88. **Rodrigues, J., C. M. Thomazini, A. Morelli, and G. C. M. de Batista.** 2004. Reduced etiological role for enteropathogenic *Escherichia coli* in cases of diarrhea in Brazilian infants. *J. Clin. Microbiol.* **42:**398–400.

89. **Roels, T. H., M. E. Proctor, L. C. Robinson, K. Hulbert, C. A. Bopp, and J. P. Davis.** 1998. Clinical features of infections due to *Escherichia coli* producing heat-stable toxin during an outbreak in Wisconsin: a rarely suspected cause of diarrhea in the United States. *Clin. Infect. Dis.* **26:**898–902.

90. **Seo, K. H., and J. F. Frank.** 1999. Attachment of *Escherichia coli* O157:H7 to lettuce leaf surface and bacterial viability in response to chlorine treatment as demonstrated by using confocal scanning laser microscopy. *J. Food Prot.* **62:**3–9.

91. **Slaghuis, B. A., M. C. Te Giffel, R. R. Beumer, and G. André.** 1997. Effect of pasturing on the incidence of *Bacillus cereus* spores in raw milk. *Int. Dairy J.* **7:**201–205.

92. **Solomon, E. B., C. J. Potenski, and K. R. Matthews.** 2002. Effect of irrigation method on transmission to and persistence of *Escherichia coli* O157:H7 on lettuce. *J. Food Prot.* **65:**673–676.

93. **Solomon, E. B., S. Yaron, and K. R. Matthews.** 2002. Transmission of *Escherichia coli* O157:H7 from contaminated manure and irrigation water to lettuce plant tissue and its subsequent internalization. *Appl. Environ. Microbiol.* **68:**397–400.

94. **Steele, B. T., N. Murphy, G. S. Arbus, and C. P. Rance.** 1982. An outbreak of hemolytic uremic syndrome associated with ingestion of fresh apple juice. *J. Pediatr.* **101:**963–965.

95. **Steele, M., and J. Odumeru.** 2004. Irrigation water as source of foodborne pathogens on fruit and vegetables. *J. Food Prot.* **67:**2839–2849.

96. **Takeuchi, K., and J. F. Frank.** 2000. Penetration of *Escherichia coli* O157:H7 into lettuce tissues as affected by inoculum size and temperature and the effect of chlorine treatment on cell viability. *J. Food Prot.* **63:**434–440.

97. **Tarr, C. L., T. M. Large, C. L. Moeller, D. W. Lacher, P. I. Tarr, D. W. Acheson, and T. S. Whittam.** 2002. Molecular characterization of a serotype O121:H19 clone, a distinct shiga toxin-producing clone of pathogenic *Escherichia coli*. *Infect. Immun.* **70:**6853–6859.

98. **Valcour, J. E., P. Michel, S. A. McEwen, and J. B. Wilson.** 2002. Associations between indicators of livestock farming intensity and incidence of human shiga toxin-producing *Escherichia coli* infection. *Emerg. Infect. Dis.* **8:**252–257.

99. **Van Beneden, C. A., W. E. Keene, R. A. Strang, D. H. Werker, A. S. King, B. Mahon, K. Hedberg, A. Bell, M. T. Kelly, V. K. Balan, W. R. MacKenzie, and D. Fleming.** 1999. Multinational outbreak of *Salmonella enterica* serotype Newport infections due to contaminated alfalfa sprouts. *JAMA* **281:**158–162.

100. **Van Duynhoven, Y. T. H. P., M.-A. Widdowson, C. M. de Jager, T. Fernandes, S. Neppelenbroek, W. van den Brandhof, W. J. B. Wannet, J. A. van Kooij, H. J. M. Rietveld, and W. van Pelt.** 2002. *Salmonella enterica* serotype Enteritidis phage type 4b outbreak associated with bean sprouts. *Emerg. Infect. Dis.* **8:**440–443.

101. **Wachtel, M. R., L. C. Whitehand, and R. E. Mandrell.** 2002. Prevalence of *Escherichia coli* associated with a cabbage crop inadvertently irrigated with partially treated sewage wastewater. *J. Food Prot.* **65:**471–475.

102. **Wang, G., and M. P. Doyle.** 1998. Survival of enterohemorrhagic *Escherichia coli* O157:H7 in water. *J. Food Prot.* **61:**662–667.

103. **Wang, G., T. Zhao, and M. P. Doyle.** 1996. Fate of enterohemorrhagic *Escherichia coli* O157:H7 in bovine feces. *Appl. Environ. Microbiol.* **62:**2567–2570.

104. **Watanabe, Y., K. Ozasa, J. H. Mermin, P. M. Griffin, K. Masuda, S. Imashuku, and T. Sawada.** 1999. Factory outbreak of *Escherichia coli* O157:H7 infection in Japan. *Emerg. Infect. Dis.* **5:**424–428.

105. **Werber, D., A. Fruth, A. Liesegang, M. Littmann, U. Buchholz, R. Prager, H. Karch, T. Breuer, H. Tschäpe, and A. Ammon.** 2002. A multistate outbreak of shiga toxin–producing *Escherichia coli* O26:H11 infections in Germany, detected by molecular subtyping surveillance. *J. Infect. Dis.* **186:**419–422.

106. **Winthrop, K. L., M. S. Palumbo, J. A. Farrar, J. C. Mohle-Boetani, S. Abbott, M. E. Beatty, G. Inami, and S. B. Werner.** 2003. Alfalfa sprouts and *Salmonella* Kottbus infection: a multistate outbreak following inadequate seed disinfection with heat and chlorine. *J. Food Prot.* **66:**13–17.

107. **Xiao, L., K. Alderisio, J. Limor, M. Royer, and A. A. Lal.** 2000. Identification of species and sources of *Cryptosporidium* oocysts in storm waters with a small-subunit rRNA-based diagnostic and genotyping tool. *Appl. Environ. Microbiol.* **66:**5492–5498.

108. **Zhao, T., M. P. Doyle, and R. E. Besser.** 1993. Fate of enterohemorrhagic *Escherichia coli* O157:H7 in apple cider with and without preservatives. *Appl. Environ. Microbiol.* **59:**2526–2530.

109. **Zimmerman, S.** 12 July 1996. Outbreak traced to red leaf lettuce. *Chicago Sun-Times*, Chicago, Ill.

chapter 9

When the Well Runs Dry

AUGUST 31, 1854, MARKED THE ONSET of a cholera epidemic in the Broad Street neighborhood of London, England, the worst one yet to hit the city. Residents fled the area in order to avoid the "miasma" of cholera. Within 6 days, at least three-quarters of the neighborhood population had departed or died. The epidemic killed more than 500 people in just 10 days (89).

Cholera had already caught the attention of John Snow, a London physician and surgeon, even before the 1854 outbreak. By studying the history of its journey from Asia to England and the disease's pattern of spread during an outbreak, he deduced that cholera was carried by people. But, unlike most of his contemporaries, Snow did not believe that the disease was transmitted via an airborne effluvium or contagion emanating from a sick person. Rather, he concluded that cholera was spread through contaminated food or water.

Snow had an opportunity to test his theory during the Broad Street epidemic. The pattern of contagion led him to suspect the neighborhood water pump as the source of the outbreak. He presented his case to the area's Board of Guardians on September 7, 1854. The Board disabled the pump by removing its handle the next day. The epidemic petered out and the last cholera death occurred on September 29. Snow never claimed that this incident proved his theory—the outbreak had already begun to decline before the pump handle was removed (76). Instead, he sought additional supporting evidence by studying another cholera outbreak, one that had taken place earlier that same year (89).

The outbreak that hit the Kennington district of London began in July 1854. Two companies, Lambeth Company and Southward and Vauxhall Company, supplied water to this district. Lambeth drew its water from a part of the Thames River upstream from London's sewage outflow, whereas Southward and Vauxhall obtained its water supply from an area farther downstream, below the city's raw sewage outflow. As part of his investigation, Snow reviewed the 1853 cholera death statistics from parts of the city served by the two companies, sorting the data according to whether each household was supplied with water by Lambeth or by Southward and Vauxhall. He discovered that the cholera death rate during the 1853 outbreak was 8.5 times as high in Southward and Vauxhall households (315 per 10,000) as in Lambeth Company households (37 per 10,000). Snow's statistical study of these and other earlier cholera outbreaks led him to conclude unequivocally that "all the evidence proving the communication of cholera through the medium of water, confirms that with which I set out, of its communication in the crowded habitations of the poor, in coal mines and other places, by the hands getting soiled with the evacuations of the patients, and by small quantities of these evacuations being swallowed with the food, as paint is swallowed by house painters of uncleanly habits, who contract lead-colic in this way."

The delivery of potable water to city and town residents has improved immeasurably since the days of John Snow. Thanks to his efforts, and those of Robert Koch beginning in 1883 (23), citizens of the developed world no longer need to worry about the safety of their drinking water—most of the time.

Opening the Floodgates

There was a time when water from wells, lakes, and rivers was relatively free from fecal contamination and the pathogens associated with it. But population pressure from humans and domesticated animals has changed the safety profile of surface and ground waters (81, 102). Since the late 1980s, outbreaks of gastrointestinal disease due to *Salmonella*, *Shigella*, *Plesiomonas*, *Escherichia coli* (O157:H7 and O121:H19), *Giardia*, *Cryptosporidium*, and norovirus have been traced to drinking untreated well water and to recreational water activities such as lake swimming, water parks, swimming pools, and interactive water fountains (Table 9.1) (33). But these incidents are dwarfed by the *Cryptosporidium* outbreak that hit Milwaukee, Wisc., in 1993.

Milwaukee residents had every reason to believe that their drinking water supply was safe. The city's water system, which served more than 1.6 million residents in the greater Milwaukee area, was fed by two water treatment plants, one in the northern part of the city and one in the south (5, 67). Either plant was capable on its own of supplying water for the entire city, and both facilities were operating in compliance with all federal and state regulations (1).

The northern and southern plants obtained their raw water from different locations, but both followed the same procedure to purify water for the city's residents (Fig. 9.1). First, chlorine and polyaluminum chloride (a coagulant) were mixed into the water—the chlorine for disinfection and the polyaluminum chloride to encourage particles to stick together.

Table 9.1 Other outbreaks of gastrointestinal illness traced to drinking or recreational use of untreated and inadequately treated water

Year	Country	Pathogen	Source of outbreak[a]	No. of cases (deaths)	Reference(s)
1988	United States	*Cryptosporidium* sp.	Inadequate maintenance of municipal swimming pool filtration system, combined with resistance of *Cryptosporidium* to chlorine	44	90, 91
1989/90	United States	*E. coli* O157:H7	Contaminated unchlorinated potable water supply	243 (4)	93
1990	Japan	*E. coli* O157:H7	Contaminated well water	174	2
1991	United States	*E. coli* O157:H7 and *Shigella sonnei*	Swimming in contaminated lake	59[b]	61
1992	Scotland	*E. coli* O157 phage type 49	Children's paddling pool contaminated by child with diarrhea	6	22
1993	United States	*Salmonella enterica* serotype Typhimurium	Unchlorinated community water supply	>650 (7)	4
1993	United States	*C. parvum*	Contaminated swimming pool	51	68
1994	United States	*C. parvum*	Swimming in contaminated lake	2,070	62
1995	Canada	*E. coli* O157:H7	Well water on dairy farm contaminated by manure-contaminated surface water	1	57
1995	United States	*E. coli* O157:H7	Swimming in contaminated lake	12	103
1995	United States	*C. parvum*	Contaminated hose nozzle used to obtain drinking water at day camp	77	85
1995	United States	*S. sonnei*	Contaminated well water	82	10
1995	Canada	*Toxoplasma gondii*	Contamination following heavy rainfall of water reservoir serving municipality; water chlorinated but unfiltered	100	20, 56
1996	United States	*E. coli* O157:H7	Improper and inadequate chlorination of swimming pool water	18	42
1996	United States	*Plesiomonas shigelloides* and *S. enterica* serotype Hartford	Contaminated well water	56	99
1997	United States	*Cryptosporidium* sp.	Playing in water sprinkler fountain	4	32
1998	United States	*E. coli* O157:H7	Unchlorinated municipal water supply	157	80
1998	Finland	Norovirus	Inadequate chlorination of municipal water supply	1,655	63
1999	United States	*C. parvum* and *S. sonnei*	Interactive water fountain	38	75
1999	United States	*E. coli* O121:H19	Swimming in contaminated lake	11	71
1999	United States	*E. coli* O157:H7 and *Campylobacter jejuni*	Contaminated water distribution system at county fair	775	19
2001	Canada	*E. coli* O157:H7	Swimming in contaminated water at public beach	4	25
2001	United States	Norovirus	Contaminated well water	20	3
2004	United States	Norovirus	Improper pool maintenance, resulting in inadequate chlorination	53	18

[a] Those outbreaks involving *S. sonnei* were most likely due to human fecal contamination of the water, since *Shigella* is a human-specific pathogen and not a zoonotic.
[b] Twenty-one cases due to *E. coli* O157:H7 and 38 cases due to *S. sonnei*.

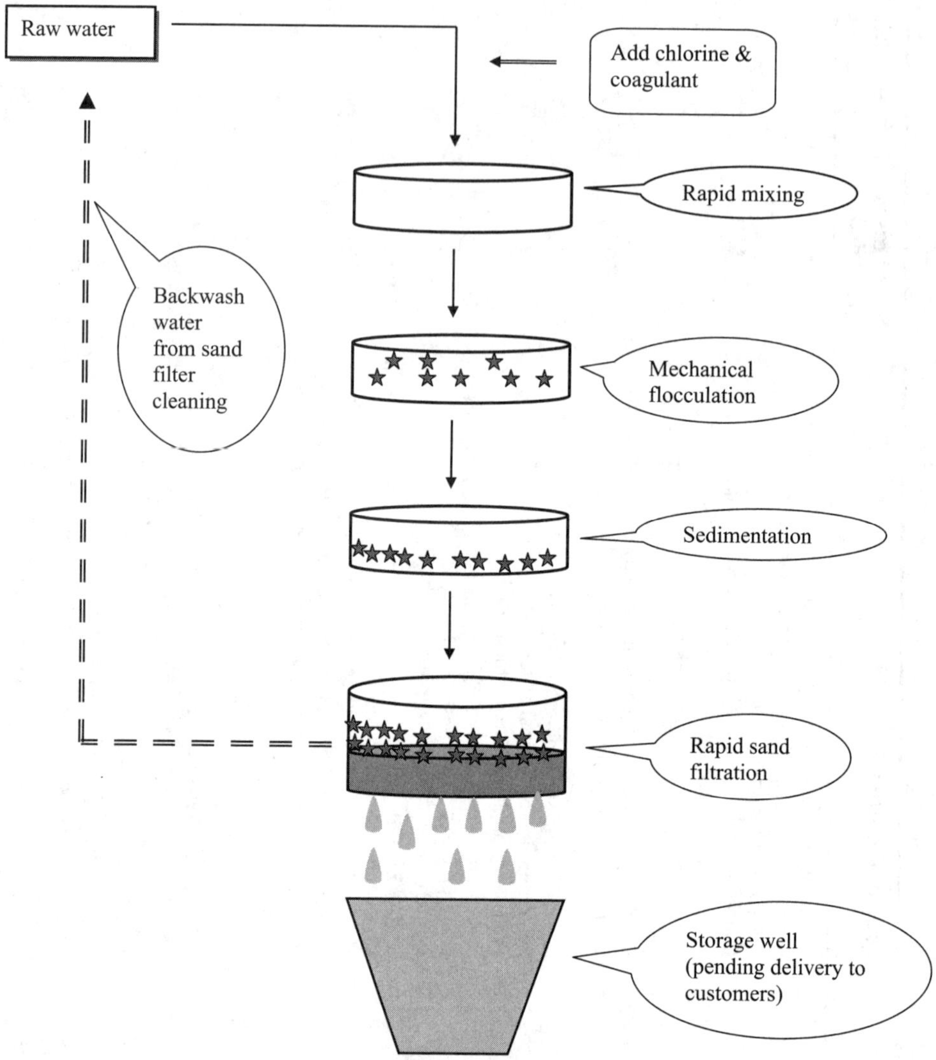

Figure 9.1 Water treatment in Milwaukee, Wis., at the time of the 1993 *Cryptosporidium* outbreak (67).

Then, a mechanical flocculation step caused the sticky particles to adhere to each other, forming floc (large, fluffy particles) that could be filtered out of the water easily. Next, a sedimentation step allowed most of the floc to settle out. Finally, the water was filtered through sand to remove the remaining floc. The sand filters, which needed periodic cleaning to remove accumulated floc, were backwashed at regular intervals with water that was then cycled back through the entire treatment process. Consistently, both plants produced water that met all state water quality standards in place at the time (67).

On March 18, 1993, the cloudiness of the treated water from the southern plant increased from a consistent turbidity reading of not more than 0.25 nephelometric turbidity unit (NTU) to 0.35 NTU. Between March 23 and April 1, the turbidity remained consistently at or above 0.45 NTU, reaching a maximum level of 1.7 NTU twice during this period. The southern plant operators corrected the problem by increasing the concentration of polyaluminum chloride. Then, on April 2, they decided to switch from polyaluminum chloride to alum, a different coagulant, in order to make the treated water less corrosive and reduce the risk of leaching lead out of lead pipes in customers' household plumbing systems (77). By April 5, the turbidity of the treated water had shot back up to 1.5 NTU (67). The plant's flocculation and filtration system was allowing suspended particles—and something more hazardous—to remain in the treated water. *Cryptosporidium parvum* was defeating Milwaukee's water treatment procedure.

The spring rains had swollen the two rivers that emptied into Milwaukee harbor on Lake Michigan. Along with the increased volume of water came snow runoff from farms and slaughterhouses along the banks of those rivers, runoff that most likely contained *C. parvum* oocysts. The oocyst-contaminated water found its way into Milwaukee's southern water treatment plant (67).

The first victims of the *C. parvum* outbreak began to experience diarrhea on March 1, 1993. By April 5, the Milwaukee Department of Health realized that a major outbreak of gastrointestinal illness was under way, and they notified the Wisconsin Division of Health of the problem. On April 7, *Cryptosporidium* oocysts were isolated from stool samples in the Milwaukee area for the first time. That same evening, suspecting that the pathogen was being spread in the public water supply, Milwaukee issued a "boil water" advisory to city residents (5). Two days later, with evidence pointing increasingly to the southern water treatment plant as the source of the outbreak, officials decided to close that plant temporarily (6). Laboratory testing of ice blocks produced between March 25 and April 9 using water from the southern treatment plant confirmed that the water contained *Cryptosporidium* oocysts.

The "boil water" advisory and closure of the southern water treatment plant had their desired effect. The outbreak subsided by the middle of April, but not before claiming an estimated 403,000 victims, including 12 deaths (7, 67).

The city of Milwaukee established a committee of investigation to determine the cause of the outbreak and to recommend corrective measures. The committee blamed the outbreak on a water supply fouled by heavy rainfall and seasonal lake turbidity, calling it a "natural disaster." The members also concluded that chemists at the southern water treatment plant were inexperienced in the correct use of the new coagulating agent that had been introduced into the plant's treatment protocol around the time the outbreak began. The committee report included 24 recommendations to improve the safety and reliability of Milwaukee's drinking water supply (7).

The magnitude of the outbreak, and the fact that it occurred in a water district that was in full compliance with all federal and state regulatory standards, prompted the Centers for Disease Control and Prevention (CDC) to act. In September 1994, CDC organized a workshop to address the health issues raised by the outbreak. Workshop participants studied four areas of concern: (i) surveillance and epidemiological systems; (ii) appropriate public health responses to detection of *Cryptosporidium* oocysts in drinking water; (iii) cryptosporidiosis and the immunocompromised; and (iv) research priorities, sampling methods, and interpretation of results. The workshop produced several recommendations, and CDC impaneled the Working Group on Waterborne Cryptosporidiosis, made up of representatives from various agencies and organizations, to pursue the concerns highlighted by the workshop report (1). CDC initiated coverage of cryptosporidiosis as a nationally reportable disease in 1995 (52).

CDC wasn't the only agency to take the Milwaukee outbreak to heart. The Environmental Protection Agency (EPA), which participated in the

CDC Workshop on Waterborne Cryptosporidiosis: Concerns, Recommendations, and Conclusions (1)

1. Develop and initiate surveillance for *Cryptosporidium*, possibly including making cryptosporidiosis a reportable disease to CDC, monitoring sales of antidiarrheal medications, monitoring incidence of diarrheal complaints in hospitals and nursing homes, monitoring lab data for reports of *Cryptosporidium*, and monitoring tap water in selected cities for incidence of *Cryptosporidium* oocysts. This program would require cooperation of federal, state, county, and municipal officials.
2. Establish rapid response teams at CDC and EPA to provide immediate support for epidemiological investigation of a suspected outbreak of cryptosporidiosis, including lab analysis of stool samples, serological tests for *Cryptosporidium*, case-control studies, and prospective epidemiological studies.
3. Develop clear guidelines for implementing and lifting "boil water" advisories.
4. Educate the general public about *Cryptosporidium*, its health significance, and EPA's plan to require monitoring of drinking water for the presence of oocytes.
5. Educate the immunocompromised population (e.g., human immunodeficiency virus/AIDS community, cancer patients undergoing chemotherapy, organ transplant patients) about their elevated risk of developing severe or life-threatening illness as a result of *Cryptosporidium* infection, even in the absence of an outbreak, and suggest possible water use strategies to avoid sporadic infections, including boiling or filtering their drinking water or drinking only bottled water.
6. Develop and validate reliable, less labor-intensive lab methods to detect and enumerate *Cryptosporidium* oocysts from drinking water.

CDC-sponsored workshop, strengthened its turbidity standards following the 1993 incident. EPA already knew from previous research studies that small numbers of protozoans such as *Cryptosporidium* spp. and *Giardia* spp. could pass through water treatment filters (64). Therefore, to determine the magnitude of the risk to the nation's drinking water supply, the agency began to collect data on the presence and concentration of several microbes—including *Giardia* and *Cryptosporidium*—in untreated and treated water from utilities serving communities larger than 10,000 people (37). EPA used the information it collected to update the U.S. drinking water regulations in 1998 (for communities of more than 10,000 people) and in 2002 (for communities of fewer than 10,000 people), adding measures to reduce the risk of *Cryptosporidium* spp. contamination in treated drinking water (38, 39).

The CDC and EPA reacted constructively and logically to the Milwaukee outbreak. They identified the problem, determined its magnitude by gathering the necessary data and expert advice, and implemented protective measures. But the regulatory process has a very long incubation period. EPA's Final Rule covering drinking water for small communities took effect in February 2002, almost 9 years after the 1993 *Cryptosporidium* outbreak (39).

Troubled Waters

While U.S. regulators were spending time and money to improve drinking water safety, something very different was happening across the border in the Canadian province of Ontario. Prior to 1993, all testing of municipal drinking water in the province was carried out in the laboratories of the Ontario Ministry of the Environment (MOE). Any unsatisfactory results were automatically reported to the local Medical Officer of Health, who would then decide whether or not to issue a "boil water" advisory. In 1993, a few municipalities began to submit their drinking water samples to private laboratories instead of using the government labs. Then, in 1996, the Ontario government decided to save money by withdrawing completely from municipal water testing. Instead, municipalities now had to send all of their water samples to private environmental testing labs, none of which were licensed, subject to inspection, or regulated in any way (24). And because government labs weren't involved in the water testing, Medical Officers of Health were no longer advised routinely about unsatisfactory results. There was no law or regulation in place requiring that they be notified (43, 78).

The privatization of water testing in Ontario put the detection of microbiological contamination into hands that were both unregulated and less experienced than the government-run labs. It also created confusion over who was responsible for notifying health authorities when results were out of compliance with the province's drinking water standards. In doing so, it opened the door to Walkerton.

When it rains …

April showers are reputed to bring forth May flowers, but May downpours sometimes elicit a much nastier crop—contamination of well water by surface water runoff. In 2000, the 5-day period from May 8 to May 12 was especially stormy in parts of Ontario. One hundred thirty-four millimeters (5¼ inches) of rain fell in the Walkerton area, more than half of it—70 mm (2¾ inches)—on May 12. The heavy rainfall washed away manure from recently fertilized fields, contaminating surface waters with *E. coli* O157:H7 and *Campylobacter* spp. Some of that contaminated water found its way into one of Walkerton's wells. And Walkerton's residents began to fall ill.

It was May 18 before people started to realize that something was seriously wrong. A large number of children missed school that day, and two were admitted to the hospital suffering from bloody diarrhea. Even more cases turned up the following day, and the local health unit began to investigate. Health officials contacted Stan Koebel, the general manager of the town's Public Utility Commission (PUC), told him of the outbreak, and asked about the water supply. Koebel assured the health unit that the town's water was safe. Despite those assurances, 2 days later, the local Medical Officer of Health decided to issue a "boil water" advisory to Walkerton's residents.

By May 24, 600 people (out of a population of 4,800) were suffering symptoms of *E. coli* O157:H7 infection. Schools and day care centers were closed as a precaution; area hospitals were flooded with calls from people suffering from bloody diarrhea, vomiting, fever, and cramps; 11 people were hospitalized locally and 4 more were airlifted to a larger regional hospital because of the critical nature of their symptoms (11). By the time the outbreak finally ended, 7 people had died and 2,300—almost half of the town's population—had been infected with either *E. coli* O157:H7 or *Campylobacter jejuni* (29, 78).

Walkerton was served by a municipal chlorinated water supply overseen by the town's PUC. The water was drawn from three wells. Well 5 was shallow, just 15 meters (49 feet) below the surface (78). Wells 6 and 7 were deeper—72 meters (234 feet) and 76 meters (247 feet), respectively (101). Water pumped from the wells was chlorinated before entering the distribution system; flow rates and chlorine levels were monitored daily, and samples were taken weekly for microbiological testing (78). As far as they knew, the town's residents had no reason to question the safety of their drinking water. But Walkerton's water safety was an illusion, and the PUC, like the emperor in the fairy tale, had no clothes.

Well 5 was drilled in 1978. In a reversal of the usual sequence, the town drilled the well before applying to the MOE for a Certificate of Approval. And even when newly constructed, the well was vulnerable. Within 24 hours after the start of its initial pump test, water from Well 5 tested positive for fecal coliforms, an indication that the well may have become contaminated with surface water. Although recommendations were made that land use in the area surrounding Well 5 be controlled, nothing was done to reduce the risk of contamination. In accordance with the standard practice at the time, no specific operating conditions or requirements were attached to the MOE's approval of the well.

Well 5 was inspected several times by the province between 1978 and 1980. Each time, the inspector repeated warnings about the likelihood of surface water contamination. The lab results confirmed that the risks were real (31). In 1980, 4 of 42 samples—nearly 10%—contained unacceptable quantities of coliform and fecal coliform bacteria. Yet Walkerton ignored the warnings, and the well was not inspected at all during the 1980s (78).

Well 6 was drilled in 1982 and approved in 1983 (78). Nearly five times as deep as Well 5, it was thought to be more secure from surface contamination. Even so, in 1983, the MOE alerted the Walkerton PUC to test results showing that Well 6 was probably being contaminated by agricultural runoff (31). Walkerton's official response to the warning from the provincial government has been lost; the town discards its water records after 5 years (101). But in 1987, perhaps in response to concerns about Well 5 and Well 6, Walkerton drilled one more well.

Well 7 was located 357 meters (1,171 feet) from Well 6 and was drilled to approximately the same depth, 76 meters (247 feet). It was approved for operation by the MOE in October 1987 on two conditions: that the town monitor the impact of Wells 6 and 7 on the status and condition of the shallow and deep aquifers feeding these wells, and that the results of the monitoring program be reported to the MOE within 15 months after the Well 7 Certificate of Approval was issued. Consistent with its approach to drinking water safety oversight over the years, the Walkerton PUC failed to satisfy either condition. At the end of 15 months, no monitoring program was in place and no report was forthcoming. And nobody at the MOE noticed until the Walkerton system was inspected in 1991 (78).

On November 19, 1991, Brian Jaffray conducted a routine inspection of the Walkerton water system. During his review of the Certificates of Approval for the wells, he noted that the conditions for approval of Well 7 had never been satisfied. Jaffray gave the Walkerton PUC until June 1, 1992, to submit its monitoring program report. The MOE finally received a report from Walkerton on September 26, 1994—5 years late and more than 2 years after Jaffray's deadline. This situation was not unusual; the MOE simply did not have enough personnel or resources to monitor

Coliforms, Fecal Coliforms, and *E. coli*: What Do They Indicate?

Coliforms are a collection of gram-negative, oxidase-negative, non-spore-forming, rod-shaped bacteria that ferment lactose with the production of acid and usually—but not always—gas (65, 96). *Salmonella* spp. and *Shigella* spp. are gram-negative, non-spore-forming, rod-shaped bacteria, but they don't ferment lactose (except for a few atypical strains). Therefore, *Salmonella* spp. and *Shigella* spp. are not coliforms. *Bacillus cereus*, *Staphylococcus aureus*, *Clostridium botulinum*, and *Clostridium perfringens* are gram-positive bacteria. They are not coliforms. Neither are *Vibrio* spp., *Yersinia* spp., *Campylobacter* spp., hepatitis A virus, nor norovirus. The only pathogens that are coliforms are the pathogenic *E. coli*. Why, then, would anyone want to look for coliforms in water?

Coliform bacteria are members of the family *Enterobacteriaceae*, the same family that includes *Salmonella*, *Shigella*, and *E. coli*. The group shares with its family members similar degrees of susceptibility to heat, cold, and disinfectants such as chlorine. It also shares similar requirements for growth. This makes the coliform group a useful surrogate for these waterborne bacterial pathogens. Environmental conditions that permit and encourage the survival and growth of coliforms may also support survival of their pathogenic cousins (96).

Coliforms (also sometimes referred to as the "coli-aerogenes group") were chosen to be indicators of water pollution (from sewage or other sources of fecal material), because they were thought to reside in the intestinal tract of animals and to be present in water in a relatively fixed and predictable proportion to *Salmonella enterica* serotype Typhi (26, 94, 96). In time, however, scientists realized that many coliforms were naturally resident in the plant world—in algal mats, wastes from the pulp and paper industry, and soil, for example—and were not always associated with fecal contamination. They determined that a more heat-tolerant subset of coliforms, which they dubbed the fecal coliform group (also known as thermotolerant coliforms), was more closely tied to pollution of rivers, streams, lakes, oceans, and other waters with fecal material from warm-blooded animals (96). Fecal coliforms were defined as those coliforms that could grow and ferment lactose when incubated at a temperature of $44.5 \pm 0.2°C$ [(~112°F)], and became a tool to detect possible fecal contamination of drinking water, recreational water, surface water, groundwater, and seawater in shellfish growing beds and to monitor "fecal" bacterial levels in treated sewage effluents (9, 27, 35, 36, 40, 50, 55, 70, 72–74, 83, 104, 105).

Members of the fecal coliform group include some *Klebsiella* spp. and *Citrobacter* spp. in addition to *E. coli* (58). Of these, *E. coli* is associated most closely with fecal contamination (96). Until the development of several new one-step methods in the 1970s and 1980s, detecting and confirming *E. coli* in water samples was a laborious and expensive procedure (13, 34, 47, 59, 60, 69, 88, 97, 100). Once these new methods gained acceptance, *E. coli* supplanted the fecal coliform group as the preferred indicator of fecal contamination in drinking water (51, 96, 105).

The absence of *E. coli*, or even the coliform group, from a water sample does not guarantee that the water is safe to drink or swim in (95). Some pathogens are better able to survive in very cold or very hot water than *E. coli*, and several pathogens, including viruses and protozoa in addition to some bacterial pathogens, react much differently to chlorine or salinity (28, 45, 46, 49, 54, 66). Nevertheless, the presence of *E. coli* is a useful indicator of fecal contamination of untreated or treated water (14, 36, 41, 53, 87, 104). And finding coliforms or, more specifically, *E. coli* in a drinking water sample is a red flag that the microbiological safety of the water has been compromised (86, 96, 100).

the reporting deadlines that it imposed on the province's water system operators (78).

The MOE inspected Walkerton's water system again in February 1998, but its inspection did not include an assessment of Well 5. The inspector simply assumed that the well's casing was secure and that Well 5 was not subject to potential surface water contamination. The inspector failed to detect two significant problems at Walkerton—improper chlorination and poor monitoring practices (78). Both of these issues were major factors in the May 2000 outbreak.

Who was minding the store?

The Walkerton PUC was responsible for the town's water distribution system. The town council and the mayor relied on the PUC to monitor and protect the safety of the municipality's drinking water. The commission's general manager, Stan Koebel, was responsible for overseeing the operation of the water system while the commissioners focused on financial issues (8). Unfortunately, Koebel was out of his depth (43, 44, 78).

Stan Koebel joined the staff of the PUC in 1972. At first, he worked on the water system; then, in 1976, he completed a lineman apprentice program and shifted his orientation to electricity. In 1981, Koebel was made PUC foreman, responsible for the employees taking care of both the water and the electrical systems for the town. Koebel became general manager of the Walkerton PUC in 1988, when the former general manager retired.

Koebel's knowledge of the principles of drinking water safety was minimal. He had never read the section of the Ontario Drinking Water Objectives dealing with indicators of water safety; he did not know that the presence of *E. coli* or fecal coliforms in water was a health hazard; he was unfamiliar with the major concepts of water chlorination; and he did not understand some of the technical terms (e.g., "turbidity" and "organic nitrogen") that appeared regularly in MOE inspection reports and correspondence (78).

During his tenure with the PUC, Koebel never was required to take any courses or pass even a single exam relating to water system management or water safety. In 1988, he was certified as a "class 2 water operator" under a voluntary grandparenting program introduced the previous year by the MOE. His certificate was upgraded—once more without an examination—to a "class 3" water distribution license in 1996 when Walkerton's water system was reclassified to "class 3" from "class 2." Provincial regulations promulgated in 1993 required that every water operator receive 40 h of technical training each year. Koebel did not accumulate anywhere near the required 40 h, even though money had been set aside in the PUC budget for that purpose.

A Litany of Lapses: Walkerton PUC's Faulty Procedures and Practices (78)

1. The province required that 13 water samples be tested for bacterial contamination every month. Most months, Walkerton tested only eight or nine samples.
2. Water samples were often deliberately misidentified; it was common practice to take several samples from a single site and label each one differently.
3. Water was not adequately chlorinated. The PUC staff believed their well water was of high quality and chose to ignore the province's chlorination instructions.
4. Residual chlorine levels were not monitored daily, and the method that PUC staff often used to monitor chlorine was inaccurate.
5. Beginning in 1979, PUC staff falsified the daily residual chlorine test results. The residual chlorine levels recorded for the water samples would have been impossible to attain, based on the amount of chlorine added to the system.
6. Stan Koebel submitted misleading annual reports to the MOE, incorporating the fictitious residual chlorine data.
7. Stan Koebel sometimes allowed a well to pump unchlorinated water into the distribution system. Frank and Stan Koebel falsified daily operating records for Well 7 to conceal that this well had operated twice during May 2000 (May 3 to 9 and May 15 to 19) without chlorination.

The year that Stan Koebel became general manager of the PUC, his brother Frank Koebel took over the foreman's position. Frank received his water distribution certificate under the same grandparenting program as Stan. Like his brother, he was completely unequipped for the technical aspects of his job. Frank also failed to comply with the province's 40-hours-per-year technical training requirement. In his 25 years with the PUC, he never even took a course on water chlorination.

The Koebel brothers inherited a number of routine operating procedures from their predecessors at the PUC. Some of those procedures were dangerously misguided—even deceitful. Stan Koebel ignored warnings from the MOE about deficiencies in PUC sampling and monitoring procedures. And in May 2000, with the *E. coli* O157:H7 outbreak gaining momentum, he and his brother Frank tried to cover their tracks by falsifying or altering records (30, 78).

A tragedy of errors

On April 22, 2000, David Biesenthal, a local farmer and veterinarian, spread fresh manure on his field and harrowed it into the soil. He followed the government's recommended practices for handling, storing, and spreading fresh manure to the letter (78). The manure was spread to within 81 meters (263 feet) of Well 5. The Ontario government was unconcerned about the proximity of fresh manure to a well. Indeed, Ontario's Minister of Agriculture, Ernie Hardeman, later boasted about both the quality and the safety of manure used in the province (15).

On May 2, 2000, Well 7, which had been off-line since March 10, was activated. It was equipped with an automatic chlorinator. The following day, May 3, Stan Koebel instructed his brother Frank to remove the chlorinator from Well 7. Stan expected Frank to replace the old chlorinator with a new one, which had been stored on the PUC premises since December 1998. Stan left town for a conference on May 5, leaving Frank in charge. But Frank did not install the new chlorinator in Stan's absence. Instead, he allowed Well 7 to pump untreated water through the town's distribution system. On May 9, Well 7—still without its required chlorinator—was taken off-line and Wells 5 and 6 were turned on. Well 7 was reactivated on May 15 and ran until May 19, when it was deactivated once more to be fitted with its new chlorinator. Ironically, Well 7 was not the source of the Walkerton outbreak; that honor was reserved for Well 5. But operating Well 7 without chlorination for several days ensured that there was no residual chlorine remaining in Walkerton's water distribution system (12, 78).

Well 5 operated nonstop until May 15 (except for a 15-hour period beginning at 10:45 p.m. on May 12); Well 6 cycled on and off during this time, depending on demand. The heavy rains that fell in the Walkerton area between May 8 and May 12 produced some flooding in the area of Well 5 on May 12, contaminating the well with surface water from the adjacent—recently manured—field. Without proper adjustment of the amount of chlorine added to the well water to compensate for the additional organic material (soil and manure) introduced with the surface water, there would not have been enough active chlorine in the water to inactivate bacterial contaminants.

On May 13 and 14, Frank Koebel recorded his daily residual chlorine test results for Wells 5 and 6. As he and his predecessors had done for years, Frank entered a fictitious result for both wells (30). Had he actually carried out the required tests, he most likely would have learned that there was no residual chlorine in the water pumped from Well 5 (78).

Water from the Walkerton distribution system was sampled for bacteriological analysis on May 1 and again on May 8. Most of the May 1 samples were negative; the samples labeled "Well 5 raw" and "Well 5 treated" contained coliforms, but no *E. coli*. The samples taken on May 8, while Wells 5 and 6 were off-line and the system was being fed exclusively by Well 7, were all negative.

The next samples were taken on May 15 and consisted of the usual weekly samples, as well as an extra set of samples from a nearby highway construction project. Several samples from both sets were positive for coliforms and *E. coli*. This result triggered a resampling of water at the construction project on May 18. The following day, the results came back—positive for coliforms and *E. coli*. On learning the news, Stan Koebel began to flush and superchlorinate the water supply. But his response was too little and too late. The outbreak was under way, and the superchlorination failed to disinfect the distribution system. There were just too many dead ends and too much old, worn pipe in the network (16, 21). Samples taken on May 21, 22, and 23 at two dead-end locations in the system were still contaminated with coliforms and *E. coli* in spite of Koebel's efforts (78).

By May 19, the day that Koebel began flushing the system, Walkerton's outbreak already had begun. The day before (May 18), some town residents called the water department to find out whether there was any problem with the drinking water. Even though he already had received news of the *E. coli*-positive samples, Koebel assured the callers that the town's water was safe. That same evening, Stan Koebel made his regular report at the monthly meeting of the PUC. He made no mention of the positive samples. The next day, even as he was attempting to disinfect the system, Koebel assured the local public health inspector that the water was "okay." Again on May 20, Koebel was asked about the water, and again he sidestepped questions from both the MOE and the local health unit. And on May 21, even after the "boil water" advisory had been issued, Koebel chose not to mention the positive *E. coli* test results to the mayor or to the MOE (78).

The people of Walkerton paid a severe price for the Koebel brothers' errors of omission and commission. But Stan and Frank Koebel did not escape unscathed. Both pled guilty to criminal charges relating to their actions (17). Stan Koebel was sentenced to 1 year in jail, and Frank to 9 months of house arrest (48). Nevertheless, the blame for the outbreak should not be laid entirely at the feet of the Koebel brothers.

Breaking the safety barrier

Protecting the food and water supply from contamination is an exercise in erecting hurdles or barriers against the microbes. In the case of Walkerton, those barriers did not exist. Wells had been drilled in close proximity to farm fields that were fertilized with manure. For more than 2 decades, the PUC staff routinely skimped on chlorination procedures and testing requirements—even falsifying records. And the water system was overseen by individuals who were blissfully ignorant of the basic principles of drinking water safety and hygiene. The final protective barrier against a waterborne disease outbreak was dismantled when the Ontario government closed its water testing labs in 1996—forcing municipalities to send their samples to private labs—without providing adequate oversight, and without ensuring that local health officials would automatically be notified of unsatisfactory test results. With that decision, the stage was set for the Walkerton outbreak. All that was needed was the contaminant.

The Commission of Inquiry that was established to study the Walkerton outbreak recognized the importance of erecting barriers to ensure safety of the water supply throughout Ontario. In Part II of his study report, Commissioner Dennis O'Connor wrote (79):

> Putting in place a series of measures, each independently acting as a barrier to passing water-borne contaminants through the system to consumers, achieves a greater overall level of protection than does relying exclusively on a single barrier (e.g., treatment alone or source protection alone). A failure in any given barrier will not cause a failure of the entire system. The challenge is to ensure that each of the barriers is functioning properly, so that together they constitute the highest level of protection that is reasonably and practically available.

Even a state-of-the-art water distribution system can't guarantee safe drinking water. Milwaukee found that out in 1993. But 20-plus years of inadequate procedures and risky old habits, and a provincial government bent on cutting costs, erased whatever protective barriers against microbial contamination that the Walkerton water distribution system might have had. And Walkerton's residents still are suffering the consequences—in long-term health effects, loss of business, and loss of confidence in the ability of their government agencies to deliver on basic infrastructure needs (82, 84, 92, 98).

References

1. **Addiss, D. G., M. J. Arrowood, M. E. Bartlett, D. G. Colley, D. D. Juranek, J. E. Kaplan, R. Perciasepe, J. R. Elder, S. E. Regli, and P. S. Berger.** 1995. Assessing the public health threat associated with waterborne cryptosporidiosis: report of a workshop. *Morb. Mortal. Wkly. Rep.* **44**(RR-6):1–16.
2. **Akashi, S., K. Joh, A. Tsuji, H. Ito, H. Hoshi, T. Hayakawa, J. Ihara, T. Abe, M. Hatori, and T. Mori.** 1994. A severe outbreak of haemorrhagic colitis and haemolytic uraemic syndrome associated with *Escherichia coli* O157:H7 in Japan. *Eur. J. Pediatr.* **153:**650–655.
3. **Anderson, A. D., A. G. Heryford, J. P. Sarisky, C. Higgins, S. S. Monroe, R. S. Beard, C. M. Newport, J. L. Cashdollar, G. S. Fout, D. E. Robbins, S. A. Seys, K. J. Musgrave, C. Medus, J. Vinjé, J. S. Bresee, H. M. Mainzer, and R. I. Glass.** 2003. A waterborne outbreak of Norwalk-like virus among snowmobilers—Wyoming, 2001. *J. Infect. Dis.* **187:**303–306.
4. **Angulo, F. J., S. Tippen, D. J. Sharp, B. J. Payne, C. Collier, J. E. Hill, T. J. Barrett, R. M. Clark, E. E. Geldreich, H. D. Donnell, Jr., and D. L. Swerdlow.** 1997. A community waterborne outbreak of salmonellosis and the effectiveness of a boil water order. *Am. J. Public Health* **87:**580–584.
5. **Anonymous.** 8 April 1993. Illness prompts water alert in Milwaukee. *Chicago Tribune*, Chicago, Ill.
6. **Anonymous.** 9 April 1993. Milwaukee fights water-borne illness. *Seattle Times*, Seattle, Wash.
7. **Anonymous.** 25 June 1993. Milwaukee report: water crisis was a natural disaster. *Chicago Tribune*, Chicago, Ill.

8. **Anonymous.** 28 November 2000. I wasn't interested in safety of water: Walkerton official. *National Post* (National edition), Don Mills, Ontario, Canada.
9. **Apte, S. C., C. M. Davies, and S. M. Peterson.** 1995. Rapid detection of faecal coliforms in sewage using a colorimetric assay of β-D-galactosidase. *Water Res.* **29:**1803–1806.
10. **Arnell, B., J. Bennett, R. Chehey, and J. Greenblatt.** 1996. *Shigella sonnei* outbreak associated with contaminated drinking water—Island Park, Idaho, August 1995. *Morb. Mortal. Wkly. Rep.* **45:**229–231.
11. **Avery, R.** 24 May 2000. *E. coli* 'epidemic' hits almost 600; schools closed near Owen Sound in outbreak called 'unprecedented'. *Toronto Star* (Ontario edition), Toronto, Ontario, Canada.
12. **Avery, R.** 21 October 2000. *E. coli* source confirmed; study looked at Walkerton water. *Toronto Star* (Ontario edition), Toronto, Ontario, Canada.
13. **Barrell, R. A. E.** 1992. A comparison between tryptone bile agar and membrane lauryl sulphate broth for the enumeration of presumptive *Escherichia coli* in water. *Water Res.* **26:**677–681.
14. **Bergstein-Ben Dan, T., D. Wynne, and Y. Manor.** 1997. Survival of enteric bacteria and viruses in Lake Kinneret, Israel. *Water Res.* **31:**2755–2760.
15. **Blackwell, T.** 1 June 2000. Don't blame "quality" manure for tragedy, minister says. *The Ottawa Citizen*, Ottawa, Ontario, Canada.
16. **Blackwell, T.** 18 October 2000. Water system legal despite defects: inquiry. *The Gazette*, Montreal, Quebec, Canada.
17. **Blackwell, T.** 3 December 2004. Walkerton deaths haunt Stan Koebel, his lawyer says: Crown urges close to maximum penalty. *National Post* (National edition), Don Mills, Ontario, Canada.
18. **Blevins, L. Z., D. Itani, A. Burns, C. Lohff, S. Schoenfeld, W. Knight, N. Thayer, J. Oetjen, N. Pugsley, C. Otto, M. Beach, M.-A. Widdowson, J. Bresee, R. Glass, S. Monroe, L. Browne, S. Adams, M. Amundson, and L. J. Podewils.** 2004. An outbreak of norovirus gastroenteritis at a swimming club—Vermont, 2004. *Morb. Mortal. Wkly. Rep.* **53:**793–795.
19. **Bopp, D. J., B. D. Sauders, A. L. Waring, J. Ackelsberg, N. Dumas, E. Braun-Howland, D. Dziewulski, B. J. Wallace, M. Kelly, T. Halse, K. Aruda Musser, P. F. Smith, D. L. Morse, and R. J. Limberger.** 2003. Detection, isolation, and molecular subtyping of *Escherichia coli* O157:H7 and *Campylobacter jejuni* associated with a large waterborne outbreak. *J. Clin. Microbiol.* **41:**174–180.
20. **Bowie, W. R., A. S. King, D. H. Werker, J. L. Isaac-Renton, A. Bell, S. B. Eng, and S. A. Marion for the BC *Toxoplasma* Investigation Team.** 1997. Outbreak of toxoplasmosis associated with municipal drinking water. *Lancet* **350:**173–177.
21. **Brennan, R.** 25 July 2000. Walkerton water pipes to be replaced; province to scrap aging section of main still infested with *E. coli*. *Toronto Star* (Ontario edition), Toronto, Ontario, Canada.
22. **Brewster, D. H., M. I. Brown, D. Robertson, G. L. Houghton, J. Bimson, and J. C. M. Sharp.** 1994. An outbreak of *Escherichia coli* O157 associated with a children's paddling pool. *Epidemiol. Infect.* **112:**441–447.
23. **Brock, T. D.** 1999. *Robert Koch. A Life in Medicine and Bacteriology*. ASM Press, Washington, D.C.

24. **Brooke, J.** 10 July 2000. Few left untouched after deadly *E. coli* flows through an Ontario town's water. *New York Times*, New York, N.Y.

25. **Bruneau, A., H. Rodrigue, J. Ismäel, R. Dion, and R. Allard.** 2004. Outbreak of *E. coli* O157:H7 associated with bathing at a public beach in the Montreal-centre region. *Can. Communic. Dis. Rep.* **30:**133–136. [Online.] http://www.phac-aspc.gc.ca/publicat/ccdr-rmtc/04vol30/dr3015ea.html. Accessed 14 June 2006.

26. **Burman, N. P., and C. W. Oliver.** 1952. A comparative study of Folpmers' glutamic acid medium for the detection of *Bact. coli* in water. *Proc. Soc. Appl. Bacteriol.* **15:**1–7.

27. **Cabelli, V. J., F. T. Brezenski, and D. Pedersen.** 1982. Inaccuracy of the preincubation modified M-FC method for estimating fecal coliform densities in marine waters. *J. Water Pollut. Control Fed.* **54:**1237–1240.

28. **Carrillo, M., E. Estrada, and T. C. Hazen.** 1985. Survival and enumeration of the fecal indicators *Bifidobacterium adolescentis* and *Escherichia coli* in a tropical rain forest watershed. *Appl. Environ. Microbiol.* **50:**468–476.

29. **Clark, C. G., L. Price, R. Ahmed, D. L. Woodward, P. L. Melito, F. G. Rodgers, F. Jamieson, B. Ciebin, A. Li, and A. Ellis.** 2003. Characterization of waterborne outbreak-associated *Campylobacter jejuni*, Walkerton, Ontario. *Emerg. Infect. Dis.* **9:**1232–1241.

30. **Cohen, T.** 8 December 2000. Workers admit altering records in *E. coli* inquiry. *Times Union*, Albany, N.Y.

31. **Crosby, D.** 10 August 2000. Walkerton's water contaminated for 21 years, report suggests: 1979 tests found coliform bacteria in one of town's wells. *National Post* (National edition), Don Mills, Ontario, Canada.

32. **Deneen, V. C., P. A. Belle-Isle, C. M. Taylor, L. L. Gabriel, J. B. Bender, J. H. Wicklund, C. W. Hedberg, and M. T. Osterholm.** 1998. Outbreak of cryptosporidiosis associated with a water sprinkler fountain—Minnesota, 1997. *Morb. Mortal. Wkly. Rep.* **47:**856–860.

33. **Dennis, D. T., R. P. Smith, J. J. Welch, C. G. Chute, B. Anderson, J. L. Herndon, and C. F. von Reyn.** 1993. Endemic giardiasis in New Hampshire: a case-control study of environmental risks. *J. Infect. Dis.* **167:**1391–1395.

34. **Edberg, S. C., M. J. Allen, and D. B. Smith.** 1991. Defined substrate technology method for rapid and specific simultaneous enumeration of total coliforms and *Escherichia coli* from water: collaborative study. *J. Assoc. Off. Anal. Chem.* **74:**526–529.

35. **Efstratiou, M. A., A. Mavridou, S. C. Richardson, and J. A. Papadakis.** 1998. Correlation of bacterial indicator organisms with *Salmonella* spp., *Staphylococcus aureus* and *Candida albicans* in sea water. *Lett. Appl. Microbiol.* **26:**342–346.

36. **Espigares, M., C. Coca, M. Fernández-Crehuet, O. Moreno, and R. Gálvez.** 1996. Chemical and microbiologic indicators of faecal contamination in the Guadalquivir (Spain). *Eur. Water Pollut. Control* **6:**7–13.

37. **Federal Register.** 1996. National primary drinking water regulations: monitoring requirements for public drinking water supplies; Final Rule. *Fed. Regist.* **61:**24353–24388.

38. **Federal Register.** 1998. National primary drinking water regulations: interim enhanced surface water treatment; Final Rule. *Fed. Regist.* **63:**69477–69521.

39. **Federal Register.** 2002. National primary drinking water regulations: long term enhanced surface water treatment rule; Final Rule. *Fed. Regist.* **67:**1811–1844.

40. **Ferguson, C. M., B. G. Coote, N. J. Ashbolt, and I. M. Stevenson.** 1996. Relationships between indicators, pathogens and water quality in an estuarine system. *Water Res.* **30:**2045–2054.

41. **Figueras, M. J., F. Polo, I. Inza, and J. Guarro.** 1994. Poor specificity of m-Endo and m-FC culture media for the enumeration of coliform bacteria in sea water. *Lett. Appl. Microbiol.* **19:**446–450.

42. **Friedman, M. S., T. Roels, J. E. Koehler, L. Feldman, W. F. Bibb, and P. Blake.** 1999. *Escherichia coli* O157:H7 outbreak associated with an improperly chlorinated swimming pool. *Clin. Infect. Dis.* **29:**298–303.

43. **Gillis, C.** 19 October 2000. Water manager had 'chaotic' approach to tests: *E. coli* deaths inquiry: mislabelled samples and argued over laboratory costs. *National Post* (National edition), Don Mills, Ontario, Canada.

44. **Gillis, C.** 27 October 2000. Official admits he let contaminated water results slide: Walkerton inquiry: utility manager's readings were false, lawyer reveals. *National Post* (National edition), Don Mills, Ontario, Canada.

45. **Grabow, W. O. K., V. Gauss-Müller, O. W. Prozesky, and F. Deinhardt.** 1983. Inactivation of hepatitis A virus and indicator organisms in water by free chlorine residuals. *Appl. Environ. Microbiol.* **46:**619–624.

46. **Griffin, D. W., C. J. Gibson III, E. K. Lipp, K. Riley, J. H. Paul III, and J. B. Rose.** 1999. Detection of viral pathogens by reverse transcriptase PCR and of microbial indicators by standard methods in the canals of the Florida Keys. *Appl. Environ. Microbiol.* **65:**4118–4125.

47. **Halls, S., and P. A. Ayers.** 1974. A membrane filtration technique for the enumeration of *Escherichia coli* in seawater. *J. Appl. Bacteriol.* **37:**105–109.

48. **Harries, K.** 21 December 2004. Stan Koebel gets one year in jail, 'I hope you can forgive me sometime'; Brother gets 9 months of house arrest, 'No sentence … can assuage the enormous losses'. *Toronto Star* (Ontario edition), Toronto, Ontario, Canada.

49. **Harwood, V. J., A. D. Levine, T. M. Scott, V. Chivukula, J. Lukasik, S. R. Farrah, and J. B. Rose.** 2005. Validity of the indicator organism paradigm for pathogen reduction in reclaimed water and public health protection. *Appl. Environ. Microbiol.* **71:**3163–3170.

50. **Hassani, L., L. Rafouk, and A. A. Alla.** 1999. Antibiotic resistance among faecal coliform bacteria isolated from wastewater before and after treatment by an experimental sand filter. *World J. Microbiol. Biotechnol.* **15:**277–279.

51. **Health Canada.** 2006. Guidelines for Canadian drinking water quality—summary table. Guidelines for microbiological parameters. Bacteriological guidelines. Health Canada. [Online.] http://www.hc-sc.gc.ca/ewh-semt/pubs/water-eau/doc_sup-appui/sum_guide-res_recom/micro_e.html#1. Accessed 16 June 2006.

52. **Hlavsa, M. C., J. C. Watson, and M. J. Beach.** 2005. Cryptosporidiosis surveillance—United States 1999–2002. *Morb. Mortal. Wkly. Rep.* **54**(SS01): 1–8.

53. **Ho, B. S. W., and T.-Y. Tam.** 1997. Enumeration of *E. coli* in environmental waters and wastewater using a chromogenic medium. *Water Sci. Tech.* **35:** 409–413.

54. **Hörman, A., R. Rimhanen-Finne, L. Maunula, C.-H. von Bonsdorff, N. Torvela, A. Heikinheimo, and M.-L. Hänninen.** 2004. *Campylobacter* spp., *Giardia* spp., *Cryptosporidium* spp., noroviruses, and indicator organisms in surface water in southwestern Finland, 2000–2001. *Appl. Environ. Microbiol.* **70:**87–95.

55. **Hunt, D. A., and J. Springer.** 1978. Comparison of two rapid test procedures with the standard EC test for recovery of fecal coliform bacteria from shellfish-growing waters. *J. Assoc. Off. Anal. Chem.* **61:**1317–1323.

56. **Isaac-Renton, J., W. R. Bowie, A. King, G. S. Irwin, C. S. Ong, C. P. Fung, M. O. Shokeir, and J. P. Dubey.** 1998. Detection of *Toxoplasma gondii* oocysts in drinking water. *Appl. Environ. Microbiol.* **64:**2278–2280.

57. **Jackson, S. G., R. B. Goodbrand, R. P. Johnson, V. G. Odorico, D. Alves, K. Rahn, J. B. Wilson, M. K. Welch, and R. Khakhria.** 1998. *Escherichia coli* O157:H7 diarrhoea associated with well water and infected cattle on an Ontario farm. *Epidemiol. Infect.* **120:**17–20.

58. **Jay, J. M.** 2000. *Modern Food Microbiology*, 6th ed. Aspen Publishers, Inc., Gaithersburg, Md.

59. **Jermini, M., F. Domeniconi, and M. Jäggli.** 1994. Evaluation of C-EC-agar, a modified mFC-agar for the simultaneous enumeration of faecal coliforms and *Escherichia coli* in water samples. *Lett. Appl. Microbiol.* **19:**332–335.

60. **Juck, D., J. Ingram, M. Prévost, J. Coallier, and C. Greer.** 1996. Nested PCR protocol for the rapid detection of *Escherichia coli* in potable water. *Can. J. Microbiol.* **42:**862–866.

61. **Keene, W. E., J. M. McAnulty, F. C. Hoesly, L. P. Williams, Jr., K. Hedberg, G. L. Oxman, T. J. Barrett, M. A. Pfaller, and D. W. Fleming.** 1994. A swimming-associated outbreak of hemorrhagic colitis caused by *Escherichia coli* O157:H7 and *Shigella sonnei*. *N. Engl. J. Med.* **331:**579–584.

62. **Kramer, M. H., F. E. Sorhage, S. T. Goldstein, E. Dalley, S. P. Wahlquist, and B. L. Herwaldt.** 1998. First reported outbreak in the United States of cryptosporidiosis associated with a recreational lake. *Clin. Infect. Dis.* **26:**27–33.

63. **Kukkula, M., L. Maunula, E. Silvennoinen, and C.-H. von Bonsdorff.** 1999. Outbreak of viral gastroenteritis due to drinking water contaminated by Norwalk-like viruses. *J. Infect. Dis.* **180:**1771–1776.

64. **LeChevallier, M. W., W. D. Norton, and R. G. Lee.** 1991. *Giardia* and *Cryptosporidium* spp. in filtered drinking water supplies. *Appl. Environ. Microbiol.* **57:**2617–2621.

65. **Ley, A., S. Barr, D. Fredenburgh, M. Taylor, and N. Walker.** 1993. Use of 5-bromo-4-chloro-3-indolyl-β-D-galactopyranoside for the isolation of β-galactosidase-positive bacteria from municipal water supplies. *Can. J. Microbiol.* **39:**821–825.

66. **Lund, V.** 1996. Evaluation of *E. coli* as an indicator for the presence of *Campylobacter jejuni* and *Yersinia enterocolitica* in chlorinated and untreated oligotrophic lake water. *Water Res.* **30:**1528–1534.

67. **MacKenzie, W. R., N. J. Hoxie, M. E. Proctor, M. S. Gradus, K. A. Blair, D. E. Peterson, J. J. Kazmierczak, D. G. Addiss, K. R. Fox, J. B. Rose, and J. P. Davis.** 1994. A massive outbreak in Milwaukee of *Cryptosporidium* infection transmitted through the public water supply. *N. Engl. J. Med.* **331:**161–167.

68. **MacKenzie, W. R., J. J. Kazmierczak, and J. P. Davis.** 1995. An outbreak of cryptosporidiosis associated with a resort swimming pool. *Epidemiol. Infect.* **115:**545–553.

69. **Mates, A., and M. Shaffer.** 1989. Membrane filtration differentiation of *E. coli* from coliforms in the examination of water. *J. Appl. Bacteriol.* **67:** 343–346.

70. **Mates, A., and M. Shaffer.** 1992. Quantitative determination of *Escherichia coli* from coliforms and faecal coliforms in sea water. *Microbios* **71:**27–32.

71. **McCarthy, T. A., N. L. Barrett, J. L. Hadler, B. Salsbury, R. T. Howard, D. W. Dingman, C. D. Brinkman, W. F. Bibb, and M. L. Cartter.** 2001. Hemolytic-uremic syndrome and *Escherichia coli* O121 at a lake in Connecticut, 1999. *Pediatrics* **108:**59–65.

72. **McFeters, G. A., M. W. LeChevallier, and M. J. Domek.** 1984. Project summary. Injury and the improved recovery of coliform bacteria in drinking water. U.S. Environmental Protection Agency, Water Engineering Research Laboratory, Cincinnati, Ohio.

73. **Medema, G. J., I. A. van Asperen, and A. H. Havelaar.** 1997. Assessment of the exposure of swimmers to microbiological contaminants in fresh waters. *Water Sci. Tech.* **35:**157–163.

74. **Miescier, J. J., V. E. Carr, J. F. Musselman, and S. A. Furfari.** 1978. Fecal coliform methods for examination of sea water: interlaboratory evaluation of split sample analysis. *J. Assoc. Off. Anal. Chem.* **61:**772–778.

75. **Minshew, P., K. Ward, Z. Mulla, R. Hammond, D. Johnson, S. Heber, and R. Hopkins.** 2000. Outbreak of gastroenteritis associated with an interactive water fountain at a beachside park—Florida, 1999. *Morb. Mortal. Wkly. Rep.* **49:**565–568.

76. **Molenda, J. R.** 1992. Cholera, John Snow and the pump handle. *Dairy Food Environ. Sanit.* **12:**12–15.

77. **Nash, J. M.** 19 April 1993. The waterworks flu. A tiny parasite gets the blame for making thousands of Milwaukeeans miserable. *TIME*, New York, N.Y.

78. **O'Connor, D. R.** 2002. *Part One: Report of the Walkerton Inquiry: The Events of May 2000 and Related Issues.* Ontario Ministry of the Attorney General, Toronto, Ontario, Canada. [Online.] http://www.attorneygeneral.jus.gov.on.ca/english/about/pubs/walkerton/part1/. Accessed 14 June 2006.

79. **O'Connor, D. R.** 2002. *Part Two: Report of the Walkerton Inquiry: A Strategy for Safe Drinking Water.* Ontario Ministry of the Attorney General, Toronto, Ontario, Canada. [Online.] http://www.attorneygeneral.jus.gov.on.ca/english/about/pubs/walkerton/part2/. Accessed 14 June 2006.

80. **Olsen, S. J., G. Miller, T. Breuer, M. Kennedy, C. Higgins, J. Walford, G. McKee, K. Fox, W. Bibb, and P. Mead.** 2002. A waterborne outbreak of *Escherichia coli* O157:H7 infections and hemolytic uremic syndrome: implications for rural water systems. *Emerg. Infect. Dis.* **8:**370–375.

81. **Ong, C., W. Moorehead, A. Ross, and J. Isaac-Renton.** 1996. Studies of *Giardia* spp. and *Cryptosporidium* spp. in two adjacent watersheds. *Appl. Environ. Microbiol.* **62:**2798–2805.

82. **Palmer, K.** 19 January 2002. Inquiry leaves students jaded; water scandal teaches Walkerton class a lesson in political games. *Toronto Star* (Ontario edition), Toronto, Ontario, Canada.

83. **Patrick, F. M., and K. Callaghan.** 1982. Comparison of Sartorius nutrient pad sets with standard membrane filtration methods for enumerating total and faecal coliforms in water. *NZ J. Sci.* **25:**377–378.

84. **Perkel, C.** 26 November 2001. Walkerton's water bill; *E. coli* tragedy cost $155 million, study concludes. *Toronto Star* (Ontario edition), Toronto, Ontario, Canada.

85. **Regan, J., R. McVay, M. McEvoy, J. Gilbert, R. Hughes, T. Tougaw, E. Parker, W. Crawford, J. Johnson, J. Rose, S. Boutros, S. Roush, T. Belcuore, C. Rains, J. Munden, L. Stark, E. Hartwig, M. Pawlowicz, R. Hammond, D. Windham, and R. Hopkins.** 1996. Outbreak of cryptosporidiosis at a day camp—Florida, July–August 1995. *Morb. Mortal. Wkly. Rep.* **45:**442–444.

86. **Rice, E. W., M. J. Allen, D. J. Brenner, and S. C. Edberg.** 1991. Assay for β-glucuronidase in species of the genus *Escherichia* and its applications for drinking-water analysis. *Appl. Environ. Microbiol.* **57:**592–593.

87. **Rice, E. W., T. C. Covert, S. A. Johnson, C. H. Johnson, and D. J. Reasoner.** 1995. Detection of *Escherichia coli* in water using a colorimetric gene probe assay. *J. Environ. Sci. Health* **A30:**1059–1067.

88. **Sartory, D. P., and L. Howard.** 1992. A medium detecting β-glucuronidase for the simultaneous membrane filtration enumeration of *Escherichia coli* and coliforms from drinking water. *Lett. Appl. Microbiol.* **15:**273–276.

89. **Snow, J.** 1855. *On the Mode of Communication of Cholera*, 2nd ed. John Churchill, London, England. [Online.] http://www.ph.ucla.edu/epi/snow/snowbook.html. Accessed 14 June 2006.

90. **Sorvillo, F. J., K. Fujioka, B. Nahlen, M. P. Tormey, R. Kebabjian, and L. Mascola.** 1992. Swimming-associated cryptosporidiosis. *Am. J. Public Health* **82:**742–744.

91. **Sorvillo, F. J., K. Fujioka, M. Tormey, R. Kebabjian, W. Tokushige, L. Mascola, S. Schweid, M. Hillario, and S. H. Waterman.** 1990. Swimming-associated cryptosporidiosis—Los Angeles County. *Morb. Mortal. Wkly. Rep.* **39:**343–345.

92. **Spears, T.** 22 November 2003. Tainted water caused kidney damage. *Ottawa Citizen*, Ottawa, Ontario, Canada.

93. **Swerdlow, D. L., B. A. Woodruff, R. C. Brady, P. M. Griffin, S. Tippen, H. D. Donnell, Jr., E. Geldreich, B. J. Payne, A. Meyer, Jr., J. G. Wells, K. D. Greene, M. Bright, N. H. Bean, and P. A. Blake.** 1992. A waterborne outbreak in Missouri of *Escherichia coli* O157:H7 associated with bloody diarrhea and death. *Ann. Intern. Med.* **117:**812–819.

94. **Thomas, H. A., Jr.** 1955. Statistical analysis of coliform data. *Sewage Ind. Wastes* **27:**212 221.

95. **Thurman, R., B. Faulkner, D. Veal, G. Cramer, and M. Meiklejohn.** 1998. Water quality in rural Australia. *J. Appl. Microbiol.* **84:**627–632.

96. **Toranzos, G. A., and G. A. McFeters.** 1997. Detection of indicator microorganisms in environmental freshwaters and drinking waters, p. 184–194. *In* C. J. Hurst, G. R. Knudsen, M. J. McInerney, L. D. Stetzenbach, and M. V. Walter (ed.). *Manual of Environmental Microbiology*. ASM Press, Washington, D.C.

97. **Tortorello, M. L., and K. F. Reineke.** 2000. Direct enumeration of *Escherichia coli* and enteric bacteria in water, beverages and sprouts by 16S rRNA *in situ* hybridization. *Food Microbiol.* **17:**305–313.

98. **Vallis, M.** 19 January 2002. Family struggling with legacy of pain: 'For the rest of his life, I'm going to be scared'. *National Post* (National edition), Don Mills, Ontario, Canada.

99. **Van Houten, R., D. Farberman, J. Norton, J. Ellison, J. Kiehlbauch, T. Morris, and P. Smith.** 1998. *Plesiomonas shigelloides* and *Salmonella* serotype Hartford infections associated with a contaminated water supply—Livingston County, New York, 1996. *Morb. Mortal. Wkly. Rep.* **47:**394–396.

100. **Van Poucke, S. O., and H. J. Nelis.** 2000. A 210-min solid phase cytometry test for the enumeration of *Escherichia coli* in drinking water. *J. Appl. Microbiol.* **89:**390–396.

101. **Verma, S., and K. Donovan.** 5 August 2000. Walkerton was warned in '83; well discovered drawing water from farm ponds. *Toronto Star* (Ontario edition), Toronto, Ontario, Canada.

102. **Wallis, P. M., S. L. Erlandsen, J. L. Isaac-Renton, M. E. Olson, W. J. Robertson, and H. van Keulen.** 1996. Prevalence of *Giardia* cysts and *Cryptosporidium* oocysts and characterization of *Giardia* spp. isolated from drinking water in Canada. *Appl. Environ. Microbiol.* **62:**2789–2797.

103. **Warrner, M., K. Kuo, L. Williams, B. Adam, C. Langkop, R. Ruden, B. Francis, and T. Haupt.** 1996. Lake-associated outbreak of *Escherichia coli* O157:H7—Illinois, 1995. *Morb. Mortal. Wkly. Rep.* **45:**437–439.

104. **Whitman, R. L., M. B. Nevers, and P. J. Gerovac.** 1999. Interaction of ambient conditions and fecal coliform bacteria in southern Lake Michigan beach waters: monitoring program implications. *Natural Areas J.* **19:**166–171.

105. **World Health Organization.** 2004. *Guidelines for Drinking-Water Quality*, 3rd ed. Volume 1, *Recommendations*. World Health Organization, Geneva, Switzerland. [Online.] http://www.who.int/water_sanitation_health/dwq/gdwq3/en/index.html. Accessed 16 June 2006.

chapter 10

Mad Cows and Englishmen

BOVINE SPONGIFORM ENCEPHALOPATHY (BSE), or "mad cow disease," is just one of several syndromes that are known collectively as transmissible spongiform encephalopathies (TSEs). Members of this family of diseases include chronic wasting disease of deer and elk, transmissible mink encephalopathy, Creutzfeldt-Jakob disease (CJD), kuru, Gerstmann-Sträussler-Scheinker syndrome, fatal familial insomnia, Huntington's disease, and other miscellaneous encephalopathies (9, 13, 45, 46, 49, 86). Spongiform encephalopathies are recognized by the characteristic sponge-like appearance of the victim's brain. Scrapie, a disease that affects sheep and goats, was one of the earliest TSEs to be recognized (13).

Scrapie was first described in the mid-1700s, although it was not understood to be a transmissible disease until 1936 (13). Sheep that suffer from the disease scratch and scrape themselves against posts—hence the name—and bite at their legs and feet. They also stop eating, essentially starving themselves to death. In the decades since the French researchers Cuillé and Chelle first demonstrated that the agent causing scrapie could be passed from animal to animal, the disease has been transmitted—naturally or through lab intervention—to sheep, goats, mice, and hamsters (13, 19, 40). And some experts believe that BSE may have started when scrapie jumped from sheep to cows through the practice of feeding to cattle rendered meat and bone meal derived from sheep (13, 39).

Mad Cows

There is no better symbol of the historic importance of sheep and of the wool trade to England than the woolsack on which the Lord Chancellor sits when he presides over the British House of Lords, a practice that began in the 14th century during the reign of King Edward III (54). In 1755, Britain's prime minister alerted the House of Commons to the existence of scrapie in the country's sheep population as an economic threat to the prosperity of the country's wool industry. The incidence of scrapie in herds of British sheep increased during the 18th century and into the early 19th century as inbreeding of the sheep intensified. Although the incidence of the disease gradually declined toward the end of the 19th century along with the practice of inbreeding, scrapie never disappeared completely from the British Isles (13).

Dairy farming and cattle rearing are as much a staple of British agriculture as the raising of sheep, and have been for just as long. But cattle weren't thought to be susceptible to scrapie or to any other TSE (55).

The first hint that something was wrong (although no one knew it at the time) surfaced in 1984 at Pitsham Farm, the Stent family farm in Midhurst, West Sussex, a county in the southeast of England (Table 10.1). A cow began acting strangely shortly before Christmas; she died in February 1985, after developing an unsteady gait and head tremors (80). Other cows in Stent's herd—nine in all—soon developed similar symptoms. Stent and his veterinarian sent the brain from one of the afflicted cows to the local Veterinary Investigation Centre (Vet Centre) for examination by Carol Richardson, a pathologist from the Central Veterinary Laboratory (Central Lab). On examining the brain microscopically, Richardson diagnosed an acute moderate spongiform encephalopathy, suggesting "...a toxicity of some description." It was only in retrospect, after bovine spongiform encephalopathy was recognized as a transmissible disease of cattle in 1986, that veterinary scientists realized Pitsham Farm was the site of the first disease cluster in what was to become a major epidemic (56).

In the spring of 1985, cows at Plurenden Manor Farm in Kent, another county in southeastern England, began to exhibit unusual symptoms: nervousness, aggression, and lack of coordination. In November 1986, after more than a year of trying to treat the symptoms of an increasing number of ill cattle, the veterinarian called on an officer from the local Vet Center office for help. The officer submitted the brains from three of the cows to the Central Lab for diagnosis. Coincidentally, that laboratory received a brain from yet another location around the same time. On December 19, following pathological examination of all four brains, the head of the Central Lab's pathology department reported that "...the lesions observed have similarities to spongiform encephalopathies of other species and in particular scrapie of sheep." The Central Lab, after reviewing archived microscopic brain sections from the Pitsham Farm cow, concluded that Stent's cattle most likely also had been victims of BSE (56).

Once the British government realized that the country's farmers were faced with an epidemic of sick and dying cattle, it moved to find answers to four major questions:

1. What was the source of the outbreak;
2. How was it spread;
3. How could the BSE outbreak be stopped; and
4. What was the risk that the disease could be transmitted to humans?

Table 10.1 Major milestones in the United Kingdom BSE/vCJD outbreak and the government's efforts to contain and control it (5, 77)

Date	Event
November 1984	First cow becomes ill with what is later determined to be BSE.
November 1986	First identification of BSE.
15 December 1987	Initial epidemiological studies point to meat and bone meal as BSE vehicle.
14 June 1988	British government promulgates Bovine Spongiform Encephalopathy Order 1988; Article 7 of the Order—to take effect 18 July 1988—bans feeding of ruminant protein to ruminants.
7 July 1988	British government announces decision to mandate slaughter of all BSE-affected cattle.
18 July 1988	First ruminant feed ban takes effect; ban is scheduled to expire 31 December 1988.
28 November 1988	Northern Ireland makes BSE a notifiable disease.
30 December 1988	British government prolongs the ruminant feed ban (originally set to expire December 31) and prohibits the use of milk from affected or suspected cattle for any purpose other than feeding to the cow's own calf.
11 January 1989	Northern Ireland bans use of animal protein in ruminant feed.
27 February 1989	Southwood report published; Tyrrell Committee on BSE research established.
10 June 1989	Tyrrell Committee submits its report.
13 November 1989	British government bans use of specified bovine offals (SBO) for human consumption in England and Wales.
31 December 1989	Ruminant feed ban made permanent by British government.
30 January 1990	Scotland and Northern Ireland ban use of SBO for human consumption.
3 April 1990	British government announces establishment of permanent Spongiform Encephalopathy Advisory Committee.
24 September 1990	Lab transmission of BSE to a pig is announced.
25 September 1990	Ban on use of SBO is extended to its use in any animal feeds.
15 February 1991	Expert Group on Animal Feedingstuffs (Lamming Committee) holds first meeting.
27 March 1991	First case of BSE in cow born after the ruminant feed ban went into effect.
15 June 1992	Lamming Committee issues its report.
2 November 1994	British government extends ban on the use of SBO in animal feed and bans use of mammalian protein in ruminant feedingstuffs.
15 December 1995	Amended SBO order takes effect. Use of bovine vertebral column is prohibited in the manufacture of all mechanically recovered meat and also in production of some other products for human consumption; use of bovine mechanically recovered meat made from the vertebral column is prohibited in food for humans. All plants producing bovine mechanically recovered meat are required to register with government.
20 March 1996	First cases of vCJD are detected and connection to BSE is suggested.
19 April 1996	Meat and bone meal are prohibited as fertilizer for agricultural land.
10 June 1996	Recall is initiated of all animal feeds containing meat and bone meal.
October 1996	Meat and bone meal animal feed recall is completed.
16 December 1997	Mandatory deboning of all beef from cattle over 6 months at slaughter before it can be sold to consumers (order rescinded December 1999).
12 January 1998	BSE Inquiry initiated under Lord Phillips of Worth Matravers.
26 October 2000	BSE Inquiry Report published.

Determining the likely source and mode of transmission for the BSE outbreak was fundamental to finding a way to stop the spread of the disease. By the end of 1987, epidemiological evidence was pointing to the conclusion that BSE was spread through contaminated meat and bone meal that was used as a cattle-feeding supplement. John Wilesmith, the epidemiologist who headed the Central Veterinary Laboratory's epidemiology department, became convinced that rendered offal of scrapie-infected sheep had introduced the scrapie agent into meat and bone meal (56).

In February 1988, the government established the Southwood Working Party, an expert panel charged with reviewing the BSE situation and recommending appropriate actions. In its June 1988 interim report, the Working Party recommended compulsory notification and slaughter of all cattle showing symptoms of BSE and also urged that a committee be established to advise on BSE research activities. At about the same time that the interim report was issued, the government introduced a ruminant feed ban—a prohibition on feeding rendered ruminant offal to ruminants—in an attempt to halt the spread of the disease throughout the country's cattle herds. That ban went into effect on July 18, 1988. The following month, the British government instituted compulsory slaughter of BSE-affected cattle (56).

The Tyrrell Committee, established in February 1989 in response to one of the recommendations contained in the Working Party report, identified key areas of research activity that needed to be addressed and set priorities for BSE research. In April 1990, the Spongiform Encephalopathy Advisory Committee, a permanent committee also chaired by David Tyrrell, replaced the Tyrrell Committee. Its role was to advise the government, specifically, the Ministry of Agriculture, Fisheries and Food, on matters relating to spongiform encephalopathies.

The government hoped that the ruminant feed ban would first contain and then quash the BSE epidemic. But at the time the ban was announced, no test had yet been developed that could detect the presence of ruminant protein in animal feed (38, 47, 51). In the absence of a diagnostic test, the ban was not enforceable; it had to operate on the honor system.

Despite the ruminant feed ban, the incidence of BSE continued to increase (Fig. 10.1). By the end of 1990, some weeks saw more than 400 new cases. In March 1991, the British government learned that BSE was being found in cattle that were born after the ruminant feed ban had gone into effect. By September 1992, more than 200 cases of BSE had been confirmed in cattle born after the ban. One year later, that number had surpassed 4,000, and by the end of June 2000 there were more than 41,000 confirmed cases of BSE in cows born after the ruminant feed ban went into effect (56).

The July 1988 ruminant feed ban contained a major loophole. Ruminant protein was banned from animal feed destined for feeding ruminants—mainly cattle and sheep—but it was permitted in animal feeds for nonruminant farm animals, such as pigs and poultry. No one knew at the time that the BSE prion infective dose was small; just 1 gram was enough to transmit BSE orally to a cow (56). Inadvertent mixing and cross-contamination of animal feeds destined for ruminants with poultry or pig feed together with the small infectious dose opened the door for the epidemic of BSE in cattle to continue despite the ruminant feed ban. In consequence, while the number of new cases of BSE in cattle peaked in 1992, it was only when the use of any mammalian protein was banned from all animal feed in 1996 that the British finally locked the barn door on BSE (55). By then, however, BSE prions had found a new, nonruminant target—*Homo sapiens* (56).

The possible risk of BSE transmission to humans had been raised early on by British officials, and the Southwood Working Party was asked to consider this possibility as part of its mandate. The Working Party concluded in 1989 that the risk of BSE transmission to humans was remote. They assumed, perhaps not unreasonably, that BSE would behave like scrapie—humans were not susceptible to scrapie (56). During the following 6 years, however, evidence began to accumulate that BSE could jump from cattle to other nonruminant species, notably cats and other felines (56). Increasingly, scientists and other medical experts voiced their concerns as the data supporting cross-species infection mounted (25, 28, 78). Nevertheless, British government spokespeople at all levels—civil servants, the minister of agriculture, the health secretary, and even Prime Minister John Major—continued to insist that British beef was safe to eat (21, 28). A scientist with the U.S. Department of Agriculture (USDA), observing that there was no proof of BSE transmission to humans, came to a similar conclusion in a 1996 report (48).

Unfortunately, both the British government officials and the USDA scientist were wrong. As a precaution, in 1990, the United Kingdom had instituted a surveillance program for CJD, the human counterpart to BSE (53). On March 20, 1996, the British government announced that 10 people under 40 years of age (including two adolescents) had contracted what appeared to be a new variant of CJD (vCJD), possibly as a result of having eaten prion-contaminated beef (55, 85). In just a few years, researchers confirmed that BSE could be transmitted orally to primates and that the same prion strain caused both BSE and vCJD (10, 23, 24, 34, 57, 69, 84). By 2002, despite residual skepticism in some quarters (77, 79), researchers also concluded that vCJD had not been always lurking unrecognized in the background but was a brand new disease, one that made its debut in humans in 1994–1995 (35, 83).

In time, it became clear that vCJD differed in many ways from the known forms of CJD (sporadic, inherited, and iatrogenic CJD). Variant CJD targeted a younger population (median age of 26 years and mean age of 29 years versus a mean age of 66 years for sporadic CJD) and had a longer duration (14 months versus 4 months) than sporadic CJD (72). The presentation and pathology of vCJD were also different from that of the other forms; patients with vCJD were much more likely than patients with sporadic CJD to exhibit psychiatric symptoms, and the brains of vCJD victims looked different under the microscope (37).

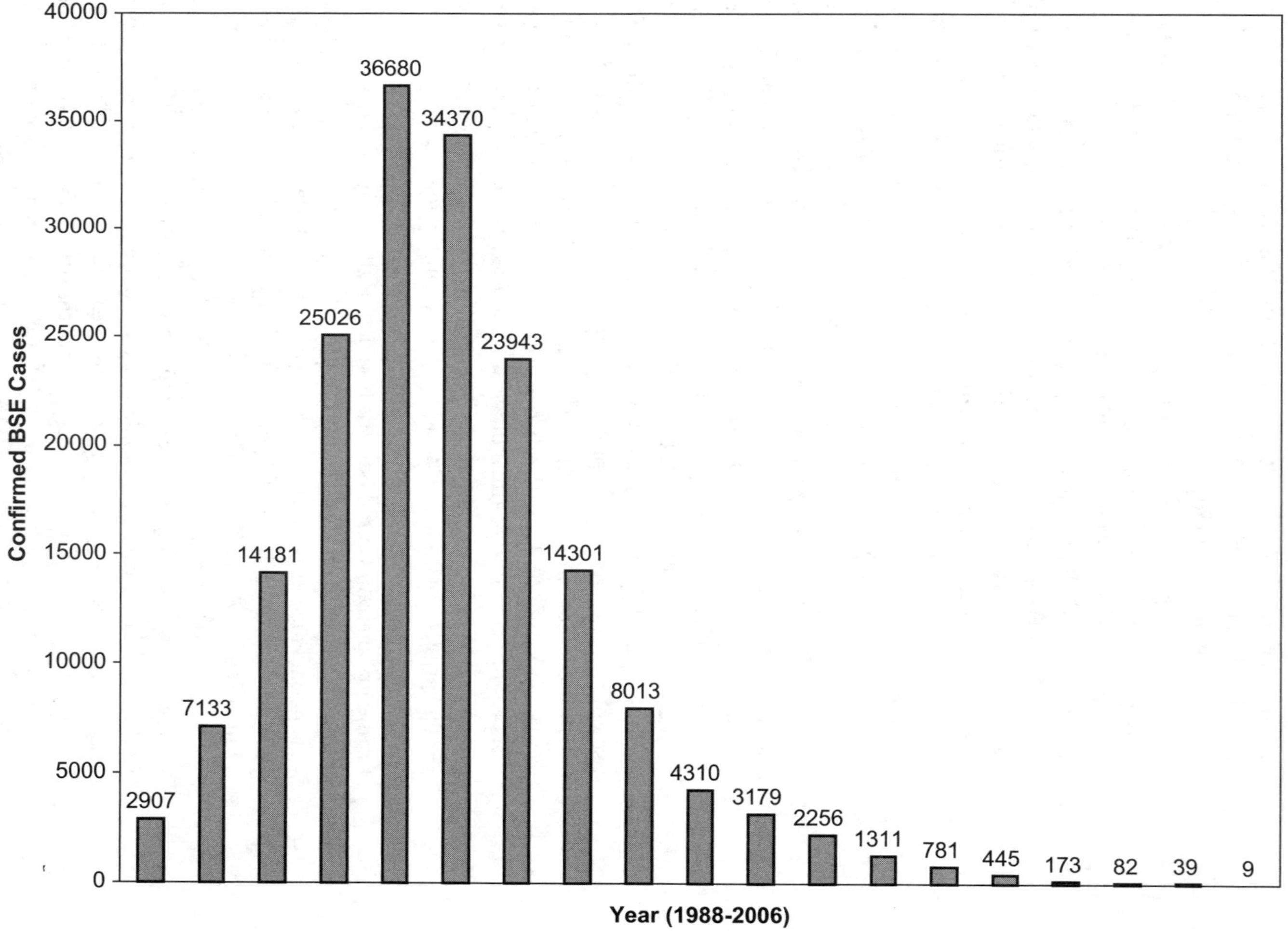

Figure 10.1 Annual incidence of confirmed BSE cases in the United Kingdom, 1988–June 2006 (6). Data for 1988 include accumulated data from earlier years. Data from 2006 are current to 9 June 2006.

Transmissible Spongiform Encephalopathies—Moving Beyond Cattle

It wasn't long before BSE displayed its talent for infecting other ruminant and nonruminant hosts. The first case of feline spongiform encephalopathy was reported in May 1990, just 1 month after the Spongiform Encephalopathy Advisory Committee was formed. Other cases soon followed, more than 80 in all. The cats most likely became infected through being fed commercial cat food that contained bovine meat and bone meal. BSE also infected exotic zoo ruminants such as nyala, elands, bison, and gemsbok. And some wildcats housed in British zoos—ocelots, pumas, cheetahs, and a tiger—developed a disease that appeared indistinguishable from BSE. These animals were fed raw meat as their regular diet. This meat would have contained bovine central nervous tissue, a prime vector for transmitting BSE. On the other hand, dogs, which were also often fed raw meat, were resistant to the disease agent's onslaught (55).

Transmission of the disease from cattle to humans presented the government and the public with the frightening prospect of a large-scale epidemic of an incurable disease, one that slowly robbed its victims, first of full control over their minds, then of control over their bodies (72). Initially, epidemiologists and clinicians predicted that the impact of the disease would be huge, with as many as 136,000 cases of vCJD expected worldwide by 2040 (33). Then, as time went on and it became evident that the susceptible population was much smaller than originally feared, experts revised their predictions downward. By 2003, the "best estimates" had been revised to between 40 and 100 future cases of vCJD—more than a thousandfold reduction from the original projections (32, 75). So far, the lower predictions have been borne out by the data. As of January 2006, a total of 159 cases of vCJD had been diagnosed in the United Kingdom since the start of the outbreak; only six victims were alive at the time of the report (4).

Though the relatively small size of the human outbreak is occasion for large sighs of relief, it cannot—and must not—be taken for an "all clear" signal. We still don't know how the BSE outbreak began in cattle, nor are we completely certain of how it was transmitted to humans. And there is still no effective treatment or cure for the disease, either in cattle or in humans.

Rendering unto Ruminants

The practice in the United Kingdom of feeding rendered animal protein to cattle and sheep was given a major boost during World War II as a result of wartime restrictions imposed on the agricultural industry. The government ordered that animal feed must contain at least 5% by weight of animal-derived protein feed such as meat and bone meal, an order that remained in effect until 1953, when all wartime restrictions were finally lifted. The increased demand for rendered animal protein stimulated growth and consolidation of the rendering industry, which continued to

Prusiner's Prions

Scientific pioneers usually are given a hard time, even when they're right. Galileo was excommunicated by the Church for daring to suggest that the Earth revolved around the sun. People scoffed at the idea that the Earth was round, until Columbus proved them wrong. Pasteur's elegant experiments disproving the theory of spontaneous generation were met with skepticism on the part of many experts (27). And even though Prusiner was awarded the Nobel Prize in 1997 for his discovery of prions (3), his theory that prions are the cause of several neurological diseases—including "mad cow disease"—still is not universally accepted.

Stanley Prusiner earned his medical degree in 1968 from the University of Pennsylvania. He moved to California for his residency in neurology and, in the fall of 1972, lost a patient to CJD (3). As he explained in his Nobel lecture, Prusiner was "...most impressed by a disease process that could kill my patient in 2 months by destroying her brain while her body remained unaffected by this process." He had difficulty believing that CJD was due to an infection with a "slow virus" (the hypothesis of the day)—the lack of fever or immune response seemed to be inconsistent with this notion—and became determined to find the correct explanation (59).

Prusiner began his quest by working with scrapie. In 1982, after 10 years of painstaking lab work—first purifying and then characterizing the scrapie infectious agent—he published his controversial theory (8, 61, 62, 64–66). According to Prusiner, neither slow viruses nor virions nor bacteria were the pathogens behind TSEs; scrapie and other TSEs were caused by prions—proteinaceous infectious particles lacking any nucleic acid component (26, 58).

An infectious protein particle with no nucleic acid component was a difficult hypothesis for many scientists to accept. In the years following the publication of Prusiner's prion theory, researchers continued to look for alternative explanations. Manuelidis proposed a retrovirus (50, 70), whereas Aiken and Marsh pursued host mitochondrial DNA (1, 2). The reluctance of these and other researchers to embrace the prion as an infectious agent was partly due to the normal presence of nearly identical proteins in the brains of animals (18). The only apparent difference between the normal, noninfectious protein (referred to by Prusiner and others as PrP^{C}) and the infectious prion (PrP^{Sc}) was the way in which the protein molecule was folded (59, 73). Prusiner and his colleagues believed that when PrP^{Sc} (misfolded) prion proteins were introduced into a mouse brain and encountered the normal Prp^{C} protein, the misfolded prion caused the normal protein molecule to refold into the PrP^{Sc} form (11, 74). If they were right, a mouse that produced no PrP^{C} protein at all would be resistant to TSE.

Prusiner's team decided to test this hypothesis. They engineered a genetic line of mice that lacked the gene controlling the production of PrP^{C}, the normal protein. Then they tried to transmit scrapie to the genetically altered mice and to normal (control) mice. In 1993, Prusiner's group in San Francisco and a team of Swiss researchers reported a breakthrough. Both groups of researchers, working independently, demonstrated that when mice were genetically engineered to remove the gene that controlled production of PrP^{C}, those mice were completely resistant to infection with TSE prions (17, 63). The Nobel Prize committee recognized the importance of this finding and awarded Prusiner the 1997 Nobel Prize for medicine for his discovery of the prion and its significance (3).

Even so, skeptics of the prion theory continued to propound their alternative explanations for the cause of TSEs. Ebringer proposed that both sporadic CJD and multiple sclerosis might be triggered by infection with *Acinetobacter* (29, 30). Broxmeyer also suggested that mad cow had a bacterial origin, opining that BSE, CJD, and Alzheimer's disease were all caused by *Mycobacterium bovis* (15, 16). Manuelidis and her coworkers insisted that the data supported the existence of a virus, stating in a 2004 research article that "... over the last 20 years no form of PrP itself has ever reproduced infection in normal animals" (7, 41–43). And in the absence of a clear demonstration that the prion theory fulfilled Koch's postulates, it is easy to understand the scientists' deeply held skepticism.

Robert Koch, a founding father of bacteriology, evolved a series of tests, which he used to

Prusiner's Prions *(continued)*

determine whether a particular microbe was the cause of a specific disease. These tests became known as "Koch's postulates" and have guided medical microbiologists in their study of infectious microbes for more than 100 years (Koch never published his postulates; he left that task to one of his disciples, Friederich Loeffler). While Koch's postulates were an invaluable tool in proving the etiology of bacterial diseases, viral pathogens presented more of a problem, since they don't reproduce in the absence of host cells. And applying the postulates to prions seemed next to impossible, until a group of researchers found a way to produce infectious prion particles in a test tube.

Claudio Soto and his colleagues at the University of Texas realized that until Koch's postulates were met, there would always be the shadow of a doubt cast over the prion theory (71). They set out to reproduce purified prion protein in vitro, using a technique called cyclic amplification. Once they succeeded in producing a high enough concentration of the infectious protein particles, they inoculated the prions into hamsters and succeeded where Manuelidis had failed; they reproduced a scrapie disease in the hamsters that was identical in symptoms and appearance to the disease produced by inoculating hamsters directly with infectious material obtained from hamster brains (20). This report, which went a long way toward fulfilling Koch's postulates, has provided the strongest evidence to date that TSEs are caused by a naked protein molecule—Prusiner's prion.

find a healthy market for its meat and bone meal even after the wartime restriction had been lifted (56).

Improvements to the meat and bone meal production process were introduced in the 1960s; a continuous rendering process gradually replaced the less efficient batch process, and in 1970 a solvent-extraction step was eliminated. The introduction of large-scale continuous rendering, in turn, spurred industry consolidation and bigger became better. Large rendering operations needed to travel farther afield to obtain enough raw materials—by-products of the sheep and cattle slaughterhouses—to feed their production lines. And renderers began to sell their meat and bone meal through a distribution network that disseminated the product throughout the country instead of just locally, unknowingly setting the stage for the nationwide spread of BSE (56).

Epidemiologists at the British government's Central Veterinary Laboratory were quick to identify animal feeds containing ruminant meat and bone meal as the probable vehicle for the BSE outbreak, a conclusion that was confirmed by the results of a subsequent case-control study (81, 82). In June 1988, the government promulgated Bovine Spongiform Encephalopathy Order 1988, part of which included a temporary ban on feeding ruminant protein to ruminants—the ruminant feed ban. This temporary ban, which was extended over the years to exclude the feeding of any mammalian protein to animals, has been credited with bringing the United Kingdom's BSE outbreak largely under control (56). But how did meat and bone meal become the vehicle for this outbreak to begin with?

Koch's Postulates, as described by Loeffler in 1883 and reproduced by Brock in his 1999 biography of Robert Koch (12)

"1. The organism must be shown to be constantly present in characteristic form and arrangement in the diseased tissue.

"2. The organism which, from its behavior appears to be responsible for the disease, must be isolated and grown in pure culture.

"3. The pure culture must be shown to induce the disease experimentally."

Some of the changes to meat and bone meal processing—elimination of solvent extraction and introduction of lower-temperature processing—coincided roughly with the time period when BSE was thought to have originated. It was logical for investigators to consider the possibility that one or more of these changes were responsible for allowing the BSE agent to survive in the meal. But the timing of the change from batch to continuous processing and the accompanying reduction in processing temperature were inconsistent with the onset of the outbreak (82). And on further study, the solvent extraction step proved to be ineffective in inactivating the BSE agent (56).

How Did BSE Begin?

Several theories have been proposed to explain the origin of BSE. The disease might have jumped to cattle from scrapie-infected sheep (13, 14); a chemical contaminant (such as a pesticide) might have caused normal brain prion protein to refold into an infectious prion (56, 67, 68); the disease might have always existed in cattle at a low level and gone undetected by animal health experts until a change in cattle-rearing or rendering practices caused them to take notice (36, 44); the first case of BSE might have been generated by a spontaneous mutation (56); or BSE might be an autoimmune disease triggered by a bacterial infection, akin to multiple sclerosis in cattle (29, 30).

Some of these hypotheses were discounted quickly. The British Central Veterinary Laboratory carried out an epidemiological study of BSE in 1987, soon after the outbreak was detected. They concluded that the disease was not inherited, nor was it imported into the United Kingdom from abroad. And the epidemiological evidence contradicted the theory that chemical contaminants were the trigger. There was no association between the use of therapeutic agents (such as antibiotics) or agricultural chemicals and the development of BSE (82).

The autoimmune hypothesis was based largely on the presence of elevated levels of antibodies to *Acinetobacter* spp., a common soil bacterium, in patients with multiple sclerosis and in cows affected with BSE (29, 30, 76). But this theory failed to answer several questions: why the disease suddenly appeared, why the epidemic was limited to the United Kingdom and to countries that imported cattle from the United Kingdom, and why the epidemic waned once controls were placed on the use of animal protein in animal feed (56).

Could BSE have begun as sporadic, spontaneous mutations—either somatic or chromosomal—in individual cows, lurking unbeknownst to farmers, veterinarians, or government regulators until a change in animal husbandry procedures triggered an outbreak? There is precedent for this hypothesis in sporadic and familial forms of CJD (9). Familial CJD arises from autosomal mutations, which are passed down to subsequent generations. Sporadic CJD is thought by many to arise as a result of spontaneous somatic mutations in an individual brain cell, giving rise to production of a small quantity of abnormally folded prion protein, PrP^{Sc}. The abnormal protein then serves as a template for the refolding of normal PrP^{C} protein molecules with which the PrP^{Sc} protein comes into contact (9).

The suggestion that BSE derived from scrapie was based on the chronic presence—for more than 2½ centuries—of scrapie-infected sheep in the United Kingdom and the use of rendered sheep by-products in cattle feed (13, 14). The theory appeared plausible when it was first proposed; there is 98% agreement between the protein sequences of scrapie and BSE prions (60). But when 10 cows were inoculated with brain tissue homogenate from scrapie-infected sheep (5 cows) or goats (5 cows), only 3 of the 10 cows became ill. Furthermore, the symptoms they developed and the microscopic appearance of their brain tissue differed from those of BSE (22). On the other hand, some observations argue in favor of this theory. The disease appears in two different forms in sheep: Type I (the classic scrapie syndrome) and the much less common Type II, or trembling syndrome. The symptoms and neuropathology of Type II scrapie closely resemble those of BSE, CJD, and kuru (52). And when BSE prions were introduced into sheep, the infected sheep exhibited symptoms similar to those shown by BSE-affected cattle (31, 52).

Lord Phillips, who chaired the BSE Inquiry initiated by the British government after the first cases of vCJD were diagnosed, favored the chance mutation theory (56). In the Inquiry Report, issued in 2000, he said "...BSE probably originated from a novel source early in the 1970s, possibly a cow or other animal that developed disease as a consequence of a gene mutation. The origin of the disease will probably never be known with certainty."

Lord Phillips' statement is probably correct. We shall likely never know for sure when, how, or where BSE was born. But, like a meteor flashing across the sky, we can follow its path as it spreads to other countries around the world and trace its impact on the international cattle and beef industry and on world trade.

References

1. **Aiken, J. M., J. L. Williamson, L. M. Borchardt, and R. F. Marsh.** 1990. Presence of mitochondrial D-loop DNA in scrapie-infected brain preparations enriched for the prion protein. *J. Virol.* **64:**3265–3268.
2. **Aiken, J. M., J. L. Williamson, and R. F. Marsh.** 1989. Evidence of mitochondrial involvement in scrapie infection. *J. Virol.* **63:**1686–1694.
3. **Altman, L. K.** 7 October 1997. U.S. scientist wins Nobel for controversial work. *New York Times*, New York, N.Y.
4. **Andrews, N. J.** 19 January 2006. Incidence of variant Creutzfeldt-Jakob disease onsets and deaths in the UK. January 1994–December 2005. The National Creutzfeldt-Jakob Disease Surveillance Unit, University of Edinburgh, Edinburgh, United Kingdom. [Online.] http://www.cjd.ed.ac.uk/vcjdqdec05.htm. Accessed 19 June 2006.
5. **Anonymous.** 2004. Bovine spongiform encephalopathy. Chronology of events (as at 20 July 2004). Department for Environment, Food and Rural Affairs, United Kingdom. [Online.] http://www.defra.gov.uk/animalh/bse/directory.html. Accessed 19 June 2006.
6. **Anonymous.** 16 June 2006. BSE: Statistics—BSE—GB weekly cumulative statistics. Department for Environment, Food and Rural Affairs, United Kingdom. [Online.] http://www.defra.gov.uk/animalh/bse/statistics/weeklystats.html#pass. Accessed 19 June 2006.
7. **Arjona, A., L. Simarro, F. Islinger, N. Nishida, and L. Manuelidis.** 2004. Two Creutzfeldt-Jakob disease agents reproduce prion protein-independent identities in cell cultures. *Proc. Natl. Acad. Sci. USA* **101:**8768–8773.
8. **Baringer, J. R., and S. B. Prusiner.** 1978. Experimental scrapie in mice: ultrastructural observations. *Ann. Neurol.* **4:**205–211.
9. **Belay, E. D.** 1999. Transmissible spongiform encephalopathies in humans. *Annu. Rev. Microbiol.* **53:**283–314.
10. **Bons, N., N. Mestre-Frances, P. Belli, F. O. Cathala, D. C. Gajdusek, and P. Brown.** 1999. Natural and experimental oral infection of nonhuman primates by bovine spongiform encephalopathy agents. *Proc. Natl. Acad. Sci. USA* **96:**4046–4051.
11. **Borchelt, D. R., M. Scott, A. Taraboulos, N. Stahl, and S. B. Prusiner.** 1990. Scrapie and cellular prion proteins differ in their kinetics of synthesis and topology in cultured cells. *J. Cell Biol.* **110:**743–752.
12. **Brock, T. D.** 1999. *Robert Koch. A Life in Medicine and Bacteriology*. ASM Press, Washington, D.C.
13. **Brown, P., and R. Bradley.** 1998. 1755 and all that: a historical primer of transmissible spongiform encephalopathy. *BMJ* **317:**1688–1692.
14. **Brown, P., R. G. Will, R. Bradley, D. M. Asher, and L. Detwiler.** 2001. Bovine spongiform encephalopathy and variant Creutzfeldt-Jakob disease: background, evolution, and current concerns. *Emerg. Infect. Dis.* **7:**6–16.
15. **Broxmeyer, L.** 2004. Is mad cow disease caused by a bacteria? *Med. Hypotheses* **63:**731–739.
16. **Broxmeyer, L.** 2005. Thinking the unthinkable: Alzheimer's, Creutzfeldt-Jakob and mad cow disease: the age-related reemergence of virulent, foodborne,

bovine tuberculosis or losing your mind for the sake of a shake or burger. *Med. Hypotheses* **64:**699–705.

17. **Bueler, H., A. Aguzzi, A. Sailer, R. A. Greiner, P. Autenried, M. Aguet, and C. Weissmann.** 1993. Mice devoid of PrP are resistant to scrapie. *Cell* **73:**1339–1347.
18. **Carlson, G. A., D. Westaway, S. J. DeArmond, M. Peterson-Torchia, and S. B. Prusiner.** 1989. Primary structure of prion protein may modify scrapie isolate properties. *Proc. Natl. Acad. Sci. USA* **86:**7475–7479.
19. **Carp, R. I., G. S. Merz, and P. C. Licursi.** 1976. Scrapie in vitro: agent replication and reduced cell yield. *Infect. Immun.* **14:**163–167.
20. **Castilla, J., P. Saá, C. Hetz, and C. Soto.** 2005. *In vitro* generation of infectious scrapie prions. *Cell* **121:**195–206.
21. **Chaudhary, V.** 8 December 1995. Major stops short of endorsing 'safe' beef. 'There is currently no evidence that BSE can be transmitted to humans.' *The Guardian*, Manchester, United Kingdom.
22. **Clark, W. W., J. L. Hourrigan, and W. J. Hadlow.** 1995. Encephalopathy in cattle experimentally infected with the scrapie agent. *Am. J. Vet. Res.* **56:** 606–612.
23. **Collinge, J.** 1997. Human prion diseases and bovine spongiform encephalopathy (BSE). *Hum. Mol. Gen.* **6:**1699–1705.
24. **Collinge, J., K. C. Sidle, J. Meads, J. Ironside, and A. F. Hill.** 1996. Molecular analysis of prion strain variation and the aetiology of 'new variant' CJD. *Nature* **383:**685–690.
25. **Dealler, S., and R. Lacey.** 1991. Beef and bovine spongiform encephalopathy: the risk persists. *Nutr. Health* **7:**117–133.
26. **Diener, T. O., M. P. McKinley, and S. B. Prusiner.** 1982. Viroids and prions. *Proc. Natl. Acad. Sci. USA* **79:**5220–5224.
27. **Dubos, R. J.** 1998. T. D. Brock (ed.). *Pasteur and Modern Science*. ASM Press, Washington, D.C.
28. **Durham, M.** 10 December 1995. Public health. Safe to eat? We'll be the judge of that. Nobody believes the Government when it says we have nothing to fear from beef. A food watchdog may help. *The Guardian*, Manchester, United Kingdom.
29. **Ebringer, A., T. Rashid, and C. Wilson.** 2004. Bovine spongiform encephalopathy as an autoimmune disease evoked by *Acinetobacter*: implications for multiple sclerosis and Creutzfeldt-Jakob disease, p. 383–394. *In* Y. Schoenfeld and N. R. Rose (ed.), *Infection and Autoimmunity*, Elsevier B.V., Amsterdam, The Netherlands.
30. **Ebringer, A., T. Rashid, C. Wilson, R. Boden, and E. Thompson.** 2005. A possible link between multiple sclerosis and Creutzfeldt-Jakob disease based on clinical, genetic, pathological and immunological evidence involving *Acinetobacter* bacteria. *Med. Hypotheses* **64:**487–494.
31. **Foster, J. D., D. W. Parnham, N. Hunter, and M. Bruce.** 2001. Distribution of the prion protein in sheep terminally affected with BSE following experimental oral transmission. *J. Gen. Virol.* **82:**2319–2326.
32. **Ghani, A. C., C. A. Donnelly, N. M. Ferguson, and R. M. Anderson.** 2003. Updated projections of future vCJD deaths in the UK. *BMC Infect. Dis.*

3:4. [Online.] http://www.biomedcentral.com/1471-2334/3/4. Accessed 20 June 2006.

33. **Ghani, A. C., N. M. Ferguson, C. A. Donnelly, and R. M. Anderson.** 2000. Predicted vCJD mortality in Great Britain. *Nature* **406:**583–584.

34. **Hill, A. F., M. Desbruslais, S. Joiner, K. C. L. Sidle, I. Gowland, J. Collinge, L. J. Doey, and P. Lantos.** 1997. The same prion strain causes vCJD and BSE. *Nature* **389:**448–450.

35. **Hillier, C. E. M., R. L. Salmon, J. W. Neal, and D. A. Hilton.** 2002. Possible underascertainment of variant Creutzfeldt-Jakob disease: a systematic study. *J. Neurol. Neurosurg. Psychiatry* **72:**304–309.

36. **Institute of Food Science & Technology.** 1996. BSE: a review of the scientific evidence. *Br. Food J.* **98:**4–16.

37. **Knight, R. S. G., and R. G. Will.** 2004. Prion diseases. *J. Neurol. Neurosurg. Psychiatry* **75:**i36–i42.

38. **Krcmár, P., and E. Rencová.** 2001. Identification of bovine-specific DNA in feedstuffs. *J. Food Prot.* **64:**117–119.

39. **MacKnight, C.** 2001. Clinical implications of bovine spongiform encephalopathy. *Clin. Infect. Dis.* **32:**1726–1731.

40. **Malone, T. G., R. F. Marsh, R. P. Hanson, and J. S. Semancik.** 1978. Membrane-free scrapie activity. *J. Virol.* **25:**933–935.

41. **Manuelidis, L.** 1998. Vaccination with an attenuated Creutzfeldt-Jakob disease strain prevents expression of a virulent agent. *Proc. Natl. Acad. Sci. USA* **95:**2520–2525.

42. **Manuelidis, L., and Z. Y. Lu.** 2003. Virus-like interference in the latency and prevention of Creutzfeldt-Jakob disease. *Proc. Natl. Acad. Sci. USA* **100:** 5360–5365.

43. **Manuelidis, E. E., and L. Manuelidis.** 1993. A transmissible Creutzfeldt-Jakob disease-like agent is prevalent in the human population. *Proc. Natl. Acad. Sci. USA* **90:**7724–7728.

44. **Marsh, R. F.** 1993. Bovine spongiform encephalopathy: a new disease of cattle? *Arch. Virol. Suppl.* **7:**255–259.

45. **Marsh, R. F., J. M. Miller, and R. P. Hanson.** 1973. Transmissible mink encephalopathy: studies on the peripheral lymphocyte. *Infect. Immun.* **7:** 352–355.

46. **Miller, M. W., E. S. Williams, C. W. McCarty, T. R. Spraker, T. J. Kreeger, C. T. Larsen, and E. T. Thorne.** 2000. Epizootiology of chronic wasting disease in freeranging cervids in Colorado and Wyoming. *J. Wildl. Dis.* **36:**676–690.

47. **Momcilovic, D., and A. Rasooly.** 2000. Detection and analysis of animal materials in food and feed. *J. Food Prot.* **63:**1602–1609.

48. **Moon, H. W.** 1996. Bovine spongiform encephalopathy: hypothetical risk of emergence as a zoonotic foodborne epidemic. *J. Food Prot.* **59:**1106–1111.

49. **Moore, R. C., F. Xiang, J. Monaghan, D. Han, Z. Zhang, L. Edström, M. Anvret, and S. B. Prusiner.** 2001. Huntington disease phenocopy is a familial prion disease. *Am. J. Hum. Genet.* **69:**1385–1388.

50. **Murdoch, G. H., T. Sklaviadis, E. E. Manuelidis, and L. Manuelidis.** 1990. Potential retroviral RNAs in Creutzfeldt-Jakob disease. *J. Virol.* **64:**1477–1486.

51. **Myers, M. J., S. L. Friedman, D. E. Farrell, D. A. Dove-Pettit, M. F. Bucker, S. Kelly, S. Madzo, W. Campbell, R.-F. Wang, D. Paine, and C. E. Cerniglia.** 2001. Validation of a polymerase chain reaction method for the detection of rendered bovine-derived materials in feedstuffs. *J. Food Prot.* **64:**564–566.

52. **Narang, H. K.** 2001. Lingering doubts about spongiform encephalopathy and Creutzfeldt-Jakob disease. *Exp. Biol. Med.* **226:**640–652.

53. **National CJD Surveillance Unit.** 2004. Thirteenth annual report. 2004. Creutzfeldt-Jakob disease surveillance in the UK. The National CJD Surveillance Unit, Western General Hospital, Edinburgh, United Kingdom. [Online.] http://www.cjd.ed.ac.uk/report13.htm. Accessed 20 June 2006.

54. **Parliamentary Education Unit.** 1999. Explore Parliament. The UK Parliament's education website. [Online.] http://www.explore.parliament.uk. Accessed 20 June 2006.

55. **Pattison, J.** 1998. The emergence of bovine spongiform encephalopathy and related diseases. *Emerg. Infect. Dis.* **4:**390–394.

56. **Phillips, L., J. Bridgeman, and M. Ferguson-Smith.** 2000. The BSE Inquiry—16 volumes. *Inquiry into the Emergence and Identification of Bovine Spongiform Encephalopathy (BSE) and Variant Creutzfeldt-Jakob Disease (vCJD) and the Action Taken in Response to It up to 20 March 1996.* Stationery Office, London, England, United Kingdom. [Online.] http://www.bseinquiry.gov.uk/report/index.htm. Accessed 19 June 2006.

57. **Prinz, M., G. Huber, A. J. S. Macpherson, F. L. Heppner, M. Glatzel, H.-P. Eugster, N. Wagner, and A. Aguzzi.** 2003. Oral prion infection requires normal numbers of Peyer's patches but not of enteric lymphocytes. *Am. J. Pathol.* **162:**1103–1111.

58. **Prusiner, S. B.** 1982. Novel proteinaceous infectious particles cause scrapie. *Science* **216:**136–144.

59. **Prusiner, S. B.** 1998. Prions. *Proc. Natl. Acad. Sci. USA* **95:**13363–13383.

60. **Prusiner, S. B., M. Fuzi, M. Scott, D. Serban, H. Serban, A. Taraboulos, J. M. Gabriel, G. A. Wells, J. W. Wilesmith, R. Bradley, et al.** 1993. Immunologic and molecular biologic studies of prion proteins in bovine spongiform encephalopathy. *J. Infect. Dis.* **167:**602–613.

61. **Prusiner, S. B., D. F. Groth, C. Bildstein, F. R. Masiarz, M. P. McKinley, and S. P Cochran.** 1980. Electrophoretic properties of the scrapie agent in agarose gels. *Proc. Natl. Acad. Sci. USA* **77:**2984–2988.

62. **Prusiner, S. B., D. F. Groth, M. P. McKinley, S. P. Cochran, K. A. Bowman, and K. C. Kasper.** 1981. Thiocyanate and hydroxyl ions inactivate the scrapie agent. *Proc. Natl. Acad. Sci. USA* **78:**4606–4610.

63. **Prusiner, S. B., D. Groth, A. Serban, R. Koehler, D. Foster, M. Torchia, D. Burton, S.-L. Yang, and S. J. DeArmond.** 1993. Ablation of the prion protein (PrP) gene in mice prevents scrapie and facilitates production of anti-PrP antibodies. *Proc. Natl. Acad. Sci. USA* **90:**10608–10612.

64. **Prusiner, S. B., W. J. Hadlow, C. M. Eklund, and R. E. Race.** 1977. Sedimentation properties of the scrapie agent. *Proc. Natl. Acad. Sci. USA* **74:**4656–4660.

65. **Prusiner, S. B., W. J. Hadlow, C. M. Eklund, R. E. Race, and S. P. Cochran.** 1978. Sedimentation characteristics of the scrapie agent from murine spleen and brain. *Biochemistry* **17:**4987–4992.
66. **Prusiner, S. B., M. P. McKinley, D. F. Groth, K. A. Bowman, N. I. Mock, S. P. Cochran, and F. R. Masiarz.** 1981. Scrapie agent contains a hydrophobic protein. *Proc. Natl. Acad. Sci. USA* **78:**6675–6679.
67. **Purdey, M.** 1996. The UK epidemic of BSE: slow virus or chronic pesticide-initiated modification of the prion protein? Part 1: Mechanisms for a chemically induced pathogenesis/transmissibility. *Med. Hypotheses* **46:**429–443.
68. **Purdey, M.** 1996. The UK epidemic of BSE: slow virus or chronic pesticide-initiated modification of the prion protein? Part 2: An epidemiological perspective. *Med. Hypotheses* **46:**445–454.
69. **Scott, M. R., R. Will, J. Ironside, H.-O. B. Nguyen, P. Tremblay, S. J. DeArmond, and S. B. Prusiner.** 1999. Compelling transgenetic evidence for transmission of bovine spongiform encephalopathy prions to humans. *Proc. Natl. Acad. Sci. USA* **96:**15137–15142.
70. **Sklaviadis, T. K., L. Manuelidis, and E. E. Manuelidis.** 1989. Physical properties of the Creutzfeldt-Jakob disease agent. *J. Virol.* **63:**1212–1222.
71. **Soto, C., and J. Castilla.** 2004. The controversial protein-only hypothesis of prion propagation. *Nat. Med.* **10:**S63–S67.
72. **Spencer, M. D., R. S. G. Knight, and R. G. Will.** 2002. First hundred cases of variant Creutzfeldt-Jakob disease: retrospective case note review of early psychiatric and neurological features. *BMJ* **324:**1479–1482.
73. **Stahl, N., and S. B. Prusiner.** 1991. Prions and prion proteins. *FASEB J.* **5:**2799–2807.
74. **Taraboulos, A., D. Serban, and S. B. Prusiner.** 1990. Scrapie prion proteins accumulate in the cytoplasm of persistently infected cultured cells. *J. Cell Biol.* **110:**2117–2132.
75. **Thomas, P. J., M. J. Newby, and R. Zwissler.** 2003. New predictions for vCJD numbers. *Br. Food J.* **105:**420–433.
76. **Tiwana, H., C. Wilson, J. Pirt, W. Cartmell, and A. Ebringer.** 1999. Autoantibodies to brain components and antibodies to *Acinetobacter calcoaceticus* are present in bovine spongiform encephalopathy. *Infect. Immun.* **67:**6591–6595.
77. **Tucker, A.** 30 July 1998. Science and technology. Mad cows and wild professors. *The Guardian*, Manchester, United Kingdom.
78. **Turner, T., and T. Kelsey.** 22 October 1995. Experts fear 1.5m 'mad cows' eaten. Mad cow disease. *The Times*, London, United Kingdom.
79. **Venters, G. A.** 2001. New variant Creutzfeldt-Jakob disease: the epidemic that never was. *BMJ* **323:**858–861.
80. **Walters, M. J.** 2003. *Six Modern Plagues and How We Are Causing Them.* Island Press, Washington, D.C.
81. **Wilesmith, J. W., J. B. Ryan, and W. D. Hueston.** 1992. Bovine spongiform encephalopathy: case-control studies of calf feeding practices and meat and bonemeal inclusion in proprietary concentrates. *Res. Vet. Sci.* **52:**325–331.
82. **Wilesmith, J. W., G. A. Wells, M. P. Cranwell, and J. B. Ryan.** 1988. Bovine spongiform encephalopathy: epidemiological studies. *Vet. Rec.* **123:**638–644.

83. **Will, R. G.** 2002. Variant Creutzfeldt-Jakob disease. How new is new? *J. Neurol. Neurosurg. Psychiatry* **72:**285–286.

84. **Will, R. G., and J. W. Ironside.** 1999. Oral infection by the bovine spongiform encephalopathy prion. *Proc. Natl. Acad. Sci. USA* **96:**4738–4739.

85. **Will, R. G., J. W. Ironside, M. Zeidler, S. N. Cousens, K. Estibeiro, A. Alperovitch, S. Poser, M. Pocchiari, A. Hofman, and P. G. Smith.** 1996. A new variant of Creutzfeldt-Jakob disease in the UK. *Lancet* **347:**921–925.

86. **Williams, E. S., and S. Young.** 1980. Chronic wasting disease of captive mule deer: a spongiform encephalopathy. *J. Wildl. Dis.* **16:**89–98.

chapter 11

The Politics of Prions—BSE and World Trade

BY THE END OF 2004, bovine spongiform encephalopathy (BSE) had spread far beyond the United Kingdom, invading 20 European countries, Canada, the United States, and Japan (Table 11.1). The rest of the world, through good luck or good management, remained BSE free (78).

The European Union (EU), formerly the European Economic Community, maintains a common agricultural policy among its member states. This policy ensures the free passage of live animals and other agricultural products between members of the EU. In 1986, this included trade in live cattle, beef, bovine semen, and bovine embryos. It wasn't long, however, before some members of the EU began to question the wisdom of importing cattle, beef, and other bovine products from the United Kingdom (68).

Locking the Barn Door

The first report of BSE in 1986 didn't cause much of a stir outside Great Britain. But that began to change in December 1987, when it became clear that the disease had been transmitted in animal feed supplemented with ruminant meat and bone meal (68). In January 1988, a British farmer living in Portugal, having learned about the existence of BSE in British herds, decided against importing British cattle into Portugal. Six months later, in July 1988, the Dutch government voiced its concerns about the safety of British cattle to the EU Standing Veterinary Committee. In response, the United Kingdom negotiated a bilateral agreement with The Netherlands that prohibited the export from the United Kingdom to The Netherlands even of asymptomatic cattle from BSE-positive herds. The United Kingdom reached similar understandings with Germany, Ireland, and Denmark (68). In July 1989, the EU banned export from the United Kingdom to all EU member countries of all live cattle born before July 18, 1988, the date the British ruminant feed ban went into effect. The EU also barred United Kingdom export of cattle born to cows that were either suspected or confirmed to be BSE positive. But, unlike the bilateral agreements reached by

Table 11.1 Countries reporting cases of bovine spongiform encephalopathy (BSE) as of December 31, 2004[a]

Country	Year of first reported BSE case (reference)	Total no. of confirmed cases of BSE
Austria	2001 (76)	1
Belgium	1997 (30)	128
Canada	2003 (78)	2
Czech Republic	2001 (30)	15
Denmark	1992 (30)	13
Finland	2001 (30)	1
France	1991 (30)	946
Germany	1992 (30)	362
Greece	2001 (30)	1
Ireland	1989 (30)	1,470
Israel	2002 (78)	1
Italy	1994 (30)	125
Japan	2001 (30, 75)	14
Liechtenstein	1998 (30)	2
Luxembourg	1997 (30)	2
Netherlands	1997 (30)	77
Poland	2002 (78)	20
Portugal	1990 (30)	949
Slovakia	2001 (30)	20
Slovenia	2001 (30)	4
Spain	2000 (30)	532
Switzerland	1990 (30)	456
United Kingdom	1986 (30)	184,138
United States	2003 (78)	1

[a] Adapted from *ProMED-mail*, June 6, 2005, and *Eurosurveillance Weekly*, vol. 5, no. 50 , with additional information from other sources (30, 78).

the United Kingdom, Germany, The Netherlands, and Ireland, the EU permitted the export of cattle from herds in which the presence of BSE had been confirmed (68).

The European export ban was tightened at the beginning of March 1990, when the EU restricted United Kingdom live cattle exports to calves younger than 6 months and specified that those calves must be slaughtered (in their country of destination) before they reached the age of 6 months. At the end of March 1990, the EU also banned the use of specified bovine offal—parts of a slaughtered cow that were believed to be most susceptible to carrying the BSE infectious agent—in animal feeds.

The EU wasn't alone in responding to the presence of BSE in British cattle. In July 1989, the United States banned the importation of live cattle, embryos, and bull semen from the United Kingdom. By the end of April 1990, Australia, Finland, Israel, Sweden, West Germany, and New Zealand had followed suit (26). Japan, Morocco, Canada, and South Africa stipulated that live cattle imported from the United Kingdom must be from herds certified to be BSE free (68). France and Germany, both EU members, went even further, banning all imports of British beef. Although the British government and the EU protested the French and German actions strenuously, by February 1991, Algeria, Bahrain, Brazil, Canada, China, Egypt, Iran, Iraq, Jordan, Morocco, Saudi Arabia, Syria, Tunisia, Turkey, United Arab Emirates, and the USSR joined Germany and France in imposing outright bans on the importation of British beef (55). Several other countries, including Cyprus, Hong Kong, the Ivory Coast, Austria, Mexico, and Malta, imposed less severe restrictions (68).

The EU's agricultural policy addressed live cattle, beef, and beef products but ignored rendered animal products such as beef tallow or meat and bone meal (MBM). When, in June 1988, the United Kingdom adopted its ruminant feed ban (prohibiting the feeding to ruminants of animal feed containing rendered ruminant protein), EU member nations did not do likewise, nor did most of the United Kingdom's other customers. By 1990, countries as far apart as Norway and Nigeria, Saudi Arabia and South Korea, and Turkey and Taiwan were still importing MBM from the United Kingdom (68). And all of these countries were equally far apart in their reaction to a letter issued to 25 countries by a British official in February 1990 reminding the United Kingdom's customers of Great Britain's continuing ruminant feed ban.

The United States and Canada did not institute their own ruminant feed bans until 1997, 7 years after receiving the British advisory, and only after the World Health Organization (WHO) recommended the ban (49, 79). Other countries—Australia and New Zealand, for example—beat the WHO to the starting gate; Australia banned the importation of all animal feed in 1989, and New Zealand initiated a voluntary ruminant feed ban in 1996 (71, 72). Japan, on the other hand, waited until 2001 to prohibit the feeding of ruminant MBM to other ruminants—and reported its first confirmed case of BSE later that same year (75).

EU member countries also varied in their response to the MBM issue. France had banned the importation of MBM from the United Kingdom in 1989, 5 years before the EU took similar action. But the French ban was ineffective; British MBM continued to enter France illegally through Belgium until the EU implemented a ruling banning the sale of British MBM to other EU member countries in 1994 (74). And while the EU also banned ruminant-to-ruminant feeding of MBM that same year, it continued to allow ruminant MBM to be used in pig and poultry feed. In March 1996, the United Kingdom banned the incorporation of mammalian MBM into feed destined for all farm animals, even including horses and farmed fish. The British went even further only 1 month later, prohibiting mammalian MBM in fertilizers sold for use on agricultural lands, although still allowing its use in private gardens or greenhouses (17).

By the time the EU acted, British MBM was no longer the only problem. The presence of BSE in European cattle raised the fear that MBM from continental Europe might have become contaminated by the BSE prion,

"It's Politics, Not Science"—A Tale of Three Countries (69)

In ancient Rome, it was called *dignitas*. To augment one's *dignitas* was the aim of every Roman nobleman; to see that *dignitas* diminished or destroyed was a punishment worse than death.

"Face" is to Japan what *dignitas* was to ancient Rome. "Losing face" is the worst thing that can happen to an individual (especially a public figure), a corporation, or the country as a whole. And the detection of BSE in a slaughtered cow in Japan in September 2001 caused a massive loss of face to Prime Minister Junichiro Koizumi, his government, and the country's cattle and beef industry (75).

Unlike many other countries, including the United States and Canada, Japan waited until 2001 to ban the importation of cattle feed from European countries (70). In June 2001, the EU's Scientific Steering Committee, after reviewing Japan's BSE risk status at that country's request, concluded that there was an "objective risk" of BSE in Japanese cattle. The Japanese government, at the reported behest of the country's cattle and beef industry, tried to have the committee's conclusions suppressed. A government representative claimed that the risk rating was unfair and expressed concern about a "panic reaction." The report was never formally published (67).

When news of Japan's first reported case of BSE was released in September 2001, and vague assurances and contradictory information began to emerge from the offices of the Ministry of Agriculture and Ministry of Health, consumers reacted with shock and expressions of distrust in their government's continued assurances that the country's beef was safe to eat (45, 58, 65). In an attempt to regain consumer confidence, and its own face, the government went overboard. It announced an immediate inspection of Japan's entire cattle population for "clinical signs" of BSE and instituted an intensive BSE screening program (8, 9, 93). From October 18, 2001, onward, the brain of every slaughtered cow would be tested for the disease—even though there was no scientifically sound justification for such a comprehensive testing program (51, 64). Within a month of the start of the intensive screening, the new program produced its first positive result; the brain of an asymptomatic cow tested positive for BSE (10). A third cow was reported BSE positive just 10 days later (11, 12).

On learning that BSE was detected in a cow in Canada in May 2003, and in another cow in the United States in December 2003, Japan joined several other countries in banning the importation first of Canadian, then of U.S., cattle and beef (16, 46). But it's much easier to close a border than to reopen one, and the insistence of Japan on testing 100% of slaughtered cattle for BSE made reestablishing even a partial beef trade among the three countries especially difficult. Japan's borders remained closed to U.S. and Canadian beef until December 2005, reopening partway only after Japan finally abandoned its insistence that cows 20 months and younger be tested for BSE (23, 114).

The cattle and beef industry in all three countries was, and still is, large and influential. In 2003, Canadian cattle and beef exports to the United States alone were worth \$2.2 billion per year (46). Japan's domestic beef market in 2001 was valued at \$2.1 billion per year, and the U.S. beef industry lost more than \$6 billion in export sales to Japan during the 2 years that country refused to allow entry to U.S. beef (64, 114). With such a large amount of money at stake, it's no wonder that beef became a political football. Unfortunately, a combination of consumer fears, industry pressures, and government pratfalls in all three countries precluded unemotional, thoughtful, and science-based policies from being put into place to resolve what quickly disintegrated into a major trade irritant (20, 24, 44, 45, 54, 65, 84).

especially since BSE-positive cattle in both Belgium and Germany had been rendered for use in animal feeds (46). Ireland, Italy, Belgium, France, Portugal, Germany, Spain, and Switzerland—all EU member countries—have each experienced more than 100 cases of BSE (Table 11.1). In contrast, Canada, which banned the importation of MBM and blood meal from all countries except the United States in 1988, has experienced only five cases of BSE—one in a breeder cow imported from the United Kingdom in 1987 (32, 33, 35). Australia (which banned the import of MBM for animal feed in 1989), New Zealand (which never allowed the importation of MBM for animal feed), and Argentina (which never imported MBM and never used MBM in cattle feed) have all remained BSE free (29, 71–73, 85).

Not-So-Strange Bedfellows

The 1989 Canada–United States Free Trade Agreement (CUSTA) opened the way to a free flow of agricultural products, including livestock, meat, and poultry, between the two countries by providing for a phased elimination of tariffs and the removal of nontariff trade barriers (2). In 1994, the North American Free Trade Agreement (NAFTA) replaced CUSTA, creating an enlarged free trade zone that encompassed Canada, Mexico, and the United States (41).

Even before CUSTA, there was extensive cooperation and sharing of information between the United States and Canadian food safety and health agencies. Under CUSTA, and later under NAFTA, the trend toward harmonizing many food safety procedures and protocols intensified (33). The two countries worked to coordinate their response to the threat of BSE from imported live cattle, beef, and animal feeds (32, 33) (see Table 11.2).

Between 1980 and 1989, both the United States and Canada imported a small number of live cattle from the United Kingdom. In 1989, once the risk of BSE became known, the United States banned further importation of ruminants from Britain. Canada followed suit with its own import ban in 1990; it traced all of the cattle of United Kingdom origin that it had imported during the 1980s and began to monitor those cattle for any clinical signs of BSE. In 1993, one of the British cows (imported into Canada in 1987) showed symptoms of BSE. It was slaughtered and confirmed to be BSE positive. The Canadian government ordered the slaughter of all the remaining animals that had been imported from the United Kingdom and tested each one for BSE. All of them proved to be BSE free (33).

As the threat posed by BSE to the cattle and beef industry—and concern for its possible spread to humans—increased, the United States and Canada tightened their import controls. Both countries implemented progressively stricter bans on the importation of beef, edible beef products, and rendered beef products, including MBM from countries either already affected by or at risk of becoming affected by BSE. In 1997, following the WHO recommendation, Canada and the United States banned the feeding of ruminant MBM to other ruminants (the "ruminant feed ban"). And as part of their overall risk management efforts, the United States and Canadian governments began a routine BSE testing program (33).

Table 11.2 Sequence of events in the BSE-related trade disputes among the United States, Canada, and Japan

Date	Event
8 December 1993	Canada reports its first confirmed case of BSE; the affected cow had been imported from the United Kingdom (6, 34).
22 December 2000	Japan announces it will ban import of European beef, effective January 1, 2001 (7).
1 January 2001	Canada initiates nationwide electronic tag system to track all cattle; compulsory system to take effect on 1 July 2002 (1).
10 September 2001	Japan reports its first suspected case of BSE (70, 75).
18 September 2001	United States imposes import restrictions on beef from Japan in response to announcement of first suspected case of BSE in Japan (95).
18 September 2001	Japan announces ban on use of meat and bone meal in animal feed for cattle (82).
22 September 2001	Japan confirms its first case of BSE (75).
20 May 2003	Canada confirms its first case of BSE in a Canadian-born cow. United States and Japan halt importation of Canadian beef (31, 46).
30 July 2003	Japan announces increase in tariffs on imported beef and pork (87).
8 August 2003	United States announces partial lifting of ban on importation of Canadian beef (96).
31 October 2003	USDA proposes reopening U.S. border to importation of live cattle from Canada (3, 97).
23 December 2003	United States announces its first suspected case of BSE (98).
26 December 2003	Japan bans importation of U.S. beef (16).
28 December 2003	U.S. cow suspected of being BSE positive tentatively traced to Canada (42).
30 December 2003	USDA announces new policies to strengthen protection against spread of BSE (99).
2 March 2005	U.S. District Court judge issues temporary injunction to block importation of live Canadian cattle into the United States (54).
24 June 2005	USDA announces confirmed positive BSE test result in U.S.-born cow; Canada announces it will not ban importation of U.S. beef (43, 110).
14 July 2005	U.S. federal appeals court lifts injunction preventing importation of live cattle from Canada into the United States (27).
11 December 2005	Japan announces partial lifting of bans on importation of beef from United States and Canada; United States announces reciprocal resumption of imports from Japan (23, 111, 114).
20 January 2006	Japan reinstates its ban on U.S. beef after finding spinal column material in shipment from one supplier (89).
21 June 2006	Japan announces that it will lift ban on imports of U.S. beef once it has completed spot inspections of exporting plants as agreed to between itself and the United States (83).
26 June 2006	Canada institutes ban on use of specified risk materials in all animal feed, pet foods, and fertilizers (36).

The testing programs started small: just 53 tests per 1 million head of cattle in Canada and 26 tests per 1 million in the United States in the first year. By 2003, both countries had increased their testing frequency, to 276 in Canada and 213 in the United States. And in 2004, the United States performed 1,859 tests per 1 million head of cattle and Canada carried out 1,591. The more than fivefold increase in testing frequency in Canada and the eightfold increase in the United States between 2003 and 2004 were triggered by the detection, in 2003, of a BSE-positive cow in Canada (33).

On January 31, 2003, an Alberta cow sent for slaughter was identified as a "downer," unable to walk. Her head was removed at the slaughterhouse for BSE testing; the rest of the carcass, declared unfit for human consumption due to pneumonia, was rendered, and the rendered material was used in animal feed. Because Canada believed itself to be BSE free, the downer's head was given a low priority in the lab. It wasn't until May 16 that the regional lab completed its testing and reported finding evidence of BSE in the sample. On May 20, the international BSE reference laboratory in the United Kingdom confirmed the Canadian lab's initial finding. Canada had identified its first homegrown BSE-positive cow (31).

International reaction to the Canadian finding was swift. The U.S. government immediately banned all beef and all live cattle imports from Canada (62). Other customers of Canadian beef, including Japan, South Korea, Australia, New Zealand, and Mexico, quickly announced their own bans, with Singapore and Indonesia not far behind (13, 14, 28, 50). Meanwhile, Canadian officials began tracing the origin of the affected cow and imposed quarantines on any suspect herds as they were identified.

Canada had implemented a mandatory identification tag for all cattle in 2001, but the BSE-positive cow was born before the program went into effect, making tracing its origins more difficult (92). Nevertheless, by the beginning of July, Canada had determined the history of the affected cow, including the herds with which it had been in contact during its 8-year life. As a precaution, the government ordered the destruction of more than 2,700 head of cattle. All cattle over 24 months old in the affected herds—more than 2,000 animals in all—were tested for BSE, with uniformly negative results (31). Canada also submitted, in June, to an international inspection of its BSE safeguards. The inspection team praised Canada's response to the finding of a BSE-positive cow but recommended several improvements to the country's BSE firewall, including the elimination of specified risk materials—those tissues likely to contain the BSE prion—from cattle at slaughter (56). Canada responded in July 2003 by instituting mandatory removal of the brain and spinal cord at slaughter from all cattle over 30 months old (48). And in August, the United States and Mexico announced that they would reopen their borders to low-risk Canadian meat, such as certain cuts of boneless beef (60, 66, 96). The first shipments of Canadian beef started rolling across the Canada–United States border in early September 2003 (61). But just as the country's beef industry began to breathe a little more easily, another Canadian-born BSE-positive cow was identified—this time in the United States.

On December 23, 2003, U.S. Department of Agriculture (USDA) Secretary Ann Veneman announced that a possible case of BSE had been detected in a downer cow slaughtered in the state of Washington 2 weeks earlier. Unlike the Canadian situation, meat from the Washington cow had entered the human food supply. The slaughterhouse recalled more than 10,000 pounds of ground beef—meat from the suspect cow having been ground together with meat from 20 other cows (40, 57). The United States' international beef customers reacted swiftly, just as they had done after the May 2003 Canadian BSE finding. Japan and South Korea, two of its largest customers, immediately halted the importation of U.S. beef products (88). Other countries, including Australia, Singapore, Thailand, Malaysia, Hong Kong, Russia, and Taiwan, also closed their doors to U.S. beef, at least temporarily (15, 91). Canada, on the other hand, reserved judgment and kept its borders open (39).

USDA officials immediately began the process of tracing the affected cow's origins as soon as they learned of the positive BSE test result. Thanks in part to Canada's mandatory ear tag identification system, it took just 4 days to determine that the cow was born in Alberta, Canada (113). It took another 7 weeks for USDA to locate 28 of the 80 cattle that had accompanied the BSE-positive cow across the border. Due to poor record-keeping and the lack of even a voluntary national cattle-tagging system in the United States, the remaining 52 animals were never traced (86, 101).

Even though the BSE-positive cow was not native born, the incident was a wake-up call to the USDA. After years of permitting the use for human consumption of meat from cattle unable to walk on their own (so-called "downer cattle" or "downers"), Agriculture Secretary Veneman announced on December 30 that meat from downers would no longer be considered fit for human consumption. She also introduced several other new policies meant to safeguard the U.S. meat supply from BSE: holding meat from suspect cows until BSE testing was complete (rather than releasing meat from the cow immediately after physical inspection); establishing an expanded definition of "specified risk materials"; prohibiting air injection stunning; banning meat from Advanced Meat Recovery systems (systems using hydraulic pressure to separate meat from bones) for human consumption; and promising to speed up development and implementation of a mandatory nationwide cattle-tagging system (99).

The National Animal Identification System ran into roadblocks, not the least of which was the "speeded up" pace at which USDA tackled the problem. Part of the reason for the lack of progress was inadequate funding of the program; for example, no funds were appropriated for the project at all in 2004 (25). The USDA's Animal and Plant Health Inspection Service finally published a discussion draft of its proposed mandatory tracking system in May 2005 (108). USDA modified its proposals based on feedback it received from industry and farming representatives, and the resulting nationwide *voluntary* tracking system should be in place by 2009 (5, 112).

Although the U.S. cattle and beef industries supported some form of identification system, it was not long before they began to lobby against the downer cattle ban. Industry representatives argued that the majority of downer cows suffered from physical disabilities such as broken legs rather than from disease and were perfectly safe to eat (94). USDA, however, held its ground on this issue.

Born in the USA

On June 25, 2004, USDA announced that a downer cow had tested positive for BSE in a preliminary screening test (102). On June 29, a second cow was reported BSE positive in a preliminary test (103). A third positive screening result was announced on November 18 (106). Thanks to the downer ban that was still in force, none of the meat from the three suspect cows was allowed to enter the food supply, nor was any of it used for animal feed (102, 103, 106). None of the three suspect cows were confirmed BSE positive by USDA's immunohistochemistry (IHC) confirming test (104, 105, 107), but the story did not end there.

In June 2005, USDA's Inspector General expressed concern over the conflicting results obtained using the rapid tests and confirming tests for the three downer cows (22). At her insistence, USDA retested samples from all three cows using a different, and more sensitive, confirming test, the Western blot (38). The November downer cow produced a positive Western blot test result (109). To resolve the discrepancy between the IHC and Western blot results, USDA sent a tissue sample from the positive animal to the international BSE reference lab in Weybridge, England. On June 24, USDA announced that the Weybridge lab had confirmed a BSE-positive result in the sample from the November cow—the first U.S.-born cow to be confirmed BSE positive (110).

Until the June 2005 episode, USDA staunchly defended its confirming test against detractors, calling it the "gold standard" despite concerns that the IHC test might be less sensitive than the Western blot procedure (19, 59). But following the release of the Weybridge results, Agriculture Secretary Mike Johanns announced that, henceforth, the USDA lab would confirm all positive screening results using both the IHC and the Western blot methods and would consider a sample confirmed as BSE positive if either of these tests produced a positive result (110).

Science-Based Policies?

In 2003, officials from the USDA criticized the long delay between slaughter and testing of the first BSE-positive Canadian cow, commenting that a testing backlog of nearly 4 months would not be tolerated in the United States (47). Whenever its own BSE policies or programs have been questioned, USDA has been quick to defend its actions as "science-based" (100). But something clearly went wrong with the tests carried out on the November downer cow, resulting in a 7-month lag between the positive screening test result and the eventual confirmation that the cow was BSE positive (21).

Consumer groups and some food safety experts often have expressed skepticism over the logic and effectiveness of portions of USDA's BSE surveillance program. Some of their concerns have been well founded. The Office of Inspector General (OIG), as part of its audit of the BSE program, reviewed the basis for USDA's policy on sampling cows for testing. Despite USDA's insistence that its sampling program was based on solid science, the OIG report questioned the validity of the assumptions on which the statistical sampling program was based (63). And the events leading up to and following the retesting of tissue samples from the November cow show clearly that there were gaps in USDA's lab methods as well.

When USDA found a BSE-positive cow (the Canadian-born cow) in 2003, it confirmed the results of its rapid screening test by running the IHC test. When that test was positive, it double-checked the IHC result with the Western blot test, which also produced a positive result (52). Because the IHC and Western blot both yielded the same positive result, USDA scientists apparently concluded that the Western blot test was superfluous. Unfortunately, they failed to consider the possibility that the IHC test might prove less sensitive, producing a false-negative result at prion concentrations that could still be detected by Western blot.

Despite public pronouncements in 2004 by USDA officials expressing confidence in the reliability of the IHC test, the agency's scientists were working behind the scenes to improve the sensitivity of the test. In addition to the two negative IHC tests that they ran immediately following the positive screening test result on the November cow, the USDA lab ran a third IHC test using a modified procedure that was designed to increase the sensitivity of the test. The enhanced IHC test produced a positive result, but the information was not used or reported because the newer procedure had not been formally validated. Although a test result from an unvalidated method should never be the sole basis for taking (or not taking) regulatory action, it would have been prudent, nevertheless, to retest the sample using the Western blot technique. Instead, the data were simply filed away (53).

It was the existence of this third IHC test result, found during an OIG audit of the USDA's BSE surveillance program, that ultimately triggered the retesting of brain tissue from the November cow in June 2005 (22). The results of the retesting raised questions, in turn, about the reliability of the USDA lab's handling even of the validated IHC test protocol. The June 2005 retest of the November cow was carried out simultaneously by the USDA lab and by the international BSE reference lab in Weybridge. Both the Weybridge lab and the USDA lab (the same one that reported the negative IHC result the previous year) confirmed the presence of BSE prions in the November cow using both the IHC and the Western blot tests (37).

The discrepancy in test results was not the only irregularity in USDA procedures exposed by the OIG audit and the subsequent retesting. Other lapses and oversights complicated attempts to identify unequivocally the cow that had yielded the BSE-positive brain sample and the herd from which it had come. An ID tag indicating the breed of the infected cow was mislabeled, leading to confusion as to its herd of origin (53); workers froze the brain sample (which can harm the sensitivity of the test); the USDA lab that ran the BSE tests neglected to complete its paperwork; and the infected cow's carcass was stored with four other carcasses while testing was being carried out, making it difficult to determine which of the five carcasses needed to be destroyed (37).

BSE and vCJD—The Bottom Line

Much has been made of the need to control the spread of BSE and variant Creutzfeldt-Jakob disease (vCJD). The worldwide cost of surveillance and control of BSE internationally is in the billions of dollars (18). But, it has become increasingly evident that the risk of an individual contracting vCJD through having eaten BSE-affected beef is extremely low (90). Notwithstanding the human tragedies associated with 159 confirmed cases of vCJD in the United Kingdom and additional cases in several countries, including, among others, France, Ireland, The Netherlands, Canada, and Italy (Table 11.1), BSE, like scrapie, has turned out to be primarily an economic issue rather than a significant risk to human health (4, 77, 80, 81, 90). Why, then, should we be concerned about USDA's handling of its follow-up investigation after finding a positive screening result in November 2004?

Aside from the economic impact of short- and long-term closures of a number of international borders to U.S. live cattle and beef products, the events surrounding the eventual confirmation of the first U.S.-born BSE-positive cow opened the door to questions about the reliability of internally generated data on which USDA may be basing some of its regulatory and enforcement policies—not only for BSE but for other food safety issues as well. It is incumbent on a regulatory agency to use the best available laboratory and field investigation methods and to document all of its actions and data completely and according to clearly established procedures. Science-based policies, to be effective, must be based on the rational application of sound science.

References

1. **Acharya, M.** 12 February 2001. Canada's herd goes digital; an old idea that worked has been adapted with new technology to protect animals and consumers. *Toronto Star*, Toronto, Ontario, Canada.
2. **Agriculture and Agri-Food Canada.** 2003. Agri-food trade policy. Canada–United States Free Trade Agreement (FTA). Chapter Seven: Agriculture. Agriculture and Agri-Food Canada. [Online.] http://www.agr.gc.ca/itpd-dpci/english/trade_agr/fta_agr.htm. Accessed 24 June 2006.

3. **Alberts, S., and L. Pynn.** 21 October 2003. Americans ready to make decision on live-cattle ban. *The Vancouver Sun*, Vancouver, B.C., Canada.
4. **Andrews, N. J.** 19 January 2006. Incidence of variant Creutzfeldt-Jakob disease onsets and deaths in the UK, January 1994–December 2005. The National Creutzfeldt-Jakob Disease Surveillance Unit, University of Edinburgh, Edinburgh, United Kingdom. [Online.] http://www.cjd.ed.ac.uk/vcjdqdec05.htm. Accessed 19 June 2006.
5. **Animal and Plant Health Inspection Service.** 2006. National animal identification system. Animal and Plant Health Inspection Service, U.S. Department of Agriculture. [Online.] http://animalid.aphis.usda.gov/nais/index.shtml. Accessed 24 June 2006.
6. **Anonymous.** 17 December 1993. Trade reputation at stake in cow disease scare. *The Vancouver Sun*, Vancouver, Canada.
7. **Anonymous.** 26 December 2000. Japan says no to EU beef. *China Daily (North American edition)*, New York, N.Y.
8. **Anonymous.** 14 September 2001. Officials probe for mad-cow disease in Japan. *The Asian Wall Street Journal*, New York, N.Y.
9. **Anonymous.** 19 September 2001. 1 million cattle to undergo mad cow tests. *Mainichi Daily News*, Tokyo, Japan.
10. **Anonymous.** 21 November 2001. Diseased cow showed no signs of 'madness'. *Mainichi Daily News*, Tokyo, Japan.
11. **Anonymous.** 1 December 2001. Inspectors turn up link between 3 mad cow cases. *Mainichi Daily News*, Tokyo, Japan.
12. **Anonymous.** 3 December 2001. Japan confirms 3rd case of mad cow disease. *Jiji Press English News Service*, Tokyo, Japan.
13. **Anonymous.** 22 May 2003. Canada beef business on freeze. *Oakland Tribune*, Oakland, Calif.
14. **Anonymous.** 22 May 2003. Canadian beef imports banned. *Korea Herald*, Seoul, South Korea.
15. **Anonymous.** 24 December 2003. Nations ban U.S. beef; mad cow discovery scares Asian countries. *Madison Capital Times*, Madison, Wis.
16. **Anonymous.** 26 December 2003. Japan officially bans US beef imports over BSE fears. *BBC Monitoring Asia Pacific*, London, United Kingdom.
17. **Anonymous.** 2004. Bovine spongiform encephalopathy. Chronology of events (as at 20 July 2004). [Online.] http://www.defra.gov.uk/animalh/bse/directory.html. Accessed 26 July 2005.
18. **Anonymous.** 18 October 2004. EU budget firms up for BSE. Foodnavigator.com/Europe. [Online.] http://www.foodnavigator.com/news/news-ng.asp?id=55459-eu-budget-firms. Accessed 24 June 2006.
19. **Anonymous.** 24 February 2005. Consumers Union asks feds to retest suspect mad cow after crucial test omitted; USDA urged to follow intl recognized procedures. *U.S. Newswire*, Washington, D.C.
20. **Anonymous.** 27 May 2005. Time is ripe to resume imports of US beef. *The Daily Yomiuri*, Tokyo, Japan.
21. **Anonymous.** 21 June 2005. Consumers Union calls on USDA to go the extra mile to protect consumers, cattle industry. *U.S. Newswire*, Washington, D.C.

22. **Anonymous.** 1 July 2005. Mad beef policy. *Los Angeles Times*, Los Angeles, Calif.

23. **Anonymous.** 13 December 2005. Wary Japan resumes beef imports. *Toronto Star*, Toronto, Ontario, Canada.

24. **Anonymous.** 23 March 2006. Meatpacker may sue USDA over testing; the firm wants to check all its cattle for mad cow disease, but the department says that wouldn't ensure safety. *Los Angeles Times*, Los Angeles, Calif.

25. **Antosh, N.** 6 March 2004. Tracking of cattle becomes key goal. Mad cow issue prods USDA. *Houston Chronicle*, Houston, Tex.

26. **Barr, R.** 22 April 1990. 'Mad cow disease' kills 10,000 cattle in Britain livestock: the government sees only a remote risk to humans. The malady may be spread through cattle feed. *Los Angeles Times*, Los Angeles, Calif.

27. **Barrionuevo, A.** 15 July 2005. Despite concerns on mad cow, court allows Canada imports. *New York Times*, New York, N.Y.

28. **Betts, M.** 22 May 2003. Mad cow ban put on beef from Canada. *Dominion Post*, Wellington, New Zealand.

29. **Biosecurity New Zealand.** 2005. Transmissible spongiform encephalopathies. Biosecurity New Zealand. [Online.] http://www.biosecurity.govt.nz/pests-diseases/animals/tse/. Accessed 25 June 2006.

30. **Bryant, G.** 13 December 2001. BSE in Finland and the rest of Europe. *Eur. Surveill. Wkly.* **5**. [Online.] http://www.eurosurveillance.org/ew/2001/011213.asp. Accessed 25 June 2006.

31. **Canadian Food Inspection Agency.** 2 July 2003. BSE in North America. Summary of the report of the investigation of bovine spongiform encephalopathy (BSE) in Alberta, Canada. Canadian Food Inspection Agency. [Online.] http://www.inspection.gc.ca/english/anima/heasan/disemala/bseesb/ab2003/evalsume.shtml. Accessed 25 June 2006.

32. **Canadian Food Inspection Agency.** 2005. Bovine spongiform encephalopathy (BSE). Canadian Food Inspection Agency, Animal Products, Animal Health and Production Division. [Online.] http://www.inspection.gc.ca/english/anima/heasan/disemala/bseesb/bseesbfse.shtml. Accessed 25 June 2006.

33. **Canadian Food Inspection Agency.** 2005. Technical overview of BSE in Canada—March 2005. Canadian Food Inspection Agency. [Online.] http://www.inspection.gc.ca/english/anima/heasan/disemala/bseesb/200503canadae.shtml. Accessed 25 June 2006.

34. **Canadian Food Inspection Agency.** 2006. Chronology of BSE events. Canadian Food Inspection Agency. [Online.] http://www.inspection.gc.ca/english/anima/heasan/disemala/bseesb/chronoe.shtml. Accessed 24 June 2006.

35. **Canadian Food Inspection Agency.** 2006. Completed investigations. Canadian Food Inspection Agency. [Online.] http://www.inspection.gc.ca/english/anima/heasan/disemala/bseesb/comenqe.shtml. Accessed 25 June 2006.

36. **Canadian Food Inspection Agency.** 26 June 2006. Canada strengthens feed controls. Canadian Food Inspection Agency. [Online.] http://www.inspection.gc.ca/english/corpaffr/newcom/2006/20060626e.shtml. Accessed 27 June 2006.

37. **Dworkin, A.** 25 June 2005. New mad cow case takes U.S. testing up another notch. *The Oregonian*, Portland, Oreg.

38. **Fabi, R.** 12 June 2005. Mad cow tests may last 2 weeks. USDA scientists need more results to confirm case. *Houston Chronicle*, Houston, Tex.

39. **Fabi, R., and R. Cowan.** 24 December 2003. U.S. hit by first case of mad cow. Single animal in Washington state tests positive. Recall of possibly affected beef could begin today. *Toronto Star*, Toronto, Ontario, Canada.

40. **Food Safety and Inspection Service.** 23 December 2003. Washington firm recalls beef products following presumptive BSE determination. Recall Release FSIS-RC-067-2003. Food Safety and Inspection Service, U.S. Department of Agriculture. [Online.] http://www.fsis.usda.gov/OA/recalls/prelease/pr067-2003.htm. Accessed 26 July 2005.

41. **Foreign Affairs and International Trade Canada.** 2003. Canada and the North American Free Trade Agreement. Foreign Affairs and International Trade Canada. [Online.] http://www.dfait-maeci.gc.ca/nafta-alena/menu-en.asp. Accessed 25 June 2006.

42. **Gersema, E.** 28 December 2003. Mad cow traced to Canada; origin outside U.S. could save exports for beef industry. *Chicago Sun-Times*, Chicago, Ill.

43. **Gillie, R.** 26 June 2005. Mad cow: US beef welcome in Canada; Taiwan resets ban, Japan concerned after second case found in America. *Sunday Gazette-Mail*, Charleston, W.Va.

44. **Grant, J.** 4 May 2005. Cattlemen issue threat to US over meat processing. *Financial Times*, London, United Kingdom.

45. **Hadfield, P.** 23 September 2001. Ministry bungle puts Japan at risk of BSE. Carcass of infected cow was processed into animal feed. *The Sunday Telegraph*, London, United Kingdom.

46. **Hallinan, J. T., and M. Heinzl.** 22 May 2003. Mad-cow disease shakes up markets—new case of illness leads US, Japan, Australia, South Korea to halt Canadian cattle imports. *The Asian Wall Street Journal*, New York, N.Y.

47. **Harper, T.** 23 May 2003. Testing delay queried by U.S. officials; backlog called 'intolerable'; isolating imports considered. *Toronto Star*, Toronto, Ontario, Canada.

48. **Health Canada.** 18 July 2003. Government of Canada announces new BSE measure. Health Canada. [Online.] http://www.hc-sc.gc.ca/ahc-asc/media/nr-cp/2003/bse-esb_e.html. Accessed 25 June 2006.

49. **Health Canada.** 15 July 2005. Overview of Canada's BSE safeguards. Health Canada. [Online.] http://www.hc-sc.gc.ca/fn-an/securit/animal/bse-esb/safe-prot_e.html. Accessed 25 June 2006.

50. **Heinzl, M., T. Carlisle, and J. T. Hallinan.** 23 May 2003. Singapore, Indonesia join Canadian beef ban. *The Asian Wall Street Journal*, New York, N.Y.

51. **Hutton, B.** 19 October 2001. Mad cow fears as Japan says beef is safe. *Financial Times*, London, United Kingdom.

52. **Ivanovich, D.** 17 June 2005. Feds skipped key mad cow disease test in 2004 case. USDA changes its protocols after animal initially had been cleared. *Houston Chronicle*, Houston, Tex.

53. **Kaufman, M.** 25 June 2005. Retesting reveals mad cow case; USDA criticized for first clearing animal of disease. *The Washington Post*, Washington, D.C.

54. **Kawar, M.** 3 March 2005. Court ban yields new debates. A federal judge cites health concerns in an order against reopening the US border to Canadian cattle. *Omaha World-Herald*, Omaha, Nebr.

55. **Kaye, J.** 4 June 1990. One symptom of 'mad cow' disease is trade friction. *Los Angeles Times*, Los Angeles, Calif.

56. **Kihm, U., W. Hueston, D. Heim, and S. MacDiarmid.** 26 June 2003. Report on actions taken by Canada in response to the confirmation of an indigenous case of BSE. [Online.] http://www.inspection.gc.ca/english/anima/heasan/disemala/bseesb/ab2003/internate.shtml. Accessed 25 June 2006.

57. **Kilman, S., S. Leung, and T. Carlisle.** 24 December 2003. Case of mad cow found in the U.S. for first time; discovery expected to jolt American cattle industry; food supply called safe. *The Wall Street Journal*, New York, N.Y.

58. **Magnier, M.** 12 September 2001. Japan shocked by reported case of 'mad cow' disease. Asia: shares of dairy and meat businesses fall in wake of announcement about afflicted animal. Nation's neighbors impose beef bans. *Los Angeles Times*, Los Angeles, Calif.

59. **McNeil, D. G., Jr.** 24 November 2004. Suspicion of a mad cow case proves unfounded, tests find. *New York Times*, New York, N.Y.

60. **Monchuk, J.** 25 August 2003. Beef delivery to U.S. market delayed; Sept. 1 target date won't be met. Permits coming after Labour Day. *Toronto Star*, Toronto, Ontario, Canada.

61. **Monchuk, J.** 10 September 2003. Alberta beef heading for U.S. *Toronto Star*, Toronto, Ontario, Canada.

62. **Neergaard, L.** 21 May 2003. U.S. bans Canadian beef after mad cow discovery. Canadian health officials say the infected 8-year-old cow did not make it to the food chain. *The Grand Rapids Press*, Grand Rapids, Mich.

63. **Office of Inspector General.** August 2004. *Audit Report. Animal and Plant Health Inspection Service and Food Safety and Inspection Service Bovine Spongiform Encephalopathy (BSE) Surveillance Program—Phase I.* Report No. 50601-9-KC. U.S. Department of Agriculture, Office of Inspector General, Great Plains Region. [Online.] http://www.usda.gov/oig/rptsauditsaphis.htm. Accessed 25 June 2006.

64. **Ono, Y., and S. Stecklow.** 15 October 2001. Japan plans to examine all cattle for BSE—move is seen as bid to save country's $2.1 billion domestic beef market. *The Wall Street Journal*, New York, N.Y.

65. **Ono, Y., and S. Stecklow.** 30 November 2001. Mad-cow disease finds Japan unprepared as Tokyo fumbles in bid to reassure public. *The Wall Street Journal*, New York, N.Y.

66. **Paraskevas, J.** 19 August 2003. NAFTA trio seeks to revise criteria for BSE threats. *The Ottawa Citizen*, Ottawa, Ontario, Canada.

67. **Parry, R. L.** 19 June 2001. Japan puts pressure on Brussels to keep BSE warning under wraps. *The Independent*, London, United Kingdom.

68. **Phillips, L., J. Bridgeman, and M. Ferguson-Smith.** 2000. The BSE Inquiry—16 Volumes. *Inquiry into the Emergence and Identification of Bovine Spongiform Encephalopathy (BSE) and Variant Creutzfeldt-Jakob Disease (vCJD) and the Action Taken in Response to It Up to 20 March 1996.* Stationery Office, London, England, United Kingdom. [Online.] http://www.bseinquiry.gov.uk/report/index.htm. Accessed 21 June 2006.

69. **Price, R.** 6 February 2005. Cattlemen's group fights to keep ban on Canadian beef. Ohioans warn of risk to US consumers if border reopens; meeting set Monday. *The Columbus Dispatch*, Columbus, Ohio.

70. **Prideaux, E.** 11 September 2001. Japan reports first suspected case of mad cow disease. *Wisconsin State Journal*, Madison, Wis.

71. **ProMED-Ahead.** 24 April 1996. Spongiform encephalopathy (43): UK export restrictions. *ProMED Ahead* 19960424.0783. [Online.] http://www.promedmail.org/pls/askus/f?p=2400:1202:8169964021050853870::NO::F2400_P1202_CHECK_DISPLAY,F2400_P1202_PUB_MAIL_ID:X,844. Accessed 25 June 2006.

72. **ProMED-Ahead.** 25 July 1996. BSE & scrapie—New Zealand: free. *ProMED-Ahead* 19960725.1327. [Online.] http://www.promedmail.org/pls/askus/f?p=2400:1202:8169964021050853870::NO::F2400_P1202_CHECK_DISPLAY,F2400_P1202_PUB_MAIL_ID:X,302. Accessed 25 June 2006.

73. **ProMED-Ahead.** 26 July 1996. BSE & scrapie—Australia: free. *ProMED-Ahead* 19960726.1333. [Online.] http://www.promedmail.org/pls/askus/f?p=2400:1202:8169964021050853870::NO::F2400_P1202_CHECK_DISPLAY,F2400_P1202_PUB_MAIL_ID:X,294. Accessed 25 June 2006.

74. **ProMED-mail.** 24 March 1998. BSE, current status—European Union. *ProMED-mail* 19980324.0541. [Online.] http://www.promedmail.org/pls/askus/f?p=2400:1202:8169964021050853870::NO::F2400_P1202_CHECK_DISPLAY,F2400_P1202_PUB_MAIL_ID:X,6126. Accessed 25 June 2006.

75. **ProMED-mail.** 22 September 2001. BSE?—Japan: confirmed. *ProMED-mail* 20010923.2303. [Online.] http://www.promedmail.org/pls/promed/f?p=2400:1202:3369683606479121243::NO::F2400_P1202_CHECK_DISPLAY,F2400_P1202_PUB_MAIL_ID:X,14519. Accessed 24 June 2006.

76. **ProMED-mail.** 16 December 2001. BSE—Austria, Finland: first cases. *ProMED-mail* 20011216.3043. [Online.] http://www.promedmail.org/pls/askus/f?p=2400:1202:8169964021050853870::NO::F2400_P1202_CHECK_DISPLAY,F2400_P1202_PUB_MAIL_ID:X,17003. Accessed 25 June 2006.

77. **ProMED-mail.** 23 November 2004. CJD (new var.)—France: 9th case. *ProMED-mail* 20041123.3138. [Online.] http://www.promedmail.org/pls/askus/f?p=2400:1001:8169964021050853870::NO::F2400_P1001_BACK_PAGE,F2400_P1001_ARCHIVE_NUMBER,F2400_P1001_USE_ARCHIVE:1001,20041123.3138,Y. Accessed 25 June 2006.

78. **ProMED-mail.** 6 June 2005. BSE update 2005 (02). *ProMED-mail* 20050606.1574. [Online.] http://www.promedmail.org/pls/askus/f?p=2400:1001:8169964021050853870::NO::F2400_P1001_BACK_PAGE,F2400_P1001_PUB_MAIL_ID:1010,29208. Accessed 25 June 2006.

79. **ProMED-mail.** 11 June 2005. BSE, bovine USA: suspected. *ProMED-mail* 20050611.1625. [Online.] http://www.promedmail.org/pls/askus/f?p=2400:1001:8169964021050853870::NO::F2400_P1001_BACK_PAGE,F2400_P1001_PUB_MAIL_ID:1000,29260. Accessed 25 June 2006.

80. **ProMED-mail.** 3 July 2005. CJD (new variant) update 2005 (07). *ProMED-mail* 20050703.1889. [Online.] http://www.promedmail.org/pls/askus/f?p=2400:1001:8169964021050853870::NO::F2400_P1001_BACK_PAGE,F2400_P1001_PUB_MAIL_ID:1010,29520. Accessed 25 June 2006.

81. **Public Health Agency of Canada.** May 2003. First Canadian case of variant Creutzfeldt-Jakob disease (variant CJD). Public Health Agency of Canada. [Online.] http://www.phac-aspc.gc.ca/cjd-mcj/vcjd-ca_e.html. Accessed 25 June 2006.

82. **Reitman, V.** 19 September 2001. Japan outlaws feeding of bone meal to cattle. Agriculture: decision follows news that a plant ground up the carcass of an animal suspected of 'mad cow' disease and sold it as fertilizer and food for chicken, pigs. *Los Angeles Times*, Los Angeles, Calif.

83. **Ruff, J.** 22 June 2006. Japan prodded to move quickly. Some US lawmakers and cattle producers will seek reprisals if the country is slow to implement a deal on beef imports. *Omaha World-Herald*, Omaha, Nebr.

84. **Schmidt, L.** 14 January 2004. Japan agrees to reopen talks on Canadian beef: 'one step in a number of steps'. *National Post*, Don Mills, Ontario, Canada.

85. **Schudel, A. A., B. J. Carrillo, E. L. Weber, J. Blanco Viera, E. J. Gimeno, C. Van Gelderen, E. Ulloa, A. Nader, B. G. Cané, and R. H. Kimberlin.** 1996. Analysis of risk factors and active surveillance for BSE in Argentina, p. 138–145. *In* C. Gibbs, Jr. (ed.), *Bovine Spongiform Encephalopathy: the BSE Dilemma.* Proceedings of the Serono Symposia. Springer-Verlag New York Inc., New York, N.Y.

86. **Simon, S.** 11 January 2004. USDA plans to beef up livestock ID system; millions of cattle would be given electronic tags. Some ranchers object to the cost and fear the tracking could expose them to lawsuits. *Los Angeles Times*, Los Angeles, Calif.

87. **Soble, J.** 30 July 2003. Japan boosts beef, pork tariffs amid a steep rise in imports. *The Asian Wall Street Journal*, New York, N.Y.

88. **Sparshott, J.** 24 December 2003. U.S. gets first case of mad cow disease. *The Washington Times*, Washington, D.C.

89. **Sparshott, J.** 21 January 2006. Japan reimposes ban on US beef imports; Brooklyn meatpacker included parts of spinal column in shipment. *The Washington Times*, Washington, D.C.

90. **Thomas, P. J., M. J. Newby, and R. Zwissler.** 2003. New predictions for vCJD numbers. *Br. Food J.* **105:**420–433.

91. **Thompson, J.** 24 December 2003. Risk from mad cow case is rated low. Officials downplay the health threat from the first U.S. infection as Japan, South Korea and other nations ban American beef. *Omaha World-Herald*, Omaha, Nebr.

92. **Tibbetts, J.** 21 May 2003. Tracking the infected cow's history. *The Ottawa Citizen*, Ottawa, Ontario, Canada.

93. **Tolbert, K.** 20 September 2001. Japan to test 1 million cattle for 'mad cow'; concerns grow after first case botched. *The Washington Post*, Washington, D.C.

94. **Tomson, B.** 12 April 2004. Strict US ban on lame cattle from meat supply is opposed. *The Wall Street Journal*, New York, N.Y.

95. **United States Department of Agriculture.** 18 September 2001. Release No. 0179.01. USDA imposes import restrictions due to BSE in Japan. U.S. Department of Agriculture. [Online.] http://www.usda.gov/wps/portal/!ut/p/_s.7_0_A/7_0_1OB/.cmd/ad/.ar/sa.retrievecontent/.c/6_2_1UH/.ce/7_2_5JM/.p/5_2_4TQ/.d/2/_th/J_2_9D/_s.7_0_A/7_0_1OB?PC_7_2_5JM_contentid=2001%2F09%2F0179.html&PC_7_2_5JM_parentnav=LATEST_RELEASES&PC_7_2_5JM_navid=NEWS_RELEASE#7_2_5JM. Accessed 24 June 2006.

96. **United States Department of Agriculture.** 8 August 2003. Release No. 0281.03. Veneman announces that import permit applications for certain ruminant products from Canada will be accepted. U.S. Department of

Agriculture. [Online.] http://www.usda.gov/wps/portal/!ut/p/_s.7_0_A/7_0_1OB/.cmd/ad/.ar/sa.retrievecontent/.c/6_2_1UH/.ce/7_2_5JM/.p/5_2_4TQ/.d/3/_th/J_2_9D/_s.7_0_A/7_0_1OB?PC_7_2_5JM_contentid=2003%2F08%2F0281.html&PC_7_2_5JM_navtype=RT&PC_7_2_5JM_parentnav=LATEST_RELEASES&PC_7_2_5JM_navid=NEWS_RELEASE#7_2_5JM. Accessed 25 June 2006.

97. **United States Department of Agriculture.** 31 October 2003. Release No. 0372.03. USDA issues proposed rule to allow live animal imports from Canada. Releases risk assessment by Harvard Center for Risk Analysis. U.S. Department of Agriculture. [Online.] http://www.usda.gov/wps/portal/!ut/p/_s.7_0_A/7_0_1OB/.cmd/ad/.ar/sa.retrievecontent/.c/6_2_1UH/.ce/7_2_5JM/.p/5_2_4TQ/.d/5/_th/J_2_9D/_s.7_0_A/7_0_1OB?PC_7_2_5JM_contentid=2003%2F10%2F0372.html&PC_7_2_5JM_parentnav=LATEST_RELEASES&PC_7_2_5JM_navid=NEWS_RELEASE#7_2_5JM. Accessed 24 June 2006.

98. **United States Department of Agriculture.** 23 December 2003. Release No. 0432.02. USDA makes preliminary diagnosis of BSE. U.S. Department of Agriculture. [Online.] http://www.usda.gov/wps/portal/!ut/p/_s.7_0_A/7_0_1OB/.cmd/ad/.ar/sa.retrievecontent/.c/6_2_1UH/.ce/7_2_5JM/.p/5_2_4TQ/.d/2/_th/J_2_9D/_s.7_0_A/7_0_1OB?PC_7_2_5JM_contentid=2003%2F12%2F0432.html&PC_7_2_5JM_parentnav=LATEST_RELEASES&PC_7_2_5JM_navid=NEWS_RELEASE#7_2_5JM. Accessed 24 June 2006.

99. **United States Department of Agriculture.** 30 December 2003. Release No. 0449.03. Veneman announces additional protection measures to guard against BSE. U.S. Department of Agriculture. [Online.] http://www.usda.gov/wps/portal/!ut/p/_s.7_0_A/7_0_1OB/.cmd/ad/.ar/sa.retrievecontent/.c/6_2_1UH/.ce/7_2_5JM/.p/5_2_4TQ/.d/6/_th/J_2_9D/_s.7_0_A/7_0_1OB?PC_7_2_5JM_contentid=2003%2F12%2F0449.html&PC_7_2_5JM_navtype=RT&PC_7_2_5JM_parentnav=LATEST_RELEASES&PC_7_2_5JM_navid=NEWS_RELEASE#7_2_5JM. Accessed 25 June 2006.

100. **United States Department of Agriculture.** 22 January 2004. Release No. 0035.04. Dr. J. B. Penn, Under Secretary of Agriculture for Farm and Foreign Agricultural Services and other USDA and FDA officials press briefing on BSE at the American Embassy. U.S. Department of Agriculture. [Online.] http://www.usda.gov/wps/portal/!ut/p/_s.7_0_A/7_0_1OB/.cmd/ad/.ar/sa.retrievecontent/.c/6_2_1UH/.ce/7_2_5JM/.p/5_2_4TQ/.d/0/_th/J_2_9D/_s.7_0_A/7_0_1OB?PC_7_2_5JM_contentid=2004%2F01%2F0035.html&PC_7_2_5JM_navtype=RT&PC_7_2_5JM_parentnav=LATEST_RELEASES&PC_7_2_5JM_navid=NEWS_RELEASE#7_2_5JM. Accessed 25 June 2006.

101. **United States Department of Agriculture.** 9 February 2004. Release No. 0074.04. Final BSE update—Monday, February 9, 2004. U.S. Department of Agriculture. [Online.] http://www.usda.gov/wps/portal/!ut/p/_s.7_0_A/7_0_1OB/.cmd/ad/.ar/sa.retrievecontent/.c/6_2_1UH/.ce/7_2_5JM/.p/5_2_4TQ/.d/6/_th/J_2_9D/_s.7_0_A/7_0_1OB?PC_7_2_5JM_contentid=2004%2F02%2F0074.html&PC_7_2_5JM_navtype=RT&PC_7_2_5JM_parentnav=LATEST_RELEASES&PC_7_2_5JM_navid=NEWS_RELEASE#7_2_5JM. Accessed 25 June 2006.

102. **United States Department of Agriculture.** 25 June 2004. Release No. 0263.04. Statement by Deputy Administrator Dr. John Clifford for the

Animal and Plant Health Inspection Service. U.S. Department of Agriculture. [Online.] http://www.usda.gov/Newsroom/0263.04.html. Accessed 25 June 2006.

103. **United States Department of Agriculture.** 29 June 2004. Release No. 0266.04. Statement by Deputy Administrator Dr. John Clifford for the Animal and Plant Health Inspection Service. U.S. Department of Agriculture. [Online.] http://www.usda.gov/Newsroom/0266.04.html. Accessed 25 June 2006.

104. **United States Department of Agriculture.** 30 June 2004. Release No. 0272.04. Statement by Deputy Administrator Dr. John Clifford for the Animal and Plant Health Inspection Service. U.S. Department of Agriculture. [Online.] http://www.usda.gov/Newsroom/0272.04.html. Accessed 25 June 2006.

105. **United States Department of Agriculture.** 2 July 2004. Release No. 0275.04. Statement by Deputy Administrator Dr. John Clifford for the Animal and Plant Health Inspection Service. U.S. Department of Agriculture. [Online.] http://www.usda.gov/Newsroom/0275.04.html. Accessed 25 June 2006.

106. **United States Department of Agriculture.** 18 November 2004. Release No. 0501.04. Statement by Andrea Morgan, Associate Deputy Administrator Animal and Plant Health Inspection Service. U.S. Department of Agriculture. [Online.] http://www.usda.gov/wps/portal/!ut/p/_s.7_0_A/7_0_1OB?contentidonly=true&contentid=2004/11/0501.xml. Accessed 25 June 2006.

107. **United States Department of Agriculture.** 23 November 2004. Release No. 0508.04. Statement by John Clifford, Deputy Administrator Animal and Plant Health Inspection Service. U.S. Department of Agriculture. [Online.] http://www.usda.gov/wps/portal/!ut/p/_s.7_0_A/7_0_1OB?contentidonly=true&contentid=2004/11/0508.xml. Accessed 25 June 2006.

108. **United States Department of Agriculture.** 5 May 2005. Release No. 0149.05. USDA unveils multi-year draft strategic plan for the national animal identification system. Requests input from industry. U.S. Department of Agriculture. [Online.] http://www.usda.gov/wps/portal/!ut/p/_s.7_0_A/7_0_1OB?contentidonly=true&contentid=2005/05/0149.xml. Accessed 24 June 2006.

109. **United States Department of Agriculture.** 10 June 2005. Release No. 0206.05. Statement by Dr. John Clifford regarding further analysis of BSE inconclusive test results. U.S. Department of Agriculture. [Online.] http://www.usda.gov/wps/portal/!ut/p/_s.7_0_A/7_0_1OB/.cmd/ad/.ar/sa.retrievecontent/.c/6_2_1UH/.ce/7_2_5JM/.p/5_2_4TQ/.d/9/_th/J_2_9D/_s.7_0_A/7_0_1OB?PC_7_2_5JM_contentid=2005%2F06%2F0206.xml&PC_7_2_5JM_navtype=RT&PC_7_2_5JM_parentnav=LATEST_RELEASES&PC_7_2_5JM_navid=NEWS_RELEASE#7_2_5JM. Accessed 25 June 2006.

110. **United States Department of Agriculture.** 24 June 2005. Release No. 0232.05. USDA announces BSE test results and new BSE confirmatory testing protocol. U.S. Department of Agriculture. [Online.] http://www.usda.gov/wps/portal/!ut/p/_s.7_0_A/7_0_1OB?contentidonly=true&contentid=2005/06/0232.xml. Accessed 25 June 2006.

111. **United States Department of Agriculture.** 11 December 2005. Release No. 0544.05. Statement by Agriculture Secretary Mike Johanns regarding the opening of the Japanese market to US beef. U.S. Department of Agriculture. [Online.] http://www.usda.gov/wps/portal/!ut/p/_s.7_0_A/7_0_1OB/.

cmd/ad/.ar/sa.retrievecontent/.c/6_2_1UH/.ce/7_2_5JM/.p/5_2_4TQ/.d/3/_th/J_2_9D/_s.7_0_A/7_0_1OB?PC_7_2_5JM_contentid=2005%2F12%2F0544.xml&PC_7_2_5JM_parentnav=LATEST_RELEASES&PC_7_2_5JM_navid=NEWS_RELEASE#7_2_5JM. Accessed 24 June 2006.

112. **United States Department of Agriculture.** 6 April 2006. Release No. 0120.06. Johanns releases national animal identification system implementation plan. USDA's general standards for database integration also available. U.S. Department of Agriculture. [Online.] http://www.usda.gov/wps/portal/!ut/p/_s.7_0_A/7_0_1OB/.cmd/ad/.ar/sa.retrievecontent/.c/6_2_1UH/.ce/7_2_5JM/.p/5_2_4TQ/.d/1/_th/J_2_9D/_s.7_0_A/7_0_1OB?PC_7_2_5JM_contentid=2006%2F04%2F0120.xml&PC_7_2_5JM_parentnav=LATEST_RELEASES&PC_7_2_5JM_navid=NEWS_RELEASE#7_2_5JM. Accessed 24 June 2006.

113. **Vedantam, S., and B. Harden.** 28 December 2003. Officials now convinced cow came from Canada. *Oakland Tribune*, Oakland, Calif.

114. **Zhang, J.** 13 December 2005. US, Japan partially lift trade restrictions on beef. *The Wall Street Journal*, New York, N.Y.

chapter 12

Asymptomatic Carriers and Captive Audiences

MARY MALLON WAS A SERIAL KILLER. During the course of a career spanning roughly 15 years, she was responsible for the deaths of 3 people and the illness of an additional 44. Though never brought to trial, she was arrested twice and held in isolation for periods of 3 and 23 years. To her dying day in 1938 at the age of 69, she vehemently protested her innocence. The public knew her as Typhoid Mary (84).

Mary was born in Ireland in 1869 and emigrated to the United States when she was just 14 years old. She found work as a cook and by 1906 had worked for several wealthy families in the New York City area. In 1906 she was hired through an employment agency by the Warren family to cook for them at their summer cottage in Oyster Bay on the north shore of Long Island, N.Y. (113). Doubtless, Mary brought her personal belongings along with her to Oyster Bay on August 4, 1906. Unfortunately, she also brought an uninvited guest—*Salmonella enterica* serotype Typhi (134).

On August 27, members of the Warren household began to experience symptoms of typhoid fever. By September 3, 6 of the 11-member household were affected, including Mrs. Warren, two of her daughters, and three servants (113). Mary Mallon did not become ill, and she left Oyster Bay just 3 weeks after the typhoid outbreak began (134).

The owner of the Oyster Bay cottage, George Thompson, was concerned that an unexplained typhoid outbreak would make it impossible for him to rent his property the following summer. Although Thompson suspected that the drinking water supply to the cottage might have become contaminated with sewage, analysis of drinking water samples by two independent chemists failed to uncover any contamination. Still concerned, Thompson hired George Soper, a sanitary engineer, to find, and correct, the source of the typhoid (113). At first, Soper thought that the Warren family and their servants might have been exposed to typhoid by eating raw clams. On further investigation, however, he established that no clams had been eaten for 6 weeks before the outbreak—far too long for clams to be the source of the typhoid, which has a 10- to 14-day incubation period (15, 134).

Then Soper's attention was drawn to the history of the house and the medical history of its tenants. He quickly established that there had been no previous cases of typhoid in the Oyster Bay cottage in the 13 years before the Warren family's occupancy. Indeed, typhoid fever was not a common ailment in Oyster Bay. And none of the six people who developed typhoid fever had traveled anywhere in the several weeks preceding the onset of their symptoms (134). But if the source of the typhoid wasn't the raw clams, the Oyster Bay environment, the water supply, or the result of travel, where did the infection originate?

In 1902 Paul Frosch, a colleague of Robert Koch, was sent to investigate a major outbreak of typhoid fever in Trier, Germany. Frosch and his investigation team determined that *S. enterica* serotype Typhi (then known as *Bacillus typhosus*) had contaminated the drinking water supply. But although the measures they took to clean up the water supply reduced the spread of typhoid, those measures did not eliminate the disease completely. After analyzing the incidence of new typhoid cases in one village, Koch concluded that otherwise-healthy people must be the source of the new cases. He called these unwitting typhoid accomplices "carriers" (26).

When Soper learned that Mary Mallon had joined the Warren staff just 3 weeks before the first household member fell ill with typhoid, he began to suspect that she was a carrier of the disease. He confronted Mary with his suspicions in the hope of learning details of her work history and medical background. He succeeded only in angering her. She vehemently denied ever having typhoid fever and would give Soper no additional information. Fortunately, he was able to obtain a partial list of Mary's previous employers from the agency through which she found work (134). When he contacted her previous employers, Soper discovered that seven of the eight families for which she worked during the previous 10 years had suffered an outbreak of typhoid while Mary was their cook (113).

Soper presented his data to Hermann Briggs, the New York City Department of Health's medical officer (134). Briggs sent an emissary, Josephine Baker, to speak to Mary and convince her to cooperate with the health department by furnishing stool and urine samples for testing. But Baker was no more successful with Mary than Soper had been. Finally, in March 1907, the police were called, and Mary was taken forcibly to Willard Park Hospital where she was tested and found to be shedding serotype Typhi in her stool (59). She was detained by the Department of Health for 3 years and was released in 1910 after promising not to work as a cook and not to engage in any work that would bring her into contact with food in any way. She was also required to report periodically to the Department of Health, much like a convicted criminal on parole. But after a while, Mary broke parole and disappeared (134).

In January and February of 1915, the Sloane Hospital for Women in New York City suffered an outbreak of 25 cases of typhoid fever, mostly among its nurses and attendants. Three months before the outbreak began, the hospital had hired a new cook, known to them as Mrs. Brown. Mary Mallon had resurfaced. Before she could be stopped, Mary disappeared again, but was soon found and detained. Mary Mallon remained an involuntary resident of Riverside Hospital on North Brother Island, N.Y., until her death in 1938 (84).

Typhoid Mary wasn't unique, except in her notoriety. She may have been the first healthy carrier of serotype Typhi identified in the United States (134), but she was not the only one. In 1919, New York City was estimated to have 25,000 typhoid carriers (84). Although typhoid fever is far less common today than it was a century ago, serotype Typhi is still with us. In the first 7 months of 2005, typhoid outbreaks were reported in Algeria, Congo, Fiji, Gabon, Kenya, Malaysia, Russia, the Philippines, and Turkmenistan (6,–11, 120–122). And the United States has not been immune from the occasional spurt of typhoid cases either. There have been at least five U.S. outbreaks of typhoid fever

Salmonella and Asymptomatic Carriers

No one can say when Mary Mallon became infected with *Salmonella enterica* serotype Typhi. But it's clear that, once a carrier, she continued to shed the pathogen in her stool for many years—perhaps even for the rest of her life (113).

Long-term shedding of *Salmonella* is the rule, rather than the exception, for healthy carriers. The pathogen hides out in the gallbladder and is released, apparently at random intervals, into the intestines. It's not uncommon for someone who has suffered from salmonellosis to continue shedding the microbe for several weeks after his or her symptoms have disappeared (132). But this is different from chronic carriage. An individual can, unsuspectingly, become a chronic carrier and shedder of *S. enterica* serotype Typhi or *S. enterica* serotype Paratyphi for months or even years, without ever having shown any symptoms of the disease at all (58, 87, 138).

Detecting an asymptomatic carrier of *Salmonella* can be difficult. Shedding is sporadic and unpredictable. Therefore, a suspected carrier must sometimes submit stool sample after stool sample in the hope (or fear) of detecting the pathogen. Once an individual is found to be a chronic carrier, the battle has just begun. Curing a carrier of his or her chronic infection is not an easy task.

For many years, although the incidence of *Salmonella* carriage was known to be higher among those individuals who suffered from a diseased gallbladder, the mechanism behind the association between the gallbladder and chronic carriage of *Salmonella* was not well understood. Then, in 1992, a group of researchers at the Chinese University of Hong Kong postulated that serotype Typhi formed a biofilm on the surface of gallstones lodged in a patient's common bile duct. When the gallstones were removed, the patient ceased shedding the pathogen (82). Ten years later, researchers in San Antonio, Tex., confirmed independently that *Salmonella* spp. could form biofilms when inoculated onto the surface of gallstones. They established several conditions—including the presence of bile, flagella, and quorum sensing—that affected the efficiency of biofilm formation on the stones. Biofilm formation by *Salmonella*, according to these researchers, most likely was essential to the development of a carrier state (123). And understanding the mechanism of the biofilm formation may help the medical profession to develop effective methods to eliminate or prevent asymptomatic carriage of *Salmonella*.

in the last 25 years, one each in Oakland, Calif.; Jackson, Mich.; New York, N.Y.; Skagit County, Wash.; and Silver Spring, Md. (18, 19, 87, 124, 138).

Carrying On

Although outbreaks of typhoid fever that occur in underdeveloped countries are usually linked to contaminated water, the most common mode of serotype Typhi transmission in the United States is the asymptomatic carrier (18, 19, 87, 124, 138). *S. enterica* serotype Typhi, however, is just one of several gastrointestinal pathogens that can be spread by human contact. *Shigella* spp., hepatitis A virus, norovirus, *Campylobacter*, *Cryptosporidium*, various *Escherichia coli* serotypes, and other *S. enterica* serotypes are often transmitted by unsuspecting carriers (36, 40, 41, 49, 51, 58, 67, 110, 115, 132, 144). And, as several dozen customers and employees of a Hardee's fast-food restaurant outlet in Winona, Minn., found out in 1989, one asymptomatic carrier working in a restaurant or cafeteria can trigger a chain-reaction *Salmonella* outbreak (2, 65).

The outbreak, as far as investigators could tell, began on September 18, 1989, when a food handler (designated in the investigation report as "employee A") working for a Hardee's fast-food restaurant in Winona started to experience abdominal cramps. The employee continued to work for several days despite the discomfort but stayed home starting on September 23, after developing diarrhea. A sample of the employee's stool, taken on September 27, contained *S. enterica* serotype Enteritidis (65).

The Minnesota Department of Health first received word of the possible outbreak on September 27, after four patients with symptoms of salmonellosis visited the local hospital emergency room between September 23 and 25. All four patients reported having eaten at Hardee's during the 3 days before they began to develop their symptoms. An additional 10 people were also identified by the hospital and a local clinic as suffering from diarrhea, and *Salmonella* was isolated from their stools. When contacted by the health department investigators, all 10 reported that they had eaten at the same Hardee's restaurant outlet as the first four patients. On visiting the restaurant, investigators learned that several employees were also ill. On September 27, the same day that the health department learned of the outbreak, restaurant management agreed to close until the source of the *Salmonella* outbreak was found and the situation could be remedied (65).

In all, 89 people, including both restaurant patrons and employees, were identified as probably (symptoms of diarrhea and fever or chills) or certainly (confirmed isolation of *Salmonella* from the patient's stool in addition to the symptoms) suffering from salmonellosis. Of those 89 victims, the health department confirmed that 50, including 17 restaurant employees, were infected with serotype Enteritidis (65).

Table 12.1 Hardee's 1989 *Salmonella* outbreak: probable timeline (65)

Date	Event
September 18	Employee A begins to experience abdominal cramps; continues working through September 22.
September 20	Second employee becomes ill.
September 21	First patrons become ill.
September 23	Employee A develops diarrhea; stays home.
September 23	Patients begin to visit hospital emergency room.
September 27	Hospital notifies health department of cluster of four *Salmonella* cases.
September 27	Minnesota Department of Health investigators visit restaurant and determine that several employees had suffered from diarrhea during the preceding 2 weeks.
September 27	Hardee's closes voluntarily until the source of the outbreak can be determined and corrected.
September 29	Last outbreak victims begin to experience symptoms.

The investigation team reconstructed the probable series of events that culminated in the outbreak, based on the results of interviews, lab analyses, and inspection of the restaurant facilities (Table 12.1). They concluded that employee A was the probable source of the outbreak. This employee, who continued to handle food while suffering from abdominal cramps, likely transmitted *Salmonella* to coworkers and restaurant patrons; the infected coworkers, in their turn, passed the pathogen to additional patrons, even after employee A became ill with diarrhea and stayed home from work. The investigators noted that just a few of the infected food handlers reported experiencing symptoms of salmonellosis; asymptomatic or mildly symptomatic food handlers were actively shedding *Salmonella* in their stool and—probably as a result of the inadequate hand washing practices observed during a routine inspection of the restaurant less than 1 week before the outbreak—transferred *Salmonella* onto various food items (65).

Coffee, Tea, or Hepatitis?

While Mary Mallon put a face to the concept of asymptomatic carriage of disease, the scope of the problem extends far beyond carriage of *Salmonella*, or of other bacterial pathogens. Viruses such as norovirus and hepatitis A can also be spread rapidly and broadly by carriers who either are asymptomatic, are in the early, presymptomatic stages of illness, or are convalescent but still shedding infectious virus particles (16, 25, 60). Some parasites, such as *Entamoeba histolytica*, are also spread in this way (137).

Unlike *Salmonella*, hepatitis A virus (HAV) has a long incubation period; 15 to 50 days can elapse between the time a victim becomes infected and the onset of the first symptoms (40). The virus particles can appear in a victim's stool as early as 10 days after exposure—several days or weeks before symptoms first appear (15). The long asymptomatic period during which a victim can unwittingly transmit hepatitis to others makes detecting an outbreak and tracing its source a slow and difficult process, as happened in Bristol County, Mass., in 2001 (83).

On October 17, 2001, a restaurant employee in Swansea, Mass., began to experience symptoms of hepatitis. On October 26, the restaurant's owners notified the state's public health authorities that one of its employees had been diagnosed with HAV infection. That same day, the owners closed the restaurant for an extensive cleaning. The next day, state public health inspectors visited the restaurant; they determined that no other workers were complaining of symptoms of hepatitis and that the restaurant was in complete compliance with the state's sanitary requirements. Nineteen restaurant workers were treated with an injection of immunoglobulin as a precaution and the restaurant was allowed to reopen (83).

The ill employee was a manager; he handled food only to a limited extent. When interviewed by the health authorities, the employee stated that he wore gloves whenever handling food and that he was careful and diligent about washing his hands. The employee wore an ostomy bag but said that he never changed the bag at work. On the basis of the employee's limited contact with food, and his apparently good personal hygiene practices, the Massachusetts Department of Public Health (MDPH) decided that it would not be necessary to notify the public of the incident (83).

The general public first learned of the problem on November 24, almost a full month later, when the *Providence Journal* and the *Boston Globe* broke the story that an outbreak was under way (3, 125). Twenty-five cases of hepatitis A had been diagnosed in Bristol County, Mass., between late October and November 24, and health authorities were scrambling to find the source of the virus, to identify more victims, and to treat with immunoglobulin those who had come into contact with hepatitis victims (125).

On November 27, 3 days after the story first appeared in local newspapers, the MDPH made it official: 33 cases of HAV infection had been confirmed in Bristol County since November 8 (94). By the time the outbreak was over, 46 people—three-fourths of whom had eaten at the D'Angelo Sandwich Shop in Swansea, Mass., in October or early November—had contracted hepatitis (83, 126). Two of the victims were food handlers at another restaurant, Rudy's Country Store, raising the specter of a secondary outbreak. Massachusetts health officials recommended that everyone who had been exposed to a hepatitis victim or who had eaten at Rudy's between November 14 and November 23 should receive an injection of immunoglobulin as a precaution. More than 1,700 people followed the state's advice, receiving free injections at an area hospital clinic (126); no cases of hepatitis were ever traced to Rudy's (83).

The Hepatitis Viruses

Although sometimes lumped casually together in an introductory virology course, in reality, hepatitis viruses A through G have very little in common with each other, except their propensity for attacking the liver. They belong to at least four different virus families. Hepatitis B is a DNA-containing virus, whereas A, C, D, E, and G rely on RNA as their genetic makeup (81); and hepatitis B, C, D, and G are transmitted parenterally (e.g., through transfusions or sharing of contaminated needles by intravenous drug users), while hepatitis A and E are enteric viruses (40, 81). Hepatitis F appears to be no more than a researcher's hypothesis; the existence of this virus, first proposed in 1994, was still unconfirmed in 2002 (76).

Hepatitis A and E viruses infect their victims through the ingestion of contaminated food or water or by direct person-to-person contact (fecal-oral route). Hepatitis E is found most commonly in parts of Asia (including the Indian subcontinent and the Middle East) and Africa, but it also has been reported in California (131, 133, 141). Hepatitis A, on the other hand, is distributed worldwide (40).

Hepatitis A virus is a survivor, which increases its potential for causing mischief (98). It is resistant, except at high levels, to most disinfectants, including hypochlorite, glutaraldehyde, and quaternary ammonium compounds; is partially protected from heat inactivation in higher-fat dairy products; can persist in live oysters for up to 6 weeks after they have been removed from contaminated growing beds; and is able to survive in dried feces for a month or more (17, 70, 79, 97, 101). Hepatitis A virus can remain alive for several hours on a person's hands and is transferred easily from the hands of an infected food handler onto the food being handled (16, 96).

In the early 1990s, researchers in Ottawa, Canada studied the survival of hepatitis A virus on the fingertips of volunteers. They smeared the fingertips of five volunteers with a fecal suspension containing a measured quantity of hepatitis A virus and found that up to 30% of the virus could still be detected after 4 hours. And even after drying for 4 hours on a volunteer's hand, those living virus particles could be transferred from the inoculated fingertips to metal disks by simple pressure (96). The researchers also had volunteers test the effectiveness of various hand washing products—regular hand soap, medicated hand soaps, and germicidal soaps—at removing or killing hepatitis A virus. They determined that most of the products were inadequate; live virus particles could still be transferred to metal disks even after the volunteers washed their hands with plain soap or with many of the medicated products. Only using hand cleaners that contained high levels of alcohol prevented the transfer of live hepatitis A particles (99).

In view of how difficult it can be to wash away or kill the virus, vaccination of food service workers against hepatitis A virus has been proposed as a way to prevent outbreaks. According to a study published in 2000, vaccinating 100,000 U.S. food service workers would prevent 2,500 symptomatic infections, 93,000 days of illness, and 8 deaths. The study's authors estimated that a compulsory vaccination program would cost $8.1 million, which would be offset by savings due to reduced loss of work days and reduced need for outbreak investigations (69). The city of St. Louis adopted compulsory vaccination of food service workers in 2001 (37).

Many health authorities, though, disagree with this approach. The New York State's Department of Health, for example, does not require vaccination of food service workers on the basis that only 2 to 3% of hepatitis A cases have been traced back to restaurant food and that high employee turnover makes compulsory vaccination of food service workers impractical (111).

The MDPH handling of this outbreak was both a failure and a success. The department's initial evaluation that the D'Angelo employee was unlikely to transmit his hepatitis was flawed. It was based on the employee's evaluation of his own personal hygiene practices and on the relatively small portion of his time spent handling food. By the time the state realized that an outbreak was under way, it was too late to warn the sandwich shop's patrons or to offer protective immunoglobulin injections to them (83, 127). On the other hand, the MDPH acted promptly to warn the public on finding out that two outbreak victims were food handlers at another restaurant. By doing so, and by offering immunoglobulin to the patrons of the second restaurant and to anyone who had direct contact with hepatitis victims from the initial outbreak, the Massachusetts health officials short-circuited a possible chain-reaction epidemic.

Shipping Out

Being on the receiving end of a gastrointestinal ailment transmitted by an asymptomatic food handler is always unpleasant, and sometimes dangerous, for the victim. But fast action on the part of public health workers can put an abrupt stop to a restaurant-based outbreak. If a restaurant must be closed, the population won't go hungry or thirsty. Things are different, however, on board a ship (42, 105). Passengers and crew alike are captives of a cruise ship's kitchens and food service staff. Except for excursions ashore while the ship is in port, a cruise ship's population relies on the ship's restaurants, room service, bars, and dining rooms for all of its meals and snacks. A single carrier of a pathogen such as norovirus can cause an explosive outbreak of gastrointestinal illness among the ship's passengers and crew, resulting in quarantine of passengers in their quarters or even in an early return of a cruise liner to its home port. And, sometimes, the problem carries over to subsequent cruises.

The U.S. Center for Disease Control (CDC) established a Vessel Sanitation Program (VSP) in the 1970s, in response to several disease outbreaks on cruise ships (33). VSP personnel conduct sanitation inspections of ships that call at U.S. ports, investigate disease outbreaks, train ship employees to prepare and handle food safely, review the design of new ships, and inspect newly constructed ships. CDC publishes on its website the results of its cruise ship inspections, making this information easily available to the public (33, 34).

For many years, CDC's efforts appeared to be successful. Only a handful of outbreaks per year (mean, 5.5; median, 5) were reported on cruise ships between 1994 and 2001 (35). Then in 2002, the annual number of outbreaks increased dramatically, rising nearly fivefold to an average of 28 (median, 27) outbreaks per year between 2002 and 2004. The pattern continued into 2005, with 16 outbreaks reported in the first half of the year (35). And more than half (56%) of the outbreaks that occurred from 1994 on were due to norovirus (Fig. 12.1).

One of the worst shipboard outbreaks investigated by CDC began in October 2002. It encompassed six consecutive cruises of the Holland

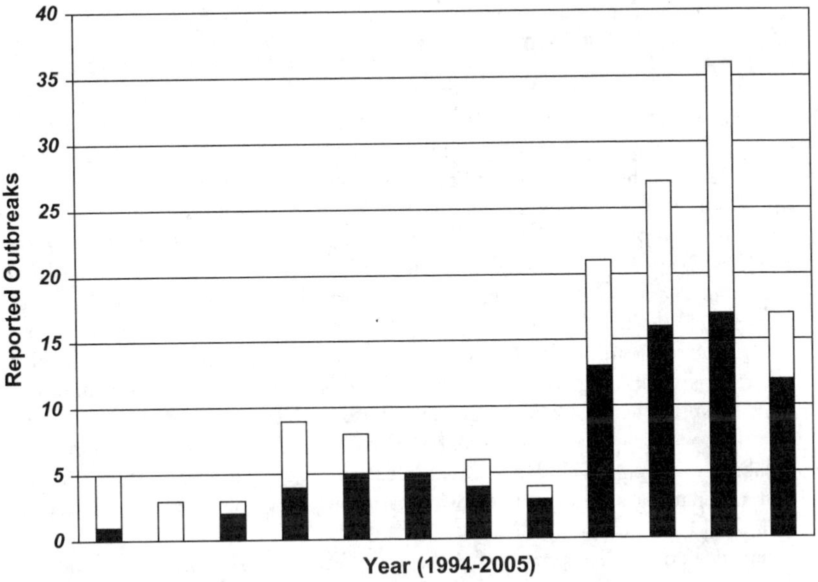

Figure 12.1 Frequency of reported outbreaks of gastroenteritis due to norovirus on cruise ships: 1994–2005 (35). Solid columns, norovirus; open columns, other gastroenteritis outbreaks.

America Line's passenger cruise ship *Amsterdam*. Several hundred people, passengers and crew members, became infected with norovirus, and the outbreak also spread to a nursing home when a resident who had contracted norovirus on one of the cruises returned to the facility (68).

North to Alaska

If real life were a screenplay, the norovirus outbreak story would have begun in Vancouver, B.C. Between April and June 2002, Canada's third-largest city experienced a series of norovirus gastroenteritis outbreaks that were traced to two restaurants and a hotel banquet (103). Investigators determined, on the basis of nucleic acid sequencing, that the same strain of norovirus caused both the hotel outbreak and a series of outbreaks that were traced to a restaurant located just a few blocks from the hotel. The virus's profile also matched norovirus outbreaks that had taken place in nursing homes located in three towns outside of, but near to, Vancouver (103).

Vancouver, Canada, is the port of departure for many of the passenger liners that cruise the Alaska coast. On July 18, 2002, the Holland America ship *Ryndam* stopped at Vancouver to take on passengers for its cruise to Ketchikan, Alaska. The next day, five passengers suffering from gastrointestinal illness reported to the infirmary (38). On July 23, 5 days into the cruise, Holland

Norovirus—What's in a Name?

Acute infectious nonbacterial gastroenteritis, characterized by a short (24 to 48 hours) incubation period, vomiting, diarrhea, and a relatively short (12 to 60 hours) duration, has been recognized as a disease syndrome since the 1940s (45, 75). But it was only in 1972 that researchers with the U.S. National Institutes of Health (NIH) were able to determine its cause—a 27-nm virus particle (75).

In 1968, an outbreak of infectious gastroenteritis struck a school in Norwalk, Ohio (61). Despite the best efforts of investigators, no microorganism could be linked to the outbreak. In time, NIH researchers, realizing that the pathogen they sought was fastidious and would be extremely difficult to cultivate, eschewed conventional isolation techniques. Instead, they used immune electron microscopy, mixing a sample with antiserum and examining the mixture for the presence of aggregated virus particles in suspensions of fecal material from victims of the outbreak (75). Serological tests and human volunteers helped confirm that the 27-nm virus was the long-sought cause of at least some unexplained cases of infectious gastroenteritis (74, 75). The newly discovered pathogen became known as the "Norwalk virus," in recognition of the 1968 outbreak (106).

In time, other small human viruses that were similar in morphology to the Norwalk virus were discovered. The names of each of these "Norwalk-like viruses" were based on the geographic location where they were first discovered, for example, Sapporo, Snow Mountain, and Hawaii. Ultimately, the Norwalk-like viruses (also known as the Norwalk family of viruses) were assigned to the family *Caliciviridae* and given the genus names of *Sapovirus* and *Norovirus* (21, 62, 95).

America notified the CDC of an outbreak of gastrointestinal disease on board the *Ryndam*. By the time the ship docked in Vancouver on July 25, 167 passengers (out of 1,318) and 9 of the crew of 564 had fallen ill with norovirus infection (32).

After the passengers disembarked on July 25, the crew disinfected the ship. That same afternoon, the *Ryndam* took on a new complement of passengers and sailed from Vancouver for another 7-day cruise to Alaska. By the end of the sixth day, 189 passengers and 30 crew members were complaining of gastrointestinal illness. This time, when the ship returned to Vancouver, Holland America cancelled its next scheduled departure and carried out an extensive cleaning and sanitizing of the ship (32). These efforts appeared to work; no more shipboard outbreaks were reported to CDC from the *Ryndam* until 2004 (35).

On October 1, approximately 2 months after the *Ryndam*'s illness-marred cruises, the *Amsterdam* set sail from Vancouver on a 21-day cruise down the west coast of North America and through the Panama Canal to Florida (29). By October 16, 101 passengers (8%) and 14 members of the ship's crew (2%) had reported to the infirmary with symptoms of gastrointestinal illness (38). By the time the ship arrived in Fort Lauderdale, Fla., at the end of the 21-day cruise, 30% of the passengers and 4% of the crew had suffered from norovirus-caused gastroenteritis (145).

South to the Sunbelt

The *Amsterdam* set sail again on October 22, almost immediately after its arrival in Fort Lauderdale—this time for a 7-day Caribbean cruise, during which 33 passengers (3%) and 7 crew (1%) became infected with norovirus. Its second Caribbean cruise was worse: 155 passengers (12%) and 18 members of the crew (3%) were stricken with the virus (38). When the *Amsterdam* also experienced a norovirus outbreak during its third consecutive Caribbean cruise, Holland America finally decided to take the *Amsterdam* off-line for 10 days in order to scrub and sanitize the ship and its contents (29). Everything was cleaned—even the Bibles (4). The 10-day sanitizing program worked; no further outbreaks were reported on the *Amsterdam* during the 2002–2003 Caribbean cruise season (38). But this wasn't the end of the norovirus outbreaks. Even Walt Disney's Imagineers could not protect that company's cruise ship from the ubiquitous microbe.

Disney's *Magic* left Port Canaveral, Fla., on November 16, 2002, for a 7-day Caribbean cruise (38, 72). The first ill passenger surfaced on the ship's second day at sea (68), and by November 20, 28 passengers and 7 crew members had been infected by norovirus (38). By November 23, when the ship returned to port, 260 passengers (12% of the passenger list) and 17 crew members (2% of the crew) had reported gastrointestinal symptoms. All passengers and crew of the *Magic* were handed a CDC questionnaire while still at sea. Of the passengers who completed and returned the questionnaire, 492 (21%) reported symptoms that were consistent with norovirus infection. As soon as the ship reached port, Disney staff began a 14-hour marathon of cleaning and sanitizing, and CDC inspectors from the VSP boarded the *Magic* to try to determine what might have contributed to the outbreak (38, 72). The *Magic* set sail with a new set of passengers that same evening (73).

Unfortunately, the 14-hour cleaning session was not enough to eliminate the virus. Symptoms of norovirus infection appeared among both passengers and crew soon after the *Magic* sailed on the evening of November 23. Twenty-three crew members (2%) and 195 passengers had reported being ill by the time the ship returned to Port Canaveral on November 30 (38). This time, while the ship was still out at sea, the Disney Cruise Line chose to cancel the *Magic*'s scheduled December 1 cruise and subject the ship to an intensive cleaning and decontamination instead (107). The weeklong cleanup reduced the presence of norovirus without eliminating the problem completely (145). The number of people who fell ill from norovirus during the December 7 cruise dropped to just 61 passengers and crew. And the numbers decreased to fewer than 40 patients with norovirus infection in the following two cruises—well below the 3% incidence rate that constitutes an outbreak according to CDC (33, 145).

The Canadian Connection

It's tempting to link the outbreaks that took place in 2002 on the *Ryndam*, the *Amsterdam*, and the *Magic*. The same strain of norovirus—the Farmington Hills (FH) strain (named for the town in Michigan where it was first identified as the source of an outbreak in 2002)—was responsible for the outbreaks on all three ships (145). And it's equally a temptation to blame the contamination of the *Ryndam* and the *Amsterdam* on norovirus that was endemic in Vancouver and Alaska at the time of the shipboard outbreaks. Although no one has compared the strains prevalent in Vancouver in 2002 to the FH strain, there were at least two FH norovirus outbreaks in Alaska that year, one in October and the other in November (145). A passenger or member of the crew of the *Ryndam* or the *Amsterdam* might have picked up the virus in port and spread it through the ship. It's even possible—though unlikely, given the 2-month gap between outbreaks—that the virus traveled from the *Ryndam* to the *Amsterdam*, carried by a crew member or hitching a ride on a piece of equipment.

The possibility that the *Amsterdam* and *Magic* outbreaks were linked is even more remote. The two ships are owned and operated by different corporations; Holland America operates the *Amsterdam* and Disney Cruise Lines owns the *Magic*. The ships also sailed from different ports; the *Amsterdam* used Fort Lauderdale's Port Everglades as its Florida base and the *Magic* sailed out of Port Canaveral, Fla. (29, 72).

The trail of epidemiological evidence also argues against a sequential transmission of the virus from ship to ship. The *Ryndam*, *Amsterdam*, and *Magic* were all hit by the FH norovirus strain. Before the year was out, FH had been responsible for 7 shipboard outbreaks and at least 10 additional land-based outbreaks in eight U.S. states (145). Investigators of the outbreaks that occurred on the *Ryndam* and the *Amsterdam* could find no common links pointing to a sequential transfer of the virus from ship to ship. There were no crew transfers between the two ships, and no food or water supply in common, even though Holland America operated both ships and both sailed from the same port until the *Amsterdam* left for Florida on October 1 (145). The CDC concluded that one or more infected passengers probably brought the norovirus onto the ship. Once on board, the virus spread rapidly and widely, foiling attempts to disinfect the ships between scheduled sailings. And even when the disinfection was effective, the virus was sometimes introduced anew. The outbreaks that were reported on the *Magic* after the ship had been scrubbed down and returned to service were due to norovirus strains with two slightly different genetic makeups; the original outbreak strain, which had survived attempts to disinfect the ship, had been joined by a second strain, introduced by an infected passenger or crew member (145).

Norovirus continues to be a problem wherever large numbers of people gather. Day care facilities, nursing homes, military barracks, Las Vegas casinos, and hurricane evacuation shelters—have all suffered norovirus outbreaks (1, 12, 14, 27, 43, 55, 60, 63, 93, 100, 102, 136). Even the Boy Scouts are not immune (86). The virus has an incubation period of 24 to 48 hours and a relatively short (usually 48 hours or less), but explosive, symptomatic stage, during which it can be spread easily by person-to-person contact or through contaminated food, water, and surfaces (13, 146), and victims can continue to shed the virus for as long as 2 weeks after their symptoms have disappeared (61). Yet norovirus is benign compared to some of the gastrointestinal pathogens that can be found in hospitals.

The Nosocomial Nightmare

Hospitals are dangerous places. Even the briefest, most routine hospital stay carries risks, not the least of which is the risk of contracting a nosocomial (hospital-acquired) infection. Patients are completely dependent on the hospital staff for their food, beverages, and medication and must rely on the hospital's attention to proper food handling procedures, cleaning and sanitation, sterilization of instruments, and personal hygiene. An asymptomatic carrier—whether patient, staff member, or visitor—of a gastrointestinal pathogen in a hospital is an invitation to an outbreak.

Outbreaks of gastroenteritis can be particularly severe in a hospital or nursing home due to the reduced resistance to infection that is associated with hospital patients and nursing home residents. *Salmonella* spp., *Shigella* spp., norovirus, rotavirus, *Campylobacter*, and several other pathogens have been implicated in many such outbreaks (Table 12.2). Most nosocomial outbreaks last no more than a few weeks or months and affect a relatively small number of patients and staff. But there is the occasional exception.

Recently, the Canadian province of Quebec experienced a 2-year-long nosocomial outbreak caused by a virulent strain of *Clostridium difficile*. The outbreak, which began in late 2002 or early 2003, encompassed at least 12 hospitals in Montreal and one in Sherbrooke and, in 2004 alone, sickened at least 44,000 patients, killing more than 274 of them (48, 50, 53, 54, 118).

C. difficile is an important cause of nosocomial infections worldwide and has been responsible for diarrheal illnesses in Europe, Asia, Africa, the Americas, and Australia (56, 57, 66, 77, 90, 104, 108, 109, 112, 114, 117, 130, 135, 139, 140). It targets the immunocompromised, the elderly, and people being treated with broad-spectrum antibiotics such as cephalosporins (139). Some individuals who become infected with *C. difficile* remain asymptomatic; others develop a watery, sometimes recurrent, diarrhea. And 1 to 3% of patients are stricken with a life-threatening form of colitis (119).

In mid-2002, researchers at the Centre Hospitalier Universitaire de Sherbrooke (University of Sherbrooke Medical Centre) noticed an increased

Table 12.2 Examples of nosocomial outbreaks of gastrointestinal disease

Pathogen	Venue	Country	Year	No. of cases (deaths)	Reference
Campylobacter jejuni	Hospital	Spain	2000	2	88
Clostridium difficile	Hospital	France	1993–94	37	57
C. difficile	Hospital	Spain	2000–01	35	117
C. difficile	Hospital	United States	2000–01	253 (18)	109
C. difficile	Hospital	Japan	2001	15	130
Escherichia coli O157:H7	Psychiatric hospital	Canada	2002	109 (2)	20
Hepatitis A virus	Hospital	United States	1991	26	28
Hepatitis A virus	Hospital	Norway	1996	15	71
Norovirus	Multiple hospitals	New Zealand	2004	>424	93
Norovirus	Hospital	United Kingdom	2002	99	60
Pseudomonas aeruginosa	Hospital	Korea	1997–99	7	78
Salmonella enterica serotype Enteritidis	Hospital	United Kingdom	1993	29	51
S. enterica serotype Livingstone	Hospital	Tunisia	2002	16	23
S. enterica serotype Reading	Hospital	United States	1990	≥27	129
S. enterica serotype Saintpaul	Hospital	United States	2001	11	22
S. enterica serotype Senftenberg	Hospital	United States	1993–94	22	85
Shigella dysenteriae	Orphanage	Guatemala	1991	4 (2)	39
S. dysenteriae	Hospital	United States	1983	107	89

incidence of what they called "severe *C. difficile*-associated diarrhea," or CDAD. They reviewed all cases of CDAD that had been diagnosed from January 1, 1991, to December 31, 2003, in their 683-bed hospital—1,721 in all. They discovered that the incidence of CDAD had increased more than fourfold during the 12 years covered by their review; the proportion of patients who suffered complications more than doubled from 7.1 to 18.2%, and the death rate among infected patients jumped from 4.7% in 1991–92 to 13.8% in 2003. The researchers offered three possible explanations for the heightened incidence and severity of *C. difficile* infections in their hospital, any or all of which might have played a role: introduction of a more virulent strain of *C. difficile*; a higher proportion of older patients than in previous years; or the evolution of a less sanitary hospital environment due to cutbacks in the hospital's budget (118).

The 12-hospital Montreal epidemic was caused by the same strain of *C. difficile* as the dominant strain responsible for the Sherbrooke outbreak. Hospitals in both cities suffered from budget limitations that forced cutbacks in the size and training of their cleaning staffs. Many of the affected hospitals had been built around the turn of the 20th century, lacked bedside hand washing stations, and had too few private rooms with private bathrooms for isolating *C. difficile*-infected patients. Karl Weiss, the infectious disease specialist and microbiologist at Montreal's Hôpital

Clostridium difficile—the Aptly Named Pathogen

C. difficile is a gram-positive, anaerobic, spore-forming bacterium. When Hall and O'Toole first found it in the stools of healthy infants in 1935, they named their new bacterium *Bacillus difficile*, a testimony to the difficulty they had in isolating it (92). Since then, it has truly lived up to its initial reputation. *C. difficile* is tough to kill (being a spore-former, it is more resistant to heat and germicides than the average pathogen) and difficult to contain in a hospital setting. The syndrome it causes (pseudomembranous colitis) is hard to diagnose and can be confused with diseases such as Crohn's or ulcerative colitis (80). An outbreak of *C. difficile* can be a hospital's worst nightmare.

C. difficile is a normal inhabitant of the intestinal tracts of newborn infants and has been found in up to 63% of healthy newborns and young children (80). It occurs less frequently in healthy adults, with carriage rates of up to 15% having been reported (80). Most of the time, *C. difficile* is kept under control by the rest of the normal, healthy intestinal flora. But when that flora is erased by the use of a broad-spectrum antibiotic, *C. difficile* takes over, causing symptoms that may range from mild, watery diarrhea to a life-threatening colitis accompanied by fever (which can exceed 101°F [38°C] in 26 to 66% of patients), pseudomembrane formation, elevated leukocyte counts, and inflammation (80).

Once they have been correctly diagnosed, patients who have *C. difficile* diarrhea or colitis can be treated successfully. Often, fluid replacement combined with stopping the use of the offending antibiotic treatment is all that is needed. Other times, doctors resort to changing to a different antibiotic, such as metronidazole or vancomycin, to which *C. difficile* is sensitive (143). Probiotic therapy is also sometimes used to help repopulate the patient's intestines with "good" bacteria (44). But the best defense against a *C. difficile* outbreak is proper infection control procedure: strict adherence to correct use of cleaners and sanitizers; careful attention to washing of hands by doctors, nurses, and orderlies; and isolation of patients with diarrhea in private rooms equipped with private bathrooms (30, 143).

Maisonneuve-Rosemont, got to the heart of the problem during an interview with the *Canadian Medical Association Journal*. "We are," he explained, "practising 21st century medicine in a 19th century environment" (54).

The symptoms of *C. difficile* infection are triggered by cytotoxins produced by the microbe. The outbreaks that took place in Montreal and Sherbrooke between 2002 and 2005 apparently were caused by an especially virulent strain of *C. difficile*, one that was later shown to produce 20 times the usual amount of cytotoxin (24). This same strain was also implicated in several extended outbreaks of nosocomial diarrhea in the United States beginning sometime in 2001 and in the United Kingdom during 2003–2005 (24, 52, 64). But increased virulence alone is not enough to explain the lengthy outbreaks in Montreal and Sherbrooke. The age of hospital facilities and the limited resources usually devoted to sanitation are also key elements in the equation.

As Weiss pointed out, many of the hospitals in the province of Quebec were built around the start of the 20th century (54). Although state of the art at the time of their construction, these hospitals fell far short of modern requirements. Many rooms lacked bedside sinks at which doctors and nurses could wash their hands before and after visiting patients—a necessity to prevent the spread of infectious diseases (91). At the time of the outbreaks, these older hospitals were equipped with very few private rooms and private bathrooms; when they were built, large wards housing several patients were the norm. Even today, it's not unusual in many hospitals for four patients to share a "semiprivate" room with a single bathroom (142), and the shortage of private rooms makes it difficult to isolate patients who are infected with *C. difficile* from the rest of the hospital population (47). As Mary Vearncombe (medical director for infection prevention and control at a major Canadian hospital) pointed out in an interview in 2004, "If you are in an old, four-bed room sharing a toilet with four people, one of whom has diarrhea, you're going to get it" (30).

Infection control in Quebec's aging hospitals was made more difficult by the province's cutbacks to hospital funding in the decade before the *C. difficile* outbreak (5). Lack of funds translated into bed closures and staff reductions. And staff reductions meant fewer housekeepers and less frequent—and less thorough—cleaning of patient rooms and bathrooms (47, 116). Nor, at first, did the Quebec government increase hospital funding in the face of this outbreak. In ruling out the short-term release of additional money to help hospitals fight *C. difficile*, the health minister's press attachée explained the government's position (46). "Even if we spent thousands and thousands of dollars," she told a reporter with Montreal's *Gazette*, "if (doctors) don't wash their hands between patients, all that money would be useless." Finally, in January 2005, Quebec announced that it would increase its hospital funding by $20 million: $15 million for hospital upgrades and renovations, and $5 million to improve cleaning and infection-control programs (31).

A Clean Sweep

Cleaning and sanitation are not glamorous activities. As the Quebec hospital outbreaks demonstrate, they can often be given short shrift—even by people who understand their importance—when budgets are prepared. Yet, the relatively small amounts of money saved by cutting back on cleaning can be greatly overshadowed by the costs that are incurred as a result of a major outbreak. In the United Kingdom, for example, nosocomial gastroenteritis outbreaks cost British taxpayers an estimated £115 million ($184 million) in 2002–2003 (90). Holland America and Disney lost revenue when they were forced to cancel sailings in order to scrub and sanitize their cruise ships after repeated outbreaks of norovirus gastroenteritis (4, 107). Restaurants can often face closures or lawsuits when they are found to be the source of a food-borne outbreak (126, 127). The D'Angelo Sandwich Shop, for example, agreed to pay victims of its hepatitis A outbreak a total of $270,000 in settlement of a class action lawsuit (128). And taxpayers foot the bill for government investigations of disease outbreaks. While it's not possible to prevent the occurrence of every carrier-caused outbreak, careful attention to cleaning, sanitizing, and personal hygiene can limit their scope and their duration—saving lives, money, and reputations.

References

1. **Anderson, L., S. Miller, J. Greenblatt, L. Steiner-Sichel, V. Reddy, R. Heffernan, D. Cimini, S. Balter, D. Weiss, M. Layton, A. Agasan, M. Backer, A. Ebrahimzadeh, M. Kacica, C. Scott, N. Chatterjee, P. Smith, M. Fuschino, M. Bhat-Gregerson, E. Forsyth, K. R. Steingart, P. Pace, M. Payne, M. Johnson, J. Massey, M. A. Widdowson, J. S. Bresee, R. I. Glass, U. Parashar, S. Monroe, R. S. Beard, H. White, L. Hadley, S. Bulens, and M. Charles.** 2003. Norovirus activity—United States, 2002. *Morb. Mortal. Wkly. Rep.* **52:**41–45.
2. **Anonymous.** 5 October 1989. Winona—workers may be *Salmonella* source. *Star Tribune*, Minneapolis, Minn.
3. **Anonymous.** 24 November 2001. State investigating hepatitis A cases. *Boston Globe*, Boston, Mass.
4. **Anonymous.** 23 November 2002. Cruise line cleans ship after virus sickens 500. *New York Times*, New York, N.Y.
5. **Anonymous.** 23 October 2004. How to better control *C. difficile* outbreak. *The Gazette*, Montreal, Quebec, Canada.
6. **Anonymous.** 1 February 2005. Typhoid fever cases. *BusinessWorld*, Manila, The Philippines.
7. **Anonymous.** 13 March 2005. Uzbek typhoid outbreak traced to Turkmenistan—Kazakh paper. *BBC Monitoring Central Asia*, London, United Kingdom.
8. **Anonymous.** 24 April 2005. 'Stop blaming people for typhoid outbreak'. *New Straits Times*, Kuala Lumpur, Malaysia.
9. **Anonymous.** 20 July 2005. Thirteen typhoid fever cases reported in western Algeria. *BBC Monitoring Middle East*, London, United Kingdom.
10. **Anonymous.** 26 July 2005. Typhoid, hepatitis A rates growing in Russia—Onishchenko. *Daily News Bulletin (English)*, Moscow, Russia.
11. **Anonymous.** 29 July 2005. First fatality in Fiji typhoid outbreak. *BBC Monitoring Asia Pacific*, London, United Kingdom.
12. **Arness, M., M. Canham, B. Feighner, E. Hoedebecke, J. Cuthie, C. Polyak, D. R. Skillman, J. English, C. Jenkins, T. Barker, T. Cieslak, and D. N. Taylor.** 1999. Norwalk-like viral gastroenteritis in U.S. army trainees—Texas, 1998. *Morb. Mortal. Wkly. Rep.* **48:**225–227.
13. **Arness, M. K., B. H. Feighner, M. L. Canham, D. N. Taylor, S. S. Monroe, T. J. Cieslak, E. L. Hoedebecke, C. S. Polyak, J. C. Cuthie, R. L. Fankhauser, C. D. Humphrey, T. L. Barker, C. D. Jenkins, and D. R. Skillman.** 2000. Norwalk-like viral gastroenteritis outbreak in U.S. army trainees. *Emerg. Infect. Dis.* **6:**204–207.
14. **Bailey, M. S., C. J. Boos, G. Vautier, A. D. Green, H. Appleton, C. I. Gallimore, J. J. Gray, and N. J. Beeching.** 2005. Gastroenteritis outbreak in British troops, Iraq. *Emerg. Infect. Dis.* **11:**1625–1628.

15. **Baron, S. (ed.).** 1996. *Medical Microbiology*, 4th ed. The University of Texas Medical Branch at Galveston, Galveston, Tex. [Online.] http://www.ncbi.nlm.nih.gov/books/bv.fcgi?rid=mmed. Accessed 27 June 2006.

16. **Bidawid, S., J. M. Farber, and S. A. Sattar.** 2000. Contamination of foods by food handlers: experiments on hepatitis A virus transfer to food and its interruption. *Appl. Environ. Microbiol.* **66:**2759–2763.

17. **Bidawid, S., J. M. Farber, S. A. Sattar, and S. Hayward.** 2000. Heat inactivation of hepatitis A virus in dairy foods. *J. Food Prot.* **63:**522–528.

18. **Birkhead, G. S., D. L. Morse, W. C. Levine, J. K. Fudala, S. F. Kondracki, H. G. Chang, M. Shayegani, L. Novick, and P. A. Blake.** 1993. Typhoid fever at a resort hotel in New York: a large outbreak with an unusual vehicle. *J. Infect. Dis.* **167:**1228–1232.

19. **Bodovitz, K.** 26 August 1988. East Bay officials seek source of typhoid cases. *San Francisco Chronicle*, San Francisco, Calif.

20. **Bolduc, D., L. F. Srour, L. Sweet, A. Neatby, E. Galanis, S. Isaacs, and G. Lim.** 2004. Severe outbreak of *Escherichia coli* O157:H7 in health care institutions in Charlottetown, Prince Edward Island, fall, 2002. *Can. Commun. Dis. Rep.* **30:**81–88.

21. **Bon, F., H. Giraudon, C. Sancey, C. Barranger, M. Joannes, P. Pothier, and E. Kohli.** 2004. Development and evaluation of a new commercial test allowing the simultaneous detection of noroviruses and sapoviruses by reverse transcription-PCR and microplate hybridization. *J. Clin. Microbiol.* **42:** 2218–2220.

22. **Bornemann, R., D. M. Zerr, J. Heath, J. Koehler, M. Grandjean, R. Pallipamu, and J. Duchin.** 2002. An outbreak of *Salmonella* serotype Saintpaul in a children's hospital. *Infect. Control Hosp. Epidemiol.* **23:**671–676.

23. **Bouallègue-Godet, O., Y. Ben Salem, L. Fabre, M. Demartin, P. A. D. Grimont, R. Mzoughi, and F.-X. Weill.** 2005. Nosocomial outbreak caused by *Salmonella enterica* serotype Livingstone producing CTX-M-27 extended-spectrum β-lactamase in a neonatal unit in Sousse, Tunisia. *J. Clin. Microbiol.* **43:**1037–1044.

24. **Branswell, H.** 11 April 2005. *C. difficile* strain high in toxins. *The Gazette*, Montreal, Quebec, Canada.

25. **Bresee, J. S., M.-A. Widdowson, S. S. Monroe, and R. I. Glass.** 2002. Foodborne viral gastroenteritis: challenges and opportunities. *Clin. Infect. Dis.* **35:**748–753.

26. **Brock, T. D.** 1999. *Robert Koch. A Life in Medicine and Bacteriology.* ASM Press, Washington, D.C.

27. **Brown, D., J. Gray, P. MacDonald, A. Green, D. Morgan, G. Christopher, R. Glass, and R. Turcios.** 2002. Outbreak of acute gastroenteritis associated with Norwalk-like viruses among British military personnel—Afghanistan, May 2002. *Morb. Mortal. Wkly. Rep.* **51:**477–479.

28. **Burkholder, B. T., V. G. Coronado, J. Brown, J. H. Hutto, C. N. Shapiro, B. Robertson, and B. A. Woodruff.** 1995. Nosocomial transmission of hepatitis A in a pediatric hospital traced to an anti-hepatitis A virus-negative patient with immunodeficiency. *Pediatr. Infect. Dis. J.* **14:**261–266.

29. **Canedy, D.** 20 November 2002. Cruise line cancels trip after onboard sickness. *New York Times*, New York, N.Y.

30. **Carey, E.** 6 August 2004. Hospital cutbacks let virus flourish; patients are older, budgets tighter. New strains pose serious concerns. *Toronto Star*, Toronto, Ontario, Canada.

31. **Carroll, A.** 28 January 2005. $20 million injected against superbug. *The Gazette*, Montreal, Quebec, Canada.

32. **Centers for Disease Control and Prevention.** 2002. Outbreak of Norwalk virus aboard Holland America's *Ryndam* in Alaska. Centers for Disease Control and Prevention, National Center for Environmental Health, Vessel Sanitation Program. [Online.] http://www.cdc.gov/nceh/vsp/surv/outbreak/2002/ryndam.htm. Accessed 27 June 2006.

33. **Centers for Disease Control and Prevention.** 2005. About the vessel sanitation program. Centers for Disease Control and Prevention, National Center for Environmental Health, Vessel Sanitation Program. [Online.] http://www.cdc.gov/nceh/vsp/desc/aboutvsp.htm. Accessed 27 June 2006.

34. **Centers for Disease Control and Prevention.** 2005. Vessel sanitation program. Advanced scores search. Centers for Disease Control and Prevention, National Center for Environmental Health, Vessel Sanitation Program. [Online.] http://wwwn.cdc.gov/vsp/InspectionQueryTool/Forms/InspectionSearch.aspx. Accessed 27 June 2006.

35. **Centers for Disease Control and Prevention.** 2005. Updates of gastrointestinal illness among passengers and crew for international cruise lines. Centers for Disease Control and Prevention, National Center for Environmental Health, Vessel Sanitation Program. [Online.] http://www.cdc.gov/nceh/vsp/surv/GIlist.htm. Accessed 26 June 2006.

36. **Christenson, B., A. Ringner, C. Blucher, H. Billaudelle, K. N. Gundtoft, G. Eriksson, and M. Bottiger.** 1983. An outbreak of *Campylobacter* enteritis among the staff of a poultry abattoir in Sweden. *Scand. J. Infect. Dis.* **15:** 167–172.

37. **City of St. Louis.** 2005. Hepatitis A vaccination requirement for food service workers. City of St. Louis Department of Health. [Online.] http://stlouis.missouri.org/citygov/health/hepaord.html. Accessed 27 June 2006.

38. **Cramer, E. H., D. Forney, A. L. Dannenberg, M. A. Widdowson, J. S. Bresee, S. Monroe, R. S. Beard, H. White, S. Bulens, E. Mintz, C. Stover, E. Isakbaeva, J. Mullins, J. Wright, V. Hsu, W. Chege, and J. Varma.** 2002. Outbreaks of gastroenteritis associated with noroviruses on cruise ships—United States, 2002. *Morb. Mortal. Wkly. Rep.* **51:**1112–1115.

39. **Cruz, J. R., F. Cano, L. Rodriguez, C. A. Rios, P. Guerra, and Z. Leonardo.** 1991. International notes: *Shigella dysenteriae* type 1—Guatemala, 1991. *Morb. Mortal. Wkly. Rep.* **40:**421, 427–428.

40. **Cuthbert, J. A.** 2001. Hepatitis A: old and new. *Clin. Microbiol. Rev.* **14:** 38–58.

41. **Daniels, N. A., D. A. Bergmire-Sweat, K. J. Schwab, K. A. Hendricks, S. Reddy, S. M. Rowe, R. L. Fankhauser, S. S. Monroe, R. L. Atmar, R. I. Glass, and P. Mead.** 2000. A foodborne outbreak of gastroenteritis associated with Norwalk-like viruses: first molecular traceback to deli sandwiches contaminated during preparation. *J. Infect. Dis.* **181:**1467–1470.

42. **Daniels, N. A., J. Neimann, A. Karpati, U. D. Parashar, K. D. Greene, J. G. Wells, A. Srivastava, R. V. Tauxe, E. D. Mintz, and R. Quick.** 2000.

Traveler's diarrhea at sea: three outbreaks of waterborne enterotoxigenic *Escherichia coli* on cruise ships. *J. Infect. Dis.* **181:**1491–1495.

43. **DaRosa, A.** 28 March 2004. Flu-like bug hits Vegas casino; 1,475 cases seen since December. *San Diego Union-Tribune*, San Diego, Calif.
44. **Dendukuri, N., V. Costa, M. McGregor, and J. M. Brophy.** 2005. Probiotic therapy for the prevention and treatment of *Clostridium difficile*-associated diarrhea: a systematic review. *Can. Med. Assoc. J.* **173:**167–170.
45. **Deneen, V. C., J. M. Hunt, C. R. Paule, R. I. James, R. G. Johnson, M. J. Raymond, and C. W. Hedberg.** 2000. The impact of foodborne calicivirus disease: the Minnesota experience. *J. Infect. Dis.* **181:**S281–S283.
46. **Derfel, A.** 26 September 2004. More money for superbug ruled out: wait-and-see with *C. difficile*. *The Gazette*, Montreal, Quebec, Canada.
47. **Derfel, A.** 23 October 2004. Five hospitals ignore *C. difficile* guidelines. *The Gazette*, Montreal, Quebec, Canada.
48. **Derfel, A.** 5 April 2005. Patients to join fight against dangerous bug: drug targets *C. difficile*. *National Post*, Don Mills, Ontario, Canada.
49. **Doganci, T., E. Araz, A. Ensari, M. Tanyuksel, and L. Doganci.** 2002. Detection of *Cryptosporidium parvum* infection in childhood using various techniques. *Med. Sci. Monit.* **8:**MT223–MT226.
50. **Dougherty, K.** 23 June 2005. Quebec plans wiser use of antibiotics to curb hospital infections. *The Gazette*, Montreal, Quebec, Canada.
51. **Dryden, M. S., N. Keyworth, R. Gabb, and K. Stein.** 1994. Asymptomatic foodhandlers as the source of nosocomial salmonellosis. *J. Hosp. Infect.* **28:**195–208.
52. **Eggertson, L.** 2004. *C. difficile*: by the numbers. *Can. Med. Assoc. J.* **171:** 1331–1332.
53. **Eggertson, L.** 2005. Quebec reports *C. difficile* mortality statistics. *Can. Med. Assoc. J.* **173:**139.
54. **Eggertson, L., and B. Sibbald.** 2004. Hospitals battling outbreaks of *C. difficile*. *Can. Med. Assoc. J.* **171:**19–21.
55. **Evenson, B.** 27 October 2000. Virus passed on gridiron left players heaving lunch: football itself was likely carrier of the bug: study. *National Post*, Don Mills, Ontario, Canada.
56. **Fernandez Canigia, L., J. Nazar, M. Arce, J. Dadamio, J. Smayevsky, and H. Bianchini.** 2001. [*Clostridium difficile* diarrhea: frequency of detection in a medical center in Buenos Aires, Argentina] [English abstract; article in Spanish]. *Rev. Argent. Microbiol.* **33:**101–107.
57. **Ferroni, A., J. Merckx, T. Ancelle, B. Pron, E. Abachin, F. Barbut, J. Larzul, P. Rigault, P. Berche, and J. L. Gaillard.** 1997. Nosocomial outbreak of *Clostridium difficile* diarrhea in a pediatric service. *Eur. J. Clin. Microbiol. Infect. Dis.* **16:**928–933.
58. **Francis, S., J. Rowland, K. Rattenbury, D. Powell, W. N. Rogers, L. Ward, and S. R. Palmer.** 1989. An outbreak of paratyphoid fever in the UK associated with a fish-and-chip shop. *Epidemiol. Infect.* **103:**445–448.
59. **Froude, J.** 2004. Remembering Typhoid Mary. *Irish America* **20:**22.
60. **Gallimore, C. I., D. Cubitt, N. du Plessis, and J. J. Gray.** 2004. Asymptomatic and symptomatic excretion of noroviruses during a hospital outbreak of gastroenteritis. *J. Clin. Microbiol.* **42:**2271–2274.

61. **Glass, R. I., J. Noel, T. Ando, R. Fankhauser, G. Belliot, A. Mounts, U. D. Parashar, J. S. Bresee, and S. S. Monroe.** 2000. The epidemiology of enteric caliciviruses from humans: a reassessment using new diagnostics. *J. Infect. Dis.* **181:**S254–S261.

62. **Green, K. Y., T. Ando, M. S. Balayan, T. Berke, I. N. Clarke, M. K. Estes, D. O. Matson, S. Nakata, J. D. Neill, M. J. Studdert, and H.-J. Thiel.** 2000. Taxonomy of the caliciviruses. *J. Infect. Dis.* **181:**S322–S330.

63. **Harasim, P.** 5 November 2004. Visitors leave LV with virus. *Las Vegas Review-Journal,* Las Vegas, Nev.

64. **Health Protection Agency.** 2005. Outbreak of *Clostridium difficile* infection in a hospital in south east England. *CDR Weekly* **15**(24): 2–3. [Online.] http://www.hpa.org.uk/cdr/archives/archive05/News/news2405.htm. Accessed 27 June 2006.

65. **Hedberg, C. W., K. E. White, J. A. Johnson, L. M. Edmonson, J. T. Soler, J. A. Korlath, L. S. Theurer, K. L. MacDonald, and M. T. Osterholm.** 1991. An outbreak of *Salmonella enteritidis* infection at a fast-food restaurant: implications for foodhandler-associated transmission. *J. Infect. Dis.* **164:** 1135–1140.

66. **Herrera, P., A. Cotera, A. Fica, T. Galdo, and M. Alvo.** 2003. [High incidence and complications of *Clostridium difficile* diarrhea among patients with renal diseases] [English abstract; article in Spanish]. *Rev. Med. Chil.* **131:** 397–403.

67. **Hossain, M. A., K. Z. Hasan, and M. J. Albert.** 1994. *Shigella* carriers among non-diarrhoeal children in an endemic area of shigellosis in Bangladesh. *Trop. Geogr. Med.* **46:**40–42.

68. **Isakbaeva, E. T., M.-A. Widdowson, R. S. Beard, S. N. Bulens, J. Mullins, S. S. Monroe, J. Bresee, P. Sassano, E. H. Cramer, and R. I. Glass.** 2005. Norovirus transmission on cruise ship. *Emerg. Infect. Dis.* **11:**154–157.

69. **Jacobs, R. J., S. F. Grover, A. S. Meyerhoff, and T. A. Paivanas.** 2000. Cost effectiveness of vaccinating food service workers against hepatitis A infection. *J. Food Prot.* **63:**768–774.

70. **Jean, J., J.-F. Vachon, O. Moroni, A. Darveau, I. Kukavica-Ibrulj, and I. Fliss.** 2003. Effectiveness of commercial disinfectants for inactivating hepatitis A virus on agri-food surfaces. *J. Food Prot.* **66:**115–119.

71. **Jensenius, M., S. H. Ringertz, D. Berild, H. Bell, R. Espinoza, and B. Grinde.** 1998. Prolonged nosocomial outbreak of hepatitis A arising from an alcoholic with pneumonia. *Scand. J. Infect. Dis.* **30:**119–123.

72. **Johnson, R.** 23 November 2002. Inspectors to board Disney cruise ship to search for cause of stomach virus. *Pittsburgh Post-Gazette,* Pittsburgh, Pa.

73. **Johnson, R.** 27 November 2002. More fall ill on voyage of Disney ship; federal health officials could delay the Magic's next cruise for more cleaning. *Orlando Sentinel,* Orlando, Fla.

74. **Kapikian, A. Z.** 2000. The discovery of the 27-nm Norwalk virus: an historic perspective. *J. Infect. Dis.* **181:**S295–S302.

75. **Kapikian, A. Z., R. G. Wyatt, R. Dolin, T. S. Thornhill, A. R. Kalica, and R. M. Chanock.** 1972. Visualization by immune electron microscopy of a 27-nm particle associated with acute infectious nonbacterial gastroenteritis. *J. Virol.* **10:**1075–1081.

76. **Kelly, D., and S. Skidmore.** 2002. Hepatitis C-Z: recent advances. *Arch. Dis. Child.* **86:**339–343.

77. **Kim, K.-H., I.-S. Suh, J. M. Kim, C. W. Kim, and Y.-J. Cho.** 1989. Etiology of childhood diarrhea in Korea. *J. Clin. Microbiol.* **27:**1192–1196.

78. **Kim, S. W., K. R. Peck, S.-I. Jung, Y.-S. Kim, S. Kim, N. Y. Lee, and J.-H. Song.** 2001. *Pseudomonas aeruginosa* as a potential cause of antibiotic-associated diarrhea. *J. Korean Med. Sci.* **16:**742–744.

79. **Kingsley, D. H., and G. P. Richards.** 2003. Persistence of hepatitis A virus in oysters. *J. Food Prot.* **66:**331–334.

80. **Knoop, F. C., M. Owens, and I. C. Crocker.** 1993. *Clostridium difficile*: clinical disease and diagnosis. *Clin. Microbiol. Rev.* **6:**251–265.

81. **Konomi, N., C. Miyoshi, C. La Fuente Zerain, T.-C. Li, Y. Arakawa, and K. Abe.** 1999. Epidemiology of hepatitis B, C, E, and G virus infections and molecular analysis of hepatitis G virus isolates in Bolivia. *J. Clin. Microbiol.* **37:**3291–3295.

82. **Lai, C. W., R. C. Chan, A. F. Cheng, J. Y. Sung, and J. W. Leung.** 1992. Common bile duct stones: a cause of chronic salmonellosis. *Am. J. Gastroenterol.* **87:**1198–1199.

83. **LaPorte, T., D. Heisey-Grove, P. Kludt, B. T. Matyas, A. DeMaria, Jr., R. Dicker, A. De, A. Fiore, O. Nainan, and D. S. Friedman.** 2003. Foodborne transmission of hepatitis A—Massachusetts, 2001. *Morb. Mortal. Wkly. Rpt.* **52:**565–567.

84. **Leavitt, J. W.** 1996. *Typhoid Mary. Captive to the Public's Health.* Beacon Press, Boston, Mass.

85. **L'Ecuyer, P. B., J. Diego, D. Murphy, E. Trovillion, M. Jones, D. F. Sahm, and V. J. Fraser.** 1996. Nosocomial outbreak of gastroenteritis due to *Salmonella senftenberg*. *Clin. Infect. Dis.* **23:**734–742.

86. **Levitz, J.** 5 August 2005. Scrubbed clean, Camp Yawgoog reopens Sunday. *The Providence Journal*, Providence, R.I.

87. **Lin, F.-Y. C., J. M. Becke, C. Groves, B. P. Lim, E. Israel, E. F. Becker, R. M. Helfrich, D. S. Swetter, T. Cramton, and J. B. Robbins.** 1988. Restaurant-associated outbreak of typhoid fever in Maryland: identification of carrier facilitated by measurement of serum Vi antibodies. *J. Clin. Microbiol.* **26:**1194–1197.

88. **Llovo, J., E. Mateo, A. Muñoz, M. Urquijo, S. L. W. On, and A. Fernández-Astorga.** 2003. Molecular typing of *Campylobacter jejuni* isolates involved in a neonatal outbreak indicates nosocomial transmission. *J. Clin. Microbiol.* **41:**3926–3928.

89. **Longfield, R., E. Strohmer, R. Newquist, J. Longfield, J. Coberly, G. Howell, and R. Thomas.** 1983. Hospital-associated outbreak of *Shigella dysenteriae* type 2—Maryland. *Morb. Mortal. Wkly. Rep.* **32:**250–252.

90. **Lopman, B. A., M. H. Reacher, I. B. Vipond, D. Hill, C. Perry, T. Halladay, D. W. Brown, W. J. Edmunds, and J. Sarangi.** 2004. Epidemiology and cost of nosocomial gastroenteritis, Avon, England, 2002–2003. *Emerg. Infect. Dis.* **10:**1827–1834.

91. **Louie, T. J., and J. Meddings.** 2004. *Clostridium difficile* infection in hospitals: risk factors and responses. *Can. Med. Assoc. J.* **171:**45–46.

92. **Lyerly, D. M., H. C. Krivan, and T. D. Wilkins.** 1988. *Clostridium difficile*: its disease and toxins. *Clin. Microbiol. Rev.* **1:**1–18.

93. **MacDonald, N.** 30 August 2004. Virus causes mayhem in hospitals, rest homes. *Dominion Post*, Wellington, New Zealand.

94. **Massachusetts Department of Public Health.** 27 November 2001. Hepatitis A outbreak in Bristol County. The Commonwealth of Massachusetts, Executive Office of Health and Human Services, Department of Public Health, Boston, Mass. [Online.] http://www.mass.gov/dph/media/2001/pr1127.htm. Accessed 27 June 2006.

95. **Mayo, M. A.** 2002. Virus taxonomy—Houston 2002. *Arch. Virol.* **147:** 1071–1076.

96. **Mbithi, J. N., V. S. Springthorpe, J. R. Boulet, and S. A. Sattar.** 1992. Survival of hepatitis A virus on human hands and its transfer on contact with animate and inanimate surfaces. *J. Clin. Microbiol.* **30:**757–763.

97. **Mbithi, J. N., V. S. Springthorpe, and S. A. Sattar.** 1990. Chemical disinfection of hepatitis A virus on environmental surfaces. *Appl. Environ. Microbiol.* **56:**3601–3604.

98. **Mbithi, J. N., V. S. Springthorpe, and S. A. Sattar.** 1991. Effect of relative humidity and air temperature on survival of hepatitis A virus on environmental surfaces. *Appl. Environ. Microbiol.* **57:**1394–1399.

99. **Mbithi, J. N., V. S. Springthorpe, and S. A. Sattar.** 1993. Comparative *in vivo* efficiencies of hand-washing agents against hepatitis A virus (HM-175) and poliovirus type 1 (Sabin). *J. Clin. Microbiol.* **59:**3463–3469.

100. **McCarthy, M., M. K. Estes, and K. C. Hyams.** 2000. Norwalk-like virus infection in military forces: epidemic potential, sporadic disease, and the future direction of prevention and control efforts. *J. Infect. Dis.* **181:**S387–S391.

101. **McCaustland, K. A., W. W. Bond, D. W. Bradley, J. W. Ebert, and J. E. Maynard.** 1982. Survival of hepatitis A virus in feces after drying and storage for 1 month. *J. Clin. Microbiol.* **16:**957–958.

102. **McCullough, M., and L. Krieger.** 8 September 2005. Rumors of dysentery, cholera are unfounded, health officials say. Knight Ridder Tribune News Service, Washington, D.C.

103. **McIntyre, L., L. Vallaster, C. Kurzac, J. Fung, A. McNabb, M.-K. Lee, P. Daly, M. Petric, and J. Isaac-Renton.** 2002. Gastrointestinal outbreaks associated with Norwalk virus in restaurants in Vancouver, British Columbia. *Can. Commun. Dis. Rep.* **28:**197–203. [Online.] http://www.phac-aspc.gc.ca/publicat/ccdr-rmtc/02vol28/dr2824ea.html. Accessed 27 June 2006.

104. **Miller, M. A., M. Hyland, M. Ofner-Agostini, M. Gourdeau, and M. Ishak.** 2002. Morbidity, mortality, and healthcare burden of nosocomial *Clostridium difficile*-associated diarrhea in Canadian hospitals. *Infect. Control Hosp. Epidemiol.* **23:**137–140.

105. **Minooee, A., and L. S. Rickman.** 1999. Infectious diseases on cruise ships. *Clin. Infect. Dis.* **29:**737–744.

106. **Monroe, S. S., T. Ando, and R. I. Glass.** 2000. Introduction: human enteric caliciviruses—an emerging pathogen whose time has come. *J. Infect. Dis.* **181:** S249–S251.

107. **Morse, D.** 29 November 2002. Disney Cruise Line will remove ship from service for cleanup. *The Wall Street Journal*, New York, N.Y.

108. **Moshkowitz, M., E. Ben Baruch, Z. Kline, M. Gelber, Z. Shimoni, and F. Konikoff.** 2004. Clinical manifestations and outcome of pseudomembranous colitis in an elderly population in Israel. *Isr. Med. Assoc. J.* **6:**201–204.

109. **Muto, C. A., M. Pokrywka, K. Shutt, A. B. Mendelsohn, K. Nouri, K. Posey, T. Roberts, K. Croyle, S. Krystofiak, S. Patel-Brown, A. W. Pasculle, D. L. Paterson, M. Saul, and L. H. Harrison.** 2005. A large outbreak of *Clostridium difficile*-associated disease with an unexpected proportion of deaths and colectomies at a teaching hospital following increased fluoroquinolone use. *Infect. Control Hosp. Epidemiol.* **26:**273–280.

110. **Nastasi, A., C. Mammina, M. R. Villafrate, G. Dicuonzo, E. Aiello, and G. Scaglione.** 1991. Reemergence of *Shigella dysenteriae* type 2 in Sicily: an epidemiological evaluation. *Microbiologica* **14:**219–222.

111. **New York State Department of Health.** 2004. Hepatitis A and food service workers. New York State Department of Health. [Online.] http://www.health.state.ny.us/nysdoh/communicable_diseases/en/hepafood.htm. Accessed 27 June 2006.

112. **Norén, T., T. Åkerlund, E. Bäck, L. Sjöberg, I. Persson, I. Alriksson, and L. G. Burman.** 2004. Molecular epidemiology of hospital-associated and community-acquired *Clostridium difficile* infection in a Swedish county. *J. Clin. Microbiol.* **42:**3635–3643.

113. **Ochs, R.** 13 April 1998. Long Island: our story. Dinner with Typhoid Mary. A household in Oyster Bay is stricken, and the trail leads to the cook, Mary Mallon. *Newsday*, Long Island, N.Y.

114. **Oguike, J. U., and A. C. Emeruwa.** 1990. Incidence of *Clostridium difficile* in infants in rural and urban areas of Nigeria. *Microbiologica* **13:**267–271.

115. **Ojeda, A., V. Prado, J. Martinez, C. Arellano, A. Borczyk, W. Johnson, H. Lior, and M. M. Levine.** 1995. Sorbitol-negative phenotype among enterohemorrhagic *Escherichia coli* strains of different serotypes and from different sources. *J. Clin. Microbiol.* **33:**2199–2201.

116. **Parkes, D.** 6 June 2004. Dirty hospitals raise risk: doctor. Deadly infections under public spotlight. *The Gazette*, Montreal, Quebec, Canada.

117. **Pazos, R., A. Isusi, R. Fernández, L. Barbeito, A. Bravo, I. Cantón, P. Gayoso, and R. Lebrato.** 2003. Brote nosocomial de diarrea por *Clostridium difficile* en un servicio de cirugía vascular [Nosocomial diarrhea outbreak due to *Clostridium difficile* in a vascular surgery department]. *Enferm. Infecc. Microbiol. Clin.* **21:**237–241.

118. **Pépin, J., L. Valiquette, M.-E. Alary, P. Villemure, A. Pelletier, K. Forget, K. Pépin, and D. Chouinard.** 2004. *Clostridium difficile*-associated diarrhea in a region of Quebec from 1991 to 2003: a changing pattern of disease severity. *Can. Med. Assoc. J.* **171:**466–472.

119. **Poutanen, S. M., and A. E. Simor.** 2004. *Clostridium difficile*-associated diarrhea in adults. *Can. Med. Assoc. J.* **171:**51–58.

120. **ProMED-mail.** 19 January 2005. Typhoid fever—Congo DR (Kinshasa). Typhoid fever in the Democratic Republic of the Congo—update. *ProMED-mail* 20050119.0183. [Online.] http://www.promedmail.org/pls/askus/f?p=2400:1202:8169964021050853870::NO::F2400_P1202_CHECK_DISPLAY,F2400_P1202_PUB_MAIL_ID:X,27785. Accessed 27 June 2006.

121. **ProMED-mail.** 26 January 2005. Typhoid fever—Gabon (Libreville). Typhoid outbreak extends to capital hit by water supply problems. *ProMED-mail* 20050126.0281. [Online.] http://www.promedmail.org/pls/askus/f?p=2400:1202:8169964021050853870::NO::F2400_P1202_CHECK_DISPLAY,F2400_P1202_PUB_MAIL_ID:X,27882. Accessed 27 June 2006.

122. **ProMED-mail.** 27 April 2005. Typhoid fever—Kenya (Bungoma) (02). Kenyan medics worried about spread of typhoid. *ProMED-mail* 20050427.1174. [Online.] http://www.promedmail.org/pls/askus/f?p=2400:1202:8169964021050853870::NO::F2400_P1202_CHECK_DISPLAY,F2400_P1202_PUB_MAIL_ID:X,28778. Accessed 27 June 2006.

123. **Prouty, A. M., W. H. Schwesinger, and J. S. Gunn.** 2002. Biofilm formation and interaction with the surfaces of gallstones by *Salmonella* spp. *Infect. Immun.* **70:**2640–2649.

124. **Ray, D., D. Tribby, J. Eyster, S. Coopes, W. Hall, H. McGee, E. Renshaw, Jr., G. Winter, and N. Hayner.** 1982. Epidemiologic notes and reports. Typhoid fever—Michigan. *Morb. Mortal. Wkly. Rep.* **31:**544, 549–550.

125. **Reynolds, M.** 24 November 2001. Rash of hepatitis A cases hits 4 Mass. communities. *Providence Journal*, Providence, R.I.

126. **Reynolds, M.** 13 December 2001. D'Angelo sued over hepatitis A outbreak. *Providence Journal*, Providence, R.I.

127. **Reynolds, M.** 12 March 2002. Final report: D'Angelo main source of outbreak. *Providence Journal*, Providence, R.I.

128. **Reynolds, M.** 4 February 2004. Judge OKs settlement for hepatitis exposure. *The Providence Journal*, Providence, R.I.

129. **Sabetta, J. R., S. Hyman, J. Smardin, M. L. Cartter, and J. L. Hadler.** 1991. Foodborne nosocomial outbreak of *Salmonella reading*—Connecticut. *Morb. Mortal. Wkly. Rep.* **40:**804–806.

130. **Sato, H., H. Kato, K. Koiwai, and C. Sakai.** 2004. [A nosocomial outbreak of diarrhea caused by toxin A-negative, toxin B-positive *Clostridium difficile* in a cancer center hospital] [English abstract; article in Japanese]. *Kansenshogaku Zasshi* **78:**312–319.

131. **Schwartz, E., N. Piper Jenks, P. Van Damme, and E. Galun.** 1999. Hepatitis E virus infection in travelers. *Clin. Infect. Dis.* **29:**1312–1314.

132. **Sirinavin, S., L. Pokawattana, and A. Bangtrakulnondh.** 2004. Duration of nontyphoidal *Salmonella* carriage in asymptomatic adults. *Clin. Infect. Dis.* **38:**1644–1645.

133. **Smith, J. L.** 2001. A review of hepatitis E virus. *J. Food Prot.* **64:**572–586.

134. **Soper, G. A.** 1919. Typhoid Mary. *The Military Surgeon* **45**(1). [Online.] http://www.nova.edu/~kornblau/LE_class/typhoid.html. Accessed 27 June 2006.

135. **Soyletir, G., A. Eskiturk, G. Kilic, V. Korten, and N. Tozun.** 1996. *Clostridium difficile* acquisition rate and its role in nosocomial diarrhoea at a university hospital in Turkey. *Eur. J. Epidemiol.* **12:**391–394.

136. **Strausbaugh, L. J., S. R. Sukumar, and C. L. Joseph.** 2003. Infectious disease outbreaks in nursing homes: an unappreciated hazard for frail elderly persons. *Clin. Infect. Dis.* **36:**870–876.

137. **Tachibana, H., S. Kobayashi, K. Nagakura, Y. Kaneda, and T. Takeuchi.** 2000. Asymptomatic cyst passers of *Entamoeba histolytica* but not *Entamoeba*

dispar in institutions for the mentally retarded in Japan. *Parasitol. Int.* **49:** 31–35.

138. **Thayer, J., C. Milat, B. Meier, J. Hadman, S. Paciotti, R. Palmer, C. Story, D. Larson, L. Bjerkness, H. Parker, V. Jensen, R. Rognstad, P. Anderson, J. Lynne, N. Therien, B. Bartleson, L. Kentala, J. Lewis, and J. M. Kobayashi.** 1990. Epidemiologic notes and reports. Typhoid fever—Skagit County, Washington. *Morb. Mortal. Wkly. Rep.* **39:**749–751.

139. **Thomas, C., M. Stevenson, D. J. Williamson, and T. V. Riley.** 2002. *Clostridium difficile*-associated diarrhea: epidemiological data from Western Australia associated with a modified antibiotic policy. *Clin. Infect. Dis.* **35:**1457–1462.

140. **Titov, L., N. Lebedkova, A. Shabanov, Y. J. Tang, S. H. Cohen, and J. Silva, Jr.** 2000. Isolation and molecular characterization of *Clostridium difficile* strains from patients and the hospital environment in Belarus. *J. Clin. Microbiol.* **38:**1200–1202.

141. **Tsang, T. H. F., E. K. Denison, H. V. Williams, L. V. Venczel, M. M. Ginsberg, and D. J. Vugia.** 2000. Acute hepatitis E infection acquired in California. *Clin. Infect. Dis.* **30:**618–619.

142. **Valiquette, L., D. E. Low, J. Pépin, and A. McGeer.** 2004. *Clostridium difficile* infection in hospitals: a brewing storm. *Can. Med. Assoc. J.* **171:**27–29.

143. **Weir, E., and K. Flegel.** 2005. Protecting against *Clostridium difficile* illness. *Can. Med. Assoc. J.* **172:**1178.

144. **White, K. E., M. T. Osterholm, J. A. Mariotti, J. A. Korlath, D. H. Lawrence, T. L. Ristinen, and H. B. Greenberg.** 1986. A foodborne outbreak of Norwalk virus gastroenteritis. Evidence for post-recovery transmission. *Am. J. Epidemiol.* **124:**120–126.

145. **Widdowson, M.-A., E. H. Cramer, L. Hadley, J. S. Bresee, R. S. Beard, S. N. Bulens, M. Charles, W. Chege, E. Isakbaeva, J. G. Wright, E. Mintz, D. Forney, J. Massey, R. I. Glass, and S. S. Monroe.** 2004. Outbreaks of acute gastroenteritis on cruise ships and on land: identification of a predominant circulating strain of norovirus—United States, 2002. *J. Infect. Dis.* **190:**27–36.

146. **Widdowson, M.-A., A. Sulka, S. N. Bulens, R. S. Beard, S. S. Chaves, R. Hammond, E. D. P. Salehi, E. Swanson, J. Totaro, R. Woron, P. S. Mead, J. S. Bresee, S. S. Monroe, and R. I. Glass.** 2005. Norovirus and foodborne disease, United States, 1991–2000. *Emerg. Infect. Dis.* **11:**95–102.

chapter 13

Deliberately Contaminated Food

"WILL ANYONE NOTICE?" the cook wondered as he doled out portions from behind his cafeteria counter. "Will somebody complain about the taste?" For the next 2 days, the cook remained in suspense, watching for a sign that his plan was working. Then, one after another, the 13 jewelry store employees in the Elephant's Gate shopping area of Chennai, India, began to fall ill. And the cook took advantage of their absence to rob the jewelry store, making off with jewelry worth 1.2 million rupees—approximately $27,000 (8, 9).

Military strategists have used biological weapons for many centuries. In 1346, the Tatars catapulted plague-riddled corpses over the walls of the port city of Caffa, causing an epidemic and forcing the town's defenders to flee for their lives (52). The Russian army also used the corpses of plague victims as weapons, this time against opposing Swedish forces in 1710 (46). Smallpox was another favored weapon in the early days of biological warfare. Legend has it that Pizarro gave smallpox-laden clothing to the indigenous people of South America in the 16th century; and the English used variola-contaminated blankets to spread smallpox among Indian tribes that were fighting on the side of France during the French and Indian War in the mid-18th century (46). More recently, in 1993, the Aum Shinrikyo cult aerosolized a liquid suspension of *Bacillus anthracis* and dispersed it from the roof of a building in Kameido, Japan, in an attempt to cause an anthrax epidemic. Fortunately, they used a vaccine strain of *B. anthracis*—one with little or no virulence for most people—and did a poor job of aerosolizing the suspension (29, 50). And the United States was the recipient of a successful anthrax bioterror attack in the fall of 2001, the perpetrator of which remained unidentified 4 years later (45).

Unlike the use of anthrax, smallpox, or plague, the documented cases of deliberate contamination of food have not been about biological warfare or bioterrorism. Instead, the perpetrators have been driven by more mundane motives, such as malice, greed (as in the jewelry theft), or political gain (30, 51).

Salad Days in The Dalles

In 1981, Rajneesh International, an Indian cult led by the guru Bhagwan Shree Rajneesh, paid $6 million for a 130-square-mile (337 km^2) ranch in Wasco County, Oreg., and established a commune on the property consisting of 280 adherents of the Rajneesh movement. Within a year, the cult incorporated their commune as the town of Rajneeshpuram and also made their presence felt in the nearby small community of Antelope (population of 31 registered voters) by buying additional land and taking over the general store, which they renamed "Zorba the Buddha." By September 1984, they had gained political control of Antelope (2, 3). But the cult ran afoul of Wasco County commissioners who threatened to disincorporate Rajneeshpuram.

The cult's leaders, in an attempt to prevent Rajneeshpuram's disincorporation, decided to try and take over Wasco County's government in the November 1984 election (15). They began to import street people from all over the United States, housing them at Rajneeshpuram and urging them to register to vote—in favor of Rajneesh candidates, of course (in Oregon at that time, voters could register even as late as election day). Wasco's county clerk responded by turning away a busload of the cult's followers, saying that they would have to return later and submit to an examination of their qualifications to register to vote (4).

While Wasco County officials were dealing with their difficult residents, they were hit with another problem, an outbreak of *Salmonella enterica* serotype Typhimurium in The Dalles (the county seat). The outbreak of gastroenteritis, which occurred in two waves and affected 751 people, was traced to salad bars in 10 restaurants in and around the town. The first phase of the *Salmonella* outbreak began on September 9, 1984, and continued until the 18th. The second wave of the outbreak began on September 19 and ended on October 10 (51). Investigators suspected that the salad bars had been contaminated deliberately but could not find a convincing motive (1). So they concluded instead—in the absence of any compelling evidence to the contrary—that the outbreak must have been due to poor employee hygiene (5). One year later, they had reason to change their minds (51).

In October 1985, law enforcement officers found an open vial of commercial stock culture disks of serotype Typhimurium on the premises of the Rajneesh commune. It had been purchased before the onset of the salad bar outbreak, and lab tests confirmed that it matched the outbreak strain perfectly (51). In July 1986, Ma Sheela, former personal secretary to Bhagwan Shree Rajneesh, pled guilty to a charge of tampering with consumer products. Sheela explained that the cult had wanted to influence the November 1984 election results in Wasco County by causing widespread illness in the local community. She said that cult members had laced grocery store produce and restaurant salad bars and coffee creamers with

Salmonella as a dry run. If the test worked, they planned to contaminate the town's water supply shortly before the day of the election (6, 28). But, apparently, the number of illnesses resulting from their experiment was not high enough to encourage the cult to proceed with their plot, as there were no subsequent outbreaks of salmonellosis on or around election day.

The Rajneesh incident was not widely reported either in the national press or in scientific journals at the time, although the information was circulated to public health officials throughout the United States. The state and federal health officials who investigated the outbreak and uncovered the plot chose to keep the story quiet. They were concerned about the prospect of copycat crimes. Finally, in 1997, the investigators published a detailed account of the incident in the *Journal of the American Medical Association* in the hope, they said, of creating greater awareness in the scientific community of the possibility that other such incidents could occur (51). But events had overtaken them. Despite the lack of publicity received by the Rajneesh affair, a second deliberate food poisoning had already taken place in Dallas, Tex., in 1996.

Sweet Revenge

When Diane Thompson and her boyfriend split up, she decided to make him pay for her hurt. As a lab worker at Dallas's St. Paul Medical Center, she had access to a novel means of doing so. In October 1995, she infected him with a culture of *Shigella dysenteriae* type 2, a bacterium that rarely causes disease in the United States (10, 30). Her ex-boyfriend was hospitalized at St. Paul's as a result, and Diane was forced to cover her tracks by switching his stool sample so that the cause of his illness would remain undiscovered (10).

One year later, Diane struck again. This time, she inoculated muffins and doughnuts with *S. dysenteriae*, placed them in the laboratory staff room, and sent an unsigned e-mail from a supervisor's computer inviting lab employees to help themselves to the snack (7, 10). Twelve of them did, one of the lab workers bringing home a muffin to share with a family member. All 13 people who ate one of the contaminated snacks suffered nausea, abdominal discomfort, bloating, and diarrhea. Five of the workers were treated in emergency departments and four others were hospitalized for 2 to 10 days; eight of the workers required intravenous fluids. Fortunately, there were no deaths (30). Because of the suspicious nature of the outbreak, hospital management referred the case to the Federal Bureau of Investigation (7).

Following the outbreak, St. Paul Medical Center put into place new security provisions for its culture collection. The freezer in which the cultures are stored is now kept locked and can only be accessed by a supervisor. The individual culture vials are no longer labeled by name; species names have been replaced by a numerical identification system (30). Diane Thompson admitted to her guilt and was sentenced to four 20-year prison terms for having contaminated food with *Shigella*, and an additional 2 years for having tampered with her ex-boyfriend's stool specimen (10).

Confronting the Bioterrorism Threat

The incidents in The Dalles and in Dallas were two early warning indicators that bioterrorism (as opposed to state-sponsored biological or chemical warfare) could become a reality. In 1998, Johns Hopkins University professor D. A. Henderson warned that the United States was unprepared to deal with a bioterrorist attack (22). The following year, in February 1999 (more than 2½ years before the anthrax-contaminated letters reached their destinations in New York City and Washington, D.C.), Johns Hopkins University hosted the First National Symposium on Medical and Public Health Response to Bioterrorism, held in Arlington, Va. Forty-six U.S. states and 10 countries—Australia, Austria, Canada, England, Finland, France, Germany, Israel, Italy, and The Netherlands—sent representatives to the meeting (23).

In her address to the symposium, Donna Shalala (at the time, the U.S. Secretary of Health and Human Services) described terrorism as "...one of the thorniest problems of the post-Cold War era...," reminding her audience that preparing for and responding to a bioterror attack required close cooperation between levels of government and between countries (44). Other speakers summarized the biological weapons programs in the former Soviet Union and Iraq, as well as the threat posed by Aum Shinrikyo and other domestic terrorists (12, 36, 49). Epidemiology, clinical presentation of diseases caused by biological agents, availability and use of vaccines to protect civilians, and hypothetical attack scenarios were all discussed (11, 24, 25, 31, 37, 39, 43).

In November 2000, during her closing address to the Second National Symposium on Medical and Public Health Response to Bioterrorism, T. O'Toole of Johns Hopkins University mused, "A year or two or ten from now, what will we say we did about biological weapons? About the greatest threat of our era? I am hopeful that our answer will reflect honorably on our efforts..." (38). Just 1 year later, the United States was wrestling with its first domestic biological terror attack since the Rajneesh episode of 1984 (26).

The anthrax letter attack spurred the U.S. Congress to pass the Public Health Security and Bioterrorism Preparedness and Response Act of 2002 (the Bioterrorism Act). This act, which was signed into law on June 12, 2002, conferred upon the U.S. Food and Drug Administration (FDA) and the U.S. Department of Agriculture (USDA) joint responsibility for protecting the country's food supply (20). The FDA quickly hired

The Three Faces of *Bacillus anthracis*

In light of the 2001 anthrax letter bioterrorist attack in the United States, and the focus given to inhalation anthrax by the military and the medical and scientific communities, it's easy to forget that *B. anthracis* is much more than just a respiratory pathogen (32, 33, 46). In addition to its ability to cause potentially fatal respiratory disease, *B. anthracis* can also infect the skin (cutaneous anthrax) and the digestive system (gastrointestinal anthrax).

The anthrax letter attack (the source of which still has not been traced after more than 5 years) provided an object lesson in the differences in mortality between the respiratory and cutaneous forms of anthrax. Of the 22 attack victims, half suffered respiratory anthrax and the other half cutaneous anthrax. All 11 victims who contracted the cutaneous form of the disease survived; 5 of the 11 with respiratory anthrax died (26).

Gastrointestinal anthrax can be a difficult condition to diagnose. Its severity can range from a complete absence of symptoms (asymptomatic infection) to severe lesions in any part of the gastrointestinal tract from the mouth to the cecum, resulting in death (47). In some cases, it presents itself as gastroenteritis, causing nausea, loss of appetite, fever, abdominal pain, and bloody stools. The fatality rate in recognized cases of gastrointestinal anthrax is 25 to 60% (27).

Gastrointestinal anthrax is usually thought to be the least common human form of the disease (42). But many cases probably go unrecognized, either because they remain asymptomatic or because the symptoms are mistaken for simple gastroenteritis (47). The actual incidence of the gastrointestinal form of anthrax infection may be much higher than previously believed, especially in rural areas and developing countries (47).

Some scientists and historians have speculated that outbreaks of gastrointestinal anthrax have been behind certain biblical and historical events. For example, anthrax has been suggested as the cause of one of the 10 plagues of Egypt, as well as a plague that occurred in Athens in 430 B.C. More recently, in 1826, a French historian proposed anthrax as the cause of a 1770 outbreak of "charbon" in Haiti that killed 15,000 people. The symptoms and the circumstances surrounding that epidemic appear to have been consistent with gastrointestinal anthrax (35).

Today, gastrointestinal anthrax receives little attention from the media, which focus instead on the potential airborne dissemination of *B. anthracis* by terrorists. But the ability of anthrax spores to survive the conditions used to pasteurize milk (145.4°F [63°C] for 30 minutes or 161.6°F [72°C] for 15 seconds) and to endure for extended periods in soil and in food must not be forgotten (16, 35, 40, 47). The FDA, recognizing the bioterror potential of *B. anthracis* as a contaminant in food or water, has been exploring rapid detection methods for food-borne anthrax (19). Nevertheless, its current low profile as a food-borne pathogen may make *B. anthracis* a tantalizing tool for bioterrorists who, like the Rajneesh cult, wish to tamper with our food supply.

655 new field personnel and increased its surveillance of imported foods into the various U.S. ports of entry. In 2003, the agency published proposed regulations requiring all food manufacturing, processing, packing, and warehousing facilities, domestic or foreign (if exporting products to the United States), to register with FDA and to maintain records of the source of each food ingredient and the destination of each product leaving their facilities. FDA also developed guidance documents to provide information to domestic food facilities that would assist them to improve their food security systems, commissioned a threat assessment, redirected research funds to developing improved analytical methods, and cooperated with other agencies in emergency response exercises (17, 18). For its part,

in April 2002, the USDA developed and issued food security guidelines to the meat, poultry, and egg industry and has cooperated with FDA in various joint projects, including preparing and offering a food security awareness training program (13, 14, 18).

Other countries have also responded to the terror threat. Canada, member states of the European Union, Mexico, Australia, and members of the G7 states, to name a few, have taken steps both individually and in concert to review, upgrade, and coordinate their biosecurity plans (21, 41, 48). But no system is ever foolproof. The food we eat and the water we drink pass through too many different hands for us to ever hope to achieve 100% security. We might manage to shut the door on possible large-scale bioterrorism attacks on our food supply, but we will always be susceptible to more petty instances of intentional food or water contamination, such as those committed by the Rajneesh cult and the Dallas hospital lab worker. As Centers for Disease Control and Prevention epidemiologist Shellie A. Kolavic, lead investigator of the Dallas outbreak, observed in an interview with *USA Today*, "When a scientist 'decides to commit a crime... there's not much we can do' " (34).

References

1. **Altman, L. K.** 12 August 1997. Some medical puzzles lead to dark, and criminal, minds. *New York Times*, New York, N.Y.
2. **Anonymous.** 12 March 1982. Town may abolish itself to bar sect's takeover. *New York Times*, New York, N.Y.
3. **Anonymous.** 16 September 1984. Tension building over Oregon sect. Recruiting of many homeless to live at group's ranch prompts political fear. *New York Times*, New York, N.Y.
4. **Anonymous.** 11 October 1984. Voter registration halted as guru is picketed. *New York Times*, New York, N.Y.
5. **Anonymous.** 9 November 1984. *Salmonella* culprits: dirty hands. *Seattle Times*, Seattle, Wash.
6. **Anonymous.** 22 July 1986. Ma Sheela pleads guilty in poisoning. *Seattle Times*, Seattle, Wash.
7. **Anonymous.** 12 November 1996. FBI probes tainted doughnuts. *The Commercial Appeal*, Memphis, Tenn.
8. **Anonymous.** 7 August 2003. 13 fall ill after taking food. *News Today*, Chennai, India. [Online.] http://foodhaccp.com/msgboard.mv?parm_func=showmsg+parm_msgnum=1009573. Accessed 28 June 2006.
9. **Anonymous.** 15 August 2003. Cook decamps with jewels. *News Today*, Chennai, India. [Online.] http://foodhaccp.com/msgboard.mv?parm_func=showmsg+parm_msgnum=1009600. Accessed 28 June 2006.
10. **Becka, H.** 12 September 1998. 20-year sentence given in taintings. Woman gives tearful apology for sickening workers, ex-boyfriend. *Dallas Morning News*, Dallas, Tex.

11. **Cieslak, T. J., and E. M. Eitzen, Jr.** 1999. Clinical and epidemiologic principles of anthrax. *Emerg. Infect. Dis.* **5**:552–555.

12. **Davis, C. J.** 1999. Nuclear blindness: an overview of the biological weapons programs of the former Soviet Union and Iraq. *Emerg. Infect. Dis.* **5**: 509–512.

13. **Department of Agriculture.** 2002. FSIS security guidelines for food processors. U.S. Department of Agriculture, Food Safety and Inspection Service. [Online.] http://www.fsis.usda.gov/oa/topics/SecurityGuide.pdf. Accessed 28 June 2006.

14. **Department of Agriculture.** 2005. Announcing food security training program opportunities. Protecting the food supply from intentional adulteration: an introductory training session to raise awareness. U.S. Department of Agriculture, Agricultural Marketing Service. [Online.] http://www.ams.usda.gov/foodsecurity/. Accessed 28 June 2006.

15. **Dietrich, B.** 4 October 1984. Election-day showdown coming for Rajneesh. High stakes for guru followers, Oregon neighbors. *Seattle Times*, Seattle, Wash.

16. **Dragon, D. C., D. E. Bader, J. Mitchell, and N. Woollen.** 2005. Natural dissemination of *Bacillus anthracis* spores in northern Canada. *Appl. Environ. Microbiol.* **71**:1610–1615.

17. **Food and Drug Administration.** 19 March 2003. Guidance for industry: FDA issues food and cosmetic security preventive measures guidance. U.S. Food and Drug Administration. [Online.] http://www.fda.gov/oc/factsheets/foodsecurity.html. Accessed 28 June 2006.

18. **Food and Drug Administration.** 23 July 2003. Progress report to Secretary Tommy G. Thompson: ensuring the safety and security of the nation's food supply. U.S. Food and Drug Administration, Center for Food Safety and Applied Nutrition. [Online.] http://www.cfsan.fda.gov/~dms/fssrep.html. Accessed 28 June 2006.

19. **Food and Drug Administration.** 2005. Testing for rapid detection of adulteration of food. Report to Congress. Submitted to the Committee on Energy and Commerce of the House of Representatives and the Committee on Health, Education, Labor, and Pensions of the Senate. Second annual report—February 2005. U.S. Food and Drug Administration. [Online.] http://www.fda.gov/oc/bioterrorism/report_adulteration.html. Accessed 28 June 2006.

20. **Food and Drug Administration.** 2005. The Bioterrorism Act of 2002. U.S. Food and Drug Administration. [Online.] http://www.fda.gov/oc/bioterrorism/bioact.html. Accessed 28 June 2006.

21. **Health and Consumer Protection Directorate General.** 2005. European clinical guidelines for bioterror agents. European Commission, Health and Consumer Protection Directorate General. [Online.] http://europa.eu.int/comm/health/ph_threats/Bioterrorisme/clin_guidelines_en.htm. Accessed 28 June 2006.

22. **Henderson, D. A.** 1998. Bioterrorism as a public health threat. *Emerg. Infect. Dis.* **4**:488–492.

23. **Henderson, D. A.** 1999. About the First National Symposium on Medical and Public Health Response to Bioterrorism. *Emerg. Infect. Dis.* **5**:491.

24. **Henderson, D. A.** 1999. Smallpox: clinical and epidemiologic features. *Emerg. Infect. Dis.* **5**:537–539.

25. **Inglesby, T. V.** 1999. Anthrax: a possible case history. *Emerg. Infect. Dis.* **5:**556–560.

26. **Jernigan, D. B., P. L. Raghunathan, B. P. Bell, R. Brechner, E. A. Bresnitz, J. C. Butler, M. Cetron, M. Cohen, T. Doyle, M. Fischer, C. Greene, K. S. Griffith, J. Guarner, J. L. Hadler, J. A. Hayslett, R. Meyer, L. R. Petersen, M. Phillips, R. Pinner, T. Popovic, C. P. Quinn, J. Reefhuis, D. Reissman, N. Rosenstein, A. Schuchat, W.-J. Shieh, L. Siegal, D. L. Swerdlow, F. C. Tenover, M. Traeger, J. W. Ward, I. Weisfuse, S. Wiersma, K. Yeskey, S. Zaki, D. A. Ashford, B. A. Perkins, S. Ostroff, J. Hughes, D. Fleming, J. P. Koplan, J. L. Gerberding, and the National Anthrax Epidemiologic Investigation Team.** 2002. Investigation of bioterrorism-related anthrax, United States, 2001: epidemiologic findings. *Emerg. Infect. Dis.* **8:**1019–1028.

27. **Kassenborg, H., R. Danila, P. Snippes, M. Wiisanen, M. Sullivan, K. E. Smith, N. Crouch, C. Medus, R. Weber, J. Korlath, T. Ristinen, R. Lynfield, H. F. Hull, J. Pahlen, T. Boldingh, K. Elfering, G. Hoffman, T. Lewis, A. Friedlander, H. Heine, R. Culpepper, E. Henchal, G. Ludwig, C. Rossi, J. Teska, J. Ezzell, and E. Eitzen.** 2000. Human ingestion of *Bacillus anthracis*-contaminated meat—Minnesota, August 2000. *Morb. Mortal. Wkly. Rep.* **49:**813–816.

28. **Katz, I.** 1 August 1994. The mystic and the mayhem. When the Bhagwan's followers took on the people of Antelope there could only be one winner. *The Guardian*, Manchester, United Kingdom.

29. **Keim, P., K. L. Smith, C. Keys, H. Takahashi, T. Kurata, and A. Kaufmann.** 2001. Molecular investigation of the Aum Shinrikyo anthrax release in Kameido, Japan. *J. Clin. Microbiol.* **39:**4566–4567.

30. **Kolavic, S. A., A. Kimura, S. L. Simons, L. Slutsker, S. Barth, and C. E. Haley.** 1997. An outbreak of *Shigella dysenteriae* type 2 among laboratory workers due to intentional food contamination. *JAMA* **278:**396–398.

31. **Kortepeter, M. G., and G. W. Parker.** 1999. Potential biological weapons threats. *Emerg. Infect. Dis.* **5:**523–527.

32. **Laganella, V. A.** 2002. Anthrax: a primary care physician's perspective. *J. Am. Osteopath. Assoc.* **102:**37–40.

33. **Lane, H. C., and A. S. Fauci.** 2001. Bioterrorism on the home front. A new challenge for American medicine. *JAMA* **286:**2595–2597.

34. **Manning, A.** 6 August 1997. Terrorism's versatility hits home with poisoning. Two cases may not have been preventable. *USA Today*, McLean, Va.

35. **Morens, D. M.** 2002. Epidemic anthrax in the eighteenth century, the Americas. *Emerg. Infect. Dis.* **8:**1160–1162.

36. **Olson, K. B.** 1999. Aum Shinrikyo: once and future threat? *Emerg. Infect. Dis.* **5:**513–516.

37. **O'Toole, T.** 1999. Smallpox: an attack scenario. *Emerg. Infect. Dis.* **5:** 540–546.

38. **O'Toole, T.** 2000. The problem of biological weapons: next steps for the nation. Second National Symposium on Medical and Public Health Response to Bioterrorism—Speaker Transcript, Washington, D.C. [Online.] http://www.upmc-biosecurity.org/pages/events/2nd_symposia/transcripts/trans_otoo.html. Accessed 28 June 2006.

39. **Pavlin, J. A.** 1999. Epidemiology of bioterrorism. *Emerg. Infect. Dis.* **5:** 528–530.

40. **Perdue, M. L., J. Karns, J. Higgins, and J. A. van Kessel.** 2003. Detection and fate of *Bacillus anthracis* (Sterne) vegetative cells and spores added to bulk tank milk. *J. Food Prot.* **66:**2349–2354.

41. **Public Health Agency of Canada.** 2005. Bioterrorism and emergency preparedness. Public Health Agency of Canada. [Online.] http://www.phac-aspc.gc.ca/ep-mu/bioem_e.html. Accessed 28 June 2006.

42. **Riedel, S.** 2005. Anthrax: a continuing concern in the era of bioterrorism. *Proc. (Baylor Univ. Med. Cent.)* **18:**234–243.

43. **Russell, P. K.** 1999. Vaccines in civilian defense against bioterrorism. *Emerg. Infect. Dis.* **5:**531–533.

44. **Shalala, D. E.** 1999. Bioterrorism: how prepared are we? *Emerg. Infect. Dis.* **5:**492–493.

45. **Shane, S., and D. Johnston.** 17 September 2005. In 4-year anthrax hunt, F.B.I. finds itself stymied, and sued. *New York Times,* New York, N.Y.

46. **Sidell, F. R., E. T. Takafuji, and D. R. Franz (ed.).** 1997. *Medical Aspects of Chemical and Biological Warfare.* Office of the Surgeon General, Department of the Army, United States of America, Falls Church, Va. [Online.] http://www.nbc-med.org/SiteContent/HomePage/WhatsNew/MedAspects/contents.html. Accessed 28 June 2006.

47. **Sirisanthana, T., and A. E. Brown.** 2002. Anthrax of the gastrointestinal tract. *Emerg. Infect. Dis.* **8:**649–651.

48. **Smallwood, R. A., A. Merianos, and J. D. Mathews.** 2002. Bioterrorism in Australia. How real is the threat, and how prepared are we? *Med. J. Aust.* **176:**251–253.

49. **Stern, J.** 1999. The prospect of domestic bioterrorism. *Emerg. Infect. Dis.* **5:**517–522.

50. **Takahashi, H., P. Keim, A. F. Kaufmann, C. Keys, K. L. Smith, K. Taniguchi, S. Inouye, and T. Kurata.** 2004. *Bacillus anthracis* incident, Kameido, Tokyo, 1993. *Emerg. Infect. Dis.* **10:**117–120.

51. **Torok, T. J., R. V. Tauxe, R. P. Wise, J. R. Livengood, R. Sokolow, S. Mauvais, K. A. Birkness, M. R. Skeels, J. M. Horan, and L. R. Foster.** 1997. A large community outbreak of salmonellosis caused by intentional contamination of restaurant salad bars. *JAMA* **278:**389–395.

52. **Wheelis, M.** 2002. Biological warfare at the 1346 siege of Caffa. *Emerg. Infect. Dis.* **8:**971–975.

chapter 14

The Impact of Imports

QUEBEC TAKES ITS FRENCH HERITAGE VERY SERIOUSLY—including French cuisine, such as the delicacy described on restaurant menus as "cuisses de grenouille" (frog legs). In the 1970s, most of the frog legs that found their way onto the plates and palates of discerning diners came from countries in the area of the Indian subcontinent, notably, India, Pakistan, and Bangladesh. Because *Salmonella* often hitched a ride on the frog legs, Canada held them in import quarantine until the Health Protection Branch lab had finished checking each batch for the pathogen. Only once a shipment was pronounced satisfactory were the frog legs allowed into the country. *Salmonella*-positive shipments were either destroyed or returned to the country of origin.

Canada wasn't the only country dealing with *Salmonella* contamination in imported frog legs in the 1970s. During the first half of the decade, *Salmonella*-contaminated frog legs from countries such as Mexico, Japan, Bangladesh, Pakistan, India, and Indonesia accounted for the highest rate of violations of all foods regulated by the U.S. Food and Drug Administration (FDA) (3). But the contaminated frog legs were no more than a nuisance compared to the complex issues of food contamination that have evolved over recent decades as a result of the ever-expanding international market for foods.

Trade Is a Two-Way Street

In 2004, international trade in agricultural products (including food) totaled $783 billion. The European Union (EU) and the United States were the largest importers of agricultural products, buying $374 billion and $88 billion, respectively. At the same time, the EU also accounted for 44% of agricultural exports (just under $345 billion); the United States exported nearly $80 billion of agricultural products and Canada exported $40 billion (87, 88).

Many countries have been both the source and the recipient of contaminated food and, as Table 14.1 illustrates, just about any type of imported food is a potential vehicle for food-borne illness. Canada exported *Salmonella*-contaminated chocolate to the United States in 1972 and imported *Salmonella*-contaminated chocolate from Belgium in 1985 (22, 53); EU member countries imported contaminated snack foods from Israel and exported contaminated meat, dairy products, and produce to their fellow members (50, 60, 70); the United States has imported contaminated produce from Mexico, Honduras, and Guatemala and—twice in the last 5 years—exported contaminated raw almonds to several countries around the world, including Mexico and Canada (10, 55, 59).

A Tough Nut To Crack

In December 2000, officials at the Ontario (Canada) Central Public Health Laboratory noticed an increase in the incidence of *Salmonella enterica* serotype Enteritidis (19). The next month, Canada's National Laboratory for Enteric Pathogens identified the isolates as serotype Enteritidis phage type 30 (PT30)—the first reported isolation of PT30 in Canada since 1992. By March 2001, PT30 had also been found in the provinces of New Brunswick and Nova Scotia. One month later, Canadian epidemiologists identified raw almonds sold by a bulk food retail chain as the probable source of the *Salmonella* (19).

On April 12, 2001, the Canadian Food Inspection Agency announced that a single batch of contaminated almonds was responsible for several cases of *Salmonella* gastroenteritis in three provinces: Ontario, New Brunswick, and Nova Scotia (15). The almonds were produced in the United States and distributed in five Canadian provinces as well as five U.S. states (15, 55). The recall—limited at first to raw almonds that had been distributed by just a single Canadian bulk food retail chain—was expanded twice and eventually encompassed both raw almonds and a mixed nut snack product sold by two Canadian bulk food chains and in U.S. retail outlets (16, 17, 55).

Almonds grow on trees, out of contact with the soil. Their protection from soil contamination ceases during harvest, when the trees are shaken mechanically to release the nuts, which are allowed to fall to the ground. The fallen almonds are left on the ground to dry for about 4 days to 2 weeks. During this drying period, the almond shells are in contact with the soil, and with any pathogen that the soil might contain. After drying, the nuts are swept into rows, then lifted onto conveyer belts and loaded into trailers, which transport them to hulling and shelling plants. During this process, pressurized air is used to remove leaves and other light debris (55).

Once they arrive at the plant, the almonds are cleaned mechanically to remove any remaining debris, following which the nuts pass through

Table 14.1 Examples of food-borne disease outbreaks associated with imported foods

Country of origin	Food	Pathogen	Countries affected	No. of illnesses (deaths)	Reference(s)
Belgium	Chocolates	*S. enterica* serotype Nima	Canada, United States	33	53
Bosnia	Raspberries	Calicivirus	Canada	>200	41
Brazil	Mango	*S. enterica* serotype Newport	United States	78 (2)	77
Brazil (via Germany)	Black pepper	*S. enterica* serotype Oranienburg	Norway	126	45
Canada	Chocolate	*S. enterica* serotype Eastbourne	United States, Canada	119	22
China	Peanuts	*S. enterica* serotype Stanley and *S. enterica* serotype Newport	Australia, Canada	7	74
Denmark	Kebab meat	*S. enterica* serotype Typhimurium	Sweden	116	50
Egypt	Tahini	*S. enterica* serotype Montevideo	Australia, New Zealand	55	84
France	Brie cheese	*E. coli* O27:H20 (enterotoxigenic)	United States	45	39, 67
France	Cheese from unpasteurized goat milk	*S. enterica* serotype Stourbridge	Switzerland, Sweden, Germany, Austria, United Kingdom (England), The Netherlands, France	52	29
Germany	Garlic cloves packed in oil	*Clostridium botulinum*	Denmark	1	64
Guatemala	Raspberries	*Cyclospora* sp.	Canada, United States	1,465	48
Guatemala	Snow peas	*C. cayetanensis*	United States	96	23
Guatemala, Honduras	Frozen fruit	*S. enterica* serotype Typhi	United States	16	58
Israel	Savory snack food	*S. enterica* serotype Agona	United Kingdom, United States	37	60
Italy	Ruccola salad	*S. enterica* serotype Thompson	Norway	20	70
Italy	Raw beef	*S. enterica* serotype Typhimurium	Denmark	22	30
Italy	Salami	*S. enterica* serotype Typhimurium	Italy, Sweden	23	51
Mexico	Frozen strawberries	Hepatitis A virus	United States	258	54
Mexico	Cantaloupes	*S. enterica* serotype Saphra	United States	24	69
Mexico	Cantaloupes	*S. enterica* serotype Poona	Canada, United States	>400	40
Mexico	Green onions	Hepatitis A virus	United States	≥555 (3)	26
The Netherlands	Alfalfa sprouts (from imported seed)	*S. enterica* serotype Stanley	Finland, United States	≥242	68

(Continued)

Table 14.1 *(Continued)*

Country of origin	Food	Pathogen	Countries affected	No. of illnesses (deaths)	Reference(s)
Poland	Frozen raspberries	Norovirus	Denmark	>1,000	31
Poland	Raw minced beef	*S. enterica* serotype Typhimurium DT104	Norway	4	56
Spain	Iceberg lettuce	*Shigella sonnei*	Norway, Sweden, United Kingdom	≥163	57
Thailand	Frozen fresh coconut milk	*Vibrio cholerae* O1, biotype El Tor	United States	3	66, 83
Turkey	Aniseed (herbal tea)	*S. enterica* serotype Agona	Germany	42	62
United States	Raw almonds	*S. enterica* serotype Enteritidis	Canada, United States	168	55
United States	Ground beef	*E. coli* O157:H7	Japan	3	65
United States	Pet treat	*S. enterica* serotype Thompson	Canada, United States	9	24

a series of rollers to break the hulls and shells. The almond kernels are separated from the hulls and shells and are stored in bulk bins or in silos. The kernels are cleaned of residual debris, fumigated to protect against insect infestation, and sorted by size and quality (55).

All of the almonds implicated in the 2001 outbreaks in Canada and the United States were traced back to a single U.S. processor, Hughson Nut, Inc., located in California's San Joaquin Valley (11, 15). On inspecting the processor's facility, investigators found *S. enterica* serotype Enteritidis PT30—a rare phage type and the same one isolated from victims of the outbreaks—in 25% (two of eight swab samples) of the environmental samples. They also recovered PT30 from one of the two lots of almonds in storage at the facility. With the cooperation of the processor, investigators identified four possible huller/sheller plants that might have supplied the contaminated almonds. Environmental swab samples (but not finished product) taken from one of the four huller/sheller operations yielded PT30 (55).

Investigators next traced the almonds back to the growers that had supplied nuts to the implicated huller/sheller plant. They identified 22 orchards, belonging to four growers, from which the almonds might have come. Soil samples from 10 of the 22 orchards contained PT30. Three of the four growers each had at least one contaminated orchard. The broad distribution of a rare *Salmonella* phage type through the soil of several orchards belonging to three growers led the investigators to suspect a common source of contamination. But that source was never identified. None of the growers used manure to fertilize their orchards, nor were there any cattle or poultry being raised or grazed nearby. And the water that was

A Blessing in Disguise

Salmonella can sometimes be found in unexpected places. In a 2-month period beginning in early December 1994, British health authorities received reports of 27 cases of *Salmonella enterica* serotype Agona in England and Wales—more than twice the number of cases that had been reported during the same period the year before. Many of the victims were children, most with Jewish surnames, and the cases were clustered in certain geographic areas. Epidemiologists tracked the source of the outbreak to a kosher peanut-flavored snack that had been manufactured in Israel and exported to the United Kingdom, Canada, and the United States (60).

The snack food was also distributed very widely in Israel, where health authorities had noticed a sharp increase in the incidence of serotype Agona throughout the country between October 1994 and January 1995. The 2,200-case outbreak was nationwide and—in part because the snack was widely distributed throughout the country—Israeli health authorities were unable to pinpoint the source of the disease until they learned about the British outbreak. With the British information in hand, Israeli health inspectors were able to follow up with the snack food manufacturer and have the implicated products recalled. Investigators from the Israeli Ministry of Health also reviewed the manufacturer's production procedures and recommended several changes (76).

This outbreak illustrates clearly the importance of international cooperation in food safety investigations—and an unexpected benefit of international trade. If the British had not detected and found the source of the serotype Agona outbreak that hit England and Wales, Israeli health authorities might never have been able to track the source of their own much larger outbreak (82).

used to irrigate all 22 orchards—both *Salmonella*-positive and *Salmonella*-negative—was drawn exclusively from the California aqueduct system (55).

The almond industry took several actions in response to this first-ever almond-related outbreak of *Salmonella* gastroenteritis. The Almond Board of California, an industry trade association, funded research to determine the potential sources of *Salmonella* in orchards, to understand the ability of *Salmonella* to survive in the environment, and to develop and validate treatments to kill *Salmonella* and other pathogens that might contaminate the nuts during harvest or processing. The industry also cooperated with government agencies to develop guidelines designed to reduce the risk of contaminating almonds during growing, harvesting, transporting, shelling, and processing of the nuts (55, 75). Unfortunately, their efforts were not in time to prevent a second outbreak.

On May 18, 2004, 3 years after the Hughson Nut outbreak, Paramount Farms recalled several production lots of whole raw almonds that had been linked to several cases of *Salmonella* gastroenteritis (34). The initial recall mushroomed as investigators gained a better understanding of the scope of the outbreak and the breadth of the product's distribution. In all, Paramount Farms and its customers issued 39 recalls between May 18 and July 9, 2004, seeking the return of at least 13 million pounds of raw almonds (36, 61). The almonds had been distributed to about 50 commercial customers and were sold in the United States, Canada, France, Italy, Japan, Korea, Malaysia, Mexico, Taiwan, and the United Kingdom (59).

The outbreak, which lasted for several months, affected 29 people in 12 U.S. states and Canada. No illnesses were reported from the eight other countries to which the almonds had been exported. The victims were infected with *S. enterica* serotype Enteritidis phage type 9c, a rare strain but different from the one that caused the 2001 outbreak. The outbreak strain was never isolated from the implicated batches of almonds, although it was detected in one environmental sample that was collected at the Paramount facility and in three samples from two of the huller/shellers that supplied Paramount (59). As a precaution, Paramount announced on May 18, 2004, that it would begin pasteurizing its raw almonds to eliminate any *Salmonella* that might possibly contaminate the product (35).

During the 3 years spanned by these two outbreaks, the almond industry responded in a responsible fashion—reacting constructively to the first problem and attempting to forestall another occurrence. But their experience shows that the best of intentions is not always enough. And if a highly industrialized country such as the United States can accidentally sell contaminated food to its international customers, how much more likely is it for a less-developed country to export food-borne pathogens?

Single-Celled Stowaways

Once seasonal items, fresh strawberries and raspberries can be found in U.S. and Canadian supermarkets most months of the year. Guatemala, Mexico, Honduras, Brazil, and Chile are just some of the countries that supply out-of-season fruit, including berries, mangoes, melons, and grapes, to their North American neighbors (13, 58, 69, 77). While international traffic in produce from less-developed countries allows U.S. and Canadian consumers access to a wide variety of fruit year-round, it also increases the chances that food-borne pathogens will spread beyond the borders of the countries in which they are endemic.

In June 1995, a Florida hospital found *Cyclospora cayetanensis* in the stools of six patients. None of the patients had traveled outside the United States recently, and there was nothing in their case histories that pointed to a common exposure to potentially contaminated food or water. By August 22, 87 cases of *Cyclospora* infection had been reported to the health departments of Florida's Palm Beach and Martin counties. When the Florida Department of Health and the U.S. Centers for Disease Control and Prevention (CDC) investigated, they discovered that the most probable source of the outbreak was imported raspberries (63). The outbreak was small; it didn't even merit an article in *Morbidity and Mortality Weekly Reports*. But it was a sign of things to come.

The *Cyclospora* problem appeared in earnest in May 1996, with outbreaks reported in 20 U.S. states (including Palm Beach County in Florida)

Cyclospora cayetanensis: Another New Kid on the Block

Discovered in 1870, *Cyclospora* was first associated with human illness in 1977 and 1978 (47, 79). But it was only in the 1990s that this parasite emerged as a significant human pathogen.

Unlike other species of *Cyclospora*, which have been found in a range of insect eaters, myriapods (centipedes, millipedes, and related animals), reptiles, and rodents, *C. cayetanensis* appears to limit itself to human hosts, although closely related species of *Cyclospora* have been found in other primates. The parasite is ingested by its host, replicates in the digestive tract, and is excreted in the stool in the form of immature, noninfectious oocysts. The oocysts mature outside the human body for an unknown number of days or weeks, during which they sporulate and become infectious. Because the freshly excreted oocysts are not infectious, cyclosporiasis is not spread through person-to-person contact (47).

Cyclosporiasis has a relatively long incubation period of about a week. Its infectious dose is still unknown (1). *C. cayetanensis* causes a range of symptoms, which can be severe in immunocompromised individuals: watery—sometimes explosive—diarrhea, loss of weight, loss of appetite, nausea, vomiting, fatigue, abdominal bloating, cramps, and muscle aches. The most common symptoms are diarrhea, anorexia, fatigue, and weight loss. The symptoms are usually self-limiting, but can be prolonged; diarrhea lasted for a median of 7 days in one study, whereas fatigue and malaise can endure for several weeks or even months (47). Acute symptoms of cyclosporiasis (diarrhea, etc.) can be treated, if necessary, with trimethoprim-sulfamethoxazole (73).

and the U.S. District of Columbia, as well as two Canadian provinces (12, 18, 46, 48). At first, investigators suspected that California strawberries were to blame for most or all of the 1,465 cases of cyclosporiasis (13, 48). But they were wrong. Within a few weeks, their attention had shifted to raspberries imported from Guatemala (46, 80, 81).

Once imported raspberries were identified as the most probable culprit, federal officials began the painstaking process of tracing the berries back to their source. CDC and FDA investigators established that the implicated raspberries had arrived by air in the United States through four ports of entry—Miami, Houston, Los Angeles, and the District of Columbia—and were purchased from as many as seven Guatemalan exporters who, in turn, obtained the berries from at least five farms. Investigators visited the Guatemalan farms, with the cooperation of the local growers, but were unable to detect *Cyclospora* either in the berries or the irrigation water supply. Their investigation, however, suffered from two handicaps: the testing methods available to them were not capable of detecting low levels of contamination, and the tests took place on samples obtained after the outbreaks had subsided (48).

At the recommendation of the U.S. investigation team, the Guatemalan raspberry industry switched to potable water (for mixing solutions that are sprayed onto berries, cleaning equipment surfaces that contact the berries, hand washing, and drinking), improved sanitary facilities for workers, identified possible sources of cross-contamination, and developed and implemented a Hazard Assessment and Critical Control Points (HACCP) program (48). The Guatemala Berry Commission also limited raspberry exports for the 1997 season to farms that were identified as "low risk" for contamination with *Cyclospora* (47). But a deep-seated contamination problem is not cured in just one growing season. *Cyclospora* outbreaks reappeared in the United States and Canada in 1997 (49). In response, the Guatemalans suspended raspberry exports to the U.S. and Canadian markets, and the 1997 outbreak ended soon after it began (47).

The CDC sent a team of researchers to Guatemala in the spring of 1997 to gain a better understanding of the pathogen's ecology and epidemiology. They conducted environmental studies and also tracked *Cyclospora* infection rates among Guatemalan raspberry farm workers and their families. Six out of 185 farm workers and their families tested during April and May of 1997 during the CDC study were infected with *Cyclospora*. Sixty-two out of 68 *Cyclospora*-infected people (91%) included in a separate case-control study had drunk untreated water in the two weeks before their illness, compared to 73% of uninfected individuals. Young children (between 1.5 and 9 years old) were the most susceptible to infection. In addition to drinking untreated water, the type of sewage drainage, ownership of chickens and other domestic fowl, and—for children under the age of 2 years—contact with the soil were significant risk factors for infection (14).

The FDA forbade importation of Guatemalan raspberries into the United States in 1998, and there were no reported outbreaks of cyclosporiasis associated with raspberries that year (47). In contrast, Canada allowed Guatemalan raspberries to enter the country and suffered through another series of outbreaks in the spring of 1998 (8, 47).

The FDA permitted raspberries from five low-risk Guatemalan producers back into the United States in 2000 only to be faced with two more outbreaks—one in Georgia and one in Philadelphia, Pa.—before the year was halfway over. Fifty-four guests at a June 2000 wedding reception in Philadelphia complained of symptoms consistent with cyclosporiasis; *Cyclospora* oocysts were found in the stools of five of the victims. The guests had been served a cake with cream filling, and the cream contained pieces of raspberry. The raspberries were traced back to one of several distributors, who had purchased berries that originated in the United States, Guatemala, and Mexico. It was impossible to determine whether the berries used in the cream filling were domestic or foreign and—if foreign—whether they were Mexican or Guatemalan. The FDA sent an inspection team to investigate the Guatemalan farm that was thought to be a possible source of both outbreaks and found nothing untoward. Nevertheless, the farm was embargoed from exporting raspberries to the United States in 2001, and there were no *Cyclospora*-related outbreaks that year (52).

Barring the Door

In the United States, the primary responsibility for overseeing the safety of imported foods rests with the FDA. In July 1999, U.S. President Clinton directed the FDA to take several new measures to reduce the risk of unsafe foods entering the United States. More new requirements were mandated under the Public Health Security and Bioterrorism Preparedness and Response Act of 2002 (85). As of December 12, 2003, the FDA must be notified in advance of all shipments of imported foods that fall under its jurisdiction. Imported meats, poultry, and processed egg products, which are regulated by the U.S. Department of Agriculture (USDA), are excluded from the FDA notification requirement (33).

To evaluate the effectiveness of its import control program, the FDA carries out regular surveys of fresh produce and other high-risk imported foods. In 1999, the agency examined 1,003 samples of imported produce from 21 countries for the presence of *Salmonella Shigella*, and *Escherichia coli* O157:H7. None of the 1,003 samples contained *E. coli* O157:H7, but 35 samples were contaminated with *Salmonella* and 9 contained *Shigella*. All shipments found to be contaminated with a pathogen were refused entry into the United States, but the 1,003 shipments that were sampled represent only a small fraction of the quantity of produce and other foodstuffs that enter the United States each year (32).

The FDA has never had the resources to test every single shipment of imported food. In 1992, the agency inspected only 8% of import shipments that reached U.S. ports of entry; that number dropped to 1.7% in 1997. The FDA's resources were declining steadily between 1992 and 1997 at the same time that the number of import shipments was rising (42). And the situation improved only marginally after the bioterrorism threat manifested itself. In 1999, fewer than 1 in 100 import shipments of seafood were tested for microbiological contamination. By the end of 2002, the percentage had risen to a meager 1.2% (43).

Even after a shipment of imported food has been inspected and cleared by the FDA, there is no guarantee that the entire shipment is pathogen free. Only an infinitesimal fraction of a shipment can be sampled for lab analysis; typically, a 500-g (1.1-lb) portion or less is tested for each pathogen (37). A negative report for *Salmonella*, for example, simply means that the lab did not find *Salmonella* in the sample that it analyzed. The pathogen could easily be present in a different part of the food shipment. And some pathogens are more difficult to detect than others. It is only in recent years that the FDA has had at its disposal a practical test for parasites like *Cyclospora* or viruses such as hepatitis A (44, 72). Thus, it's understandable that hepatitis A-contaminated strawberries from Mexico made it past the FDA's inspectors in 1997 (54). But the strawberries should never have entered the USDA's school lunch program.

A Free Lunch

The National School Lunch Program was established in 1946. The program, which is administered by the USDA, provides low-cost and free lunches to school children from low-income families. The USDA both gives cash subsidies to participating schools and also supplies schools with surplus food products purchased under domestic agricultural price support programs (27).

In 1997, the USDA purchased surplus frozen strawberries from three food brokers and distributed them to schools under the school lunch program (20). The USDA had been advised early in 1997 that a San Diego, Calif., company was purchasing strawberries in Mexico for fraudulent resale as domestic surplus agricultural product but failed to follow up on the report (6). The strawberries were distributed to schools in the District of Columbia and 15 states, including Maine and Michigan (2, 54). The food brokers also sold some of the strawberries in the United States and Canada through normal commercial channels (54).

The first sign of a problem arose in mid-March 1997 in Calhoun County, Mich., with the report of a case of hepatitis A. Within 2 weeks, county health officials were investigating more than 190 suspected cases of hepatitis A and had already documented 128 confirmed cases. Calhoun health officials were quick to link the outbreak to frozen strawberries that had been served to school students about 4 weeks before the outbreak began (5). All told, the hepatitis-contaminated strawberries infected 213 people in Michigan, 29 in Maine, 5 in Wisconsin, and 7 in Arizona (54).

The contaminated strawberries were harvested in Mexico and purchased, processed, and packed by the Andrew & Williamson Sales Company (AWSC) of San Diego, Calif., in 1996 (6). AWSC issued a certificate of origin indicating that the strawberries were grown in the United States and—acting through intermediaries—sold 1.7 million pounds of the frozen strawberries to the USDA for distribution to schools (20). In November 1997, AWSC and its president, Frederick Williamson, pled guilty to charges of conspiracy, making false statements, and making a false claim (7). Seven months later, AWSC agreed to pay $1.3 million in civil damages and $200,000 in criminal penalties, and was also ordered to pay $150,000 in restitution. Williamson was sentenced to 5 months in prison and 5 months of home detention (9).

Inadvertent entry of an occasional shipment of contaminated food is the price consumers pay to have access to foods from around the world. The strawberry-linked hepatitis outbreak was unusual only in that criminal behavior on the part of the supplier allowed an imported food to be distributed to schools through government channels. Otherwise, it was just one more instance of contaminated food entering a country despite the efforts of its food safety regulators.

Imported Food—Embargo or Embrace?

One way for a country to prevent imported foods from causing illness is to embargo imports from high-risk countries or to embargo high-risk foods from any country. Many countries have chosen to embargo imports of beef or live cattle from countries perceived to be at high risk for bovine spongiform encephalopathy (BSE). This has not contained the spread of BSE completely, but it has helped the beef and cattle industry by providing a security blanket to consumers who might otherwise have avoided eating beef. The EU maintained an import ban on genetically modified foods between 1999 and 2003, ostensibly to protect consumers and farmers from the risks associated with disseminating genetically modified organisms (78). And the importation of raw poultry from areas affected by the avian influenza virus H5N1 has been restricted by countries that are unaffected by the H5N1 virus (28, 86). International guidelines for risk-based inspection of imported foods (developed under the auspices of the World Health Organization and the Food and Agriculture Organization of the United Nations) have been agreed to by a number of countries (21).

The automatic detention of raw and cooked shrimp from India (pending results of testing) practiced by the United States is just one example of temporary embargoes used by the FDA to prevent high-risk foods (or foods from high-risk countries or processors) from entering the U.S. food supply. This approach reduces—but does not eliminate—the risk that a contaminated food from overseas might reach U.S. consumers. Some automatic detention orders, such as the FDA's import ban on raspberries from Guatemala, are in force for a relatively short time, whereas others, such as the detention of shrimp from India and other countries in the region, remain active for more than a decade (38).

Another approach to preventing contaminated food from entering a country is to demand that an exporter conform to the food safety laws and protocols of the importing country. Japan has taken this approach to the renewal of beef imports from the United States. That country's government has insisted on inspecting all meat processing facilities that wish to export beef to Japan and has reserved the right to reject any facility that fails to meet its standards. The USDA requires that foreign meat processors conform to U.S. HACCP rules if they wish to export meat products to the United States. And, since 1996, the United States has exported poultry to Russia under an agreement that sets out the microbiological and other standards and specifications that the U.S. poultry processors are required to meet in order to supply their products to Russia (4).

Despite the headlines that sometimes are triggered when an imported food is found to be the cause of a food-borne disease outbreak, the great majority of food-borne illnesses are due to domestically produced foods. Of the 2,751 outbreaks of food-borne disease reported to the CDC in

1993 to 1997, only a handful ("several outbreaks" according to the report's authors) were due to imported foods (71). Similarly, of the 214 food-borne disease outbreaks reported in Australia between 1995 and 2000, just one was traced to an imported food (25). Perhaps some of the microbiological safety and sanitation requirements placed on processors of foods that are destined for export to other countries should be applied equally to foods that are meant for domestic consumption!

References

1. **Alfano-Sobsey, E. M., M. L. Eberhard, J. R. Seed, D. J. Weber, K. Y. Won, E. K. Nace, and C. L. Moe.** 2004. Human challenge pilot study with *Cyclospora cayetanensis*. *Emerg. Infect. Dis.* **10:**726–728.
2. **Allen, J. E.** 3 April 1997. Hepatitis contamination sparks scare; executive quits as investigators track suspected frozen strawberries. *Austin American Statesman*, Austin, Tex.
3. **Andrews, W. H., C. R. Wilson, P. L. Poelma, and A. Romero.** 1977. Comparison of methods for the isolation of *Salmonella* from imported frog legs. *Appl. Environ. Microbiol.* **33:**65–68.
4. **Anonymous.** 25 March 1996. Agreed minutes from the meeting of experts of the government of the United States of America and the government of the Russia Federation regarding the application of sanitary measures to exports of poultry products from the United States to Russia. U.S. Department of Agriculture, Foreign Agricultural Service. [Online.] http://www.fas.usda.gov/itp/agreements/ruspoul.html. Accessed 2 July 2006.
5. **Anonymous.** 28 March 1997. Officials target berries in hepatitis A outbreak. *The Grand Rapids Press*, Grand Rapids, Mich.
6. **Anonymous.** 18 April 1997. Illegal fruit report too late to halt outbreak. *Times Union*, Albany, N.Y.
7. **Anonymous.** 14 November 1997. Strawberry supplier pleads guilty in hepatitis. *The Oregonian*, Portland, Oreg.
8. **Anonymous.** 1998. Outbreak of cyclosporiasis—Ontario, Canada, May 1998. *Morb. Mortal. Wkly. Rep.* **47:**806–809.
9. **Anonymous.** 16 June 1998. Food executive gets prison term for lying. *New York Times*, New York, N.Y.
10. **Anonymous.** 24 May 2004. Marler Clark sues Paramount Farms over *Salmonella*-tainted almonds. *Business Wire*, Spokane, Wash.
11. **Anonymous.** 2005. Hughson Nut. [Online.] http://www.hughsonnut.com/index.html. Accessed 2 July 2006.
12. **Artero, S.** 8 June 1996. Fruit salad suspect in possible *Cyclospora* cases. *Palm Beach Post*, West Palm Beach, Fla.
13. **Artero, S.** 11 June 1996. *Cyclospora* is showing scary affinity for strawberries. *Palm Beach Post*, West Palm Beach, Fla.
14. **Bern, C., B. Hernandez, M. B. Lopez, M. J. Arrowood, M. Alvarez de Mejia, A. M. de Merida, A. W. Hightower, L. Venczel, B. L. Herwaldt, and R. E. Klein.** 1999. Epidemiologic studies of *Cyclospora cayetanensis* in Guatemala. *Emerg. Infect. Dis.* **5:**766–774.

15. **Canadian Food Inspection Agency.** 12 April 2001. Health hazard alert. Almonds may contain dangerous bacteria. Canadian Food Inspection Agency, Office of Food Safety and Recall, Ottawa, Ontario, Canada. [Online.] http://www.inspection.gc.ca/english/corpaffr/recarapp/2001/20010412be.shtml. Accessed 2 July 2006.

16. **Canadian Food Inspection Agency.** 18 April 2001. Health hazard alert. Mixed snack products containing almonds sold by Bulk Barn may contain dangerous bacteria. Canadian Food Inspection Agency, Office of Food Safety and Recall, Ottawa, Ontario, Canada. [Online.] http://www.inspection.gc.ca/english/corpaffr/recarapp/2001/20010418e.shtml. Accessed 2 July 2006.

17. **Canadian Food Inspection Agency.** 18 May 2001. Health hazard alert. Almonds may contain dangerous bacteria. Canadian Food Inspection Agency, Office of Food Safety and Recall, Ottawa, Ontario, Canada. [Online.] http://www.inspection.gc.ca/english/corpaffr/recarapp/2001/20010518de.shtml. Accessed 2 July 2006.

18. **Chambers, J., S. Somerfeldt, L. Mackey, S. Nichols, R. Ball, D. Roberts, N. Dufford, A. Reddick, and J. Gibson.** 1996. Outbreaks of *Cyclospora cayetanensis* infection—United States, 1996. *Morb. Mortal. Wkly. Rep.* **45:**549–551.

19. **Chan, E. S., J. Aramini, B. Ciebin, D. Middleton, R. Ahmed, M. Howes, I. Brophy, I. Mentis, F. Jamieson, F. Rodgers, M. Nazarowec-White, S. C. Pichette, J. Farrar, M. Gutierrez, W. J. Weis, L. Lior, A. Ellis, and S. Isaacs.** 2002. Preliminary report. Natural or raw almonds and an outbreak of a rare phage type of *Salmonella enteritidis* infection. *Can. Commun. Dis. Rep.* **28:**97–99. [Online.] http://www.phac-aspc.gc.ca/publicat/ccdr-rmtc/02vol28/dr2812ea.html. Accessed 2 July 2006.

20. **Claiborne, W.** 11 June 1997. Firm faces charges in tainted berry case. *The Oregonian*, Portland, Oreg.

21. **Codex Committee on Food Import and Export Inspection and Certification Systems.** 2004. Joint FAO/WHO food standards programme. Proposed draft guidelines for risk-based inspection of imported foods. Codex Alimentarius Commission, Food and Agriculture Organization of the United Nations, and World Health Organization. [Online.] http://www.hc-sc.gc.ca/fn-an/alt_formats/hpfb-dgpsa/pdf/intactivit/fc13_05_e.pdf. Accessed 2 July 2006.

22. **Craven, P. C., D. C. Mackel, W. B. Baine, W. H. Barker, E. J. Gangarosa, M. Goldfield, H. Rosenfeld, R. Altman, G. Lachapelle, J. W. Davies, and R. C. Swanson.** 1975. International outbreak of *Salmonella eastbourne* infection traced to contaminated chocolate. *Lancet* **i:**788–793.

23. **Crist, A., C. Morningstar, R. Chambers, T. Fitzgerald, D. Stoops, M. Deffley, Y. Reyes, T. Hiden, J. Sullivan, D. Hawk, P. Lurie, M. Moll, S. Yeager, L. Lind, J. Burkee, K. Warren, M. Marcus, J. Reeser, H. Davidson, S. Thomas, B. L. Herwaldt, M. Hlavsa, S. P. Johnston, H. Bishop, A. daSilva, A. Hightower, D. K. El Reda, and N. Flowers.** 2004. Outbreak of cyclosporiasis associated with snow peas—Pennsylvania, 2004. *Morb. Mortal. Wkly. Rep.* **53:**876–878.

24. **Crowe, L., L. Chui, D. Everett. S. Brisdon, L. Gustafson, E. Galanis, L. McIntyre, L. MacDougall, L. Wilcott, A. Paccagnella, D. MacDonald, A. Ellis, A. Drake, J. Koepsell, C. DeBolt, S. McKeirnan, J. Duchin, R. Baer, M. Leslie, M. L. Collins, J. M. Johnson, D. E. Farmer, C. E. Keys, H. Ekperigin, F. Angulo, and R. E. Colindres.** 2006. Human salmonellosis

associated with animal-derived pet treats—United States and Canada, 2005. *Morb. Mortal. Wkly. Rep.* **55:**702–705.

25. **Dalton, C. B., J. Gregory, M. D. Kirk, R. J. Stafford, E. Kraa, and D. Gould.** 2004. Foodborne disease outbreaks in Australia, 1995 to 2000. *Commun. Dis. Intell.* **28:**211–224.
26. **Dato, V., A. Weltman, K. Waller, M. A. Ruta, A. Highbaugh-Battle, C. Hembree, S. Evenson, C. Wheeler, and T. Vogt.** 2003. Hepatitis A outbreak associated with green onions at a restaurant—Monaca, Pennsylvania, 2003. *Morb. Mortal. Wkly. Rep.* **52:**1155–1157.
27. **Department of Agriculture.** 2005. National school lunch program. U.S. Department of Agriculture, Food and Nutrition Service. [Online.] http://www.fns.usda.gov/cnd/Lunch/default.htm. Accessed 2 July 2006.
28. **Elliott, V.** 3 May 2006. Russia bans British poultry imports. *The Times*, London, United Kingdom.
29. **Espié, E., and V. Vaillant.** 2005. International outbreak of *Salmonella* Stourbridge infection, April–July 2005: results of epidemiological, food and veterinary investigations in France. *Eur. Surveill. Wkly.* **10.** [Online.] http://www.eurosurveillance.org/ew/2005/050811.asp#3. Accessed 2 July 2006.
30. **Ethelberg, S.** 2005. Salmonellosis outbreak linked to carpaccio made from imported raw beef, Denmark, June–August 2005. *Eur. Surveill. Wkly.* **10.** [Online.] http://www.eurosurveillance.org/ew/2005/050922.asp#3. Accessed 2 July 2006.
31. **Falkenhorst, G., L. Krusell, M. Lisby, S. B. Madsen, B. Böttiger, and K. Mølbak.** 2005. Imported frozen raspberries cause a series of Norovirus outbreaks in Denmark, 2005. *Eur. Surveill. Wkly.* **10.** [Online.] http://www.eurosurveillance.org/ew/2005/050922.asp#2. Accessed 2 July 2006.
32. **Food and Drug Administration.** 2001. FDA survey of imported fresh produce. FY 1999 field assignment. U.S. Food and Drug Administration, Center for Food Safety and Applied Nutrition. [Online.] http://www.cfsan.fda.gov/~dms/prodsur6.html. Accessed 2 July 2006.
33. **Food and Drug Administration.** 2003. Protecting the food supply. FDA actions on new bioterrorism legislation. Fact sheet on FDA's new food bioterrorism regulation: interim Final Rule—prior notice of imported food shipments. U.S. Food and Drug Administration, Center for Food Safety and Applied Nutrition. [Online.] http://www.cfsan.fda.gov/~dms/fsbtac13.html. Accessed 2 July 2006.
34. **Food and Drug Administration.** 18 May 2004. Paramount Farms announces nationwide recall of whole natural raw almonds. U.S. Food and Drug Administration. [Online.] http://www.fda.gov/oc/po/firmrecalls/paramount05_04.html. Accessed 2 July 2006.
35. **Food and Drug Administration.** 22 May 2004. Paramount Farms expands recall of raw almonds. U.S. Food and Drug Administration. [Online.] http://www.fda.gov/oc/po/firmrecalls/paramountexp05_04.html. Accessed 2 July 2006.
36. **Food and Drug Administration.** 2004. Recalls, market withdrawals and safety alerts archive. U.S. Food and Drug Administration. [Online.] http://www.fda.gov/oc/po/firmrecalls/archive_2004.html. Accessed 2 July 2006.
37. **Food and Drug Administration.** 2005. FDA survey of imported fresh produce. Import produce assignment (DFP # 05-16) – HIGH Priority. ORA

Concurrence # 2005031601. U.S. Food and Drug Administration, Center for Food Safety and Applied Nutrition. [Online.] http://www.cfsan.fda.gov/~dms/prodsu12.html. Accessed 2 July 2006.

38. **Food and Drug Administration**. 2006. FIARS import alerts. U.S. Food and Drug Administration, Office of Regulatory Affairs. [Online.] http://www.fda.gov/ora/fiars/ora_import_alerts.html. Accessed 1 July 2006.

39. **Francis, B. J., and J. P. Davis.** 1984. Update: gastrointestinal illness associated with imported semi-soft cheese. *Morb. Mortal. Wkly. Rep.* **33:**16, 22.

40. **Francis, B. J., J. V. Altamirano, M. G. Stobierski, W. Hall, B. Robinson, S. Dietrich, R. Martin, F. Downes, K. R. Wilcox, Jr., C. Hedberg, R. Wood, M. Osterholm, C. Genese, M. J. Hung, S. Paul, K. C. Spitalny, C. Whalen, and J. Spika.** 1991. Multistate outbreak of *Salmonella poona* infections—United States and Canada, 1991. *Morb. Mortal. Wkly. Rep.* **40:**549–552.

41. **Gaulin, C. D., D. Ramsay, P. Cardinal, and M. A. D'Halevyn.** 1999. Epidemic of gastroenteritis of viral origin associated with eating imported raspberries [Article in French; English abstract]. *Can. J. Public Health* **90:**37–40.

42. **General Accounting Office.** 2000. FDA's use of faster tests to assess the safety of imported foods. U.S. General Accounting Office, Washington, D.C. [Online.] http://www.gao.gov/new.items/rc00065.pdf. Accessed 2 July 2006.

43. **General Accounting Office.** 2004. FDA's imported seafood safety program shows some progress, but further improvements are needed. U.S. General Accounting Office, Washington, D.C. [Online.] http://www.gao.gov/new.items/d04246.pdf. Accessed 2 July 2006.

44. **Goswami, B. B.** 2001. Detection and quantitation of hepatitis A virus in shellfish by the polymerase chain reaction, p. 26.01–26.05. *In* G. J. Jackson, R. I. Merker, and R. Bandler, BAM Project Coordinators, *Bacteriological Analytical Manual Online*. U.S. Food and Drug Administration, Center for Food Safety and Applied Nutrition. [Online.] http://www.cfsan.fda.gov/~ebam/bam-26.html. Accessed 2 July 2006.

45. **Gustavsen, S., and O. Breen.** 1984. Investigation of an outbreak of *Salmonella oranienburg* infections in Norway, caused by contaminated black pepper. *Am. J. Epidemiol.* **119:**806–812.

46. **Health Protection Branch, J. Hofmann, Z. Liu, C. Genese, G. Wolf, W. Manley, K. Pilot, E. Dalley, and L. Finelli.** 1996. Update: outbreaks of *Cyclospora cayetanensis* infection—United States and Canada, 1996. *Morb. Mortal. Wkly. Rep.* **45:**611–612.

47. **Herwaldt, B. L.** 2000. *Cyclospora cayetanensis*: a review, focusing on the outbreaks of cyclosporiasis in the 1990s. *Clin. Infect. Dis.* **31:**1040–1057.

48. **Herwaldt, B. L., M.-L. Ackers, and The Cyclospora Working Group.** 1997. An outbreak in 1996 of cyclosporiasis associated with imported raspberries. *N. Engl. J. Med.* **336:**1548–1556.

49. **Herwaldt, B. L., M. J. Beach, and The Cyclospora Working Group.** 1999. The return of *Cyclospora* in 1997: another outbreak of cyclosporiasis in North America associated with imported raspberries. *Ann. Intern. Med.* **130:** 210–220.

50. **Hjertqvist, M., R. Eitrem, R. Wollin, L. Plym-Forshell, and J. Giesecke.** 2003. Outbreak of *Salmonella* Typhimurium DT 108 (or DT 170) due to imported contaminated kebab meat in Sweden. *Eur. Surveill. Wkly.* **7.**

[Online.] http://www.eurosurveillance.org/ew/2003/030814.asp#1. Accessed 2 July 2006.

51. **Hjertqvist, M., I. Luzzi, S. Löfdahl, A. Olsson, J. Rådal, and Y. Andersson.** 2006. Unusual phage pattern of *Salmonella* Typhimurium isolated from Swedish patients and Italian salami. *Eur. Surveill. Wkly.* **11.** [Online.] http://www.eurosurveillance.org/ew/2006/060209.asp#3. Accessed 2 July 2006.

52. **Ho, A. Y., A. S. Lopez, M. G. Eberhart, R. Levenson, B. S. Finkel, A. J. da Silva, J. M. Roberts, P. A. Orlandi, C. C. Johnson, and B. L. Herwaldt.** 2002. Outbreak of cyclosporiasis associated with imported raspberries, Philadelphia, Pennsylvania, 2000. *Emerg. Infect. Dis.* **8:**783–788.

53. **Hockin, J. C., J.-Y. D'Aoust, D. Bowering, J. H. Jessop, B. Khanna, H. Lior, and M. E. Milling.** 1989. An international outbreak of *Salmonella nima* from imported chocolate. *J. Food Prot.* **52:**51–54.

54. **Hutin, Y. J. F., V. Pool, E. H. Cramer, O. V. Nainan, J. Weth, I. T. Williams, S. T. Goldstein, K. F. Gensheimer, B. P. Bell, C. N. Shapiro, M. J. Alter, and H. S. Margolis.** 1999. A multistate, foodborne outbreak of hepatitis A. *N. Engl. J. Med.* **340:**595–602.

55. **Isaacs, S., J. Aramini, B. Ciebin, J. A. Farrar, R. Ahmed, D. Middleton, A. U. Chandran, L. J. Harris, M. Howes, E. Chan, A. S. Pichette, K. Campbell, A. Gupta, L. Y. Lior, M. Pearce, C. Clark, F. Rodgers, F. Jamieson, I. Brophy, and A. Ellis.** 2005. An international outbreak of salmonellosis associated with raw almonds contaminated with a rare phage type of *Salmonella* Enteritidis. *J. Food Prot.* **68:**191–198.

56. **Isakbaeva, E., B.-A. Lindstedt, B. Schimmer, T. Vardund, T.-L. Stavnes, K. Hauge, B. Gondrosen, H. Blystad, H. Kløvstad, P. Aavitsland, K. Nygård, and G. Kapperud.** 2005. *Salmonella* Typhimurium DT104 outbreak linked to imported minced beef, Norway, October–November 2005. *Eur. Surveill. Wkly.* **10.** [Online.] http://www.eurosurveillance.org/ew/2005/051110.asp#1. Accessed 30 June 2006.

57. **Kapperud, G., L. M. Rørvik, V. Hasseltvedt, E. A. Høiby, B. G. Iversen, K. Staveland, G. Johnsen, J. Leitao, H. Herikstad, Y. Andersson, G. Langeland, B. Gondrosen, and J. Lassen.** 1995. Outbreak of *Shigella sonnei* infection traced to imported iceberg lettuce. *J. Clin. Microbiol.* **33:**609–614.

58. **Katz, D. J., M. A. Cruz, M. J. Trepka, J. A. Suarez, P. D. Fiorella, and R. M. Hammond.** 2002. An outbreak of typhoid fever in Florida associated with an imported frozen fruit. *J. Infect. Dis.* **186:**234–239.

59. **Keady, S., G. Briggs, J. Farrar, J. C. Mohle-Boetani, J. O'Connell, S. B. Werner, D. Anderson, L. Tenglesen, S. Bidols, B. Albanese, C. Gordan, E. DeBess, J. Hatch, W. E. Keene, M. Plantenga, J. Tierheimer, A. L. Hackman, C. E. Rinehardt, C. H. Sandt, A. Ingram, S. Hansen, S. Hurt, M. Poulson, R. Pallipamu, J. Wicklund, C. Braden, J. Lockett, S. Van Duyne, A. Dechet, and C. Smelser.** 2004. Outbreak of *Salmonella* serotype Enteritidis infections associated with raw almonds—United States and Canada, 2003–2004. *Morb. Mortal. Wkly. Rep.* **53:**484–487.

60. **Killalea, D., L. R. Ward, D. Roberts, J. de Louvois, F. Sufi, J. M. Stuart, P. G. Wall, M. Susman, M. Schwieger, P. J. Sanderson, I. S. T. Fisher, P. S. Mead, O. N. Gill, C. L. R. Bartlett, and B. Rowe.** 1996. International epidemiological and microbiological study of outbreak of *Salmonella agona*

infection from a ready to eat savoury snack—I: England and Wales and the United States. *BMJ* **313:**1105–1107.

61. **King, W.** 29 May 2004. Tainted raw almonds made 4 ill in state. Recall covers 13 million pounds sold under various brands. *Seattle Times*, Seattle, Wash.
62. **Koch, J., A. Schrauder, K. Alpers, D. Werber, C. Frank, R. Prager, W. Rabsch, S. Broll, F. Feil, P. Roggentin, J. Bockemühl, H. Tschäpe, A. Ammon, and K. Stark.** 2005. *Salmonella agona* outbreak from contaminated aniseed, Germany. *Emerg. Infect. Dis.* **11:**1124–1127.
63. **Koumans, E. H. A., D. J. Katz, J. M. Malecki, S. Kumar, S. P. Wahlquist, M. J. Arrowood, A. W. Hightower, and B. L. Herwaldt.** 1998. An outbreak of cyclosporiasis in Florida in 1995: a harbinger of multistate outbreaks in 1996 and 1997. *Am. J. Trop. Med. Hyg.* **159:**235–242.
64. **Krusell, L., and N. Lohse.** 2003. A case of human botulism in Denmark after consumption of garlic in chili oil dressing produced in Germany. *Eur. Surveill. Wkly.* **7.** [Online.] http://www.eurosurveillance.org/ew/2003/030213.asp#1. Accessed 2 July 2006.
65. **Kudaka, J., R. Asato, K. Itokazu, M. Nakamura, K. Taira, H. Kuniyosi, Y. Kinjo, J. Terajima, H. Watanabe, J. Kobayashi, B. Swaminathan, C. R. Braden, and J. R. Dunn.** 2005. *Escherichia coli* O157:H7 infections associated with ground beef from a U.S. military installation—Okinawa, Japan, February 2004. *Morb. Mortal. Wkly. Rep.* **54:**40–42.
66. **Lacey, C., R. Talbot, J. Taylor, D. Dwyer, B. Jolbitado, C. Morrison, E. Butler-Senkel, S. Strauss, D. Murphy-Baxam, J. Libonati, E. Israel, N. Ridley, M. Smith, J. Zingeser, G. Miller, Jr., and K. Ungchusak.** 1991. Cholera associated with imported frozen coconut milk—Maryland, 1991. *Morb. Mortal. Wkly. Rep.* **40:**844–845.
67. **Levy, M. E.** 1983. Gastrointestinal illness associated with imported brie cheese—District of Columbia. *Morb. Mortal. Wkly. Rep.* **32:**533.
68. **Mahon, B. E., A. Pönkä, W. N. Hall, K. Komatsu, S. E. Dietrich, A. Siitonen, G. Cage, P. S. Hayes, M. A. Lambert-Fair, N. H. Bean, P. M. Griffin, and L. Slutsker.** 1997. An international outbreak of *Salmonella* infections caused by alfalfa sprouts grown from contaminated seeds. *J. Infect. Dis.* **175:**876–882.
69. **Mohle-Boetani, J. C., R. Reporter, S. B. Werner, S. Abbott, J. Farrar, S. H. Waterman, and D. J. Vugia.** 1999. An outbreak of *Salmonella* serogroup Saphra due to cantaloupes from Mexico. *J. Infect. Dis.* **180:**1361–1364.
70. **Nygård, K., J. Lassen, L. Vold, P. Aavitsland, and I. Fisher.** 2004. International outbreak of *Salmonella* Thompson caused by contaminated ruccola salad—update. *Eur. Surveill. Wkly.* **8.** [Online.] http://www.eurosurveillance.org/ew/2004/041216.asp#2. Accessed 2 July 2006.
71. **Olsen, S. J., L. C. MacKinon, J. S. Goulding, N. H. Bean, and L. Slutsker.** 2000. Surveillance for foodborne disease outbreaks—United States, 1993–1997. *Morb. Mortal. Wkly. Rep.* **49**(SS-01):1–51.
72. **Orlandi, P. A., C. Frazar, L. Carter, and D.-M. T. Chu.** 2004. Detection of *Cyclospora* and *Cryptosporidium* from fresh produce: isolation and identification by polymerase chain reaction (PCR) and microscopic analysis. *In* G. J. Jackson, R. I. Merker, and R. Bandler, BAM Project Coordinators, *Bacteriological Analytical Manual Online*, Chapter 19 A. U.S. Food and Drug Administration,

Center for Food Safety and Applied Nutrition. [Online.] http://www.cfsan.fda.gov/~ebam/bam-19a.html#authors. Accessed 2 July 2006.

73. **Pratdesaba, R. A., M. González, E. Piedrasanta, C. Mérida, K. Contreras, C. Vela, F. Culajay, L. Flores, and O. Torres.** 2001. *Cyclospora cayetanensis* in three populations at risk in Guatemala. *J. Clin. Microbiol.* **39:**2951–2953.

74. **Public Health Agency of Canada.** 19 October 2001. *Salmonella* Stanley and *Salmonella* Newport: Canada, Australia, United Kingdom. *Infectious Diseases News Brief.* [Online.] http://www.phac-aspc.gc.ca/bid-bmi/dsd-dsm/nb-ab/2001/nb4201_e.html. Accessed 2 July 2006.

75. **Rodriguez, R.** 26 May 2004. University of California researcher had been studying *Salmonella* in almonds. *Fresno Bee*, Fresno, Calif.

76. **Shohat, T., M. S. Green, D. Merom, O. N. Gill, A. Reisfeld, A. Matas, D. Blau, N. Gal, and P. E. Slater.** 1996. International epidemiological and microbiological study of outbreak of *Salmonella agona* infection from a ready to eat savoury snack—II: Israel. *BMJ* **313:**1107–1109.

77. **Sivapalasingam, S., E. Barrett, A. Kimura, S. Van Duyne, W. De Witt, M. Ying, A. Frisch, Q. Phan, E. Gould, P. Shillam, V. Reddy, T. Cooper, M. Hoekstra, C. Higgins, J. P. Sanders, R. V. Tauxe, and L. Slutsker.** 2003. A multistate outbreak of *Salmonella enterica* serotype Newport infection linked to mango consumption: impact of water-dip disinfestation technology. *Clin. Infect. Dis.* **37:**1585–1590.

78. **Smyth, J.** 9 February 2006. EU defends rules on genetically modified crops. *Irish Times*, Dublin, Ireland.

79. **Soave, R., and W. D. Johnson, Jr.** 1995. *Cyclospora*: conquest of an emerging pathogen. *Lancet* **345:**667–668.

80. **SoRelle, R.** 27 June 1996. More berries tested in *Cyclospora* cases. *Houston Chronicle*, Houston, Tex.

81. **Stout, D.** 10 July 1996. Intestinal ailment is linked to Latin American raspberries. *New York Times*, New York, N.Y.

82. **Tauxe, R. V., and J. M. Hughes.** 1996. International investigation of outbreaks of foodborne disease. Public health responds to the globalisation of food. *BMJ* **313:**1093–1094.

83. **Taylor, J. L., J. Tuttle, T. Pramukul, K. O'Brien, T. J. Barrett, B. Jolbitado, Y. L. Lim, D. Vugia, J. G. Morris, Jr., R. V. Tauxe, and D. M. Dwyer.** 1993. An outbreak of cholera in Maryland associated with imported commercial frozen fresh coconut milk. *J. Infect. Dis.* **167:**1330–1335.

84. **Unicomb, L., M. Kirk, G. Hogg, P. Jelfs, G. Simmons, J. Gregory, and C. Nicol.** 2003. *Salmonella* Montevideo in sesame seed-based products imported into Australia and New Zealand may have implications for Europe and elsewhere. *Eur. Surveill. Wkly.* **7.** [Online.] http://www.eurosurveillance.org/ew/2003/030918.asp#3. Accessed 2 July 2006.

85. **White House.** 3 July 1999. The President announces new measures to prevent unsafe food from entering our borders. White House Briefing Room. [Online.] http://www.foodsafety.gov/~dms/fs-wh17.html. Accessed 2 July 2006.

86. **World Health Organization.** 4 November 2005. Highly pathogenic H5N1 avian influenza outbreaks in poultry and in humans: food safety implications. World Health Organization, International Food Safety Authorities

Network. [Online.] http://www.fao.org/ag/againfo/subjects/documents/ai/Foodsafety.pdf. Accessed 1 July 2006.

87. **World Trade Organization.** 2005. International trade statistics 2005. Table IV.3, World trade in agricultural products, 2004, and Table IV.8, Leading exporters and importers of agricultural products, 2004. World Trade Organization, Geneva, Switzerland. [Online.] http://www.wto.org/english/res_e/statis_e/its2005_e/its05_bysector_e.htm. Accessed 2 July 2006.

88. **World Trade Organization.** 2005. World trade developments in 2004 and prospects for 2005. World Trade Organization, Geneva, Switzerland. [Online.] http://www.wto.org/english/res_e/statis_e/its2005_e/its05_toc_e.htm. Accessed 2 July 2006.

chapter 15

A Raw Deal

IT STARTED WITH A TRIP TO THE DAIRY. It ended with a trip to the hospital. And for Young's Jersey Dairy, it was the death knell for a 45-year tradition of selling raw milk to the public (116).

In November 2002, Young's was the only dairy in Ohio licensed to bottle and sell raw milk. The company's businesses included a working dairy farm, a snack bar, a restaurant, and a petting zoo. It employed 211 workers, including 16 members of the Young family (97). But that was about to change.

The outbreak began on November 30. Over the next 2½ months, 62 people reported experiencing one or more symptoms of gastroenteritis: nausea, cramps, fever, chills, aches, diarrhea, vomiting, and headache. The youngest victim was just 16 months, and the eldest was 82 years old (97, 115). Sixteen were employees of the dairy. On December 10, the local health department advised the Ohio Department of Health that two children, hospitalized since December 3, were suffering from *Salmonella* infection. By December 18, investigators had confirmed eight additional cases and had identified the pathogen as *Salmonella enterica* serotype Typhimurium (97).

Interviews of the victims and their families quickly pointed investigators to Young's dairy as the common factor. Health inspectors visited the dairy, reviewed all of the operations and practices, and obtained samples of the raw milk products, environmental samples, and fecal samples from the dairy cows. One sample of cream, one of butter (made by a consumer using raw milk purchased from Young's), and three samples of bottled skim milk tested positive for serotype Typhimurium, and the cultures recovered from those samples were a perfect match (determined by pulsed-field gel electrophoresis) for the clinical isolates obtained from victims of the outbreak. Four asymptomatic dairy barn workers were found to be carrying the outbreak strain, but samples obtained from the dairy's environment and the cows were all negative (97).

Investigators were never able to pinpoint the origin of the *Salmonella* contamination, though the dairy's procedures provided a clue to a possible source. The strain of serotype Typhimurium that was behind the outbreak was not a newcomer to Ohio. It had been found in two other counties in the state the year before (115). A dairy barn worker at Young's might have become an asymptomatic carrier of the pathogen in the months prior to the raw milk outbreak and then accidentally contaminated the milk at Young's during milking or bottling—procedures that offered several opportunities for microbial contamination.

Although the dairy's herd of cows was milked mechanically (twice a day), every subsequent step was carried out by hand. Four workers operated the barn and milked the cows; a fifth worker bottled the milk in jugs and made the ice cream. The operation comprised several steps (Fig. 15.1).

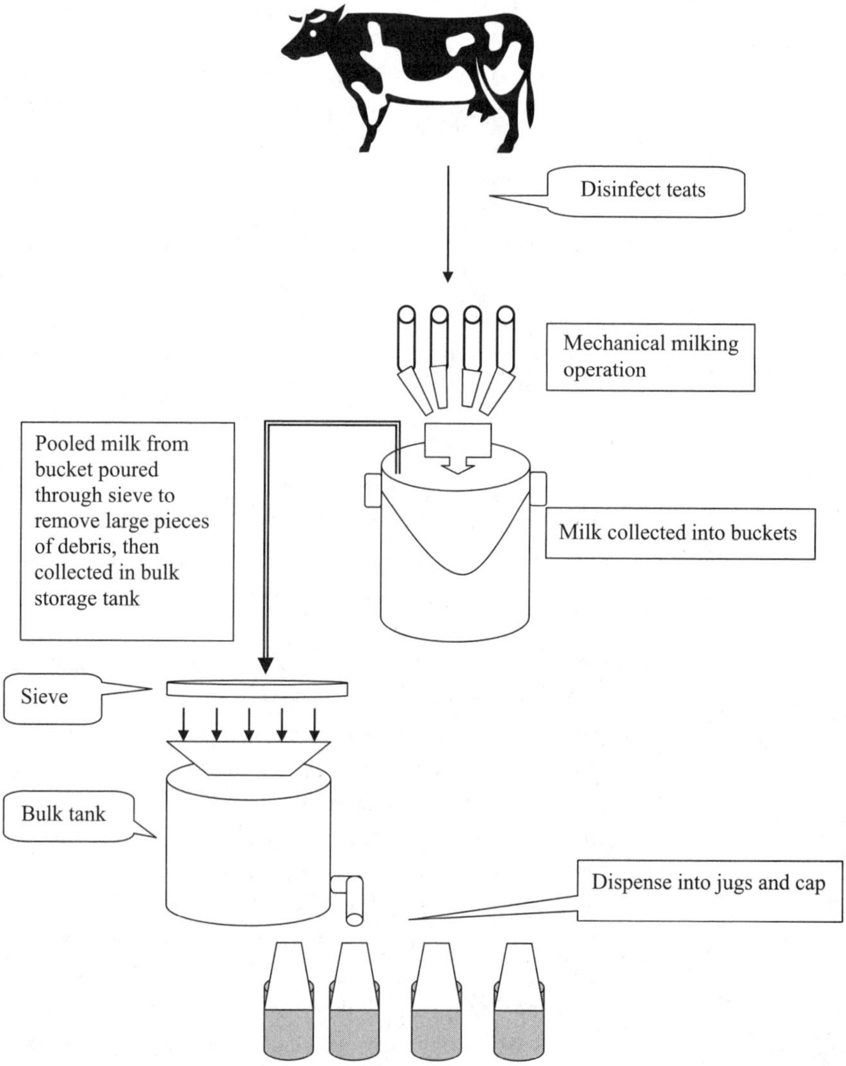

Figure 15.1 Raw milk processing and bottling at Young's Jersey Dairy (97).

After sanitizing the cows' teats, the workers milked the cows. The milk was collected in buckets, which the workers then emptied through a strainer (to remove large debris such as pieces of straw) into a milk transfer station. From there, the milk was transferred into a bulk holding tank. The contents of the holding tank were poured into jugs, which were capped by hand (97). If a dairy worker was indeed the source, rather than a victim, of the outbreak, he could have contaminated the milk at any of the several manual steps between milking and bottling.

Raw milk and unpasteurized dairy products such as cheeses have been linked to disease outbreaks caused by several pathogens, including *Salmonella*, *Shigella*, *Escherichia coli* O157:H7, *Listeria monocytogenes*, *Brucella melitensis*, *Mycobacterium bovis*, and *Campylobacter jejuni* (Table 15.1). Because of the risk of contracting a food-borne illness from drinking raw milk, the U.S. Food and Drug Administration (FDA) banned its interstate shipment for retail sale in 1987. Only a few U.S. states—California and Connecticut, for example—now allow the retail sale of raw milk (28, 34, 37). Although it is still found in stores in some European countries, others prohibit its retail sale (14, 54, 75, 108). Nevertheless, raw milk has a hardcore following among consumers in many industrialized countries around the world. Websites such as rawmilk.org and realmilk.com advocate consumption of raw milk and list places where it can be purchased in Europe, the United States, Canada, Australia, New Zealand, Israel, and Japan (8, 10). And so-called cow share programs—part ownership of a dairy cow with the right to a specified fraction of the milk it produces—have been introduced as a way around official bans on the sale of raw milk to consumers (76).

To Pasteurize or Not To Pasteurize?

Aficionados of raw milk insist that it tastes better, is more nutritious, and is safer than pasteurized milk. Is there any basis, scientific or otherwise, to their claims?

The flavor of milk and other dairy products, raw or pasteurized, is highly dependent on what a dairy herd is fed. Pasture-grazed cows produce milk of a flavor different from that of cows that have been fed grains, silage, or commercial animal feeds (18, 104, 117, 136). And the components of feed and feed supplements can have an impact on flavor, too (20).

Taste is an important criterion for some people, but the purported nutritional benefits of raw milk carry even greater weight. Raw milk proponents claim that pasteurization "... destroys enzymes, diminishes vitamin content, denatures fragile milk proteins, destroys vitamins C, B_{12}, and B_6, [and] kills beneficial bacteria...." Raw milk advocates also claim—without citing any peer-reviewed or clinical studies to back up their statements—that pasteurized milk is the culprit behind heart disease, growth problems in

Table 15.1 Examples of food-borne illness attributed to raw milk and other dairy products made from unpasteurized milk

Pathogen	Food	Country	Year	No. of cases (deaths)	Reference
B. melitensis	Unpasteurized goat cheese	United States	1983	29	112
B. melitensis	Raw goat's milk cheese	Spain	2002	11	102
Campylobacter spp.	Raw milk	United States	1982	15	83
Campylobacter spp.	Raw milk	Hungary	1998	52	80
C. jejuni	Raw milk	United States	1983	57	23
C. jejuni	Raw milk	United States	1984	12	6
C. jejuni	Raw milk	United States	1985	23	17
C. jejuni	Chocolate drink made from raw milk	Germany	1999	24	87
C. jejuni	Unpasteurized milk	United States	2001	75	74
E. coli O157	Unpasteurized cream	United Kingdom	1998	7	7
E. coli O157	Unpasteurized goat's milk cheese	United Kingdom (Scotland)	1999	30	41
E. coli O157	Raw cow's milk	United Kingdom	2000	4	110
E. coli O157	Raw cow's milk	United Kingdom	2000	2	110
E. coli O157:H	Raw cow's or goat's milk	Austria	2001	2	4
E. coli O157:H7	Raw milk	United States	1992–1993	18	81
E. coli O157:H7	Goat's milk	Czech Republic	1995	9[a]	21
E. coli O157:H7	Goat's milk	Canada	2001	5	100
L. monocytogenes	Mexican-style cheese contaminated with unpasteurized milk	United States	1985	142 (48)	89
L. monocytogenes	Unpasteurized Mexican-style cheese	United States	2000–2001	12[b]	25
M. bovis	Fresh cheese from Mexico	United States	2001–2004	35 (1)	147
S. enterica serotype Berta	Unpasteurized soft cheese	Canada	1994	82	48
S. enterica serotype Dublin	Unpasteurized cow's milk cheese	United Kingdom	1989	42	94
S. enterica serotype Dublin	Cheese made from raw cow's milk	France[c]	1995	25 (5)	137
S. enterica serotype Enteritidis	Raw milk cheese	France	2001	190	73
S. enterica serotype Enteritidis	Raw milk cheese	France	2001	25	73
S. enterica serotype Paratyphi B	Goat's milk cheese	France	1993	273 (1)	43
S. enterica serotype Stourbridge	Unpasteurized goat cheese	France[d]	2005	52	51
S. enterica serotype Typhimurium	Inadequately pasteurized milk	United States	1984	16	2
S. enterica serotype Typhimurium	Raw milk soft cheese	France	1997	113	46
S. enterica serotype Typhimurium	Raw milk	United States	2002/2003	105[e]	97
S. enterica serotype Typhimurium DT104	Milk (defective pasteurization)	United Kingdom	1998	86	12
Shigella sonnei	Unpasteurized milk curds	Lithuania	2004	41	151
S. zooepidemicus	Unpasteurized cows' milk cheese	United States	1983	16 (2)	52
Unknown[f]	Raw milk	United States	1983–1984	122	111

[a] Four of the nine cases were asymptomatic.
[b] Including 10 pregnant women. Infection resulted in five stillbirths, three premature deliveries, and two infected newborns.
[c] Cheese was manufactured in Switzerland.
[d] Twenty-seven cases in France (country of manufacture) and 25 cases in six other European countries.
[e] Sixty-two confirmed cases plus 43 suspected cases.
[f] Outbreak of chronic diarrhea syndrome of unknown etiology (111).

children, and many other ailments, including diabetes, dental caries, colic, cancer, osteoporosis, obesity, allergies, arthritis, and autism (8, 10, 103).

Proponents of drinking raw milk have been highly selective in their acknowledgment of the risks of food-borne illness. For example, when advocating the legalization of its retail sale in Los Angeles, Calif., its boosters claimed that there had not been a single outbreak of food-borne illness in California tied to raw milk between 1982 and 1997, whereas there had been 13 outbreaks in various regions of the United States that were linked to pasteurized milk and cheese (56). But they overlooked several U.S. outbreaks, in California and elsewhere, of *C. jejuni*, *Salmonella*, *E. coli* O157:H7, *B. melitensis*, *L. monocytogenes*, *Streptococcus zooepidemicus* infections and chronic diarrhea of unknown origin that were traced to raw milk or to cheese made from unpasteurized milk (Table 15.1) (146). Furthermore, their list of outbreaks due to pasteurized milk contained several inaccuracies. For example, an outbreak of *L. monocytogenes* infection was described on the list as having been linked to cheese made from pasteurized milk, whereas lab tests confirmed that the cheese either had been made from or contaminated with raw milk (89). The list also included the Jewel Dairy outbreak described in chapter 2, which had been traced, in fact, to the cross-contamination of pasteurized milk with raw milk (123).

The desire of consumers to eat raw or minimally processed food extends far beyond the realm of raw milk. Fermented dairy products made from unpasteurized milk; raw, cured, or fermented meats; raw shellfish (oysters and mussels); sushi, sashimi, and other dishes based on raw fish and seafood; and *al dente* vegetables all appear frequently on restaurant menus and on many dinner tables—sometimes with unexpected and unhappy consequences (Table 15.2 and Table 15.3).

History Repeats Itself

By 2003, consumers had become used to reading about meat recalls due to *E. coli* O157:H7 contamination. More than 2 million pounds of ground beef were recalled for that reason in 5 of the preceding 6 years—24+ million pounds in 2002 alone (chapter 7). But the recall of 739,000 pounds of meat in late June 2003 was different. It was the first time that an *E. coli* O157:H7 outbreak had been linked to undercooked steaks (62).

Stampede Meat, Inc., of Chicago, Ill., produced tenderized bacon-wrapped beefsteaks, which they sold to restaurants, institutions, and retail stores and through a door-to-door distribution network. Between May 27 and June 29, 2003, five people in Kansas, Minnesota, and Michigan became ill with *E. coli* O157:H7 infection. On June 29, the U.S. Department of Agriculture (USDA) announced that Stampede Meat was recalling vacuum-packed steaks produced between March 17 and March 22. The meat had been distributed in 16 U.S. states and in Canada (62). In all, 16 people in

Table 15.2 Examples of food-borne illness attributed to uncooked cured, fermented, or dried meat or poultry products or to raw or undercooked meat or poultry

Pathogen	Food	Country	Year	No. of cases (deaths)	Reference
C. jejuni	Undercooked barbecued chicken	United States	1982	11	78
E. coli O157:H7	Dry-cured salami	United States	1994	20	3
E. coli O157:H7	Deer meat jerky	United States	1995	11[a]	82
E. coli O157:H7	Undercooked venison from wild deer	United States	1999?	1	114
E. coli O157:H7	Undercooked nonintact steaks	United States	2003	12	84
S. enterica serotype Bovismorbificans	Raw pork	Germany	2004–2005	525 (1)	69
S. enterica serotype Coeln	Incompletely cooked hamburger	France	1998	26	72
S. enterica serotype Heidelberg	Undercooked frozen chicken nuggets and strips	Canada	2003	23	91
S. enterica serotype Newport	Raw or undercooked ground beef	United States	1975	54	59
S. enterica serotype Newport	Raw or undercooked ground beef	United States	2002	47 (1)	152
S. enterica serotype Paratyphi B	Incompletely cooked hamburger	France	1999	8	72
S. enterica serotype Thompson	Improperly cooked roast beef	United States	1996	52	127
S. enterica serotype Typhimurium	Salami sticks	United Kingdom	1987–1988	101	39
S. enterica serotype Typhimurium	Undercooked pork meat	United Kingdom	1989	206	95
S. enterica serotype Typhimurium	Raw ground beef	United States	1994	158	65
S. enterica serotype Typhimurium	Lebanon bologna	United States	1995	26	126
S. enterica serotype Typhimurium	Incompletely cooked hamburger	France	2000	35	72
S. enterica serotype Typhimurium	Raw or undercooked ground beef	United States	2004	31	40
S. enterica serotype Typhimurium DT104	Carpaccio made from raw beef	Denmark	2005	22	53
Trichinella	Incompletely cooked ground mixed meat (beef and pork)	Germany	1998	8	119
Trichinella	Mettwurst (raw smoked sausage)	Germany	1998	44	119
Trichinella	Raw horse meat	Italy	1998	24	113
Trichinella	Undercooked bear meat	France[b]	2005	17	5

[a] Six confirmed and five presumptive cases.
[b] Bear meat was imported from Canada.

Table 15.3 Examples of food-borne illness attributed to uncooked cured or dried fish and shellfish products or to raw or undercooked fish and shellfish

Pathogen	Food	Country	Year	No. of cases (deaths)	Reference(s)
Anisakis simplex	Raw anchovies	Korea	2002	1	109
A. simplex	Raw saltwater fish	Japan	1995	2	92
Clostridium botulinum Type E	Kapchunka (uneviscerated, salted, air-dried whitefish)	Israel, United States	1987	8	135
C. botulinum Type E	Uneviscerated salted fish	Egypt	1991	91	141
C. jejuni	Raw oysters	United States	1991	1	1
Gnathostoma spinigerum	Raw freshwater fish	Myanmar	2001	38	30
Hepatitis A virus[a]	Raw clams	United States	1982	74	85
Hepatitis A virus	Raw oysters	United States	1988	61	44
Hepatitis A virus	Raw oysters	Australia	1997	467	36
L. monocytogenes	Smoked mussels	New Zealand	1992	3	26
L. monocytogenes	Cold-smoked and gravad rainbow trout	Sweden	1994–1995	9 (2)	50
Norovirus[b]	Raw clams and oysters; steamed clams	United States	1982	1,017	105
Norovirus	Raw or steamed oysters	United States	1993	190	19, 38
Norovirus	Raw oysters	United States	1996	75	19
Norovirus	Raw oysters	United States	1996–1997	153	19, 57
Not identified	Raw clams and oysters	United States	1991	12	77
Pseudoterranova decipiens	Raw marine fish	Korea	1994	1	150
S. enterica serotype Singapore	Sushi rolls	Australia	2004	13	16
Vibrio cholerae non-O1	Raw clams	United States	1981	1	93
V. cholerae O1	Raw oysters	United States	1988	1	47
V. parahaemolyticus	Raw oysters	United States	1977	1	130
V. parahaemolyticus	Raw or undercooked shellfish	Canada, United States	1997	209	67
V. parahaemolyticus	Raw or undercooked oysters and clams	United States	1998	22	142
V. vulnificus	Raw oysters	United States	1996	3 (3)	96
Virus (unidentified)	Raw oysters	United Kingdom	1993	15 (1)	31

[a] Consolidation of four outbreaks of gastrointestinal illness traced to raw clams. Hepatitis A virus confirmed as pathogen responsible for illness in some of the cases.

[b] Consolidation of 103 outbreaks.

12 states became infected with *E. coli* O157:H7 as a result of the contaminated steaks (49).

The steaks produced by Stampede Meat had been treated by injection with a tenderizing and seasoning solution. Most likely, the tenderizing process transferred bacteria, including *E. coli* O157:H7, from the surface of the meat into its center (11). And some of those bacteria would have been able to survive cooking, especially if the steak was only lightly cooked (90).

The Stampede Meat outbreak had a familiar ring to it. In 1981, health officials in Pennsylvania and New Jersey noticed a spike in the incidence of salmonellosis. The increased number of cases of *Salmonella* infection proved to be part of a series of eight outbreaks that occurred in the second half of 1981. The outbreaks, which affected people living in at least four states in the U.S. Northeast, were due to contamination of a commercial precooked roast beef with three or more serotypes of *Salmonella*: serotype Newport, serotype Typhimurium, and serotype Anatum (120).

Rare roast beef has long been a popular deli sandwich meat. In 1976, the U.S. Patent Office issued a patent (assigned to the meat company E. Kahn's Sons Company of Cincinnati) for the safe preparation of rare roast beef. In their patent application, the inventors described a process for pumping a solution into the beef roast, followed by deep-fat frying of the outer surface to brown it, vacuum sealing the roast into a bag, and subjecting the roast to live steam until the internal temperature reached 128 to 134°F (53 to 57°C). After cooking, the roast was allowed to stand until its internal temperature reached 134 to 140°F (57 to 60°C) before being cooled. At the time the patent was issued, USDA regulations required that roast beef be cooked to an internal temperature of 145°F (63°C). But the inventors claimed that their process, in which the internal temperature never surpassed 140°F (60°C), resulted in roast beef that was uniformly rare and yet safe to eat (144).

Unfortunately, the inventors were wrong. Although it's unclear whether the Kahn process was used by the manufacturer of the contaminated rare roast beef, the number of different *Salmonella* serotypes isolated from unopened packages of the ready-to-eat meat, combined with the scope and duration of the outbreaks, is evidence that the cooking procedure was inadequate to ensure safety (120, 129).

The 1981 outbreaks were unique in that they were traced to a ready-to-eat product, but they are not the only instances of food-borne illnesses due to undercooked beef roasts. In 1988, an outbreak of *E. coli* O157:H7 struck 61 attendees of a university graduation banquet who had eaten undercooked (medium rare to rare) roast round of beef. An unopened package of raw frozen beef from the same production date yielded the outbreak strain of *E. coli* O157:H7 (121). Another outbreak, also traced to undercooked inside round beef roasts, affected 70 attendees of an agricultural threshing show in North Dakota in 1990 (99).

If non-intact beef—such as the steaks that were behind the Stampede Meat outbreak—is contaminated with *E. coli* O157:H7, USDA treats it as adulterated. Conversely, the agency does not consider intact beef to be adulterated when its surface is contaminated with the pathogen, because the surface bacteria should be destroyed even with no more than light cooking (58). Although this may be a reasonable expectation, USDA's policy places onto the shoulders of individual food handlers—both commercial and private individuals—the responsibility for determining whether a particular piece of meat is intact or non-intact and for applying to the meat appropriate, safe cooking parameters.

Pet Peeves

More than 30 years ago (in 1972), the U.S. government enacted a ban on interstate shipment of turtles that harbored *Salmonella* spp., and followed this up in 1975 with a ban on the sale of "... viable turtle eggs and live turtles with a carapace length of less than 4 inches ..." (35, 98). The government took these actions in response to a series of outbreaks of salmonellosis that were traced to pet turtles (145). Turtles and their eggs, however, were only a small part of the problem. Since the ban went into effect, fish aquariums, a variety of pet birds, ornamental fish, reptiles, rodents, chicks, ducklings, turtles, lizards, iguanas, hedgehogs, hamsters, and mice have been associated with cases of salmonellosis in the United States, Canada, and elsewhere (24, 45, 68, 86, 124, 128, 134, 148). Approximately 90% of reptiles carry *Salmonella* and shed it in their feces, and an estimated 3 to 5% of human salmonellosis cases can be attributed to these and other exotic pets (148).

Small or exotic animals and birds are not the only pets that carry *Salmonella* and transmit it to their owners. Cats and dogs can also be a source of this pathogen in the home. Studies have demonstrated the presence of *Salmonella*, *Campylobacter*, *Clostridium perfringens*, and other human pathogens in the feces of diarrheic and even asymptomatic dogs and cats (29, 33, 42, 55, 66, 70, 71, 79, 106, 107, 138, 139, 149). And this risk is heightened by those pet owners who engage in the currently fashionable practice of feeding raw meat and poultry to their dogs and cats (143).

The BARF Diet

The Biologically Appropriate Raw Food (BARF) diet originated in Australia in the 1980s. Ian Billinghurst, a veterinarian, came to the conclusion that commercial pet foods provided poor nutrition to domestic cats and dogs and that a raw diet was essential to promoting good health in these animals (22). His arguments struck a positive chord with a number of dog and cat breeders and owners, who began to follow Billinghurst's precepts and spread the word (9, 15).

BARF diet disciples claim that feeding raw meat and poultry to domestic cats and dogs does not carry with it a risk of infecting these animals with *Salmonella* or other food-borne pathogens. They say that the animals' immune and digestive systems are able to destroy the pathogens, rendering them harmless (15). But these claims fly in the face of several research and epidemiological studies. *Salmonella*, for example, is found often in the feces of racing greyhounds—dogs that are fed raw meat routinely (32, 101, 132). Dogs and cats both can develop clinical illness as a result of eating a *Salmonella*-contaminated diet and, even when asymptomatic, can transmit the *Salmonella* to their handlers (27, 125, 131).

In the last few years, the pet food industry has jumped onto the raw diet gravy train, producing frozen and freeze-dried raw food patties. Some pet owners have opted to use these for their convenience; others believe the manufactured patties to be safer (microbiologically). But the commercial raw food diets can promote a false sense of security. In the United States, for example, there are no specific regulations or requirements that raw pet food manufacturers must meet, just a nonbinding "guidance document" issued by the FDA's Center for Veterinary Medicine (60). A Canadian survey of raw food diets from eight manufacturers, published in 2005, found that 64% contained *E. coli* (though not *E. coli* O157); *Salmonella* was present in 20% of the samples, and *C. perfringens* was also detected in 20% (143). U.S. investigators also confirmed the presence of *Salmonella* in commercial raw food diets, though at a lower frequency than in Canada; they found *Salmonella* in just 7.1% of the raw food samples on the U.S. retail market (133). The *Salmonella* detection method used by the U.S. research team, however, was less sensitive than the method used by the Canadian researchers (133, 143). Had both groups of researchers used the same method, it's likely that the contamination rates in the U.S. and Canadian samples would have been similar.

The Raw Food Mystique

Why are raw and undercooked foods so alluring to some consumers that they are prepared to risk illness for themselves, their families, and their pets in order to enjoy them? Perhaps the answer is that many people long for a simpler time when—they perceive—their meat, produce, dairy products, and water were pure and clean. Nicols Fox, in the introduction to her 1996 book *Spoiled: The Dangerous Truth about a Food Chain Gone Haywire*, captured the spiritual nature of their nostalgia (64). "The beef I ate," she wrote, "had a personality. I had played with that steer and fed it and watched it grow, and yet I accepted its death as natural and inevitable. Then it became a part of me; its flesh became my flesh, and I was united to all life in a respectful process." But these consumers, like Nicols Fox, are chasing a mirage.

In June 2006, the United States marked the centennial of its first food safety law (61). When President Theodore Roosevelt signed the Pure Food and Drugs Act into law in 1906, consumers still were storing their perishables in iceboxes—the first home refrigerator was introduced in 1911 (13). Most perishable foods were produced and processed locally. Cold storage space was limited, the iceman and the milkman were daily visitors, fresh produce was "out of season" during the winter, and every day or two, city dwellers visited the local butcher shop to buy their meat and poultry.

On the farm, milk at the breakfast table was "fresh from the cow." Surplus milk was made into butter and cheese. All other foods were cooked, canned, or cured. Perishable foods were expected to spoil quickly, so they were eaten while the food was still fresh, not after it had been refrigerated for days or weeks. Leftovers were fed to animals or were eaten at the next meal (63). Food was eaten fresh, but—except for seasonal fruits and vegetables—it was not eaten raw.

Eating a raw food diet is a modern fad, which has evolved into a new industry (88, 118, 122). Interest in this diet fad has been fueled in part by its celebrity boosters, including Mel Gibson, Demi Moore, and Pierce Brosnan (140). But celebrity status does not equate to nutritional or food safety knowledge. Consumers who embrace raw food diets open themselves to an increased risk of becoming victims of food-borne disease agents.

References

1. **Abeyta, C., Jr., F. G. Deeter, C. A. Kaysner, R. F. Stott, and M. M. Wekell.** 1993. *Campylobacter jejuni* in a Washington State shellfish growing bed associated with illness. *J. Food Prot.* **56:**323–325.

2. **Adams, D., S. Well, B. F. Brown, S. Gregorio, L. Townsend, J. W. Skaggs, and M. W. Hinds.** 1984. Salmonellosis from inadequately pasteurized milk—Kentucky. *Morb. Mortal. Wkly. Rep.* **33:**505–506.

3. **Alexander, E. R., J. Boase, M. Davis, L. Kirchner, C. Osaki, T. Tanino, M. Samadpour, P. Tarr, M. Goldoft, S. Lankford, J. Kobyashi, P. Stehr-Green, P. Bradley, B. Hinton, P. Tighe, B. Pearson, G. R. Flores, S. Abbott, R. Bryant, S. B. Werner, and D. J. Vugia.** 1995. *Escherichia coli* O157:H7 outbreak linked to commercially distributed dry-cured salami—Washington and California, 1994. *Morb. Mortal. Wkly. Rep.* **44:**157–160.

4. **Allerberger, F., M. Wagner, P. Schweiger, H.-P. Rammer, A. Resch, M. P. Dierich, A. W. Friedrich, and H. Karch.** 2001. *Escherichia coli* O157 infections and unpasteurised milk. *Eur. Surveill. Wkly.* **6:**147–151. [Online.] http://www.eurosurveillance.org/em/v06n10/0610-222.asp. Accessed 6 July 2006.

5. **Ancelle, T., A. De Bruyne, D. Poisson, and J. Dupouy-Camet.** 2005. Outbreak of trichinellosis due to consumption of bear meat from Canada, France, September 2005. *Eur. Surveill. Wkly.* **10.** [Online.] http://www.eurosurveillance.org/ew/2005/051013.asp#3. Accessed 6 July 2006.

6. **Anonymous.** 1984. *Campylobacter* outbreak associated with certified raw milk products—California. *Morb. Mortal. Wkly. Rep.* **33:**562.

7. **Anonymous.** 1998. Cases of *Escherichia coli* O157 infection associated with unpasteurised cream in England. *Eur. Surveill. Wkly.* **2.** [Online.] http://www.eurosurveillance.org/ew/1998/981022.asp#5. Accessed 6 July 2006.

8. **Anonymous.** 2003. Preserving the people's right to choose the healthy foods they want, through education and activism. Right to Choose Healthy Food, Santa Monica, Calif. [Online.] http://www.rawmilk.org/. Accessed 6 July 2006.

9. **Anonymous.** 2004. Welcome to BONSAH. Billinghurst's Optimal Nutrition Society for Animal Health. [Online.] http://www.bonsah.com/index.php. Accessed 6 July 2006.

10. **Anonymous.** 2005. Why a campaign for real milk? Weston A. Price Foundation, Washington, D.C. [Online.] http://www.realmilk.com/why.html. Accessed 6 July 2006.

11. **Anonymous.** 9 July 2003. Coloradans told to check freezers for tainted steak. *The Pueblo Chieftain*, Pueblo, Colo.

12. **Ashraf, S., S. Gee, and S. O'Brien.** 1998. Defective pasteurisation linked to outbreak of *Salmonella typhimurium* definitive phage type 104 in England. *Eur. Surveill. Wkly.* **2.** [Online.] http://www.eurosurveillance.org/ew/1998/980917.asp#3. Accessed 6 July 2006.

13. **Association of Home Appliance Manufacturers.** 2006. History of the refrigerator. The History Channel. A&E Television Networks. [Online.] http://www.historychannel.com/exhibits/modern/fridge.html. Accessed 5 July 2006.

14. **Australia.** 2005. Australia New Zealand Food Standards Code. Standard 1.6.2: processing requirements. *Food Standards Australia New Zealand.* [Online.] http://www.foodstandards.gov.au/foodstandardscode/index.cfm#_FSCchapter1. Accessed 6 July 2006.

15. **BARF World.** 2005. Dr. Billinghurst's BARF Diet™. Biologically Appropriate Raw Foods. [Online.] http://www.barfworld.com/html/learn_more/what_is_barf.shtml. Accessed 6 July 2006.

16. **Barralet, J., R. Stafford, C. Towner, and P. Smith.** 2004. Outbreak of *Salmonella* Singapore associated with eating sushi. *Commun. Dis. Intel.* **28.**

[Online.] http://www.health.gov.au/internet/wcms/publishing.nsf/Content/cda-2004-cdi2804p.htm. Accessed 6 July 2006.

17. **Benda, B., J. Pollak, R. Benjamin, T. Livermore, H. Mitchell, S. B. Werner, and J. Chin.** 1986. *Campylobacter* outbreak associated with raw milk provided on a dairy tour—California. *Morb. Mortal. Wkly. Rep.* **35:**311–312.
18. **Bendall, J. G.** 2001. Aroma compounds of fresh milk from New Zealand cows fed different diets. *J. Agric. Food Chem.* **49:**4825–4832.
19. **Berg, D. E., M. A. Kohn, T. A. Farley, and L. M. McFarland.** 2000. Multi-state outbreaks of acute gastroenteritis traced to fecal-contaminated oysters harvested in Louisiana. *J. Infect. Dis.* **181:**S381–S386.
20. **Besong, S., J. A. Jackson, C. L. Hicks, and R. W. Hemken.** 1996. Effects of a supplemental liquid yeast product on feed intake, ruminal profiles, and yield, composition, and organoleptic characteristics of milk from lactating Holstein cows. *J. Dairy Sci.* **79:**1654–1668.
21. **Bielaszewska, M., J. Janda, K. Blahova, H. Minarikova, E. Jikova, M. A. Karmali, J. Laubova, J. Sikulova, M. A. Preston, R. Khakhria, H. Karch, H. Klazarova, and O. Nyc.** 1997. Human *Escherichia coli* O157:H7 infection associated with the consumption of unpasteurized goat's milk. *Epidemiol. Infect.* **119:**299–305.
22. **Billinghurst, I.** 2005. BARF Australia. [Online.] http://www.drianbillinghurst.com/index.html. Accessed 6 July 2006.
23. **Blessing, D. J., M. Thompson, B. Fisher, D. Schooley, M. J. Kramer, T. M. DeMelfi, M. A. McCarthy, E. J. Witte, C. W. Hays, and J. Smucker.** 1983. Campylobacteriosis associated with raw milk consumption—Pennsylvania. *Morb. Mortal. Wkly. Rep.* **32:**337–338, 344.
24. **Blythe, D., M. Goldoft, J. Lewis, P. Stehr-Green, R. Chehey, J. Greenblatt, P. R. Cieslak, K. R. Stefonek, F. C. Hoesly, and D. Fleming.** 1997. *Salmonella* serotype Montevideo infections associated with chicks—Idaho, Washington, and Oregon, spring 1995 and 1996. *Morb. Mortal. Wkly. Rep.* **46:**237–239.
25. **Boggs, J. D., R. E. Whitwam, L. M. Hale, R. P. Briscoe, S. E. Kahn, J. N. MacCormack, J.-M. Maillard, S. C. Grayson, K. S. Sigmon, J. W. Reardon, and J. R. Saah.** 2001. Outbreak of listeriosis associated with homemade Mexican-style cheese—North Carolina, October 2000–January 2001. *Morb. Mortal. Wkly. Rep.* **50:**560–562.
26. **Brett, M. S. Y., P. Short, and J. McLauchlin.** 1998. A small outbreak of listeriosis associated with smoked mussels. *Int. J. Food Microbiol.* **43:**223–229.
27. **Caldow, G. L., and M. M. Graham.** 1998. Abortion in foxhounds and a ewe flock associated with *Salmonella montevideo* infection. *Vet. Rec.* **142:** 138–139.
28. **California.** 2005. California Food and Agriculture Code, section 35781-35788. [Online.] http://www.leginfo.ca.gov/cgi-bin/displaycode?section=fac&group=35001-36000&file=35781-35788. Accessed 6 July 2006.
29. **Cave, N. J., S. L. Marks, P. H. Kass, A. C. Melli, and M. A. Brophy.** 2002. Evaluation of a routine diagnostic fecal panel for dogs with diarrhea. *J. Am. Vet. Med. Assoc.* **221:**52–59.
30. **Chai, J.-Y., E.-T. Han, E.-H. Shin, J.-H. Park, J.-P. Chu, M. Hirota, F. Nakamura-Uchiyama, and Y. Nawa.** 2003. An outbreak of gnatho-

stomiasis among Korean emigrants in Myanmar. *Am. J. Trop. Med. Hyg.* **69:** 67–73.

31. **Chalmers, J. W. T., and J. H. McMillan.** 1995. An outbreak of viral gastroenteritis associated with adequately prepared oysters. *Epidemiol. Infect.* **115:**163–167.

32. **Chengappa, M. M., J. Staats, R. D. Oberst, N. H. Gabbert, and S. McVey.** 1993. Prevalence of *Salmonella* in raw meat used in diets of racing greyhounds. *J. Vet. Diagn. Invest.* **5:**372–377.

33. **Cherry, B., A. Burns, G. S. Johnson, H. Pfeiffer, N. Dumas, D. Barrett, P. L. McDonough, and M. Eidson.** 2004. *Salmonella* Typhimurium outbreak associated with veterinary clinic. *Emerg. Infect. Dis.* **10:**2249–2251.

34. **Code of Federal Regulations.** 1992. Mandatory pasteurization for all milk and milk products in final package form intended for direct human consumption. *Code of Federal Regulations* 21CFR 1240.61. [Online.] http://www.accessdata.fda.gov/scripts/cdrh/cfdocs/cfCFR/CFRSearch.cfm?fr=1240.61. Accessed 6 July 2006.

35. **Code of Federal Regulations.** 2005. Turtles intrastate and interstate requirements. *Code of Federal Regulations* 21CFR 1240.62. [Online.] http://www.accessdata.fda.gov/scripts/cdrh/cfdocs/cfCFR/CFRSearch.cfm?fr=1240.62. Accessed 6 July 2005.

36. **Conaty, S., P. Bird, G. Bell, E. Kraa, G. Grohmann, and J. M. McAnulty.** 2000. Hepatitis A in New South Wales, Australia from consumption of oysters: the first reported outbreak. *Epidemiol. Infect.* **124:**121–130.

37. **Connecticut Department of Agriculture.** 2003. Public Act 05-175—an act concerning the revision and modernization of milk regulation statutes and the licensing of poultry dealers. State of Connecticut, Department of Agriculture, Hartford, Conn. [Online.] http://www.ct.gov/doag/cwp/view.asp?a=1366&q=296822. Accessed 6 July 2006.

38. **Conrad, C., K. Hemphill, S. Wilson, L. McFarland, K. Coulbourne, S. Qarni, S. Poster, C. Groves, C. Slemp, E. Butler, D. Matuszak, D. Dwyer, E. Israel, J. Cirino, D. Cumberland, L. Pollack, B. Brackin, M. Currier, H. Morris, M. Bissett, S. Evans, B. Respess, B. Jenkins, J. Maillard, R. Meriwether, J. N. MacCormack, B. Creasy, J. Veazey, K. Calci, S. Rippey, and G. Hoskin.** 1993. Multistate outbreak of viral gastroenteritis related to consumption of oysters—Louisiana, Maryland, Mississippi, and North Carolina, 1993. *Morb. Mortal. Wkly. Rep.* **42:**945–948.

39. **Cowden, J. M., M. O'Mahony, C. L. Bartlett, B. Rana, B. Smyth, D. Lynch, H. Tillett, L. Ward, D. Roberts, R. J. Gilbert, A. C. Baird-Parker, and D. C. Kilsby.** 1989. A national outbreak of *Salmonella typhimurium* DT 124 caused by contaminated salami sticks. *Epidemiol. Infect.* **103:**219–225.

40. **Cronquist, A., S. Wedel, B. Albanese, C. M. Sewell, D. Hoang-Johnson, T. Ihry, M. Lynch, J. Lockett, N. Kazerouni, C. O'Reilly, D. Ferguson, and EIS officers, CDC.** 2006. Multistate outbreak of *Salmonella* Typhimurium infections associated with eating ground beef—United States, 2004. *Morb. Mortal. Wkly. Rep.* **55:**180–182.

41. **Curnow, J.** 1999. *Escherichia coli* O157 outbreak in Scotland linked to unpasteurised goat's milk. *Eur. Surveill. Wkly.* **3.** [Online.] http://www.eurosurveillance.org/ew/1999/990610.asp#2. Accessed 6 July 2006.

42. **Damborg, P., K. E. P. Olsen, E. Møller Nielsen, and L. Guardabassi.** 2004. Occurrence of *Campylobacter jejuni* in pets living with human patients infected with *C. jejuni*. *J. Clin. Microbiol.* **42:**1363–1364.

43. **Desenclos, J.-C., P. Bouvet, E. Benz-Lemoine, F. Grimont, H. Desqueyroux, I. Rebiere, and P. A. Grimont.** 1996. Large outbreak of *Salmonella enterica* serotype Paratyphi B infection caused by a goats' milk cheese, France 1993: a case finding and epidemiological study. *BMJ* **312:**91–94.

44. **Desenclos, J. C., K. C. Klontz, M. H. Wilder, O. V. Nainan, H. S. Margolis, and R. A. Gunn.** 1991. A multistate outbreak of hepatitis A caused by the consumption of raw oysters. *Am. J. Public Health* **81:**1268–1272.

45. **Dessi, S., C. Sanna, and L. Paghi.** 1992. Human salmonellosis transmitted by a domestic turtle. *Eur. J. Epidemiol.* **8:**120–121.

46. **De Valk, H., E. Delarocque-Astagneau, G. Colomb, S. Ple, E. Godard, V. Vaillant, S. Haeghebaert, P. H. Bouvet, F. Grimont, P. Grimont, and J. C. Desenclos.** 2000. A community-wide outbreak of *Salmonella enterica* serotype Typhimurium infection associated with eating a raw milk soft cheese in France. *Epidemiol. Infect.* **124:**1–7.

47. **Doran, M., P. Shillam, R. E. Hoffman, and L. M. McFarland.** 1989. Toxigenic *Vibrio cholerae* O1 infection acquired in Colorado. *Morb. Mortal. Wkly. Rep.* **38:**19–20.

48. **Ellis, A., M. Preston, A. Borczyk, B. Miller, P. Stone, B. Hatton, A. Chagla, and J. Hockin.** 1998. A community outbreak of *Salmonella berta* associated with a soft cheese product. *Epidemiol. Infect.* **120:**29–35.

49. **Erickson, J.** 9 July 2003. Firm recalls frozen steaks; 16 *E. coli* infections linked to beef from Chicago business. *Rocky Mountain News*, Denver, Colo.

50. **Ericsson, H., A. Eklöw, M.-L. Danielsson-Tham, S. Loncarevic, L.-O. Mentzing, I. Persson, H. Unnerstad, and W. Tham.** 1997. An outbreak of listeriosis suspected to have been caused by rainbow trout. *J. Clin. Microbiol.* **35:**2904–2907.

51. **Espié, E., and V. Vaillant.** 2005. International outbreak of *Salmonella* Stourbridge infection, April–July 2005: results of epidemiological, food and veterinary investigations in France. *Eur. Surveill. Wkly.* **10.** [Online.] http://www.eurosurveillance.org/ew/2005/050811.asp#3. Accessed 6 July 2006.

52. **Espinosa, F. H., W. M. Ryan, P. L. Vigil, D. F. Gregory, R. B. Hilley, D. A. Romig, R. B. Stamm, E. D. Suhre, P. S. Taulbee, L. H. Zucal, R. W. Honsinger, Jr., P. J. Lindberg, M. Barcheck, J. A. Miller, R. Mitzelfelt, J. M. Montes, L. J. Nims, O. J. Rollag, N. Weber, and J. M. Mann.** 1983. Group C streptococcal infections associated with eating homemade cheese—New Mexico. *Morb. Mortal. Wkly. Rep.* **32:**510, 515–516.

53. **Ethelberg, S.** 2005. Salmonellosis outbreak linked to carpaccio made from imported raw beef, Denmark, June–August 2005. *Eur. Surveill. Wkly.* **10.** [Online.] http://www.eurosurveillance.org/ew/2005/050922.asp#3. Accessed 6 July 2006.

54. **European Communities.** 1996. S.I. No. 9/1996: European Communities (hygienic production and placing on the market of raw milk, heat-treated milk and milk-based products) regulations, 1996. European Communities. [Online.] http://www.irishstatutebook.ie/ZZSI9Y1996.html#ZZSI9Y1996. Accessed 6 July 2006.

55. **Ezell, H., B. Tramontin, R. Hudson, L. Tengelsen, C. Hahn, K. Smith, J. Bender, D. Boxrud, J. Adams, R. Frank, K. Culbertson, T. Besser, D. Rice, R. Gautom, R. Pallipamu, M. Goldoft, J. Grendon, J. Kobayashi, F. Angulo, T. Barrett, S. Rossiter, S. Sivapalasingam, and J. Wright.** 2001. Outbreaks of multidrug-resistant *Salmonella* Typhimurium associated with veterinary facilities—Idaho, Minnesota, and Washington, 1999. *Morb. Mortal. Wkly. Rep.* **50:**701–704.

56. **Fallon, S.** 2001. Is raw milk safe for babies? *Real Milk Updates.* [Online.] http://www.realmilk.com/raw-milk-babies.html. Accessed 6 July 2006.

57. **Farley, T. A., L. McFarland, M. Estes, and K. Schwab.** 1997. Viral gastroenteritis associated with eating oysters—Louisiana, December 1996–January 1997. *Morb. Mortal. Wkly. Rep.* **46:**1109–1112.

58. **Federal Register.** 1999. Beef products contaminated with *Escherichia coli* O157:H7. *Fed. Regist.* **64:**2803–2805.

59. **Fontaine, R. E., S. Arnon, W. T. Martin, T. M. Vernon, Jr., E. J. Gangarosa, J. J. Farmer III, A. B. Moran, J. H. Silliker, and D. L. Decker.** 1978. Raw hamburger: an interstate common source of human salmonellosis. *Am. J. Epidemiol.* **107:**36–45.

60. **Food and Drug Administration.** 2004. Guidance for Industry #122. Manufacture and labeling of raw meat foods for companion and captive noncompanion carnivores and omnivores. U.S. Food and Drug Administration, Center for Veterinary Medicine. [Online.] http://www.fda.gov/cvm/Guidance/Guide122.pdf. Accessed 6 July 2006.

61. **Food and Drug Administration.** 30 June 2006. FDA commemorates a century of protecting and promoting public health. U.S. Food and Drug Administration. [Online.] http://www.fda.gov/bbs/topics/NEWS/2006/NEW01403.html. Accessed 5 July 2006.

62. **Food Safety and Inspection Service.** 29 June 2003. Recall Release FSIS-RC-028-2003. Illinois firm recalls beef products for possible *E. coli* O157:H7. Food Safety and Inspection Service, United States Department of Agriculture, Washington, D.C. [Online.] http://www.fsis.usda.gov/OA/recalls/prelease/pr028-2003.htm. Accessed 6 July 2006.

63. **Foster, E. M.** 1997. Historical overview of key issues in food safety. *Emerg. Infect. Dis.* **3:**481–482.

64. **Fox, N.** 1997. *Spoiled. The Dangerous Truth about a Food Chain Gone Haywire.* BasicBooks, A Division of HarperCollins Publishers, Inc., New York, N.Y.

65. **Frazak, P. A., J. J. Kazmierczak, M. E. Proctor, J. P. Davis, J. Larson, and R. Loerke.** 1995. Outbreak of *Salmonella* serotype Typhimurium infection associated with eating raw ground beef—Wisconsin, 1994. *Morb. Mortal. Wkly. Rep.* **44:**905–909.

66. **Fukata, T., F. Naito, N. Yoshida, T. Yamaguchi, Y. Mizumura, and K. Hirai.** 2002. Incidence of *Salmonella* infection in healthy dogs in Gifu prefecture, Japan. *J. Vet. Med. Sci.* **64:**1079–1080.

67. **Fyfe, M., M. T. Kelly, S. T. Yeung, P. Daly, K. Schallie, S. Buchanan, P. Waller, J. Kobayashi, N. Therien, M. Guichard, S. Lankford, P. Stehr-Green, R. Harsch, E. DeBess, M. Cassidy, T. McGivern, S. Mauvais, D. Fleming, M. Lippmann, L. Pong, R. W. McKay, D. E. Cannon, S. B. Werner, S. Abbott, M. Hernandez, C. Wojee, J. Waddell, S. Water-**

man, J. Middaugh, D. Sasaki, P. Effler, C. Groves, N. Curtis, D. Dwyer, G. Dowdle, and C. Nichols. 1998. Outbreak of *Vibrio parahaemolyticus* infections associated with eating raw oysters—Pacific Northwest, 1997. *Morb. Mortal. Wkly. Rep.* **47:**457–462.

68. **Giljahn, L. K., and T. J. Halpin.** 1986. Turtle-associated salmonellosis—Ohio. *Morb. Mortal. Wkly. Rep.* **35:**733–734.

69. **Gilsdorf, A., A. Jansen, K. Alpers, H. Dieckmann, U. van Treeck, A. M. Hauri, G. Fell, M. Littmann, P. Rautenberg, R. Prager, W. Rabsch, P. Roggentin, A. Schroeter, A. Miko, E. Bartelt, J. Bräunig, and A. Ammon.** 2005. A nationwide outbreak of *Salmonella* Bovismorbificans PT24, Germany, December 2004–March 2005. *Eur. Surveill. Wkly.* **10.** [Online.] http://www.eurosurveillance.org/ew/2005/050324.asp#1. Accessed 6 July 2006.

70. **Gray, J. T., L. L. Hungerford, P. J. Fedorka-Cray, and M. L. Headrick.** 2004. Extended-spectrum-cephalosporin resistance in *Salmonella enterica* isolates of animal origin. *Antimicrob. Agents Chemother.* **48:**3179–3181.

71. **Guardabassi, L., S. Schwarz, and D. H. Lloyd.** 2004. Pet animals as reservoirs of antimicrobial-resistant bacteria. *J. Antimicrob. Chemother.* **54:**321–332.

72. **Haeghebaert, S., L. Duché, C. Gilles, B. Masini, M. Dubreuil, J. C. Minet, P. Bouvet, F. Grimont, E. Delarocque Astagneau, and V. Vaillant.** 2001. Minced beef and human salmonellosis: review of the investigation of three outbreaks in France. *Eur. Surveill. Wkly.* **6:**21–26. [Online.] http://www.eurosurveillance.org/em/v06n02/0602-222.asp. Accessed 6 July 2006.

73. **Haeghebaert, S., P. Sulem, L. Deroudille, E. Vanneroy-Adenot, O. Bagnis, P. Bouvet, F. Grimont, A. Brisabois, F. Le Querrec, C. Hervy, E. Espié, H. de Valk, and V. Vaillant.** 2003. Two outbreaks of *Salmonella* Enteritidis phage type 8 linked to the consumption of Cantal cheese made with raw milk, France, 2001. *Eur. Surveill. Wkly.* **8:**151–156. [Online.] http://www.eurosurveillance.org/em/v08n07/v08n07.pdf. Accessed 6 July 2006.

74. **Harrington, P., J. Archer, J. P. Davis, D. R. Croft, and J. K. Varma.** 2002. Outbreak of *Campylobacter jejuni* infections associated with drinking unpasteurized milk procured through a cow-leasing program—Wisconsin, 2001. *Morb. Mortal. Wkly. Rep.* **51:**548–549.

75. **Health Canada.** 2005. Statement from Health Canada about drinking raw milk. Health Canada, Food and Nutrition. [Online.] http://www.hc-sc.gc.ca/fn-an/securit/facts-faits/rawmilk-laitcru_e.html. Accessed 6 July 2006.

76. **Heikens, N.** 15 October 2003. Indiana residents refusing to kowtow to state milk rules. *The Indianapolis Star*, Indianapolis, Ind.

77. **Higashihara, S., B. Kanenaka, M. Ching-Lee, P. Effler, D. Akiyama, M. Sugi, E. Pon, K. Sharifzadeh, R. Waskiewicz, N. Ridley, W. Hohmann, W. Higson, M. Sobsey, and D. Wait.** 1991. Gastroenteritis associated with consumption of raw shellfish—Hawaii, 1991. *Morb. Mortal. Wkly. Rep.* **40:**303–305.

78. **Istre, G. R., M. J. Blaser, P. Shillam, and R. S. Hopkins.** 1984. *Campylobacter* enteritis associated with undercooked barbecued chicken. *Am. J. Public Health* **74:**1265–1267.

79. **Joffe, D. J., and D. P. Schlesinger.** 2002. Preliminary assessment of the risk of *Salmonella* infection in dogs fed raw chicken diets. *Can. Vet. J.* **43:** 441–442.

80. **Kálmán, M., E. Szöllösi, B. Czermann, M. Zimányi, S. Szekeres, and M. Kálmán.** 2000. Milkborne *Campylobacter* infection in Hungary. *J. Food Prot.* **63:**1426–1429.

81. **Keene, W. E., K. Hedberg, D. E. Herriott, D. D. Hancock, R. W. McKay, T. J. Barrett, and D. W. Fleming.** 1997. A prolonged outbreak of *Escherichia coli* O157:H7 infections caused by commercially distributed raw milk. *J. Infect. Dis.* **176:**815–818.

82. **Keene, W. E., E. Sazie, J. Kok, D. H. Rice, D. D. Hancock, V. K. Balan, T. Zhao, and M. P. Doyle.** 1997. An outbreak of *Escherichia coli* O157:H7 infections traced to jerky made from deer meat. *JAMA* **277:**1229–1231.

83. **Klein, B. S., J. M. Vergeront, M. J. Blaser, P. Edmonds, D. J. Brenner, D. Janssen, and J. P. Davis.** 1986. *Campylobacter* infection associated with raw milk. An outbreak of gastroenteritis due to *Campylobacter jejuni* and thermotolerant *Campylobacter fetus* subsp. fetus. *JAMA* **255:**361–364.

84. **Laine, E. S., J. M. Scheftel, D. J. Boxrud, K. J. Vought, R. N. Danila, K. M. Elfering, and K. E. Smith.** 2005. Outbreak of *Escherichia coli* O157: H7 infections associated with nonintact blade-tenderized frozen steaks sold by door-to-door vendors. *J. Food Prot.* **68:**1198–1202.

85. **Lanzillo, L., J. Reid, S. Cobb, M. DiManno, E. Podgorski, J. Lyons, D. Greenstein, N. Maher, N. Schell, V. Tulumello, D. Klotz, H. Foust, L. Bonser, J. Debbie, R. Deibel, B. Fear, J. Guzewich, K. Henry, J. Raucci, I. Loudon, D. Morse, J. Pert, M. Shayegani, P. Smith, A. Squire, R. Stricof, R. Svenson, and R. Rothenberg.** 1982. Enteric illness associated with raw clam consumption—New York. *Morb. Mortal. Wkly. Rep.* **31:**449–451.

86. **Levings, R. S., D. Lightfoot, R. M. Hall, and S. P. Djordjevic.** 2006. Aquariums as reservoirs for multidrug-resistant *Salmonella* Paratyphi B. *Emerg. Infect. Dis.* **12:**507–510.

87. **Lieftucht, A.** 1999. An outbreak of *Campylobacter* infection associated with a farm in Germany. *Eur. Surveill. Wkly.* **3.** [Online.] http://www.eurosurveillance.org/ew/1999/991125.asp#1. Accessed 6 July 2006.

88. **Life Enthusiast Co-op.** 2006. Nutrition: raw food. Primal diet testimony. Life Enthusiast Co-op. [Online.] http://www.life-enthusiast.com/index.php?Q1=Education&Q2=NutritionRaw. Accessed 6 July 2006.

89. **Linnan, M. J., L. Mascola, X. D. Lou, V. Goulet, S. May, C. Salminen, D. W. Hird, M. L. Yonekura, P. Hayes, R. Weaver, A. Audurier, B. D. Plikaytis, S. L. Fannin, A. Kleks, and C. V. Broome.** 1988. Epidemic listeriosis associated with Mexican-style cheese. *N. Engl. J. Med.* **319:**823–828.

90. **Longcore, K.** 2 July 2003. Kent health detective's work triggers meat recall; she traces illness caused by *E. coli* to tenderized steaks produced by a Chicago meat packer. *The Grand Rapids Press*, Grand Rapids, Mich.

91. **MacDougall, L., M. Fyfe, L. McIntyre, A. Paccagnella, K. Cordner, A. Kerr, and J. Aramini.** 2004. Frozen chicken nuggets and strips—a newly identified risk factor for *Salmonella* Heidelberg infection in British Columbia, Canada. *J. Food Prot.* **67:**1111–1115.

92. **Machi, T., S. Okino, Y. Saito, Y. Horita, T. Taguchi, T. Nakazawa, Y. Nakamura, H. Hirai, H. Miyamori, and S. Kitagawa.** 1997. Severe chest pain due to gastric anisakiasis. *Intern. Med.* **36:**28–30.

93. **MacRae, S., T. Clements, and J. Cournoyer.** 1982. Non-O1 *Vibrio cholerae* gastroenteritis—New Hampshire. *Morb. Mortal. Wkly. Rep.* **31:**538–539.

94. **Maguire, H., J. Cowden, M. Jacob, B. Rowe, D. Roberts, J. Bruce, and E. Mitchell.** 1992. An outbreak of *Salmonella dublin* infection in England and Wales associated with a soft unpasteurized cows' milk cheese. *Epidemiol. Infect.* **109:**389–396.

95. **Maguire, H. C., A. A. Codd, V. E. Mackay, B. Rowe, and E. Mitchell.** 1993. A large outbreak of human salmonellosis traced to a local pig farm. *Epidemiol. Infect.* **110:**239–246.

96. **Mascola, L., M. Tormey, D. Dassey, L. Kilman, S. Harvey, A. Medina, A. Tilzer, and S. Waterman.** 1996. *Vibrio vulnificus* infections associated with eating raw oysters—Los Angeles, 1996. *Morb. Mortal. Wkly. Rep.* **45:**621–624.

97. **Mazurek, J., E. Salehi, D. Propes, J. Holt, T. Bannerman, L. M. Nicholson, M. Bundesen, R. Duffy, and R. L. Moolenaar.** 2004. A multistate outbreak of *Salmonella enterica* serotype Typhimurium infection linked to raw milk consumption—Ohio, 2003. *J. Food Prot.* **67:**2165–2170.

98. **McCoy, R. H., and R. J. Seidler.** 1973. Potential pathogens in the environment: isolation, enumeration, and identification of seven genera of intestinal bacteria associated with small green pet turtles. *Appl. Microbiol.* **25:**534–538.

99. **McDonough, S., F. Heer, and L. Shireley.** 1991. Foodborne outbreak of gastroenteritis caused by *Escherichia coli* O157:H7—North Dakota, 1990. *Morb. Mortal. Wkly. Rep.* **40:**265–267.

100. **McIntyre, L., J. Fung, A. Paccagnella, J. Isaac-Renton, F. Rockwell, B. Emerson, and T. Preston.** 2002. *Escherichia coli* O157 outbreak associated with the ingestion of unpasteurized goat's milk in British Columbia, 2001. *Can. Commun. Dis. Rep.* **28:**6–8.

101. **McVey, D. S., M. M. Chengappa, D. E. Mosier, G. G. Stone, R. D. Oberst, M. J. Sylte, N. M. Gabbert, S. M. Kelly-Aehle, and R. Curtiss, III.** 2002. Immunogenicity of χ^{4127} *phoP- Salmonella enterica* serovar Typhimurium in dogs. *Vaccine* **20:**1618–1623.

102. **Méndez Martínez, C., A. Páez Jiménez, M. Cortés Blanco, E. Salmoral Chamizo, E. Mohedano Mohedano, C. Plata, A. Varo Baena, and F. Martíinez Navarro.** 2003. Brucellosis outbreak due to unpasteurized raw goat cheese in Andalucia (Spain), January–March 2002. *Eur. Surveill. Wkly.* **8:**164–168. [Online.] http://www.eurosurveillance.org/em/v08n07/0807-223.asp. Accessed 6 July 2006.

103. **Mercola, J.** 1999. Milk linked to autism, schizophrenia. [Online.] http://www.mercola.com/1999/archive/milk_linked_to_autism.htm. Accessed 6 July 2006.

104. **Moio, L., L. Rillo, A. Ledda, and F. Addeo.** 1996. Odorous constituents of ovine milk in relationship to diet. *J. Dairy Sci.* **79:**1322–1331.

105. **Morse, D. L., J. J. Guzewich, J. P. Hanrahan, R. Stricof, M. Shayegani, R. Deibel, J. C. Grabau, N. A. Nowak, J. E. Herrmann, G. Cukor, and N. R. Blacklow.** 1986. Widespread outbreaks of clam- and oyster-associated gastroenteritis. Role of Norwalk virus. *N. Engl. J. Med.* **314:**678–681.

106. **Morse, E. V., and M. A. Duncan.** 1975. Canine salmonellosis: prevalence, epizootiology, signs, and public health significance. *J. Am. Vet. Med. Assoc.* **167:**817–820.

107. **Morse, E. V., M. A. Duncan, D. A. Estep, W. A. Riggs, and B. O. Blackburn.** 1976. Canine salmonellosis: a review and report of dog to child transmission of *Salmonella enteritidis*. *Am. J. Public Health* **66:**82–83.

108. **New Zealand Minister for Food Safety.** 2002. New Zealand (milk and milk products processing) food standards 2002. New Zealand Food Safety Authority. [Online.] http://www.nzfsa.govt.nz/policy-law/legislation/food-standards/nz-food-standards-2002-milk.pdf. Accessed 6 July 2006.

109. **Noh, J. H., B.-J. Kim, S. M. Kim, M.-S. Ock, M. I. Park, and J. Y. Goo.** 2003. A case of acute gastric anisakiasis provoking severe clinical problems by multiple infection. *Korean J. Parasitol.* **41:**97–100.

110. **O'Brien, S., H. Smith, L. Lighton, and A. Mellanby.** 2000. Outbreaks of VTEC O157 infection linked to consumption of unpasteurised milk. *Eur. Surveill. Wkly.* **4.** [Online.] http://www.eurosurveillance.org/ew/2000/000608.asp#4. Accessed 6 July 2006.

111. **Osterholm, M. T., K. L. MacDonald, K. E. White, J. G. Wells, J. S. Spika, M. E. Potter, J. C. Forfang, R. M. Sorenson, P. T. Milloy, and P. A. Blake.** 1986. An outbreak of a newly recognized chronic diarrhea syndrome associated with raw milk consumption. *JAMA* **256:**484–490.

112. **Perkins, P., A. Rogers, M. Key, V. Pappas, R. Wende, J. Epstein, M. Thapar, F. Jensen, T. L. Gustafson, and E. Young.** 1983. Brucellosis—Texas. *Morb. Mortal. Wkly. Rep.* **32:**548–553.

113. **Pozio, E., D. Sacchini, P. Boni, A. Tamburrini, F. Alberici, and F. Paterlini.** 1998. Human outbreak of trichinellosis associated with the consumption of horsemeat in Italy. *Eur. Surveill. Wkly.* **3:**85–86. [Online.] http://www.eurosurveillance.org/em/v03n08/0308-222.asp. Accessed 6 July 2006.

114. **Rabatsky-Ehr, T., D. Dingman, R. Marcus, R. Howard, A. Kinney, and P. Mshar.** 2002. Deer meat as the source for a sporadic case of *Escherichia coli* O157:H7 infection, Connecticut. *Emerg. Infect. Dis.* **8:**525–527.

115. **Rahim, S.** 1 January 2003. Young's problem likely tied to human; number of *Salmonella* cases rises to 47. *Dayton Daily News*, Dayton, Ohio.

116. **Rahim, S.** 17 January 2003. Young's to stop selling, using unpasteurized milk. Health officials say *Salmonella* outbreak ended. Cox News Service, Dayton, Ohio. [Online.] http://foodhaccp.com/msgboard.mv?parm_func=showmsg+parm_msgnum=1006741. Accessed 6 July 2006.

117. **Randby, Å. T., I. Selmer-Olsen, and L. Baevre.** 1999. Effect of ethanol in feed on milk flavor and chemical composition. *J. Dairy Sci.* **82:**420–428.

118. **Rawfoodnetwork.com.** 2006. Raw and living food forums, mailing lists, and newsletters. Blue Horizon Enterprises, Wynnewood, Pa. [Online.] http://www.rawfoodnetwork.com/mailinglists.html. Accessed 6 July 2006.

119. **Rehmet, S., G. Sinn, O. Robstad, L. Petersen, A. Ammon, D. Lesser, H. David, K. Noeckler, G. Scherholz, K.-D. Erkrath, D. Pechmann, R. Kundt, G. Oltmans, R. Lange, J. Laumen, U. Nogay, M. Dixius, J. Eichenberg, F. Dinse, D. Stegemann, W. Lotz, D. Franke, P. Hagelschur, and M. Steigert.** 1999. Two outbreaks of trichinellosis in the state of Northrhine-Westfalia, Germany, 1998. *Eur. Surveill. Wkly.* **4:**78–81. [Online.] http://www.eurosurveillance.org/em/v04n08/0408-222.asp. Accessed 6 July 2006.

120. **Riley, L. W., G. T. DiFerdinando, Jr., T. M. DeMelfi, and M. L. Cohen.** 1983. Evaluation of isolated cases of salmonellosis by plasmid profile analysis: introduction and transmission of a bacterial clone by precooked roast beef. *J. Infect. Dis.* **148:**12–17.

121. **Rodrigue, D. C., E. E. Mast, K. D. Greene, J. P. Davis, M. A. Hutchinson, J. G. Wells, T. J. Barrett, and P. M. Griffin.** 1995. A university outbreak of *Escherichia coli* O157:H7 infections associated with roast beef and an unusually benign clinical course. *J. Infect. Dis.* **172:**1122–1125.

122. **Ross, R. A.** 2005. Raw news: the truth about toxins in cooked foods is making mainstream headlines. *Raw Food Life.* Adventures Marketing, LLC. [Online.] http://www.rawfoodlife.com/. Accessed 6 July 2006.

123. **Ryan, C. A., M. K. Nickels, N. T. Hargrett-Bean, M. E. Potter, T. Endo, L. Mayer, C. W. Langkop, C. Gibson, R. C. McDonald, R. T. Kenney, N. D. Puhr, P. J. McDonnell, R. J. Martin, M. L. Cohen, and P. A. Blake.** 1987. Massive outbreak of antimicrobial-resistant salmonellosis traced to pasteurized milk. *JAMA* **258:**3269–3274.

124. **Salna, B., T. Monson, T. Kurzynski, K. Gundlach, P. E. Fox, J. Kazmierczak, M. Wegner, J. P. Davis, R. Harrington, M. Dowell, R. Heald, R. Harris, W. Manley, J. Snow, A. Heryford, and S. Seys.** 2005. Salmonellosis associated with pet turtles—Wisconsin and Wyoming, 2004. *Morb. Mortal. Wkly. Rep.* **54:**223–226.

125. **Sato, Y., T. Mori, T. Koyama, and H. Nagase.** 2000. *Salmonella* Virchow infection in an infant transmitted by household dogs. *J. Vet. Med. Sci.* **62:** 767–769.

126. **Sauer, C. J., J. Majkowski, S. Green, and R. Eckel.** 1997. Foodborne illness outbreak associated with a semi-dry fermented sausage product. *J. Food Prot.* **60:**1612–1617.

127. **Shapiro, R., M.-L. Ackers, S. Lance, M. Rabbani, L. Schaefer, J. Daugherty, C. Thelen, and D. Swerdlow.** 1999. *Salmonella* Thompson associated with improper handling of roast beef at a restaurant in Sioux Falls, South Dakota. *J. Food Prot.* **62:**118–122.

128. **Smith, K., D. Boxrud, F. Leano, C. Snider, C. Braden, J. Lockett, S. Montgomery, S. Swanson, and C. O'Reilly.** 2005. Outbreak of multidrug-resistant *Salmonella* Typhimurium associated with rodents purchased at retail pet stores—United States, December 2003–October 2004. *Morb. Mortal. Wkly. Rep.* **54:**429–433.

129. **Spitalny, K. C., E. N. Okowitz, and R. L. Vogt.** 1984. Salmonellosis outbreak at a Vermont hospital. *South. Med. J.* **77:**168–172.

130. **Spite, G. T., D. F. Brown, and R. M. Twedt.** 1978. Isolation of an enteropathogenic, Kanagawa-positive strain of *Vibrio parahaemolyticus* from seafood implicated in acute gastroenteritis. *Appl. Environ. Microbiol.* **35:**1226–1227.

131. **Stiver, S. L., K. S. Frazier, M. J. Mauel, and E. L. Styer.** 2003. Septicemic salmonellosis in two cats fed a raw-meat diet. *J. Am. Anim. Hosp. Assoc.* **39:**538–542.

132. **Stone, G. G., M. M. Chengappa, R. D. Oberst, N. H. Gabbert, S. McVey, K. J. Hennessy, M. Muenzenberger, and J. Staats.** 1993. Application of polymerase chain reaction for the correlation of *Salmonella* serovars recovered from greyhound feces with their diet. *J. Vet. Diagn. Invest.* **5:**378–385.

133. **Strohmeyer, R. A., P. S. Morley, D. R. Hyatt, D. A. Dargatz, A. V. Scorza, and M. R. Lappin.** 2006. Evaluation of bacterial and protozoal contamination of commercially available raw meat diets for dogs. *J. Am. Vet. Med. Assoc.* **228:**537–542.

134. **Svitlik, C., M. Cartter, Y. McCarter, J. L. Hadler, D. Goeller, C. Groves, D. Dwyer, D. Tilghman, E. Israel, R. Housenecht, S. Yeager, and D. R. Tavris.** 1992. *Salmonella hadar* associated with pet ducklings—Connecticut, Maryland, and Pennsylvania, 1991. *Morb. Mortal. Wkly. Rep.* **41:**185–187.

135. **Telzak, E. E., E. P. Bell, D. A. Kautter, L. Crowell, L. D. Budnick, D. L. Morse, and S. Schultz.** 1990. An international outbreak of type E botulism due to uneviscerated fish. *J. Infect. Dis.* **161:**340–342.

136. **Timmons, J. S., W. P. Weiss, D. L. Palmquist, and W. J. Harper.** 2001. Relationships among dietary roasted soybeans, milk components, and spontaneous oxidized flavor of milk. *J. Dairy Sci.* **84:**2440–2449.

137. **Vaillant, V., S. Haeghebaert, J. C. Desenclos, P. Bouvet, F. Grimont, P. A. Grimont, and A. P. Burnens.** 1996. Outbreak of *Salmonella dublin* infection in France, November–December 1995. *Eur. Surveill. Wkly.* **1:**9–10. [Online.] http://www.eurosurveillance.org/em/v01n02/0102-221.asp. Accessed 6 July 2006.

138. **Van Duijkeren, E., and D. Houwers.** 2002. [*Salmonella* enteritis in dogs, not relevant?] [English abstract; article in Dutch]. *Tijdschr. Diergeneeskd.* **127:**716–717.

139. **Van Immerseel, F., F. Pasmans, J. De Buck, I. Rychlik, H. Hradecka, J.-M. Collard, C. Wildemauwe, M. Heyndrickx, R. Ducatelle, and F. Haesebrouck.** 2004. Cats as a risk for transmission of antimicrobial drug-resistant *Salmonella*. *Emerg. Infect. Dis.* **10:**2169–2174.

140. **Weaver, D.** 2003. The ecological wisdom of enjoying raw foods. River Canyon Retreat. [Online.] http://www.rivercanyonretreat.com/article1.html. Accessed 6 July 2006.

141. **Weber, J. T., R. G. Hibbs, Jr., A. Darwish, B. Mishu, A. L. Corwin, M. Rakha, C. L. Hatheway, S. El Sharkawy, S. Abd El-Rahim, M. F. S. Al-Hamd, J. E. Sarn, P. A. Blake, and R. V. Tauxe.** 1993. A massive outbreak of type E botulism associated with traditional salted fish in Cairo. *J. Infect. Dis.* **167:**451–454.

142. **Wechsler, E., C. D'Aleo, V. A. Hill, J. Hopper, D. Myers-Wiley, E. O'Keeffe, J. Jacobs, F. Guido, A. Huang, S. N. Dodt, B. Rowan, M. Sherman, A. Greenberg, D. Schneider, B. Noone, L. Fanella, B. R. Williamson, E. Dinda, M. Mayer, M. Backer, A. Agasan, L. Kornstein, F. Stavinsky, B. Neal, D. Edwards, M. Haroon, D. Hurley, L. Colbert, J. Miller, B. Mojica, E. Carloni, B. Devine, M. Cambridge, T. Root, D. Schoonmaker, M. Shayegani, W. Hastback, B. Wallace, S. Kondracki, P. Smith, S. Matiuck, K. Pilot, M. Acharya, G. Wolf, W. Manley, C. Genese, J. Brooks, Z. Dembek, and J. Hadler.** 1999. Outbreak of *Vibrio parahaemolyticus* infection associated with eating raw oysters and clams harvested from Long Island Sound—Connecticut, New Jersey, and New York, 1998. *Morb. Mortal. Wkly. Rep.* **48:**48–51.

143. **Weese, J. S., J. Rousseau, and L. Arroyo.** 2005. Bacteriological evaluation of commercial canine and feline raw diets. *Can. Vet. J.* **46:**513–516.

144. **Weiner, P. D., and J. J. Kermans.** 1976. United States Patent 3,961,090. Method of preparing rare roast beef. United States Patent Office. [Online.] http://xrint.com/patents/us/3961090. Accessed 15 March 2006.

145. **Wells, J. G., G. McConnell Clark, and G. K. Morris.** 1974. Evaluation of methods for isolating *Salmonella* and *Arizona* organisms from pet turtles. *Appl. Microbiol.* **27:**8–10.

146. **Werner, S. B., F. R. Morrison, G. L. Humphrey, R. A. Murray, and J. Chin.** 1984. *Salmonella dublin* and raw milk consumption—California. *Morb. Mortal. Wkly. Rep.* **33:**196–198.

147. **Winters, A., C. Driver, M. Macaraig, C. Clark, S. S. Munsiff, C. Pichardo, J. Driscoll, M. Salfinger, B. Kreiswirth, J. Jereb, P. LoBue, and M. Lynch.** 2005. Human tuberculosis caused by *Mycobacterium bovis*—New York City, 2001–2004. *Morb. Mortal. Wkly. Rep.* **54:**605–608.

148. **Woodward, D. L., R. Khakhria, and W. M. Johnson.** 1997. Human salmonellosis associated with exotic pets. *J. Clin. Microbiol.* **35:**2786–2790.

149. **Wright, J. G., L. A. Tengelsen, K. E. Smith, J. B. Bender, R. K. Frank, J. H. Grendon, D. H. Rice, A. M. B. Thiessen, C. J. Gilbertson, S. Sivapalasingam, T. J. Barrett, T. E. Besser, D. D. Hancock, and F. J. Angulo.** 2005. Multidrug-resistant *Salmonella* Typhimurium in four animal facilities. *Emerg. Infect. Dis.* **11:**1235–1241.

150. **Yu, J.-R., M. Seo, Y.-W. Kim, M.-H. Oh, and W.-M. Sohn.** 2001. A human case of gastric infection by *Pseudoterranova decipiens* larva. *Korean J. Parasitol.* **39:**193–196.

151. **Zagrebneviene, G., V. Jasulaitiene, B. Morkunas, S. Tarbunas, and J. Ladygaite.** 2005. *Shigella sonnei* outbreak due to consumption of unpasteurised milk curds in Vilnius, Lithuania, 2004. *Eur. Surveill. Wkly.* **10.** [Online.] http://www.eurosurveillance.org/ew/2005/051201.asp#3. Accessed 6 July 2006.

152. **Zansky, S., B. Wallace, D. Schoonmaker-Bopp, P. Smith, F. Ramsey, J. Painter, A. Gupta, P. Kalluri, and S. Noviello.** 2002. Outbreak of multidrug-resistant *Salmonella* Newport—United States, January–April 2002. *Morb. Mortal. Wkly. Rep.* **51:**545–548.

chapter 16

The Media and the Message

On March 27, 1999, readers of the *Afro-American Red Star* learned that BilMar Foods was recalling some of its hot dogs and deli meats due to the possibility that these products might have been contaminated with *Listeria* (5). Not mentioned in the article was that this recall had been announced on December 22, 1998, and had been publicized as early as December 23 or 24 in several widely circulated newspapers (2, 36, 88, 89). Consumers in cities such as Houston, Tex.; Orlando, Fla.; Pittsburgh, Pa.; and Cleveland, Ohio, learned about the recall for the first time in their local newspapers on January 1, 1999 (3, 4, 85, 86). Arguably, the media could be excused for its tardiness in reporting the recall notice, since the U.S. Department of Agriculture (USDA) only issued its own advisory on January 28, 1999—more than 1 month after BilMar first announced the recall (36).

Media performance during other recalls has been similarly uneven. For example, the *Denver Post* and the *Rocky Mountain News* (another Denver-based daily) published details of the ConAgra recall of 354,000 pounds of ground beef on July 1, 2002, the day after the company's announcement. Both newspapers continued to follow the story for several months (37, 47, 50, 61, 75, 76). In contrast, the *Chicago Tribune* ignored the story until July 19, when the recall was expanded to encompass more than 19 million pounds of ground beef (38, 52). The *New York Times*, *Los Angeles Times*, and *Seattle Times* didn't carry the recall notice until July 20 (16, 48, 59).

Fortunately, consumers no longer have to rely on newspaper reports in order to stay up-to-date on food recalls. Internet websites maintained by government agencies, consumer advocacy groups, and academic institutions routinely provide consumers with current recall information. In the United States, the Food and Drug Administration (FDA) and the USDA post detailed information on national and regional recalls; agencies in other countries, including Canada, Australia, and the United Kingdom, do likewise (19, 32, 43, 46, 80). Certain nongovernment organizations are also good sources of information for food recalls. FoodHACCP.com, for example, monitors recall announcements from several countries around the world and posts daily updates on its website (33). Most sites offer their readers automatic notification of new recalls, either by e-mail or by RSS ("Really Simple Syndication") feed or both.

Reporting on food recalls is only one facet of the media's impact on food safety. Consumers rely on news media for information on how to handle foods safely, suggestions for prevention of food-borne disease, and clear, accurate explanations of the science underlying these issues. Unfortunately, information contained in news stories is not always clear or accurate.

Headline Hysteria

"Stronger bacteria threaten nation's food, study warns," was the headline of a story that appeared in the *Houston Chronicle* in February 2002 (6). The body of the article, however, told a slightly different—and more accurate—story. The article summarized a report issued by the Institute of Food Technologists (IFT) on food-borne pathogens. According to the newspaper article, the IFT report warned about the risks of overusing antibiotics in animal feeds and discussed the possibility that using manure as a fertilizer could increase the risk of spreading pathogens. The article included quotations from two contributors to the IFT report. Nothing in the body of the *Houston Chronicle*'s summary of the IFT report supported the wording of the story's headline. *The Daily News*, reporting on a 2005 outbreak of *Escherichia coli* O157:H7 linked to raw milk, headlined one of its stories "*E. coli* deadlier than strain from years past." The body of the article implied that continued mutation of the pathogen might have been responsible for a recent increase in the incidence of *E. coli* O157:H7 disease in Clark County, Wash., in recent years, although—as the author mentioned in passing—there was no "hard evidence" of this (62).

The phenomenon of headline inflation is not limited to the U.S. media. In 2006, more than 160 people in a northern province of Thailand fell victim to an outbreak of botulism after eating locally prepared fermented bamboo shoots (13). Despite describing the source of the outbreak as "contaminated food" in the body of the article, the headline of one news story exclaimed, "U.S. expert investigating food poisoning caused by strain used in biological weapons" (12).

Happily, examples such as these are the exception. Most headlines are supported by the information contained in the accompanying articles. But journalists are not scientists. They rely on their sources for accurate information, and sometimes that information is incomplete, inaccurate, or misunderstood.

Spreading the Word

The USDA, like many government agencies around the world, issues regular reminders to the public on how to cook, handle, and store foods safely. The onset of summer barbecue season is heralded by information on cooking hamburgers, and the approach of Thanksgiving triggers reminders on how to thaw and prepare turkey and stuffing safely. Although the media are not obliged to carry these information bulletins and warnings, many news outlets do so, though sometimes in an abbreviated form.

The USDA website contains information for consumers, including information on how to use a meat thermometer to determine that food has been cooked thoroughly (41). The website states clearly that, in order to prevent cross-contamination, a meat thermometer must be washed with soap and water before and after each use. But no matter how often the USDA repeats its message (it has been doing so in news releases since at least 1998), one element—the instruction to wash the thermometer—frequently gets lost by the time the information reaches its target audience (18, 35, 53, 90). And this missing instruction can be crucial. If a thermometer isn't washed after every use, it will become a conduit for cross-contamination.

Other important information can also get lost in the translation of scientific terminology into everyday language. Occasionally, journalists use the words "virus," "bacteria," and "germ" interchangeably, referring, for example, to *E. coli* as a virus instead of a bacterium (68, 77). From time to time, they also confuse infections (in which the growth and reproduction of microbes in the victim is the trigger for symptoms) with intoxications (in which symptoms are triggered by one or more toxins produced by the pathogen) (58). Nevertheless, most of the time, the print media do a reasonable job of reporting food safety news and information.

USDA's Recommendations for Safe Cooking of Meat (41)

1. The thermometer should be inserted into the thickest part of the meat and should not touch any bone, gristle, or fat.
2. Food should be cooked until it reaches the following internal temperatures:
   (i) ground beef, pork, and egg dishes: 160°F (71°C)
   (ii) fish, steaks, and roasts: 145°F (63°C)
   (iii) chicken and poultry: 165°F (74°C)
3. The thermometer should be washed with soap and water before and after every use.

A Tangled Web

Although conventional print and broadcast media try to do a conscientious job, their impact on consumers has decreased significantly as the influence of the Internet has grown. The World Wide Web has provided a worldwide audience to anyone who has a computer and access to the Internet. This development has allowed government agencies, nongovernmental organizations, trade associations, and corporations to speak directly to consumers. It has also opened the door to the free exchange of food safety information—some of it inaccurate—between individuals.

Do-it-yourself encyclopedia

The most ambitious information-sharing project is Wikipedia, an online encyclopedia written and maintained by its users. Wikipedia contains more than 1,000,000 articles, written in over 200 different languages, though most of the entries (>920,000) are in English (10). The encyclopedia relies on voluntary contributions of articles, often written by anonymous authors. Some articles, such as the one on *Salmonella*, contain a significant amount of information; others, such as the *Listeria* article, are just a few sentences long (9, 11). Few of the articles provide the reader with a list of peer-reviewed references to corroborate the information being offered.

Wikipedia's readers are encouraged to edit articles by adding information or correcting perceived errors. This allows for the possible introduction of errors into an article that was accurate when first posted, obliging the original contributor to continue monitoring the article in order to correct any inadvertent errors that might have been inserted by a well-meaning reader. Given the multiple routes by which incorrect information can slip into Wikipedia articles, the results of a study published in *Nature* came as a pleasant surprise to prospective Wikipedia users. A 2005 survey of Wikipedia science entries found an average of four inaccuracies per Wikipedia article versus three per article in the Encyclopædia Britannica (49). The publisher of the Encyclopædia Britannica quickly challenged the results of the *Nature* study, stating that "[a]lmost everything about the journal's investigation, from the criteria for identifying inaccuracies to the discrepancy between the article text and its headline, was wrong and misleading." Britannica's 20-page rebuttal of the *Nature* study challenged the validity of the study design, the rigor with which the study was carried out, the interpretation of the data, and the appropriateness of the conclusions. Britannica also pointed out that, even based on *Nature*'s own "flawed" study, Wikipedia articles contained one-third more errors than the corresponding Encyclopædia Britannica articles with which they had been compared (27).

Going to the source

There are many ways to access timely, accurate information about food safety on the Internet. Government agencies from Australia and Argentina to Switzerland and Singapore maintain websites on which they post health and safety information for their citizens (7, 8, 80, 87). In addition, several of these government sites—notably those maintained by the FDA, the USDA, and the U.S. Centers for Disease Control and Prevention, among others—provide consumers with background information on a variety of food safety issues (20, 22, 42).

Consumer advocacy groups also maintain websites that help educate the public about food safety and good food preparation practices. While some people may disagree with one or more of the policy positions taken by these organizations, websites maintained by groups such as Safe Tables Our Priority, Consumers Union, and the Center for Science in the Public Interest are a useful supplement to government-run websites (21, 23, 83). Websites sponsored by universities and professional organizations also are good sources of reliable information (1, 54, 56, 57, 74, 79). The IFT, for example, makes available to the news media about 70 experts in various aspects of food science and food safety (55). PubMed, an online service offered by the U.S. National Library of Medicine, allows anyone with an Internet connection to search the scientific literature for articles relating to food safety that date back 30 years or more, whereas information services such as ProQuest and InfoTrac (available through public library systems) do the same for newspaper articles from 1980 on (78, 81, 91). And search engines such as Google make it easy for consumers to find a full spectrum of food safety information—accurate or not—on the Internet (51).

Blogs, etc.

A great deal of what passes for food safety information on the Internet derives from sources other than universities, government and nongovernment organizations, or professional associations. Law firms, manufacturers or distributors of "food safety products," and private individuals all add their contributions to the pot.

One of the most visible contributors in this category is the law firm of Marler Clark (67). This firm of attorneys, which first made its mark in the aftermath of the 1992–1993 *E. coli* O157:H7 outbreak in the United States, specializes in personal injury litigation and has been the lead counsel in some of the most publicized lawsuits relating to food-borne illness outbreaks in the United States (17). Marler Clark boasts of having "... litigated cases involving defective or unsafe food products in over 35 states" (65). The law firm sponsors several websites, providing basic information on the most significant food-borne pathogens, news archives relating to food-borne illness outbreaks, and a separate "blog" site for each significant food-borne pathogen (66). The scientific summaries contained on the various Marler Clark websites (the *Campylobacter* site, for example) appear to be accurate and well researched, and cite a variety of government sources and peer-reviewed scientific articles in their support (64).

Unfortunately, some of the "information" that is posted on some other blog sites is simply wrong. For example, a blogger who uses the alias "loubidy" informed her readers that mayonnaise is the most common single cause of *Salmonella* food poisoning. She based her statement on the fact that mayonnaise contains raw eggs and "... most eggs contain traces of salmonella ..." (63). Apparently, this blogger was blissfully unaware that the foods associated most frequently with *Salmonella* outbreaks are raw meats and poultry (followed by eggs, milk and dairy products, fish and seafood), that commercial mayonnaise is prepared using pasteurized eggs, and that—contrary to her categorical statement that most eggs are contaminated with *Salmonella*—the *Salmonella* contamination rate of intact shell eggs in the United States is just 1:20,000 (15, 26, 30, 31).

Unlike bloggers like "loubidy" who post false information through ignorance, other sites are more likely to mislead consumers by offering one-sided, incomplete, inaccurate, or misleading information. Often, these sites are sponsored or maintained by proponents of "natural health" diets, raw food diets, or other out-of-mainstream nutritional philosophies. Joseph Mercola, D.O., is one such practitioner (73). Mercola advocates, among other things, drinking raw milk and eating raw eggs. He states that consuming a dose of probiotic cultures "... every 30 minutes until you start to feel better ..." can cure *Salmonella* infections and that most people improve within just a few hours (72). Some of his statistics are inaccurate or outdated. The Mercola website states, without any supporting documentation, that 20% of the chicken sold in the United States is contaminated with *Campylobacter*, whereas recent studies in the United States, the United Kingdom, and Barbados have all put that level at >70% (24, 70, 71, 93). A contributing author to Mercola's website also tells readers that *Salmonella* contaminates 20% of ground meats—an incomplete and potentially misleading statement, depending on a reader's interpretation of "ground meat" (25). According to the published results of a U.S. retail study on which this information was probably based, the *Salmonella* contamination rates were 35% for ground chicken, 24% for ground turkey, 16% for ground pork, and just 6% for ground beef (92).

Playing the Percentages

In February 2005, USDA announced triumphantly that the percentage of raw ground beef samples found to be contaminated by *E. coli* O157:H7 had declined by 43.3% in 2004 when compared to the previous year, and by more than 80% since 2000 (40). The agency neglected to mention that the percentage of positive samples had increased between 1998 and 2000 (44). And USDA played down the fact that the sampling program was not designed to allow statistically valid year-to-year comparisons of data (60).

Unless samples are drawn from the same sources, in the same proportions, and analyzed using the same methods, year-over-year comparisons are meaningless. The samples included in USDA's annual reports of *E. coli* O157:H7 incidence were obtained from different locations (retail stores, federally inspected production facilities, and state-inspected facilities) and

Not *Just* the Facts

Peer-reviewed scientific journal articles are a vital source of information, and often are the foundation for news stories and for web articles such as those hosted by Wikipedia. But even these peer-reviewed articles sometimes present information in a way that can mislead the reader.

Some people believe that free-range chickens—raised in the open rather than confined in conventional commercial poultry houses—are less likely to be contaminated with *Salmonella* than their caged counterparts (94). When USDA researchers decided to test this theory by analyzing 135 free-range chicken carcasses from four producers, they reported that 31% of those carcasses were *Salmonella* positive, compared to just 9 to 13% of broiler carcasses from conventionally reared chickens (14, 45). But these statistics didn't tell the whole story.

The sampling method used for the free-range chicken study was different from the method used to obtain data for conventionally reared birds. The USDA has used the same sampling method for several years when testing broiler chickens in its routine screening program. The carcasses are always obtained immediately after they pass through a chilling bath. Each carcass is placed in a bag with 400 ml of a sterile rinse and rocked back and forth to rinse the entire surface of the bird. Then, a 30-ml portion of the rinse liquid (just 7.5% of the entire volume) is used in the *Salmonella* test (28, 39). In contrast, the analysis of free-range chicken carcasses was carried out on the entire volume of rinse liquid from each carcass—an important difference that was ignored in the conclusions contained in the research report (14).

In the USDA's initial baseline study of chicken carcass contamination, the number of *Salmonella* bacteria on a contaminated carcass ranged from fewer than 12 to more than 100,000 (from <0.03 to 280 live *Salmonella* bacteria per milliliter of carcass rinse liquid); 42% of the *Salmonella*-positive carcasses were contaminated with fewer than 12 *Salmonella* bacteria (34).

It's more difficult to detect *Salmonella* on a chicken carcass when only a few cells are present than when a carcass is heavily contaminated. And the chances of finding that contamination are lessened further when only a portion of the carcass rinse liquid is subjected to analysis. Logic tells us, and research reports confirm, that sampling a *Salmonella*-contaminated carcass with 400 ml of rinse liquid and analyzing the entire 400 ml for *Salmonella* is more likely to produce a positive result than using 400 ml of rinse liquid to sample the carcass and then only testing 30 ml for *Salmonella* (29, 84). By omitting to mention this important difference in testing methods, the researchers who reported on contamination levels in free-range poultry misled their readers into believing that free-range poultry was more highly contaminated with *Salmonella* than cage-reared poultry was.

Apples and Oranges

The free-range chicken study is not the only example of research data that are compared without reference to a consistent method. When the USDA carried out its 1995–1996 baseline study of bacterial contamination levels in poultry, inspectors selected intact broiler chicken carcasses, bagged them, placed them into insulated shipping containers, and sent them to the lab for analysis. Once at the lab, the carcasses were rinsed, and a portion of the rinse liquid was analyzed for *Salmonella* (34). Approximately 20% of those carcasses were *Salmonella* positive (34). In 1998, the very first year that the Hazard Analysis and Critical Control Points (HACCP) program was in force (for large establishments only), the

(Continued next page)

Not *Just* the Facts *(continued)*

incidence of *Salmonella*-positive broiler carcasses dropped to 10.8%, then started to climb back up as the HACCP program coverage extended to medium and small establishments, reaching 16.3% in 2005 (45).

The USDA changed its sample-handling procedures between the time it carried out the baseline study and the start of its HACCP sampling program. According to its revised sampling protocol, inspectors now rinse the carcasses at the production facility and send just the chilled rinse liquid—not the actual broiler carcass—to the lab for analysis (28). Since some bacteria might not survive an overnight stay in the chilled rinse liquid, this apparently slight difference in procedure could lead to a large difference in the ability of a lab to detect *Salmonella* (especially from a carcass that is contaminated with only a few viable cells of *Salmonella*), making it difficult to compare directly USDA's baseline data with the results obtained from the agency's ongoing HACCP sampling program (29, 69). In addition, USDA's published reports do not provide details of the analytical methods used to detect *Salmonella*, referring readers instead to the version of the agency's *Microbiology Laboratory Guidebook* that was in force at the time the study was carried out (34, 82). Unfortunately, only the most current version of the guidebook is available on the USDA website (39). Without access to archived copies of older versions of the guidebook, a reader cannot determine the details of the analytical method used to generate the data reported in some of USDA's published *Salmonella* incidence studies.

were analyzed by different methods in the lab (44). The source and nature of the data collected by USDA during the course of its enforcement activities change, inevitably, from year to year. In recent years, USDA has altered the focus of its ground meat sampling program, eliminating the sampling of meat from retail stores. The analytical methods for detecting *E. coli* O157:H7 have also changed since the program began: the sample size increased from 25 g (0.9 oz) to 325 g (11.5 oz) in 1997; a new, more sensitive detection method was introduced in 1999; and USDA labs began using a different screening method in 2005 (44). In short, USDA's year-over-year data comparison was inappropriate, even misleading. The incidence of *E. coli* O157:H7 in ground meat might well have decreased, but it was intellectually dishonest of the agency to use its enforcement data to support that conclusion.

The Buck Stops Everywhere

Journalists, if they are to succeed in their profession, must present news that grabs and holds the attention of their readers or their audiences. But no one—not journalists, not scientists, and not even web bloggers—is exempt from the responsibility of ensuring, to the extent possible, that the information he or she disseminates is accurate, clear, and not misleading. Scientists have an ethical obligation to present their research methods, data, and conclusions in an impartial and unambiguous fashion. Yet, in some ways, journalists have an even more difficult and important task. They must translate scientific and technical information, some of it highly complex, into everyday language and present the information to their readers in an accurate, readable, and entertaining way. It's to their credit that they succeed as often as they do.

References

1. **American Society for Microbiology.** 2006. ASM alerts. American Society for Microbiology, Washington, D.C. [Online.] http://www.asm.org/. Accessed 14 July 2006.
2. **Anonymous.** 23 December 1998. Food-poisoning outbreak prompts recall of meats. *Los Angeles Times*, Los Angeles, Calif.
3. **Anonymous.** 1 January 1999. Bacteria outbreak traced to food plant in Michigan. *The Plain Dealer*, Cleveland, Ohio.
4. **Anonymous.** 1 January 1999. Source of deadly outbreak confirmed. The bacterium that sickened people in 11 states was found in packages of hot dogs at a Sara Lee subsidiary. *Orlando Sentinel*, Orlando, Fla.
5. **Anonymous.** 27 March 1999. Hot dogs, meats recalled. *Afro-American Red Star*, Washington, D.C.
6. **Anonymous.** 20 February 2002. Stronger bacteria threaten nation's food, study warns. *Houston Chronicle*, Houston, Tex.
7. **Anonymous.** 2006. Health in Switzerland. Swiss Federal Office of Public Health, Consumer Protection Directorate, Bern, Switzerland. [Online.] http://www.bag.admin.ch/index.html?lang=en. Accessed 14 July 2006.
8. **Anonymous.** 2005. Your gateway to health and environment resources. Health and Environment, Singapore Government, Singapore. [Online.] http://he.ecitizen.gov.sg/. Accessed 14 July 2006.
9. **Anonymous.** 2006. *Listeria*. Wikimedia Foundation, Inc. [Online.] http://en.wikipedia.org/wiki/Listeria. Accessed 14 July 2006.
10. **Anonymous.** 2006. List of Wikipedias. Wikimedia Foundation, Inc. [Online.] http://meta.wikimedia.org/wiki/List_of_Wikipedias. Accessed 14 July 2006.
11. **Anonymous.** 2006. *Salmonella*. Wikimedia Foundation, Inc. [Online.] http://en.wikipedia.org/wiki/Salmonella. Accessed 14 July 2006.
12. **Anonymous.** 21 March 2006. U.S. expert investigating food poisoning caused by strain used in biological weapons. *OhmyNews*, Korea. [Online.] http://english.ohmynews.com/articleview/article_view.asp?at_code=318042. Accessed 14 July 2006.
13. **Anonymous.** 22 March 2006. Outbreak raises food-safety alert. A recent spate of botulism cases shows that government measures are failing to protect consumers. *The Nation*, Bangkok, Thailand. [Online.] http://www.nationmultimedia.com/2006/03/22/opinion/opinion_20003283.php. Accessed 14 July 2006.
14. **Bailey, J. S., and D. E. Cosby.** 2005. *Salmonella* prevalence in free-range and certified organic chickens. *J. Food Prot.* **68:**2451–2453.
15. **Bean, N. H., and P. M. Griffin.** 1990. Foodborne disease outbreaks in the United States, 1973–1987: pathogens, vehicles, and trends. *J. Food Prot.* **53:**804–817.

16. **Becker, E.** 20 July 2002. 19 million pounds of meat recalled after 19 fall ill. *New York Times*, New York, N.Y.

17. **Brandt, J. R., L. S. Fouser, S. L. Watkins, I. Zelikovic, P. I. Tarr, V. Nazar-Stewart, and E. D. Avner.** 1994. *Escherichia coli* O157:H7-associated hemolytic-uremic syndrome after ingestion of contaminated hamburgers. *J. Pediatr.* **125:**519–526.

18. **Brasher, P.** 4 July 2000. Meatpackers know summer is *E. coli* season. *Seattle Times*, Seattle, Wash.

19. **Canadian Food Inspection Agency.** 2006. Food recalls and allergy alerts. Canadian Food Inspection Agency. [Online.] http://www.inspection.gc.ca/english/corpaffr/recarapp/recaltoce.shtml. Accessed 14 July 2006.

20. **Center for Food Safety and Applied Nutrition.** 2006. What's new. U.S. Food and Drug Administration, Center for Food Safety and Applied Nutrition. [Online.] http://www.cfsan.fda.gov/~news/whatsnew.html. Accessed 14 July 2006.

21. **Center for Science in the Public Interest.** 2005. Food safety. Center for Science in the Public Interest, Washington, D.C. [Online.] http://www.cspinet.org/foodsafety/. Accessed 14 July 2006.

22. **Centers for Disease Control and Prevention.** 2006. Health and safety topics. Department of Health and Human Services, Centers for Disease Control and Prevention. [Online.] http://www.cdc.gov/node.do/id/0900f3ec800115e1. Accessed 14 July 2006.

23. **Consumers Union.** 2005. ConsumersUnion.org. Food. Consumers Union. [Online.] http://www.consumersunion.org/food.html. Accessed 14 July 2006.

24. **Cui, S., B. Ge, J. Zheng, and J. Meng.** 2005. Prevalence and antimicrobial resistance of *Campylobacter* spp. and *Salmonella* serovars in organic chickens from Maryland retail stores. *Appl. Environ. Microbiol.* **71:**4108–4111.

25. **Dean, C.** 2004. The bugs are winning the war. [Online.] http://www.mercola.com/2004/nov/10/bacteria_war.htm. Accessed 14 July 2006.

26. **Ebel, E., and W. Schlosser.** 2000. Estimating the annual fraction of eggs contaminated with *Salmonella enteritidis* in the United States. *Int. J. Food Microbiol.* **61:**51–62.

27. **Encyclopædia Britannica, Inc.** 2006. Fatally flawed. Refuting the recent study on encyclopedic accuracy by the journal *Nature*. Encyclopædia Britannica, Inc. [Online.] http://corporate.britannica.com/britannica_nature_response.pdf. Accessed 14 July 2006.

28. **Federal Register.** 1996. Pathogen reduction; hazard analysis and critical control point (HACCP) systems; Final Rule. *Fed. Regist.* **61:**38805–38956.

29. **Fletcher, D. L.** 2006. Influence of sampling methodology on reported incidence of *Salmonella* in poultry. *J. AOAC Int.* **89:**512–516.

30. **Food and Drug Administration.** 2005. Salad dressings and sauces (commercial). *Bad Bug Book. Foodborne Pathogenic Microorganisms and Natural Toxins Handbook.* U.S. Food and Drug Administration, Center for Food Safety and Applied Nutrition. [Online.] http://www.cfsan.fda.gov/~mow/chap1sal.html. Accessed 14 July 2006.

31. **Food and Drug Administration.** 2005. *Salmonella* spp. *Bad Bug Book. Foodborne Pathogenic Microorganisms and Natural Toxins Handbook.* U.S. Food and Drug

Administration, Center for Food Safety and Applied Nutrition. [Online.] http://www.cfsan.fda.gov/~mow/chap1.html. Accessed 14 July 2006.

32. **Food and Drug Administration.** 2006. Recalls, market withdrawals and safety alerts. U.S. Food and Drug Administration. [Online.] http://www.fda.gov/opacom/7alerts.html. Accessed 14 July 2006.

33. **FoodHACCP.com.** 2006. Food daily recall information. FoodHACCP.com. [Online.] http://www.foodhaccp.com/Recalls.htm. Accessed 14 July 2006.

34. **Food Safety and Inspection Service.** 1996. National broiler chicken microbiological baseline data collection program, July 1994–June 1995. Food Safety and Inspection Service, U.S. Department of Agriculture. [Online.] http://www.fsis.usda.gov/Science/Baseline_Data/index.asp. Accessed 14 July 2006.

35. **Food Safety and Inspection Service.** 11 August 1998. USDA urges consumers to use food thermometer when cooking ground beef patties. Food Safety and Inspection Service, U.S. Department of Agriculture. [Online.] http://www.fsis.usda.gov/oa/news/1998/colorpr.htm. Accessed 14 July 2006.

36. **Food Safety and Inspection Service.** 28 January 1999. FSIS-99-RC-06. BilMar *Listeria* recall—additional brands sold at retail. Food Safety and Inspection Service, U.S. Department of Agriculture. [Online.] http://www.fsis.usda.gov/OA/recalls/prelease/pr044-98a.htm. Accessed 14 July 2006.

37. **Food Safety and Inspection Service.** 30 June 2002. FSIS-RC-055-2002. Colorado firm recalls ground beef products for possible *E. coli* O157:H7. Food Safety and Inspection Service, U.S. Department of Agriculture. [Online.] http://www.fsis.usda.gov/OA/recalls/prelease/pr055-2002a.htm. Accessed 14 July 2006.

38. **Food Safety and Inspection Service.** 19 July 2002. FSIS-RC-055-2002. Colorado firm recalls beef trim and ground beef products for possible *E. coli* O157:H7. Food Safety and Inspection Service, U.S. Department of Agriculture. [Online.] http://www.fsis.usda.gov/OA/recalls/prelease/pr055-2002.htm. Accessed 14 July 2006.

39. **Food Safety and Inspection Service.** 2004. Laboratory guidebook. Notice of change. MLG 4.03. Isolation and identification of *Salmonella* from meat, poultry, and egg products. Food Safety and Inspection Service, U.S. Department of Agriculture. [Online.] http://www.fsis.usda.gov/PDF/MLG_4_03.pdf. Accessed 14 July 2006.

40. **Food Safety and Inspection Service.** 28 February 2005. FSIS ground beef sampling shows substantial *E. coli* O157:H7 decline in 2004. Food Safety and Inspection Service, U.S. Department of Agriculture. [Online.] http://www.fsis.usda.gov/News_&_Events/NR_022805_01/index.asp. Accessed 13 July 2006.

41. **Food Safety and Inspection Service.** 16 May 2006. Is it done yet? Thermometer placement & temperatures. Food Safety and Inspection Service, U.S. Department of Agriculture. [Online.] http://www.fsis.usda.gov/Is_It_Done_Yet/Thermometer_Placement_and_Temps/index.asp. Accessed 14 July 2006.

42. **Food Safety and Inspection Service.** 2006. Food Safety and Inspection Service (FSIS). Protecting public health through food safety and defense. Food Safety and Inspection Service, U.S. Department of Agriculture. [Online.] http://www.fsis.usda.gov/. Accessed 14 July 2006.

43. **Food Safety and Inspection Service.** 2006. FSIS recalls. Open federal cases. Food Safety and Inspection Service, U.S. Department of Agriculture. [Online.] http://www.fsis.usda.gov/Fsis_Recalls/Open_Federal_Cases/index.asp. Accessed 14 July 2006.

44. **Food Safety and Inspection Service.** 2006. Microbiological results of raw ground beef products analyzed for *Escherichia coli* O157:H7, summarized by calendar year. Food Safety and Inspection Service, U.S. Department of Agriculture. [Online.] http://www.fsis.usda.gov/Science/Ecoli_O157_Summary_Tables/index.asp. Accessed 13 July 2006.

45. **Food Safety and Inspection Service.** 2006. Progress report on *Salmonella* testing of raw meat and poultry products, 1998–2004. Food Safety and Inspection Service, U.S. Department of Agriculture. [Online.] http://www.fsis.usda.gov/Science/Progress_Report_Salmonella_Testing_Tables/index.asp. Accessed 14 July 2006.

46. **Food Standards Agency.** 2006. Food alerts. Food Standards Agency, United Kingdom. [Online.] http://www.food.gov.uk/enforcement/alerts/. Accessed 14 July 2006.

47. **Frazier, D.** 19 May 2003. Meat inspectors' clout questioned. 2002 flap at ConAgra plant helped spark two federal inquiries. *Rocky Mountain News*, Denver, Colo.

48. **Garvey, M.** 20 July 2002. Illnesses prompt federal recall of ConAgra beef. Food: sixteen confirmed cases of *E. coli* infection are linked to a plant in Colorado. The 19 million pounds of meat was shipped to 21 states. *Los Angeles Times*, Los Angeles, Calif.

49. **Giles, J.** 14 December 2005. Internet encyclopaedias go head to head. Jimmy Wales' Wikipedia comes close to Britannica in terms of the accuracy of its science entries, a *Nature* investigation finds. News@Nature.com. Nature Publishing Group. [Online.] http://www.nature.com/news/2005/051212/full/438900a.html. Accessed 14 July 2006.

50. **Good, O. S.** 1 July 2002. Greeley plant recalls *E. coli*-tainted meat. Ground beef produced on May 31 went to 13 states, says ConAgra. *Rocky Mountain News*, Denver, Colo.

51. **Google.** 2006. Google advanced search. Google, Inc. [Online.] http://www.google.com/advanced_search?hl=en. Accessed 14 July 2006.

52. **Graham, J.** 19 July 2002. *E. coli* fears spur recall of 19 million pounds of beef. *Chicago Tribune*, Chicago, Ill.

53. **Hopkins, K.** 24 May 2003. Cooked meat, clean hands stop *E. coli*; *E. coli* prevention. *Omaha World-Herald*, Omaha, Nebr.

54. **Institute of Food Technologists.** 2005. About IFT. Institute of Food Technologists, Chicago, Ill. [Online.] http://www.ift.org/cms/?pid=1000023. Accessed 14 July 2006.

55. **Institute of Food Technologists.** 2005. News media. Institute of Food Technologists, Chicago, Ill. [Online.] http://www.ift.org/cms/?pid=1000152. Accessed 13 July 2006.

56. **International Association for Food Protection.** 2006. Advancing food safety worldwide. International Association for Food Protection, Des Moines, Iowa. [Online.] http://foodprotection.org/main/default.asp. Accessed 14 July 2006.

57. **Iowa State University.** 2006. Food safety news. Food Safety Project, Iowa State University, University Extension, Ames, Iowa. [Online.] http://www.extension.iastate.edu/foodsafety/news/. Accessed 14 July 2006.

58. **Jones, T.** 2001. Guess who's coming to dinner? Unwanted guests could prove hazardous to your health. *Army Reserve Magazine.* [Online.] http://www.findarticles.com/p/articles/mi_m0KAB/is_2_47/ai_79251100. Accessed 14 July 2006.

59. **Karmon, E.** 20 July 2002. Washington state *E. coli* cases may be related to huge beef recall. *Seattle Times*, Seattle, Wash.

60. **Kawar, M.** 1 March 2005. USDA tests indicate drop in *E. coli* bacteria. *Omaha World-Herald*, Omaha, Nebr.

61. **Kirksey, J.** 1 July 2002. ConAgra recalls ground beef; meat packaged May 31 may be contaminated. *Denver Post*, Denver, Colo.

62. **LaBoe, B.** 3 January 2006. *E. coli* deadlier than strain from years past. *The Daily News*, Longview, Wash.

63. **Loubidy.** 17 March 2006. well in referance to the question i set for you people earlier... *Live Journal.* [Online.] http://loubidy.livejournal.com/91874.html. Accessed 14 July 2006.

64. **Marler Clark.** 2006. About *Campylobacter*. Your information source for *Campylobacter*. Outbreak, Inc. [Online.] http://www.about-campylobacter.com/. Accessed 14 July 2006.

65. **Marler Clark.** 2006. Food poisoning litigation. Marler Clark Attorneys at Law, L.L.P., P.S., Seattle, Wash. [Online.] http://www.marlerclark.com/foodlitigation.htm. Accessed 14 July 2006.

66. **Marler Clark.** 2006. Sponsored sites. Marler Clark Attorneys at Law, L.L.P., P.S., Seattle, Wash. [Online.] http://www.marlerclark.com/sponsored-sites.htm. Accessed 14 July 2006.

67. **Marler Clark.** 2006. Welcome. Marler Clark Attorneys at Law, L.L.P., P.S., Seattle, Wash. [Online.] http://www.marlerclark.com/. Accessed 14 July 2006.

68. **McDonald, T.** 31 March 2002. Calls for action as *E. coli* on rise again. *Sunday Herald*, Glasgow, Scotland, United Kingdom.

69. **McFeters, G. A., S. C. Cameron, and M. W. Lechevallier.** 1982. Influence of diluents, media, and membrane filters on detection of injured waterborne coliform bacteria. *Appl. Environ. Microbiol.* **43:**97–103.

70. **Meldrum, R. J., D. Tucker, and C. Edwards.** 2004. Baseline rates of *Campylobacter* and *Salmonella* in raw chicken in Wales, United Kingdom, in 2002. *J. Food Prot.* **67:**1226–1228.

71. **Meldrum, R. J., D. Tucker, R. M. M. Smith, and C. Edwards.** 2005. Survey of *Salmonella* and *Campylobacter* contamination of whole, raw poultry on retail sale in Wales in 2003. *J. Food Prot.* **68:**1447–1449.

72. **Mercola, J.** 2002. Raw eggs for your health—major update. [Online.] http://www.mercola.com/2002/nov/13/eggs.htm#. Accessed 14 July 2006.

73. **Mercola, J.** 2006. Begin your journey to independent health. [Online.] http://www.mercola.com/index.htm. Accessed 14 July 2006.

74. **Michigan State University.** 2006. Leading the global fight against foodborne illness. National Food Safety & Toxicology Center, Michigan State

University, East Lansing, Mich. [Online.] http://foodsafe.msu.edu/. Accessed 14 July 2006.

75. **Migoya, D.** 6 September 2002. ConAgra delayed beef recall for 2 days; firm waited for USDA to recheck tests. *Denver Post*, Denver, Colo.

76. **Migoya, D.** 21 November 2002. Greeley, Colo., slaughterhouse reopens. *The Denver Post*, Denver, Colo.

77. **Mills, J.** 5 October 2005. Boy, five, was 'killed by a school dinner'; meat supplier suspected as *E. coli* outbreak claims its first life. *Daily Mail*, London, England, United Kingdom.

78. **National Library of Medicine.** 2006. PubMed. A service of the National Library of Medicine and the National Institutes of Health. National Library of Medicine, U.S. Department of Health and Human Services. [Online.] http://www.ncbi.nlm.nih.gov/entrez/query.fcgi?db=PubMed. Accessed 14 July 2006.

79. **Powell, D.** 2006. Food safety network. Safe food from farm to fork. University of Guelph, Guelph, Ontario, Canada. [Online.] http://foodsafetynetwork.ca/. Accessed 14 July 2006.

80. **Product Recalls Australia.** 2006. Food [FSANZ]. Product Safety Policy Section, Compliance Strategies Branch, Australian Competition and Consumer Commission. [Online.] http://www.recalls.gov.au/view_recall_by_cat.php?recall_type=3. Accessed 14 July 2006.

81. **ProQuest Company.** 2006. ProQuest Information and Learning. ProQuest Company. [Online.] http://proquest.com/. Accessed 14 July 2006.

82. **Rose, B. E., W. E. Hill, R. Umholtz, G. M. Ransom, and W. O. James.** 2002. Testing for *Salmonella* in raw meat and poultry products collected at federally inspected establishments in the United States, 1998 through 2000. *J. Food Prot.* **65:**937–947.

83. **Safe Tables Our Priority.** 2005. Fighting foodborne illness. What's new. Safe Tables Our Priority, Burlington, Vt. [Online.] http://safetables.org/. Accessed 14 July 2006.

84. **Simmons, M., D. L. Fletcher, M. E. Berrang, and J. A. Cason.** 2003. Comparison of sampling methods for the detection of *Salmonella* on whole broiler carcasses purchased from retail outlets. *J. Food Prot.* **66:**1768–1770.

85. **Singhania, L.** 1 January 1999. Bacterial outbreak linked to meat plant in Michigan. *Houston Chronicle*, Houston, Tex.

86. **Singhania, L.** 1 January 1999. Tainted food traced to Michigan meat plant. *Pittsburgh Post-Gazette*, Pittsburgh, Pa.

87. **Sosa Estani, S.** 2004. Boletín epidemiológico nacional 2000–2001. Dirección de Epidemiología, Ministerio de Salud de la Nación, Argentina. [Online.] http://www.msal.gov.ar/htm/site/pdf/boletin2001.pdf. Accessed 14 July 2006.

88. **Sternberg, S.** 23 December 1998. Sara Lee recalls hot dogs, other meats. *USA Today*, Arlington, Va.

89. **Thomas, J.** 24 December 1998. Outbreak of food poisoning leads to warning on hot dogs and cold cuts. *New York Times*, New York, N.Y.

90. **Thompson, J.** 12 August 1998. USDA: always test burgers with thermometer. *Omaha World-Herald*, Omaha, Nebr.

91. **Thomson Gale.** 2006. InfoTrac OneFile. Thomson Gale. [Online.] http://www.gale.com/servlet/ItemDetailServlet?region=9&imprint=000&titleCode=GAL61&cf=n&type=4&id=115229. Accessed 14 July 2006.

92. **White, D. G., S. Zhao, R. Sudler, S. Ayers, S. Friedman, S. Chen, P. F. McDermott, S. McDermott, D. D. Wagner, and J. Meng.** 2001. The isolation of antibiotic-resistant *Salmonella* from retail ground meats. *N. Engl. J. Med.* **345:**1147–1154.

93. **Workman, S. N., G. E. Mathison, and M. C. Lavoie.** 2005. Pet dogs and chicken meat as reservoirs of *Campylobacter* spp. in Barbados. *J. Clin. Microbiol.* **43:**2642–2650.

94. **Wu, O.** 18 January 2006. Fowl play; how to bring out the best in America's favorite meat. *San Francisco Chronicle*, San Francisco, Calif.

chapter 17

Changing Old Habits

OVER A 6-MONTH PERIOD IN 1997 AND 1998, 40 Australian individuals and families invited video cameras into their homes—not for a reality show, but for a reality check. They were taking part in a study of consumers' food safety habits (60).

Participants began by completing a written questionnaire. Then, they were videotaped over a 2-week period while they carried out their normal kitchen activities. The video cameras captured numerous food hygiene lapses, from a household pet walking on the kitchen counter and licking margarine that was then used to prepare a sandwich, to the more usual touching of face or hair while preparing food. Typically, participants had a much more favorable view of their own food hygiene practices (based on answers to the questionnaire) than was supported by the video evidence. Most of those discrepancies related to inadequate hand washing, avenues for cross-contamination, and inadequate cleaning of kitchen work surfaces (60).

While the video study was in progress, the investigators also carried out a telephone survey of Australians' food safety practices and knowledge (59). They questioned more than 1,200 individuals, all of them the main grocery buyers for their households. Participants, who prepared food regularly in addition to purchasing it, were asked how (at what temperature and for how long) they thawed, cooled, and stored raw and cooked foods, what precautions they took against cross-contamination, and what personal hygiene measures they used (hand washing before and during food preparation). Temperature abuse proved to be a common bad habit. Eighty-six percent of the participants said that they allowed cooked leftovers to cool at room temperature before refrigerating them, 40% thawed frozen raw meat at room temperature, and fewer than 75% of respondents knew that the correct temperature for refrigerating food was 1 to 5°C (34 to 41°F). Personal hygiene was also a problem area. Eighteen percent of survey participants did not routinely wash their hands before or while they prepared food (60).

Table 17.1 Food handling safety lapses reported in two Australian studies (60, 79)

Examples of incorrect or risky food handling observed during videotaped study of consumers' behavior	Telephone survey results[a]
Cooked food allowed to cool at room temperature before refrigeration	84.5%
Unsanitary items (including pets) on kitchen work surfaces, or contaminated surfaces not cleaned correctly before being used for food preparation	70.3%
Hands not washed with soap and water after food preparation	44.2%
Frozen food thawed at room temperature	40.1%
No detergent or cleaner used to clean countertops before and/or after food preparation	38.1%
Hands not washed with soap and water before handling food	17.7%

[a] Percentage of telephone survey respondents admitting to the described incorrect or risky practice.

A second, independent, Australian food safety survey covering more than 500 households overlapped the video study. The results of this 1998 (written questionnaire) survey were similar to those from the video and telephone study (Table 17.1). Ninety-nine percent of the respondents engaged in at least one form of unsafe food handling or storage practice, including poor hand washing practices, temperature abuse of raw and cooked foods, and practices that opened the door to cross-contamination (79).

Australian consumers are not unique. Studies carried out over the last decade have uncovered similar behavior patterns in countries and regions as far apart geographically and economically as Argentina, the European Union (Germany, Italy, Ireland, Spain, and Sweden), the Caribbean islands, Uganda, and the United States (5, 8, 10, 15, 42, 55, 66, 74, 77, 87, 92, 101, 103). Consistently, investigators have found that consumers and food handlers don't understand—or understand but don't practice—safe food handling techniques such as proper cleaning of kitchen work surfaces, correct holding temperatures for raw and cooked foods, and proper separation of raw and cooked foods in the kitchen.

Other investigators have also found that consumers do not always present an accurate picture of their food handling activities when answering questionnaires. Researchers in the United States studied the differences between consumers' actual and reported behaviors in the kitchen. Seventy-two study participants were videotaped carrying out various kitchen activities and then interviewed about their food handling habits. Although at least 75% of the participants described accurately their practices with regard to thorough cooking of meat and poultry, washing their hands before beginning to prepare food, and washing knives and cutting boards after handling raw meat, fewer than 20% of the participants correctly described their use of meat thermometers and how they cleaned kitchen countertops (65).

Some studies have shown that teaching consumers about food safety improves their food handling practices measurably, and several researchers into consumer attitudes and practices have advocated promotion of safe food handling through consumer education (5, 8, 42, 55, 60, 62, 66, 77, 96). But education is not a cure-all (15). The behavior of food handlers—whether consumers or food service workers—often belies their awareness and understanding of correct procedures (8, 15, 59, 60, 69, 77, 86, 91). Knowledge is not a guarantee of performance.

Teaching Food Safety

People learn about food safety and food preparation from a variety of sources. Watching others—either in the flesh or on television—is a common way for consumers to learn how to prepare food (42, 66, 75). School courses, cookbooks, TV news items, newspapers, magazines, and the Internet are also important sources of information (66, 69, 77). Labels on packages of raw meats and shell eggs explain how to handle and store those foods correctly (106).

Food safety educators know that the way in which information is presented can make a difference to how well it is received and understood. Researchers in several countries have studied consumer food safety knowledge and attitudes to determine how best to educate consumers and food handlers in safe food practices (5, 10, 15, 30, 42, 51, 64, 65, 69, 70, 77, 96).

The World Health Organization (WHO) has taken the lead in making food safety information available to educators in its member states. The poster "Five Keys to Safer Food"—available in more than 30 languages from Arabic to Zulu—is just one of the tools that the WHO has developed to help teachers get across to school children the basic message of safe food handling procedures (104). The organization also publishes books, pamphlets, and other food safety education materials in several languages (105).

In the United States, the Department of Agriculture (USDA), the Food and Drug Administration (FDA), and the Centers for Disease Control and Prevention (CDC) all have invested in food safety education, providing information targeted to the consumer as well as resources for primary and secondary school teachers (20, 22, 33, 36). Especially noteworthy are programs such as the USDA Food Safety Mobile and Fight Bac!®, a joint government-industry consumer education effort run by the Partnership for Food Safety Education (37, 85). Other countries—Australia, Canada, New Zealand, and the United Kingdom, for example—maintain similar education programs (17, 38, 39, 83). But, as the Australian studies showed, just making information available is not enough (59, 60, 79). Consumers and food handlers must be motivated to apply what they have learned, consistently and correctly.

Selling Food Safety

In 1989, the annual cost of food-borne diseases to the national economies of Canada and the United States was estimated at $1.1 billion and $7 billion, respectively (98). In 1998–1999, food-borne illness struck Swedes at a rate of about 38 cases per thousand and cost the country approximately $138 million (71). Australians suffered an estimated 1.48 million cases of food-borne gastroenteritis in 2000; approximately 14,700 of the victims were hospitalized and 76 died (52). And, according to CDC estimates, 5,000 people die in the United States, 325,000 are hospitalized, and 76 million are made ill by food-borne pathogens every year (23, 76). Food-borne disease is not just inconvenient—it's expensive.

In 1964, the year of the first U.S. Surgeon General's report on cigarette smoking and public health, more than 42% of Americans over the age of 18 (68% of men and 32.4% of women) smoked regularly, including the Surgeon General (7, 18). Forty years later, following major educational and advertising campaigns by the U.S. government and other agencies—and in the face of massive cigarette advertising campaigns mounted by all of the cigarette manufacturers—the percentage of American adults who smoked had dropped to 22.5% (18). The percentage of adolescent smokers decreased from 36% in 1991 to 22% in 2003 (21). In contrast, the estimated number of food-borne illnesses remained essentially unchanged between 1999 and 2005 (23, 76).

Government agencies have made a serious effort to educate the consumer on how to prevent the transmission of food-borne disease, but the message is apparently not getting through. Many consumers practice risky behaviors even though they know how to handle food safely. Clearly, motivation is lacking. How else are we to explain the eating of undercooked ground meat in the face of *Escherichia coli* O157:H7 or the drinking of raw (unpasteurized) milk despite government warnings about the risks? A professionally designed and implemented media campaign to promote food safety would be costly, but an ongoing, entertaining, and highly visible food safety advertising campaign might be the only way to get the message across. And the money would be well spent if—like the results of the antismoking campaign—the incidence of food-borne disease were to be cut in half.

Sharing Information

While educating and motivating consumers to make safe, responsible decisions when purchasing, preparing, and eating food is essential, providing them with up-to-date information on which to base their actions is equally vital. Federal and state agencies inspect food manufacturing and distribution facilities; county and city health departments license and inspect restaurants and fast-food outlets (13). All of these agencies maintain records of their inspection, licensing, and enforcement activities.

A few agencies make the results of their routine investigations available to consumers. CDC publishes on its website full reports of all cruise ship inspections under its Vessel Sanitation Program (24). The FDA maintains a page on its website listing the sanitation and enforcement ratings of all interstate shippers of milk (34). Some county and city health departments require restaurants to post their most recent inspection scores prominently; others publish a list of restaurants that have been cited for unsanitary conditions (27, 28, 99). But, unless the inspection of a food production or distribution facility uncovers an immediate, direct health hazard, most inspection results only reach the public in summary form (35).

Consumers consistently express a desire for clear, accurate information and labeling, whether the topic is using irradiation to extend shelf life, country of origin labeling, or open product expiry dating (4, 19, 61, 82). Perhaps it's time to provide consumers with information on food processors' inspection scores. Companies could be encouraged, maybe even required, to disclose the score and date of their most recent federal or state inspection, much as restaurants in many jurisdictions are required to post their inspection grade. This might be done by adding the information to product labels (although those labels are already crowded with mandated information), by including it in product advertisements, or by posting the information on companies' websites. Alternatively, government agencies could follow the lead of the CDC's Vessel Sanitation Program and display inspection results on a user-friendly website. Either way, consumers would be able to choose, if they wish, to purchase food that was manufactured in high-scoring facilities.

Embracing New Technologies

Consumers represent only one link in the food safety chain, albeit a very important one. Scientists and engineers, growers and packers, processors and distributors must all do their share to reduce the incidence of food-borne diseases.

Some researchers have focused on controlling food-borne pathogens preharvest. Their approaches have included developing probiotic and vaccine treatments of poultry to reduce the incidence of *Salmonella* in poultry and eggs, determining possible sources of contamination of produce in the fields, and exploring the effects of feeding regimens and probiotic treatments on the carriage of *E. coli* O157:H7 in cattle (9, 58, 81, 97, 102, 108). Other research groups have concentrated on reducing the incidence of pathogens on fresh meat and raw produce through improved sanitation and carcass-washing methods in meat and poultry plants, and developing procedures to decontaminate fresh fruits and vegetables after harvest

(12, 47, 67). The food industry has looked to improvements in modified-atmosphere packaging, development of surface pasteurization of vacuum-packaged products, and addition of new antimicrobial agents to reduce the risk of pathogen growth in—and to extend the shelf life of—fresh and processed foods (14, 48, 56, 57, 72, 100). Finally, major efforts have gone into developing and validating alternatives to conventional heat pasteurization for both fluid and solid foods, including ultraviolet light, ultrasonic waves (with or without heat), high hydrostatic pressure (alone or combined with an antimicrobial agent), pulsed electric fields, ultrahigh-pressure homogenization, and ionizing radiation (1, 2, 6, 11, 43, 63, 73, 84, 88–90, 93, 107).

Of these advances in food technology, the adoption of ionizing radiation to reduce pathogen levels in foods has received the most public attention and debate. The food industry welcomed the approval of food irradiation, but has been slow to take advantage of the new tool due to concerns about its cost and over consumer resistance to irradiated food (3, 78, 95).

The FDA approved irradiation of a food for human consumption (wheat and wheat flour) for the first time in 1963. Since then, the agency has approved radiation treatment of foods for a variety of reasons: to control insect infestation, to delay maturation of fresh foods, and, most recently, to control pathogens in red meats, poultry, and eggs (32, 54). Canada, Australia, New Zealand, several European Union (EU) member states, and countries in Latin America, the Middle East, and the Pacific Rim all allow food irradiation for at least some purposes (16, 29, 31, 40, 53). Nevertheless, irradiated food has been a tough sell to consumers.

In 1993, a representative group of consumers were surveyed by mail to determine their knowledge and acceptance of food irradiation. One-third of those who answered the questionnaire believed that irradiated food remained radioactive; just under half of the respondents said that they would purchase irradiated beef, pork, or poultry (61). A survey of attitudes toward irradiated meat and poultry carried out in 1998 confirmed the 1993 data. Not even half of the more than 10,000 adults surveyed said that they were willing to buy irradiated meat or poultry (41). But consumer sentiments have begun to change. A 2003 survey found that only 15% of respondents thought that food remained radioactive after irradiation; 68% of those questioned would purchase irradiated meat and poultry, while 75% would buy irradiated pork (61). Surebeam, the only irradiation plant in the United States equipped with refrigeration facilities, began supplying irradiated fresh ground beef to supermarkets in 2002. Unfortunately, the company declared bankruptcy in 2004 and shut down its operations (49).

Although consumers are more ready to accept irradiated food now than in past years, they have not been won over completely. Nor are they alone. A group of Kansas restaurant managers were almost evenly split on whether or not they would buy and serve irradiated ground beef. A 2003 survey found that only 54% would do so—if it was available at the same price as regular (nonirradiated) ground beef (80). The consensus among medical practitioners is also divided. Some physicians and epidemiologists support the technology; others have serious reservations about its long-term safety, citing the potential for nutrient loss as a result of irradiation (72, 78, 94). On the other hand, the WHO determined, based on a review of 500 studies, that irradiation of food presents no significant microbiological, nutritional, or toxicological risks, and the U.S. General Accounting Office (now the Government Accountability Office) concluded that the benefits derived from irradiating food outweighed any of the concerns that have been raised (46).

Out of Many, One

In 1998, a committee of U.S. food safety experts working under the Institute of Medicine and the National Research Council concluded that "...the creation of a centralized and unified federal framework [was] critical to improve the [U.S.] food safety system," and suggested that "... a single, unified agency headed by a single administrator ..." would be a logical way to achieve this goal (25). Four countries—Canada, Denmark, Ireland, and the United Kingdom—had reached this same conclusion before the report was even released (44). By 2005, three more countries had joined the four pioneers (Table 17.2). But in the United States, the responsibility for enforcing federal food safety and labeling laws is shared by 15 agencies, under the control seven Cabinet-level departments: the Environmental Protection Agency, the Federal Trade Commission, and the Departments of Agriculture, Health and Human Services, Commerce, the Treasury, and Homeland Security (50). The bulk of regulatory responsibility rests with the FDA and the USDA's Food Safety and Inspection Service (FSIS).

The division of responsibilities and authority between FDA, FSIS, and other agencies within the USDA is confusing and apparently arbitrary. For example, FDA regulates shell eggs, whereas USDA oversees processed egg products such as liquid, frozen, or dried whole eggs, egg yolks, or egg whites (50). FSIS inspects producers of open-faced meat or poultry sandwiches, whereas FDA handles open-faced sandwiches that do not contain meat or poultry, as well as all closed-face sandwiches, regardless of the filling (45). FDA inspectors scrutinize pizza that is topped with less than 2% meat; 2% or more, and FSIS is responsible (25). Each U.S. state (and each Canadian province) is responsible for regulating foods that remain within a state's (or province's) own borders, adding another level of complexity to the regulatory system.

A fragmented inspection system is expensive, inefficient, and unfair to the industry it regulates. Many food processors, not just pizza manufacturers, find themselves the object of inspections from two federal agencies (25).

Table 17.2 Countries that have adopted a unified agency approach to food safety regulation (50)

Year enacted	Country	Rationale
1997	Canada	Cost savings
1998	Denmark	Improved effectiveness and efficiency
1998	Ireland	Restored consumer confidence; eliminated possible conflicts of interest
1999	United Kingdom	Restored consumer confidence; eliminated possible conflicts of interest
2002	New Zealand	Improved effectiveness; eliminated regulatory inconsistencies between agencies
2002	Germany	Restored consumer confidence; improved compliance with EU food safety requirements
2002	The Netherlands	Reduced regulatory agency overlap; improved compliance with EU food safety requirements

Food processors are inspected at different frequencies, depending on the agency under whose jurisdiction they happen to fall (45). And each agency has a different approach to inspection and management of the food supply it regulates, and maintains its own regulatory bureaucracy, inspection staff, and, in some cases, lab facilities—an expensive duplication of efforts. Whereas consolidating the responsibility for enforcing federal food safety laws into a single agency cannot eliminate overlap between federal and state agencies, it would at least streamline and clarify federal enforcement policies and efforts.

Countries that have consolidated their food safety system into a single agency have done so for a variety of reasons, including removing—or avoiding the appearance of—a conflict of interest between an agency's role in supporting the food industry and its regulatory responsibilities (Table 17.2). The USDA suffers from this kind of mixed mandate. On one hand, it must oversee programs designed to support and develop markets for meats, poultry, and processed egg products; on the other hand, it is expected to enforce food safety laws in those same market sectors (26). Consolidating some or all food safety regulatory activities into a single umbrella agency, one that is not part of USDA, would do more than just eliminate the inconsistencies and duplication of efforts that are now everyday events. It would remove the potential for conflict of interest that exists within one or more of the federal agencies that are now involved in food safety regulation and enforcement.

Change for the Better

We live in a world in which we are vastly outnumbered by microbes. We touch and are touched by millions of microscopic beings every day. We cannot avoid this contact—nor should we want to. After all, most microbes are either beneficial or benign. Even so, we can't ignore the health risks presented by a small minority of pathogens that may be present in our environment, including in our food and water. We owe it to ourselves, to our families, and to our communities to take all reasonable measures to ensure the safety of our food and water supply.

Some old habits are good ones; they should not be changed willy-nilly. But old methods need to be reappraised in the light of new information, and new habits should be formed as new procedures and technologies are developed and validated.

Each of us—consumer, restaurant owner, food service worker, meat or poultry producer, farmer, or food processor—must make a commitment to learn correct procedures and apply them consistently. No one (except for those few who deliberately contaminate food) intentionally causes food-borne illness, nor would food producers or processors make a conscious decision to save money by cutting corners if they knew that a food poisoning outbreak might result. But accidents and oversights happen, and we must all remain vigilant. As Arthur Liang and his coauthors said in their Conference Panel summary, *Teaming Up to Prevent Foodborne Disease* (68),

> Food safety is what society does to ensure the conditions under which people can consume food that is safe, as well as wholesome and nutritious. Safe food requires the work of producers and consumers; industry and government; local, state, federal, and, increasingly, international partners.

References

1. **Alvarez I., P. Manas, R. Virto, and S. Condon.** 2006. Inactivation of *Salmonella* Senftenberg 775W by ultrasonic waves under pressure at different water activities. *Int. J. Food Microbiol.* **108:**218–225.
2. **Alvarez I., J. Raso, A. Palop, and F. J. Sala.** 2000. Influence of different factors on the inactivation of *Salmonella senftenberg* by pulsed electric fields. *Int. J. Food Microbiol.* **55:**143–146.
3. **American Meat Institute.** 2002. AMI fact sheet: irradiation. American Meat Institute. [Online.] http://www.meatami.com/Content/NavigationMenu/PressCenter/FactSheets_InfoKits/FactSheetIrradiation.pdf. Accessed 17 July 2006.
4. **Anderson, L. J.** 26 December 1997. Food irradiation labeling will help consumers make choices about food. *Seattle Times*, Seattle, Wash.
5. **Angelillo, I. F., N. M. A. Viggiani, L. Rizzo, and A. Bianco.** 2000. Food handlers and foodborne diseases: knowledge, attitudes, and reported behavior in Italy. *J. Food Prot.* **63:**381–385.
6. **Arthur, T. M., T. L. Wheeler, S. D. Shackelford, J. M. Bosilevac, X. Nou, and M. Koohmaraie.** 2005. Effects of low-dose, low-penetration electron beam irradiation of chilled beef carcass surface cuts on *Escherichia coli* O157:H7 and meat quality. *J. Food Prot.* **68:**666–672.
7. **Bayne-Jones, S., W. J. Burdette, W. G. Cochran, E. Farber, L. F. Fieser, J. Furth, J. B. Hickam, C. LeMaistre, L. M. Schuman, and M. H. Seevers.** 1964. Smoking and health. Report of the advisory committee to the Surgeon General of the Public Health Service. PHS Publication No. 1103. U.S. Department of Health, Education, and Welfare, Public Health Service. [Online.]

http://www.cdc.gov/tobacco/sgr/sgr_1964/1964%20SGR%20Intro.pdf. Accessed 17 July 2006.

8. **Bermúdez-Millán, A., R. Pérez-Escamilla, G. Damio, A. González, and S. Segura-Pérez.** 2004. Food safety knowledge, attitudes, and behaviors among Puerto Rican caretakers living in Hartford, Connecticut. *J. Food Prot.* **67:**512–516.

9. **Betancor, L., F. Schelotto, M. Fernandez, M. Pereira, A. Rial, and J. A. Chabalgoity.** 2005. An attenuated *Salmonella* Enteritidis strain derivative of the main genotype circulating in Uruguay is an effective vaccine for chickens. *Vet. Microbiol.* **107:**81–89.

10. **Bremer, V., N. Bocter, S. Rehmet, G. Klein, T. Breuer, and A. Ammon.** 2005. Consumption, knowledge, and handling of raw meat: a representative cross-sectional survey in Germany, March 2001. *J. Food Prot.* **68:**785–789.

11. **Briñez, W. J., A. X. Roig-Sagués, M. M. Hernández Herrero, and B. Guamis López.** 2006. Inactivation of *Listeria innocua* in milk and orange juice by ultrahigh-pressure homogenization. *J. Food Prot.* **69:**86–92.

12. **Bryant, J., D. A. Brereton, and C. O. Gill.** 2003. Implementation of a validated HACCP system for the control of microbiological contamination of pig carcasses at a small abattoir. *Can. Vet. J.* **44:**51–55.

13. **Buchholz, U., G. Run, J. L. Kool, J. Fielding, and L. Mascola.** 2002. A risk-based restaurant inspection system in Los Angeles County. *J. Food Prot.* **65:**367–372.

14. **Caillet, S., M. Millette, S. Salmiéri, and M. Lacroix.** 2006. Combined effects of antimicrobial coating, modified atmosphere packaging, and gamma irradiation on *Listeria innocua* present in ready-to-use carrots (*Daucus carota*). *J. Food Prot.* **69:**80–85.

15. **Califano, A. N., G. L. De Antoni, L. Giannuzzi, and R. H. Mascheroni.** 2000. Prevalence of unsafe practices during home preparation of food in Argentina. *Dairy Food Environ. Sanit.* **20:**934–943.

16. **Canadian Food Inspection Agency.** 2006. Food irradiation. Canadian Food Inspection Agency. [Online.] http://www.inspection.gc.ca/english/fssa/concen/tipcon/irrade.shtml. Accessed 17 July 2006.

17. **Canadian Partnership for Consumer Food Safety Education.** 2006. Helping Canadians enjoy their food safely! Canadian Partnership for Consumer Food Safety Education. [Online.] http://www.canfightbac.org/cpcfse/en/. Accessed 17 July 2006.

18. **Carmona, R. H.** 2004. *The Health Consequences of Smoking: a Report of the Surgeon General, 2004.* U.S. Department of Health and Human Services. [Online.] http://www.hhs.gov/surgeongeneral/news/speeches/SgrSmoking_05272004.htm. Accessed 17 July 2006.

19. **Cates, S. C., K. M. Kosa, R. C. Post, and J. Canavan.** 2004. Consumers' attitudes toward open dating of USDA-regulated foods. *Food Prot. Trends* **24:**82–88.

20. **Centers for Disease Control and Prevention.** 2003. Infectious disease information. Food-related diseases. Centers for Disease Control and Prevention, National Center for Infectious Diseases. [Online.] http://www.cdc.gov/ncidod/diseases/food/index.htm. Accessed 17 July 2006.

21. **Centers for Disease Control and Prevention.** 2004. CDC. Protecting health for life. The state of the CDC, fiscal year 2004. Centers for Disease Control and Prevention, Atlanta, Ga. [Online.] http://www.cdc.gov/cdc.pdf. Accessed 17 July 2006.

22. **Centers for Disease Control and Prevention.** 2004. Teachers' tools. Educational resources for teachers at K–12 levels. Centers for Disease Control and Prevention, National Center for Infectious Diseases. [Online.] http://www.cdc.gov/ncidod/teachers_tools/index.htm. Accessed 17 July 2006.

23. **Centers for Disease Control and Prevention.** 2005. Food safety office. Centers for Disease Control and Prevention. [Online.] http://www.cdc.gov/foodsafety/. Accessed 17 July 2006.

24. **Centers for Disease Control and Prevention.** 2006. Vessel Sanitation Program. National Center for Environmental Health, Centers for Disease Control and Prevention, Atlanta, Ga. [Online.] http://www.cdc.gov/nceh/vsp/default.htm. Accessed 17 July 2006.

25. **Committee to Ensure Safe Food from Production to Consumption.** 1998. *Ensuring Safe Food: From Production to Consumption*. Institute of Medicine and National Research Council. National Academy Press, Washington, D.C. [Online.] http://www.nap.edu/catalog/6163.html. Accessed 17 July 2006.

26. **Department of Agriculture.** 2006. Customer statement. U.S. Department of Agriculture. [Online.] http://customerstatement.usda.gov/. Accessed 17 July 2006.

27. **Department of Environmental Health.** 2006. Food & housing: food program. County of San Diego, Department of Environmental Health, San Diego, Calif. [Online.] http://www.co.san-diego.ca.us/deh/fhd/food.html. Accessed 17 July 2006.

28. **Department of Health and Mental Hygiene.** 2006. Restaurant inspection information. New York City Department of Health and Mental Hygiene, New York, N.Y. [Online.] http://63.106.144.9/RI/web/index.do;jsessionid=4960B91064D0028053B06B952E319754?method=goldenAppleList. Accessed 17 July 2006.

29. **DoD Combat Feeding Program.** 1998. Food irradiation. U.S. Army Soldier and Biological Chemical Command, Natick RD&E Center, Natick, Mass. [Online.] http://nsc.natick.army.mil/media/print/irradiation.pdf. Accessed 17 July 2006.

30. **Edwards, Z. M., M. T. Takeuchi, V. N. Hillers, S. M. McCurdy, and M. Edlefsen.** 2006. Use of behavioral change theories in development of educational materials to promote food thermometer use. *Food Prot. Trends* **26:**82–88.

31. **Europa.** 2006. Food and feed safety. European Commission, DG Health and Consumer Protection. [Online.] http://europa.eu.int/comm/food/food/biosafety/irradiation/index_en.htm. Accessed 17 July 2006.

32. **Federal Register.** 2000. Irradiation in the production, processing and handling of food. Final Rule. *Fed. Regist.* **65:**45280–45282.

33. **Food and Drug Administration.** 2005. Consumer advice and publications on food safety, nutrition, and cosmetics. U.S. Food and Drug Administration, Center for Food Safety and Applied Nutrition. [Online.] http://www.cfsan.fda.gov/~lrd/advice.html#foodborn. Accessed 17 July 2006.

34. **Food and Drug Administration.** 2006. IMS List. Sanitation compliance and enforcement ratings of interstate milk shippers. Table of contents. U.S. Food and Drug Administration, Center for Food Safety and Applied Nutrition. [Online.] http://www.cfsan.fda.gov/~ear/ims-toc.html. Accessed 17 July 2006.

35. **Food and Drug Administration.** 17 February 2006. FDA news. US marshals seize food items from DC company. U.S. Food and Drug Administration. [Online.] http://www.fda.gov/bbs/topics/news/2006/NEW01320.html. Accessed 17 July 2006.

36. **Food Safety and Inspection Service.** 2006. Food safety education. U.S. Department of Agriculture, Food Safety and Inspection Service. [Online.] http://www.fsis.usda.gov/Food_Safety_Education/index.asp. Accessed 17 July 2006.

37. **Food Safety and Inspection Service.** 2006. Food safety education. Food safety mobile. U.S. Department of Agriculture, Food Safety and Inspection Service. [Online.] http://www.fsis.usda.gov/food_safety_education/Food_Safety_Mobile/index.asp. Accessed 17 July 2006.

38. **Food Safety Campaign.** 2006. The Food Safety Campaign's education pages. The Food Safety Campaign, Australia. [Online.] http://www.safefood.net.au/content.cfm. Accessed 17 July 2006.

39. **Food Standards Agency.** 2006. Eat well, be well. Helping you make healthier choices. Keeping food safe. Food Standards Agency, United Kingdom. [Online.] http://www.eatwell.gov.uk/keepingfoodsafe/. Accessed 17 July 2006.

40. **Food Standards Australia New Zealand.** 2003. Food irradiation. [Online.] http://www.foodstandards.gov.au/mediareleasespublications/factsheets/factsheets2003/foodirradiationupdat1943.cfm. Accessed 17 July 2006.

41. **Frenzen, P. D., E. E. DeBess, K. E. Hechemy, H. Kassenborg, M. Kennedy, K. McCombs, A. McNees, and The FoodNet Working Group.** 2001. Consumer acceptance of irradiated meat and poultry in the United States. *J. Food Prot.* **64:**2020–2026.

42. **Garayoa, R., M. Córdoba, I. García-Jalón, A. Sanchez-Villegas, and A. I. Vitas.** 2005. Relationship between consumer food safety knowledge and reported behavior among students from health sciences in one region of Spain. *J. Food Prot.* **68:**2631–2636.

43. **Gardner, D. W., and G. Shama.** 2000. Modeling UV-induced inactivation of microorganisms on surfaces. *J. Food Prot.* **63:**63–70.

44. **General Accounting Office.** 1999. GAO/RCED-99-80. Food safety. Experiences of four countries in consolidating their food safety systems. U.S. General Accounting Office. [Online.] http://www.gao.gov/archive/1999/rc99080.pdf. Accessed 17 July 2006.

45. **General Accounting Office.** 2001. GAO-02-47T. Food safety and security. Fundamental changes needed to ensure safe food. Testimony before the Subcommittee on Oversight of Government Management, Restructuring and the District of Columbia, Committee on Governmental Affairs, U.S. Senate. U.S. General Accounting Office. [Online.] http://www.gao.gov/new.items/d0247t.pdf. Accessed 17 July 2006.

46. **General Accounting Office.** 2004. GAO/RCED-00-217. Food irradiation. Available research indicates that benefits outweigh risks. U.S. General

Accounting Office. [Online.] http://www.gao.gov/new.items/rc00217.pdf. Accessed 17 July 2006.

47. **Gill, C. O., J. Bryant, and M. Badoni.** 2001. Effects of hot water pasteurizing treatments on the microbiological condition of manufacturing beef used for hamburger patty manufacture. *Int. J. Food Microbiol.* **63:**243–256.

48. **Gill, C. O., and K. H. Tan.** 1980. Effect of carbon dioxide on growth of meat spoilage bacteria. *Appl. Environ. Microbiol.* **39:**317–319.

49. **Glynn, M.** 6 February 2004. Tops, Wegmans no longer getting irradiated beef. *Buffalo News*, Buffalo, N.Y.

50. **Government Accountability Office.** 2005. GAO-05-212. Food safety. Experiences of seven countries in consolidating their food safety systems. U.S. Government Accountability Office. [Online.] http://www.gao.gov/new.items/d05212.pdf. Accessed 17 July 2006.

51. **Green, L. R., C. Selman, E. Scallan, T. F. Jones, R. Marcus, and The EHS-Net Population Survey Working Group.** 2005. Beliefs about meals eaten outside the home as sources of gastrointestinal illness. *J. Food Prot.* **68:**2184–2189.

52. **Hall, G., M. D. Kirk, N. Becker, J. E. Gregory, L. Unicomb, G. Millard, R. Stafford, K. Lalor, and the OzFoodNet Working Group.** 2005. Estimating foodborne gastroenteritis, Australia. *Emerg. Infect. Dis.* **11:**1257–1264.

53. **Health Canada.** 2006. Food irradiation. Health Canada. [Online.] http://www.hc-sc.gc.ca/fn-an/securit/irridation/index_e.html. Accessed 17 July 2006.

54. **Henkel, J.** May-June 1998. Irradiation: A safe measure for safer food. FDA Consumer. U.S. Food and Drug Administration. [Online.] http://www.fda.gov/fdac/features/1998/398_rad.html. Accessed 17 July 2006.

55. **Hillers, V. N., L. Medeiros, P. Kendall, G. Chen, and S. DiMascola.** 2003. Consumer food-handling behaviors associated with prevention of 13 foodborne illnesses. *J. Food Prot.* **66:**1893–1899.

56. **Houben, J. H., and F. Eckenhausen.** 2006. Surface pasteurization of vacuum-sealed precooked ready-to-eat meat products. *J. Food Prot.* **69:**459–468.

57. **Ingham, S. C., E. L. Schoeller, and R. A. Engel.** 2006. Pathogen reduction in unpasteurized apple cider: adding cranberry juice to enhance the lethality of warm hold and freeze-thaw steps. *J. Food Prot.* **69:**293–298.

58. **Islam, M., J. Morgan, M. P. Doyle, S. C. Phatak, P. Millner, and X. Jiang.** 2004. Fate of *Salmonella enterica* serovar Typhimurium on carrots and radishes grown in fields treated with contaminated manure composts or irrigation water. *Appl. Environ. Microbiol.* **70:**2497–2502.

59. **Jay, L. S., D. Comar, and L. D. Govenlock.** 1999. A national Australian food safety telephone survey. *J. Food Prot.* **62:**921–928.

60. **Jay, L. S., D. Comar, and L. D. Govenlock.** 1999. A video study of Australian domestic food-handling practices. *J. Food Prot.* **62:**1285–1296.

61. **Johnson, A. M., A. E. Reynolds, J. Chen, and A. V. A. Resurreccion.** 2004. Consumer attitudes towards irradiated food: 2003 vs. 1993. *Food Prot. Trends* **24:**408–418.

62. **Kain, M. L., J. A. Scanga, J. N. Sofos, K. E. Belk, J. O. Reagan, D. R. Buege, W. P. Henning, J. B. Morgan, T. P. Ringkob, G. R. Bellinger, and G. C. Smith.** 2002. Consumer behavior regarding time lapse between store-purchase

and subsequent home-storage of fresh beef retail cuts. *Dairy Food Environ. Sanit.* **22:**740–744.

63. **Kamat, A. S., S. Khare, T. Doctor, and P. M. Nair.** 1997. Control of *Yersinia enterocolitica* in raw pork and pork products by gamma-irradiation. *Int. J. Food Microbiol.* **36:**69–76.

64. **Kato, M., K. Naka, H. Yamamoto, and S. Kira.** 2004. Assessing education of food handlers and prerequisite programs in Japanese HACCP plants. *Food Prot. Trends* **24:**316–322.

65. **Kendall, P. A., A. Elsbernd, K. Sinclair, M. Schroeder, G. Chen, V. Bergmann, V. N. Hillers, and L. C. Medeiros.** 2004. Observation versus self-report: validation of a consumer food behavior questionnaire. *J. Food Prot.* **67:**2578–2586.

66. **Kennedy, J., V. Jackson, I. S. Blair, D. A. McDowell, C. Cowan, and D. J. Bolton.** 2005. Food safety knowledge of consumers and the microbiological and temperature status of their refrigerators. *J. Food Prot.* **68:**1421–1430.

67. **Kondo, N., M. Murata, and K. Isshiki.** 2006. Efficiency of sodium hypochlorite, fumaric acid, and mild heat in killing native microflora and *Escherichia coli* O157:H7, *Salmonella* Typhimurium DT104, and *Staphylococcus aureus* attached to fresh-cut lettuce. *J. Food Prot.* **69:**323–329.

68. **Liang, A. P., M. Koopmans, M. P. Doyle, D. T. Bernard, and C. E. Brewer.** 2001. Teaming up to prevent foodborne disease. *Emerg. Infect. Dis.* **7:**533–534.

69. **Li-Cohen, A. E., and C. M. Bruhn.** 2002. Safety of consumer handling of fresh produce from the time of purchase to the plate: a comprehensive consumer survey. *J. Food Prot.* **65:**1287–1296.

70. **Li-Cohen, A. E., M. Klenk, Y. Nicholson, J. Harwood, and C. M. Bruhn.** 2002. Refining consumer safe handling educational materials through focus groups. *Dairy Food Environ. Sanit.* **22:**539–551.

71. **Lindqvist, R., Y. Andersson, J. Lindbäck, M. Wegscheider, Y. Eriksson, L. Tideström, A. Lagerqvist-Widh, K.-O. Hedlund, S. Löfdahl, L. Svensson, and A. Norinder.** 2001. A one-year study of foodborne illnesses in the municipality of Uppsala, Sweden. *Emerg. Infect. Dis.* **7:**588–592.

72. **Louria, D. B.** 2001. Food irradiation: unresolved issues. *Clin. Infect. Dis.* **33:**378–380.

73. **Luchansky, J. B., G. Cocoma, and J. E. Call.** 2006. Hot water postprocess pasteurization of cook-in-bag turkey breast treated with and without potassium lactate and sodium diacetate and acidified sodium chlorite for control of *Listeria monocytogenes*. *J. Food Prot.* **69:**39–46.

74. **Marklinder, I. M., M. Lindblad, L. M. Eriksson, A. M. Finnson, and R. Lindqvist.** 2004. Home storage temperatures and consumer handling of refrigerated foods in Sweden. *J. Food Prot.* **67:**2570–2577.

75. **Mathiasen, L. A., B. J. Chapman, B. J. Lacroix, and D. A. Powell.** 2004. Spot the mistake: television cooking shows as a source of food safety information. *Food Prot. Trends* **24:**328–334.

76. **Mead, P. S., L. Slutsker, V. Dietz, L. F. McCaig, J. S. Bresee, C. Shapiro, P. M. Griffin, and R. V. Tauxe.** 1999. Food-related illness and death in the United States. *Emerg. Infect. Dis.* **5:**607–625.

77. **Meer, R. R., and S. L. Misner.** 2000. Food safety knowledge and behavior of expanded food and nutrition education program participants in Arizona. *J. Food Prot.* **63:**1725–1731.

78. **Merrill, A.** 9 February 1998. Beefing up food irradiation; producers, retailers, government know consumer acceptance is key. *Star Tribune*, Minneapolis, Minn.

79. **Mitakakis, T. Z., M. I. Sinclair, C. K. Fairley, P. K. Lightbody, K. Leder, and M. E. Hellard.** 2004. Food safety in family homes in Melbourne, Australia. *J. Food Prot.* **67:**818–822.

80. **Mulik, K., J. A. Fox, and M. A. Boland.** 2003. Acceptability of irradiated food to restaurant managers. *Food Prot. Trends* **23:**1022–1027.

81. **Nakamura, M., T. Nagata, S. Okamura, K. Takehara, and P. S. Holt.** 2004. The effect of killed *Salmonella enteritidis* vaccine prior to induced molting on the shedding of *S. enteritidis* in laying hens. *Avian Dis.* **48:**183–188.

82. **Ness, C.** 18 June 2003. Who's trying to kill food labels—and why? *San Francisco Chronicle*, San Francisco, Calif.

83. **New Zealand Food Safety Authority.** 2006. Food safety topics. New Zealand Food Safety Authority. [Online.] http://www.nzfsa.govt.nz/consumers/food-safety-topics/index.htm. Accessed 17 July 2006.

84. **Pagán, R., P. Mañas, J. Raso, and S. Condón.** 1999. Bacterial resistance to ultrasonic waves under pressure at nonlethal (manosonication) and lethal (manothermosonication) temperatures. *Appl. Environ. Microbiol.* **65:**297–300.

85. **Partnership for Food Safety Education.** 2004. Fight Bac! Partnership for Food Safety Education. [Online.] http://www.fightbac.org/main.cfm. Accessed 17 July 2006.

86. **Patil, S. R., S. Cates, and R. Morales.** 2005. Consumer food safety knowledge, practices, and demographic differences: findings from a meta-analysis. *J. Food Prot.* **68:**1884–1894.

87. **Pattron, D. D.** 2006. An observational study of the awareness of food safety practices in households in Trinidad. *Internet J. Food Safety* **8:**14–18. [Online.] http://www.foodhaccp.com/internetjournal/ijfsv8-4.pdf. Accessed 17 July 2006.

88. **Ponce, E., R. Pla, M. Mor-Mur, R. Gervilla, and B. Guamis.** 1998. Inactivation of *Listeria innocua* inoculated in liquid whole egg by high hydrostatic pressure. *J. Food Prot.* **61:**119–122.

89. **Ponce, E., R. Pla, E. Sendra, B. Guamis, and M. Mor-Mur.** 1998. Combined effect of nisin and high hydrostatic pressure on destruction of *Listeria innocua* and *Escherichia coli* in liquid whole egg. *Int. J. Food Microbiol.* **43:**15–19.

90. **Raso, J., R. Pagán, S. Condón, and F. J. Sala.** 1998. Influence of temperature and pressure on the lethality of ultrasound. *Appl. Environ. Microbiol.* **64:** 465–471.

91. **Redmond, E. C., and C. J. Griffith.** 2003. Consumer food handling in the home: a review of food safety studies. *J. Food Prot.* **66:**130–161.

92. **Shiferaw, B., S. Yang, P. Cieslak, D. Vugia, R. Marcus, J. Koehler, V. Deneen, F. Angulo, and The FoodNet Working Group.** 2000. Prevalence of high-risk food consumption and food-handling practices among adults: a multistate survey, 1996 to 1997. *J. Food Prot.* **63:**1538–1543.

93. **Sobrino-López, Á., and O. Martín-Belloso.** 2006. Enhancing inactivation of *Staphylococcus aureus* in skim milk by combining high-intensity pulsed electric fields and nisin. *J. Food Prot.* **69:**345–353.

94. **Steele, J. H.** 2001. Food irradiation: a public health challenge for the 21st century. *Clin. Infect. Dis.* **33:**376–377.

95. **Tait, N.** 6 April 1998. Hamburger medium rare, easy on the gamma rays, *The Financial Times*, London, England, United Kingdom.

96. **Takeuchi, M. T., M. Edlefsen, S. M. McCurdy, and V. N. Hillers.** 2005. Educational intervention enhances consumers' readiness to adopt food thermometer use when cooking small cuts of meat: an application of the transtheoretical model. *J. Food Prot.* **68:**1874–1883.

97. **Tkalcic, S., C. A. Brown, B. G. Harmon, A. V. Jain, E. P. Mueller, A. Parks, K. L. Jacobsen, S. A. Martin, T. Zhao, and M. P. Doyle.** 2000. Effects of diet on rumen proliferation and fecal shedding of *Escherichia coli* O157:H7 in calves. *J. Food Prot.* **63:**1630–1636.

98. **Todd, E. C.** 1989. Costs of acute bacterial foodborne disease in Canada and the United States. *Int. J. Food Microbiol.* **9:**313–326.

99. **VCH Public Health Protection.** 2006. Food establishment closures 2005. Vancouver Coastal Health. Vancouver, British Columbia, Canada. [Online.] http://www.vch.ca/environmental/docs/food/food_closures05.pdf. Accessed 17 July 2006.

100. **Venkitanarayanan, K. S., T. Zhao, and M. P. Doyle.** 1999. Antibacterial effect of lactoferricin B on *Escherichia coli* O157:H7 in ground beef. *J. Food Prot.* **62:**747–750.

101. **Wanyenya, I., C. Muyanja, and G. W. Nasinyama.** 2004. Kitchen practices used in handling broiler chickens and survival of *Campylobacter* spp. on cutting surfaces in Kampala, Uganda. *J. Food Prot.* **67:**1957–1960.

102. **Waters, S. M., R. A. Murphy, and R. F. Power.** 2005. Assessment of the effects of Nurmi-type cultures and a defined probiotic preparation on a *Salmonella typhimurium* 29E challenge in vivo. *J. Food Prot.* **68:**1222–1227.

103. **Wenrich, T., K. Cason, N. Lv, and C. Kassab.** 2003. Food safety knowledge and practices of low income adults in Pennsylvania. *Food Prot. Trends* **23:**326–335.

104. **World Health Organization.** 2006. "Five Keys to Safer Food" poster. World Health Organization. [Online.] http://www.who.int/foodsafety/publications/consumer/5keys/en/index.html. Accessed 17 July 2006.

105. **World Health Organization.** 2006. Publications related to food safety. World Health Organization. [Online.] http://www.who.int/foodsafety/publications/en/. Accessed 17 July 2006.

106. **Yang, S., F. J. Angulo, and S. F. Altekruse.** 2000. Evaluation of safe food-handling instructions on raw meat and poultry products. *J. Food Prot.* **63:**1321–1325.

107. **Yuste, J., R. Pla, M. Capellas, E. Ponce, and M. Mor-Mur.** 2000. High-pressure processing applied to cooked sausages: bacterial populations during chilled storage. *J. Food Prot.* **63:**1093–1099.

108. **Zhao, T., S. Tkalcic, M. P. Doyle, B. G. Harmon, C. A. Brown, and P. Zhao.** 2003. Pathogenicity of enterohemorrhagic *Escherichia coli* in neonatal calves and evaluation of fecal shedding by treatment with probiotic *Escherichia coli*. *J. Food Prot.* **66:**924–930.

appendix A

A Microbial Who's Who

IN 1982, JUST 34% OF FOOD-BORNE ILLNESS in the United States could be linked to a specific pathogen. The etiology of the remaining 66% of the cases reported to the Centers for Disease Control (CDC) that year could not be determined. Bacterial pathogens accounted for about one-half of the 11,050 cases for which a cause was identified; viruses (norovirus and hepatitis A virus) were responsible for most of the remaining cases. *Salmonella*, at 37%, led the list of bacterial causes of food-borne illness, whereas *Clostridium perfringens* and *Staphylococcus aureus*—at 22% and 12%, respectively—ranked second and third (193).

The only notifiable food- and waterborne diseases in the United States in 1982 were amebiasis, botulism, cholera, salmonellosis, shigellosis, typhoid fever, and hepatitis (172). By 2003, cryptosporidiosis, cyclosporiasis, giardiasis, enterohemorrhagic *Escherichia coli* (O157 and others), postdiarrheal hemolytic uremic syndrome, and listeriosis had been added to the list, and the percentage of cases of unknown etiology had dropped to less than 32% (44, 138). *Salmonella* was still the most common of the food-borne bacterial pathogens in 2005 (39% of laboratory-diagnosed cases), followed by *Campylobacter* at 34% and *Shigella* at 12.5% (303).

Profiling the Pathogens

Food-borne and waterborne pathogens encompass a broad spectrum of microbes, including viruses, bacteria, and protozoa. Some of the microorganisms described in this appendix—*Salmonella*, *S. aureus*, and *Clostridium botulinum*, among others—have been recognized as human pathogens for many decades. Others, such as *Listeria monocytogenes* and *Campylobacter*, gained their notoriety more recently. And in a few cases—*Providencia alcalifaciens*, for example—the jury is still out.

Bacillus anthracis

Microbiology: A gram-positive, nonmotile, spore-forming, rod-shaped bacterium, *B. anthracis* is relatively resistant to adverse environmental conditions such as extremes of temperature, drying, UV light, high pH, and high salt concentrations (237).

History: *B. anthracis* was first cultured in 1876 by Robert Koch, who identified it as the cause of anthrax in cattle, a conclusion that was corroborated by Louis Pasteur the following year (37, 77). A large outbreak of gastrointestinal anthrax occurred in Lebanon in the 1960s (156).

Reservoir and geographic distribution: *B. anthracis*, a worldwide zoonotic, is most prevalent in underdeveloped and developing countries. The main animal hosts are herbivores, including cattle, goats, and other livestock. *B. anthracis* spores can survive for years in soil (156, 237, 266).

Epidemiology: Gastrointestinal anthrax, a relatively uncommon form of the disease, is usually contracted through eating contaminated, undercooked meat. Some researchers believe that its incidence is underreported, as only severe cases are usually hospitalized and diagnosed (266, 287).

Incubation period: One to 7 days (237).

Symptoms and duration: Symptoms of gastrointestinal anthrax include nausea, vomiting, acute abdominal pain, and bloody diarrhea, progressing to sepsis. Symptoms can last for weeks (144, 163, 287).

Morbidity and mortality: All ages and both genders are susceptible. Infection may be asymptomatic in some cases or may produce symptoms of gastroenteritis. In severe cases, the disease can progress to bacteremia, hemorrhage, or rarely, meningitis. Mortality rates as low as 4% and as high as 60% have been reported from various outbreaks (72, 163, 237, 266).

Infective dose: No firm data are available. The infectious dose for gastrointestinal anthrax is thought to be large numbers of spores or vegetative cells (72, 144).

Bacillus cereus

Microbiology: A gram-positive, spore-forming, rod-shaped bacterium, *B. cereus* is a facultative anaerobe, able to grow either aerobically or anaerobically. Its spores can survive mild heat treatments such as pasteurization treatment of milk or the typical temperatures reached in home cooking (206, 248).

History: The first description of *B. cereus* food poisoning was published in 1950 and was linked to a contaminated vanilla sauce (206). An outbreak of *B. cereus* was documented in the United States for the first time in 1969, and the first documented outbreak in the United Kingdom took place in 1971 (151).

Reservoir and geographic distribution: *B. cereus* is widespread in soil. Not a zoonotic, it has no known animal reservoir (206, 248).

Epidemiology: *B. cereus* outbreaks are usually associated with consuming cooked, starchy foods that have been held at ambient temperature for an extended time after cooking. *B. cereus* spores survive the cooking process, then germinate and multiply in the food as it cools. Some strains of *B. cereus* produce a heat-stable emetic toxin in the food as they multiply. Other strains produce a heat-labile enterotoxin in the food and in the intestine (169, 248).

Incubation period: Symptoms can appear within 1 to 5 hours for the emetic syndrome, and 8 to 16 hours for the diarrheal syndrome (248).

Symptoms and duration: For the emetic syndrome, the main symptoms are nausea and vomiting, lasting 6 to 24 hours. Diarrhea occurs only occasionally. The main symptoms of the diarrheal syndrome are a profuse, watery diarrhea and abdominal pain. Nausea occasionally occurs, but vomiting is rare. Symptoms of the diarrheal syndrome last about 12 to 24 hours (169).

Morbidity and mortality: Both the emetic and diarrheal types of *B. cereus* food poisoning are intoxications. Victims who have eaten food contaminated with an emetic toxin-producing strain will suffer the emetic syndrome; those who ingest an enterotoxin-forming strain will exhibit symptoms of the diarrheal syndrome. Both syndromes are self-limiting (169).

Infective dose: From 10^5 to 10^8 vegetative cells or spores per gram of food (169, 206).

Campylobacter spp.

Microbiology: A genus of gram-negative, curved or spiral-shaped bacteria, *Campylobacter* spp. grow best at 108°F (42°C), a higher optimum growth temperature than for most other food-borne pathogens (8). *Campylobacter* does not grow at temperatures below 86°F (30°C) (75).

History: *Campylobacter fetus* was first determined to be a cause of diarrhea in 1957. The species was then known as *Vibrio fetus*, since its microscopic appearance resembled that of other *Vibrio* spp. The genus name *Campylobacter* was introduced in 1973. In the mid- to late 1980s, *Campylobacter* was recognized as one of the most common bacterial causes of diarrhea worldwide (8).

Reservoir and geographic distribution: *Campylobacter* enteritis is a worldwide zoonosis. Members of the genus are found in birds and mammals (29, 161).

Epidemiology: *Campylobacter* enteritis is contracted most often by eating undercooked or contaminated poultry, by drinking unpasteurized milk or untreated water, or from contact with household pets (29, 161). Most cases of *Campylobacter* enteritis are due to *C. jejuni*; *C. upsaliensis* and *C. fetus* have also been recovered from some patients (178).

Incubation period: Two to 5 days (287).

Symptoms and duration: *Campylobacter* infection produces symptoms of diarrhea (sometimes bloody), fever, and abdominal cramps, usually lasting 2 to 10 days (8, 287).

Morbidity and mortality: In most cases, the illness is acute and self-limiting. Bacteremia may result in <1% of the cases. Fewer than 1 patient in 1,000 is stricken with Guillain-Barré syndrome, an autoimmune disease of the peripheral nervous system. The mortality rate for Guillain-Barré syndrome is approximately 10% (8, 39).

Infective dose: Between 800 and 10^6 *Campylobacter* organisms must be ingested to produce illness in 10 to 50% of individuals, according to one study (8).

Clostridium botulinum

Microbiology: *C. botulinum* is a gram-positive, spore-forming, rod-shaped bacterium and a strict anaerobe (151).

History: Several outbreaks of "sausage poisoning" described in the 18th and early 19th centuries were almost certainly caused by *C. botulinum* neurotoxin. Infant botulism was first recognized in California in 1976 (151).

Reservoir and geographic distribution: *C. botulinum* is found worldwide in soil and water (151).

Epidemiology: *C. botulinum* has been associated with a variety of low-acid processed foods, including home-canned vegetables, foil-wrapped baked potatoes, and mushrooms or garlic packed in oil (287). Infants (<12 months) are susceptible to botulism through ingesting spores of *C. botulinum*, which then germinate, multiply, and produce the neurotoxin in the intestine. Consumption of honey has been associated with approximately 20% of infant botulism cases (151).

Incubation period: Twelve to 72 hours for illness caused by preformed toxin; 3 to 30 days for infant botulism, which is caused by ingestion of spores or vegetative cells (287).

Symptoms and duration: Symptoms of *C. botulinum* intoxication in adults and children (>12 months) include nausea, vomiting, difficulty swallowing, blurred vision, and muscle weakness. Lethargy, weakness, poor muscle tone, constipation, difficulty feeding, and poor gag reflex are among the symptoms of infant botulism (287).

Morbidity and mortality: Botulism can result in respiratory failure and death (287). The mortality rate is 30 to 65%. With appropriate care, the survival rate for infant botulism is close to 100% (274).

Toxic dose: The minimum lethal dose for botulinum neurotoxin varies with the toxin type; type A toxin is more lethal than types B or E. The 50% human lethal dose is approximately 1 ng/kg of body weight (151). The number of spores that must be ingested to produce infant botulism is not yet known; however, the concentration of *C. botulinum* spores in contaminated honey is usually quite low—2 to 7 spores per 25 g of honey in one study, and between 5 and 25 spores per gram in another (75, 207, 209, 274).

Clostridium difficile

Microbiology: *C. difficile* is a gram-positive, anaerobic, spore-forming, rod-shaped bacterium (205).

History: *C. difficile* was detected in several cases of human infection between 1957 and 1962, but was concluded to be a secondary infection (272). The microbe was first recognized as a cause of pseudomembranous colitis in 1978, and its role as a cause of diarrheal disease was realized in 1981 (21, 85, 105). By 1988, *C. difficile* was recognized as a significant contributor to nosocomial diarrheal disease and pseudomembranous colitis (191). A clindamycin-resistant strain of *C. difficile* emerged in the late 1980s and was the cause of four nosocomial outbreaks of diarrheal disease between 1989 and 1992 (152).

Reservoir and geographic distribution: *C. difficile* is part of the normal flora of many infants and in about 3% of adults (280). It has also been found in the intestinal tracts of dogs, elk, and pigs (11, 183, 277). It is often present in the hospital environment (191).

Epidemiology: *C. difficile* infection is almost always nosocomial, although it can also occur in long-term care or nursing facilities or even in the community (205). The microbe is spread easily in the hospital environment. Patients who share a semiprivate room or ward with an infected patient are at risk of becoming infected with *C. difficile*, especially if they are immunocompromised, elderly, or have received clindamycin therapy (152, 187).

Incubation period: Symptoms of diarrhea usual develop 4 to 9 days after a patient has begun antibiotic treatment (93).

Symptoms and duration: Diarrhea—often mild and self-limiting, but sometimes severe—is the most common symptom of *C. difficile* infection and is accompanied by abdominal pain and fever. Some patients develop severe colitis and may suffer multiple relapses (93).

Morbidity and mortality: *C. difficile* disease symptoms are the result of two toxins produced by the microbe: an enterotoxin and a cytotoxin (191). The mortality rate is variable, depending on the virulence of the strain, and has been reported to be as high as 18% in infections caused by the newly emergent strains (93). These strains have been found to produce up to 20 times as much of both toxins as previously studied strains (78, 308).

Infective dose: Unknown.

Clostridium perfringens

Microbiology: *C. perfringens* is a gram-positive, anaerobic, spore-forming, rod-shaped bacterium (247).

History: *C. perfringens* was first recognized as a cause of food-borne disease in 1945, as a result of the investigation of a series of outbreaks that took place between 1943 and early 1945. Furthermore, the investigator suggested that the symptoms were due to the action of an enterotoxin (204).

Reservoir and geographic distribution: The species is widely distributed in soil, dust, and the intestines of animals and humans (247, 254).

Epidemiology: Illness caused by *C. perfringens* usually results from ingestion of a food (typically animal protein, although other foods have also been implicated) that contains large numbers of vegetative cells of an enterotoxin-producing strain of the bacterium (211, 247). Symptoms are triggered by the enterotoxin, which is produced when ingested cells of *C. perfringens* sporulate in the small intestine (211). Although the disease is usually food borne, it has also been associated with environmental contamination in long-term care facilities (149). *C. perfringens* has also been identified as a cause of sporadic nosocomial diarrhea (305).

Incubation period: Can be 6 to 24 hours, usually 10 to 12 hours (247, 287).

Symptoms and duration: Symptoms include profuse, watery diarrhea, abdominal cramps, and nausea. They are self-limiting and typically last for 12 to 24 hours, but sometimes linger for up to 48 hours (247, 287).

Morbidity and mortality: All sectors of the population are susceptible to *C. perfringens* food poisoning. The disease is usually self-limiting, although symptoms can be more severe and of longer duration in elderly or debilitated patients; the occasional death can occur in the elderly, the debilitated, or the very young (247, 254).

Infective dose: At least 5×10^5 vegetative cells per gram of food, usually 10^6/g or more (149, 247).

Cryptosporidium spp.

Microbiology: A protozoan parasite, *Cryptosporidium* reproduces by means of environmentally stable oocysts, which are excreted in the feces of the host. When ingested, the oocysts release sporozoites, which invade epithelial cells in the intestine (181).

History: *Cryptosporidium parvum* first was recognized as a human pathogen in 1976. It emerged as a cause of life-threatening diarrhea in patients with AIDS in the early 1980s (117).

Reservoir and geographic distribution: *Cryptosporidium* is a zoonotic protozoan. Its host range varies according to species (112, 117). *C. parvum*, for example, infects mainly mammalian hosts, including humans. Waterfowl such as Canada geese can transport infectious oocysts of *C. parvum* mechanically; however, this species does not infect birds, fish, amphibians, or reptiles, although other *Cryptosporidium* spp. do (113, 114). Species of *Cryptosporidium* that are able to infect humans also include *C. hominis*, *C. muris*, *C. felis*, and *C. canis* (314). *Cryptosporidium* oocysts have been found in salt and fresh (treated and untreated) waters around the world and in shellfish harvested from contaminated waters (33, 45, 66, 96, 107, 118, 140, 166, 194, 214, 292).

Epidemiology: *Cryptosporidium* is usually waterborne, but it can also be transmitted through contaminated food or person to person (117, 181, 217, 246).

Incubation period: Five days to 2 weeks (181).

Symptoms and duration: Symptoms include watery diarrhea, often with abdominal pain and, occasionally, fever, nausea, vomiting, and loss of appetite. Some infections are asymptomatic. Symptoms usually last several days to as long as 5 weeks (181).

Morbidity and mortality: The disease is usually self-limiting, but can be prolonged and more severe, even fatal, in immunocompromised patients (181).

Infective dose: Ten to 1,000 oocysts (101, 117).

Cyclospora cayetanensis

Microbiology: *C. cayetanensis* is a protozoan parasite (6). It reproduces by means of oocysts, which are excreted in the feces of the host. The oocysts are not infectious when excreted; they sporulate in the environment, with each oocyst containing two sporocysts, each of which contains two sporozoites (131).

History: *C. cayetanensis* first was recognized as a cause of human illness in 1977 (49). The first documented outbreak in the United States occurred in 1990 (130).

Reservoir and geographic distribution: *C. cayetanensis* is most common in subtropical and tropical countries, especially those that are underdeveloped. The host range is unknown; no natural animal host for this species has been confirmed (130).

Epidemiology: Cyclosporiasis is spread through contaminated water or food, notably fresh produce (100, 131, 135, 188, 244). Since the oocysts have not yet sporulated, and thus are not infectious, at the time they are excreted, the disease does not spread person to person (259).

Incubation period: One to 11 days (287).

Symptoms and duration: Symptoms include fatigue, watery diarrhea, nausea, anorexia, weight loss, and abdominal cramps. The diarrhea may be either intermittent or protracted (282, 287). Symptoms (especially diarrhea) can linger for several weeks (287).

Morbidity and mortality: Guillain-Barré and Reiter's syndromes are occasional secondary complications of protracted bouts of cyclosporiasis (51).

Infective dose: Unknown, but thought to be low (282).

Entamoeba histolytica

Microbiology: *E. histolytica* is an invasive protozoan parasite. Its two-stage life cycle consists of trophozoites, which move by means of pseudopodia, and nonmotile, infective cysts. Ingested cysts are carried to the small intestine; once there, each cyst produces eight daughter trophozoites. The trophozoites are gradually converted to precysts, which mature into cysts as they travel through the large intestine (288).

History: *E. histolytica* was isolated from the stool of a patient with dysentery in 1875. Its infective cyst form was demonstrated for the first time in 1913, and its life cycle first described in 1925 (288).

Reservoir and geographic distribution: Humans are the principal host reservoir of *E. histolytica* (288). There have been some indications of nonhuman animal hosts, but little confirmation (150, 162, 197, 270, 281). *E. histolytica* can be found worldwide, usually in contaminated water, but is most prevalent in underdeveloped and developing countries (7, 10, 15, 22, 32, 71, 103, 104, 119, 120, 122, 203, 233, 242).

Epidemiology: Transmission is most often via contaminated water or food, although the disease can also be transmitted by direct person-to-person contact (287).

Incubation period: Two to 3 days to 1 to 4 weeks (287).

Symptoms and duration: Symptoms of *E. histolytica* infection are bloody diarrhea and lower abdominal pain. Symptoms usually last for months (287).

Morbidity and mortality: Although the majority of cases of *E. histolytica* infection are either asymptomatic or limited to symptoms of diarrhea, some patients suffer more serious illness, including inflammatory bowel disease, liver abscesses, appendicitis, and even hemolytic uremic syndrome (31, 40, 81, 294). Diabetics are at increased risk of severe symptoms (35). The case fatality rate among symptomatic individuals worldwide is approximately 0.2% (175).

Infective dose: Ten to 100 cysts (101).

Enterobacter sakazakii

Microbiology: *E. sakazakii*, a member of the *Enterobacteriaceae* family, is a gram-negative, motile, rod-shaped bacterium (226).

History: *E. sakazakii* was first thought to be a yellow-pigmented strain of *E. cloacae*. In 1961, it was linked to two cases of neonatal meningitis. The species name "*E. sakazakii*" was first introduced in 1980 (87, 226).

Reservoir and geographic distribution: *E. sakazakii* has been reported in many countries, but efforts to find an environmental reservoir for the microbe were fruitless until 2001, when it was detected in the guts of Mexican fruit flies (26, 177, 184, 220, 223). In 2003, it was also isolated from the guts of the larvae of the stable fly, a common inhabitant of cowsheds (121). However, its incidence in stable flies is low, leading some researchers to suggest that the stable fly is not a significant vector or reservoir of *E. sakazakii* (216). It has been detected in food production facilities and in households (157).

Epidemiology: *E. sakazakii* has been found at low levels in powdered infant formula in many different countries (220). The formula is reconstituted in hospitals and fed orally to newborns. Formula is sometimes prepared in advance and held under refrigeration until used. Also, due to the extended time required for feeding, the reconstituted formula can remain at room temperature for several hours, potentially allowing viable *E. sakazakii* to multiply to a critical concentration (226, 241). *E. sakazakii* outbreaks have been linked to specific production lots of powdered infant formula by detection of similar or identical strains in patient samples and in the implicated formula (133, 297).

Incubation period: One day to 3 weeks; usually less than 1 week (26, 133, 297).

Symptoms and duration: Symptoms in neonates include fever, abnormally rapid heart rate, and neurological abnormalities such as seizures (133).

Morbidity and mortality: *E. sakazakii* is an opportunistic pathogen and is a cause of neonatal necrotizing enterocolitis and neonatal meningitis, often with sepsis (133, 297). Although newborns are its most common victims, *E. sakazakii* has also infected infants, children, and adults, causing sepsis and/or respiratory infections (179). The mortality rate in newborns has been reported to be as high as 40 to 80% (226).

Infective dose: Has not yet been determined epidemiologically, but has been suggested to be high (184, 241).

Escherichia coli, enteroaggregative (EAEC)

Microbiology: A member of the family *Enterobacteriaceae*, *E. coli* is a gram-negative, facultatively anaerobic, motile, rod-shaped bacterium (82). Enteroaggregative strains of *E. coli* form aggregates when adhering to cells in HEp-2 tissue cultures (225).

History: EAEC was first described in 1987 (225).

Reservoir and geographic distribution: Although EAEC is present in both underdeveloped and industrialized countries, its host range and reservoir have not yet been determined (225).

Epidemiology: Most likely, infection results from ingestion of contaminated food or water (3, 208).

Incubation period: Eight to 18 hours in a human volunteer study (225).

Symptoms and duration: Symptoms are persistent, watery, mucoid diarrhea with little or no vomiting lasting several days, sometimes more than 2 weeks. The mean duration in one outbreak was 11 days. As many as one-third of patients experience bloody stools (225).

Morbidity and mortality: Young children are especially affected, but older children and adults are also susceptible. EAEC is a significant cause of pediatric diarrhea in underdeveloped countries and has been recognized as a cause of traveler's diarrhea (102, 167, 225, 262). The disease can be fatal; five children died in a nursery outbreak in Mexico City (225).

Infective dose: Greater than 10^8 cells (101).

Escherichia coli, enterohemorrhagic (Shiga toxin producing)

Microbiology: A member of the family *Enterobacteriaceae*, *E. coli* is a gram-negative, facultatively anaerobic, motile, rod-shaped bacterium (82). *E. coli* O157:H7 is the best-known enterohemorrhagic serotype, but some other *E. coli* serotypes also produce Shiga toxins (224, 287).

History: *E. coli* O157:H7, the best-known Shiga toxin-producing serotype, was first identified as a pathogen in 1982, when it was found to be responsible for two outbreaks related to consumption of hamburgers (295).

It was the cause of the massive outbreak of hemorrhagic colitis (bloody diarrhea with little or no fever) and hemolytic uremic syndrome (HUS) traced to the Jack in the Box fast-food chain in 1993 (42).

Reservoir and geographic distribution: A zoonotic, *E. coli* O157:H7 is found in the intestinal tracts of ruminants and some other animals (224). Dairy cattle are a significant host reservoir of *E. coli* O157:H7, which can be introduced into soil and water through contaminated feces or the spreading of manure for fertilization of crops, possibly resulting in the contamination of produce during growth or harvest (147, 176, 190, 276).

Epidemiology: *E. coli* O157:H7 can contaminate raw beef as the result of small amounts of fecal material coming into contact with a carcass during slaughter and butchering (14, 79, 158, 293). It has been found in a variety of produce, including lettuce, apples, and alfalfa sprouts. It can also be present in raw milk, as a result of fecal contamination during or after milking. Humans acquire *E. coli* O157:H7 infections by person-to-person transmission; by ingesting contaminated food (especially undercooked meat), milk, or water; by coming into contact with infected animals in petting zoos or at agricultural fairs; or by swimming or playing in contaminated water (60, 224).

Incubation period: One to 8 days (287).

Symptoms and duration: Symptoms include severe, often bloody, diarrhea, abdominal pain, and vomiting. Unless the disease progresses to HUS, symptoms last 5 to 10 days (287).

Morbidity and mortality: Most cases are self-limiting, especially in adults. Some elderly patients and young children are at risk of more serious disease. Approximately 5 to 10% of children (<10 years old) develop HUS, a severe disease affecting the kidneys, digestive system, and other organs. The fatality rate in cases of HUS is 3 to 5%, and 12 to 30% of victims will continue to suffer long-term consequences—hypertension or impairment to renal or central nervous system functions (224).

Infective dose: One hundred to 200 organisms (224).

Escherichia coli, enteroinvasive (EIEC)

Microbiology: A member of the family *Enterobacteriaceae*, *E. coli* is a gram-negative, facultatively anaerobic, motile, rod-shaped bacterium (82). EIEC strains are characterized by their ability to invade enterocytes, epithelial cells found in the intestinal mucosa of the colon and the distal end of the small intestine (88, 106).

History: EIEC, originally described as a "*Shigella*-like" microorganism, was isolated for the first time from a patient suffering from bacillary dysentery in 1946 (82, 83). The first documented outbreak due to person-to-person transmission took place in 1981 (124).

Reservoir and geographic distribution: EIEC is not zoonotic (168); it is found worldwide (24, 89, 110, 116, 124, 168, 201, 289, 291, 304).

Epidemiology: EIEC is usually transmitted through ingestion of contaminated food or water, although person-to-person transmission has also been documented (124, 168, 224).

Incubation period: Two to 48 hours (151, 304).

Symptoms and duration: The most common symptom is watery diarrhea. In a minority of cases, patients suffer from a dysentery-like syndrome, including bloody diarrhea, tenesmus, and fever (224). The median duration in one food-borne outbreak was 3 days (273).

Morbidity and mortality: Asymptomatic infections are unusual, but most victims suffer no more than a self-limiting watery diarrhea (224). The very young and elderly are especially susceptible (151). A minority of patients experience more severe symptoms, including bacillary dysentery, recurrent watery diarrhea (notably patients with human immunodeficiency virus infection or AIDS), sepsis, or bacteremia (24, 224, 304). Some cases can be fatal (304).

Infective dose: In volunteer studies, 10^6 viable cells (139).

Escherichia coli, enteropathogenic (EPEC)

Microbiology: A member of the family *Enterobacteriaceae*, *E. coli* is a gram-negative, facultatively anaerobic, motile, rod-shaped bacterium (82).

History: EPEC was recognized as a cause of infantile diarrhea as early as the 1940s (73). The term "enteropathogenic *E. coli*" was introduced in 1955 (180).

Reservoir and geographic distribution: Though found worldwide, EPEC is more significant in developing countries, where it is a significant cause of neonatal nosocomial diarrhea (73). EPEC is not a zoonotic; the reservoir for EPEC is thought to be asymptomatic or symptomatic adults and children (224).

Epidemiology: Transmission is fecal-oral in cases of infant infection, usually via contaminated hands, fomites (such as dust), or formula. Contaminated food and water have been implicated in adults (224).

Incubation period: Unknown in infants; 12 to 24 hours in adult volunteers (180).

Symptoms and duration: Symptoms of acute, watery diarrhea, vomiting, and low-grade fever, often quite severe, can be of long duration. In one outbreak, hospitalizations lasted 21 to 120 days (224).

Morbidity and mortality: Infants and children are especially susceptible, and the mortality rate can be high: 25 to 50% in developing countries and in earlier outbreaks in industrialized countries (73, 224). Modern treatment has greatly reduced the mortality rate in industrialized countries (224).

Infective dose: From 10^8 to 10^{10} in adult volunteer studies; presumed to be lower in children (224).

Escherichia coli, enterotoxigenic (ETEC)

Microbiology: A member of the family *Enterobacteriaceae*, *E. coli* is a gram-negative, facultatively anaerobic, motile, rod-shaped bacterium (82). ETEC produces two enterotoxins, one that is heat stable and one that is heat labile (151).

History: ETEC has been recognized as a cause of enteric disease since the early 1970s (212, 310).

Reservoir and geographic distribution: ETEC is endemic in many underdeveloped countries and is also found in industrialized countries (253).

Epidemiology: Transmitted through ingestion of contaminated food or water, ETEC is a significant cause of traveler's diarrhea (23, 58, 94, 253).

Incubation period: One to 3 days (287).

Symptoms and duration: Symptoms of abdominal cramps, sudden onset of watery diarrhea, and some vomiting typically last for 3 to 7 or more days (151, 287).

Morbidity and mortality: The disease is usually self-limiting, but may be life-threatening to infants (58, 224).

Infective dose: From 10^6 to 10^{10} in adults (253).

Giardia lamblia, *Giardia intestinalis*

Microbiology: *Giardia* is an anaerobic, flagellated protozoan parasite with a two-stage life cycle. Ingested cysts enter the duodenum, where they release trophozoites. The trophozoites attach to the epithelium of the duodenum, where they reproduce by binary fission, eventually resulting in the production of new infectious cysts, which are excreted (95, 267).

History: *G. lamblia* was suspected of being a human pathogen as early as 1915 (171). By the 1940s and 1950s, this suspicion had been confirmed (195, 263).

Reservoir and geographic distribution: *Giardia* is zoonotic in both domestic and wild animals (95, 141, 148, 189). The parasite has been found in treated and untreated drinking water, surface waters, cistern water, and sewage worldwide (12, 54, 90, 140, 234, 238, 292, 306). Shellfish grown in contaminated waters can also transmit *Giardia* oocysts (108).

Epidemiology: Transmission is person to person or through fecal contamination of water by infected persons or animals; there has been the occasional food-borne outbreak due to fecal contamination of food by an infected food handler (95, 111, 164, 210, 230, 239, 243, 267, 284).

Incubation period: One to 4 weeks (287).

Symptoms and duration: Symptoms, including diarrhea (acute or chronic), flatulence, and bloating, last for weeks (287).

Morbidity and mortality: Infection cures spontaneously without treatment in more than 80% of cases; the remainder respond to antigiardiasis drugs. The fatality rate is negligible (267).

Infective dose: Ten to 25 cysts (95, 101).

Hepatitis A virus

Microbiology: A small RNA virus (family *Picornaviridae*), hepatitis A is a member of the genus *Hepatovirus* (256).

History: Infectious hepatitis was attributed to a viral cause in 1950. The agent of the disease was propagated successfully in tissue cultures and then fed to human volunteers in order to reproduce the disease, thus fulfilling Koch's postulates (76, 129).

Reservoir and geographic distribution: Although the distribution of hepatitis A virus is worldwide, the virus is most prevalent in underdeveloped and developing countries, where it is widely endemic (27). Hepatitis A virus has a limited host range, consisting of humans and some nonhuman primates (19).

Epidemiology: Hepatitis A virus is transmitted through shellfish harvested from contaminated water, by drinking or swimming in contaminated water, and from contaminated foods (143, 170, 196). The food may be contaminated by an infected food handler or by having been grown or washed in contaminated water (25, 287).

Incubation period: Fifteen to 50 days; average of about 30 days (287).

Symptoms and duration: Symptoms, which include dark urine, diarrhea, jaundice, and flu-like symptoms, last 2 days to 3 months. The disease is usually self-limiting (256, 287).

Morbidity and mortality: Severity of symptoms is age associated; infected children under 6 years of age are usually asymptomatic, whereas older children and adults are more likely to develop symptoms of the infection (256).

Infective dose: Unknown.

Hepatitis E virus

Microbiology: Hepatitis E is a single-stranded, small, nonenveloped RNA virus of undetermined taxonomy (269, 307).

History: The first confirmed outbreaks of hepatitis E virus occurred in India in the 1950s. Hepatitis E virus is suspected of having caused outbreaks in the 1850s (269).

Reservoir and geographic distribution: Hepatitis E virus is zoonotic and is found in both wild and domesticated animals, notably swine (269, 278, 316). Although hepatitis E virus is more widely distributed in tropical and subtropical underdeveloped countries, it can be found in some industrialized countries as well. For example, 79% of swine in the U.S. Midwest were found to be positive for the virus in one study (269).

Epidemiology: Hepatitis E is transmitted mainly through ingestion of contaminated water or food or, to a lesser extent, by person-to-person transmission (269).

Incubation period: Two to 9 weeks (53).

Symptoms and duration: Moderate to severe jaundice is usually self-limiting (269).

Morbidity and mortality: Young adults in the range of 15 to 30 years of age are the population most at risk. Mortality is in the range of 0.5 to 3.0%, but can be as high as 15 to 25% in pregnant women (269).

Infective dose: Unknown.

Listeria monocytogenes

Microbiology: A gram-positive, psychrotrophic, aerobic to microaerophilic, motile, non-spore-forming, short rod-shaped bacterium, *L. monocytogenes* is able to survive adverse environmental conditions such as drying and high salt concentrations (151, 249, 261).

History: *L. monocytogenes* was first isolated from rabbits in 1926, and the first isolation from a human was in 1929. *L. monocytogenes* was tied to disease in newborns in 1936, but was not linked unequivocally to foodborne disease until it was shown to be the cause of a series of outbreaks in the early 1980s (92, 185, 261).

Reservoir and geographic distribution: A worldwide zoonotic, *L. monocytogenes* is widely distributed in nature, including in soil, coastal waters, and a variety of animals and birds (151, 249, 261).

Epidemiology: *L. monocytogenes* has been found in a wide variety of foods around the world, though industrialized countries have reported most of the large outbreaks of listeriosis. Individuals become infected as a result of ingesting contaminated food. Contaminated meats, dairy products, produce, and seafood have all been implicated in outbreaks of listeriosis (13, 16, 69, 92, 134, 185, 186, 192, 198, 222, 227, 232, 236, 252, 255, 257, 261, 265).

Incubation period: Nine to 48 hours for gastrointestinal symptoms; 2 to 6 weeks for invasive syndromes; within 7 days of birth for infants infected in the womb; 2 to 3 weeks for neonates infected during delivery (260, 268, 287).

Symptoms and duration: Common symptoms are fever, muscle aches, nausea, and diarrhea—mild flu-like symptoms. Duration is variable, depending on symptoms and immunocompetence of the patient (287).

Morbidity and mortality: Listeriosis is self-limiting in many cases. Pregnant women who are infected with *L. monocytogenes* may suffer miscarriage or can transmit the infection to the fetus in utero. Elderly or immunocompromised victims are at elevated risk of suffering bacteremia or meningitis. Newborns and infants can suffer sepsis or meningitis. The mortality rate (including stillbirths) for neonatal infection is approximately 50%. The occasional death can also occur among adult victims, most often those who are older or who are immunocompromised (236, 260, 287).

Infective dose: Unknown.

Norovirus

Microbiology: Norovirus, a member of the family *Caliciviridae*, is a single-stranded RNA virus (256).

History: Norovirus (then called Norwalk virus) was recognized in 1972 as a cause of "acute non-bacterial infectious gastroenteritis," a syndrome that had been known to physicians and epidemiologists since the 1940s (159).

Reservoir and geographic distribution: Norovirus gastroenteritis is primarily a disease of developed countries (86). Zoonotic transmission has been suggested, but not confirmed (173).

Epidemiology: Norovirus infection is transmitted through ingestion of contaminated food and water or by person-to-person spread through contact with contaminated surfaces or other fomites (146, 287).

Incubation period: Twenty-four to 48 hours (287).

Symptoms and duration: Symptoms include nausea, vomiting, and copious, watery diarrhea, lasting 24 to 60 hours (287).

Morbidity and mortality: An important cause of gastroenteritis among adults, norovirus has caused large outbreaks among restaurant patrons, in day care centers, in nursing homes, in hospitals, and in hotels and has been a recurring problem in recent years on cruise ships (146, 173, 311). Although usually self-limiting and of relatively short duration, symptoms can persist for a longer period of time in the elderly (173). Approximately 1% of victims suffer severely enough to warrant hospitalization (311).

Infective dose: Ten to 100 viral particles (101).

Providencia spp.

Microbiology: A member of the family *Enterobacteriaceae*, *Providencia* is a gram-negative, motile, rod-shaped bacterium (82).

History: A 1989 report suggested that *Providencia alcalifaciens* could play a role in traveler's diarrhea (5). The first food-borne outbreak linked to *P. alcalifaciens* occurred in Japan in 1996 (218).

Reservoir and geographic distribution: *Providencia* spp. are normal constituents of human and animal feces worldwide (4, 5, 218, 231, 300).

Epidemiology: Transmission is most likely through ingestion of contaminated food, but very little information is available so far.

Incubation period: The mean incubation period in the 1996 Japan outbreak was approximately 69 hours (218).

Symptoms and duration: Symptoms in the Japanese outbreak were mild and self-limiting and included diarrhea (nonbloody), abdominal pain, and fever (218). Individuals suffering from *Providentia rettgeri* infections have also complained of vomiting (315).

Morbidity and mortality: Illnesses tend to be self-limiting. There have been no reported deaths so far due to *Providencia* gastroenteritis (218, 300, 315)

Infective dose: Unknown.

Salmonella enterica (nontyphoid)

Microbiology: A member of the family *Enterobacteriaceae*, *Salmonella enterica* is a gram-negative, facultatively anaerobic, non-spore-forming, rod-shaped bacterium (82).

History: *S. enterica* serotype Choleraesuis (originally named *Bacillus choleraesuis*), the causative agent of hog cholera, was first isolated in 1885 (75). Over time, the known number of serotypes has increased, from 45 in 1933 to more than 80 in 1939 and approximately 2,200 by 2001 (123).

Reservoir and geographic distribution: Except for a small number of serotypes, most notably *S. enterica* serotype Paratyphi A and serotype

Paratyphi B (151), *Salmonella* is zoonotic worldwide. Humans, animals, insects, reptiles, and birds all host *Salmonella* spp., although some serotypes are host-adapted (151). The relative distribution of *Salmonella* serotypes varies from region to region, and even from country to country within a region (65, 98).

Epidemiology: *S. enterica* has been found in meat and poultry, dairy products, eggs, fresh produce, spices, condiments, chocolate, and other foods (151). Some household pets, especially reptiles, carry the pathogen asymptomatically and shed it in their feces (64, 299). Individuals are infected by ingesting food or water that has become contaminated with *S. enterica*, by touching a reptile or other animal that is a carrier of the microbe, by swimming or playing in contaminated water, or by person-to-person (secondary) transmission (20, 48, 128, 151, 182, 228, 271, 279, 298).

Incubation period: Usually 1 to 3 days (287).

Symptoms and duration: Typically lasting 4 to 7 days, symptoms of salmonellosis can include diarrhea, fever, cramps, and vomiting (287). Serotype Paratyphi infections produce symptoms similar to those found in typhoid fever (18, 287).

Morbidity and mortality: Salmonellosis is usually self-limiting, but long-term consequences, including bacteremia, bacterial endocarditis, and osteomyelitis, can occur in a small percentage of cases (1). Individuals who were infected with *S. enterica* serotype Dublin had a 12-fold greater risk of dying within 1 year of their initial infection than a control group in one study; those infected with other serotypes of *S. enterica* had a two- to threefold greater risk than controls did (127).

Infective dose: The infective dose varies, depending on the serotype and the susceptibility of the victim. The food that is ingested along with the *Salmonella* is also a factor, as some constituents of food help protect the microbe from the acidic conditions in the stomach. Usually ingestion of 10^7 to 10^9 cells is required to develop salmonellosis, but an estimated infective dose of just 10^2 cells was calculated for the 1972 outbreak caused by *S. enterica* serotype Eastbourne (59).

Salmonella enterica serotype Typhi

Microbiology: *S. enterica* serotype Typhi, a gram-negative, facultatively anaerobic, non-spore-forming, rod-shaped bacterium, is a member of the family *Enterobacteriaceae* (82).

History: Serotype Typhi was first viewed microscopically in 1880 and was cultured for the first time in 1884. An epidemic of typhoid fever was traced to drinking water and sewage in 1902, the same year when Koch first reported on the connection between asymptomatic carriers and epidemic typhoid (37).

Reservoir and geographic distribution: Serotype Typhi is found worldwide, but its incidence is far lower in industrialized countries than in developing countries (56). It does not cause a zoonosis (18). Approximately 3 to 5% of infected individuals become asymptomatic carriers after recovering from their acute illness, with the pathogen residing in the gallbladder and being shed sporadically in the stool (174, 296).

Epidemiology: Transmission is by the fecal-oral route, usually through ingestion of contaminated food or water, although person-to-person transmission is also possible (38, 301).

Incubation period: Seven to 14 days (18).

Symptoms and duration: Fever, malaise, nausea, anorexia, abdominal pain, and nonproductive cough are common symptoms. An abdominal rash consisting of pinkish lesions 1 to 4 mm in diameter can develop. Diarrhea and constipation are both equally likely to occur. Fever tends to rise during the first week, plateau in the second week, and fall off during the third and fourth weeks (18).

Morbidity and mortality: Complications of typhoid fever can include secondary opportunistic infection, acute gallbladder inflammation, acute renal failure, meningitis, and hepatitis, among others (50, 125, 145, 155). The mortality rate with effective antibiotic treatment is less than 1% (2).

Infective dose: Whereas the infective dose in volunteer studies is $>10^6$, a study of the incubation period of typhoid outbreaks is suggestive of a lower infective dose (30, 61, 115).

Shigella spp.

Microbiology: Members of the family *Enterobacteriaceae*, *Shigella* spp. are gram-negative, nonmotile, aerobic, rod-shaped bacteria (82). The genus *Shigella* comprises four species—*S. boydii*, *S. dysenteriae*, *S. flexneri*, and *S. sonnei*—all of which are pathogenic (151).

History: First isolated from feces in cases of dysentery in 1888, *S. dysenteriae* was the first *Shigella* species to be described (82).

Reservoir and geographic distribution: *Shigella* spp. are found in the intestinal tracts of humans and primates (250). *S. dysenteriae* remains a problem in underdeveloped countries, whereas *S. sonnei* and *S. flexneri* have replaced *S. dysenteriae* as the dominant *Shigella* species in industrialized countries (285).

Epidemiology: Transmission of *Shigella* spp. is through the fecal-oral route, either by direct person-to-person contact, fomites such as utensils, contamination of food by an infected food handler, or washing produce in contaminated water (36, 55, 62, 99, 126, 160, 221, 240, 264, 275). Drinking or swimming in contaminated water has also been the source of an outbreak of shigellosis (165).

Incubation period: Twenty-four to 48 hours (287).

Symptoms and duration: *S. dysenteriae* is the cause of dysentery, a syndrome that comprises abdominal cramps and frequent passage of diarrheic stools containing blood and mucus (75). The other three species of *Shigella* cause shigellosis, a less severe gastrointestinal disease. Symptoms of shigellosis include abdominal cramps, fever, and diarrhea, sometimes bloody or containing mucus, and usually last 4 to 7 days (289).

Morbidity and mortality: Symptoms of shigellosis tend to be more severe than those experienced in other forms of gastroenteritis (75). In developing countries, *Shigella* infections are complicated by underlying poor nutrition and poor health, and deaths from shigellosis are relatively common (285). In industrialized countries, shigellosis is largely self-limiting. Complications, however, can occur, including severe dehydration, Reiter's syndrome (a form of reactive arthritis), and hemolytic uremic syndrome (75).

Infective dose: Ten to 100 bacteria (101).

Staphylococcus aureus

Microbiology: A member of the family *Micrococcaceae*, *S. aureus* is a gram-positive coccus that grows in grape-like clusters (139).

History: Staphylococcal food poisoning was first studied in 1894. The first volunteer study was reported in 1914, when the investigator drank milk that had been contaminated with *S. aureus* cultures. The presence of an enterotoxin in *S. aureus*-contaminated food was established in 1930 (151).

Reservoir and geographic distribution: Species of *Staphylococcus* are host-adapted; that is, different species colonize different host animals. *S. aureus* colonizes humans and many domesticated animals. In humans, it is carried, usually asymptomatically, on the skin, in the nostrils, and around the perineal area (151).

Epidemiology: *S. aureus* can be found at low levels in many foods of animal origin. It can also be introduced into foods by human carriers. Extended holding of a contaminated food at improper temperature allows enterotoxin-producing strains to multiply and produce a heat-stable enterotoxin in the food (151).

Incubation period: One to 6 hours (287).

Symptoms and duration: Symptoms are sudden-onset nausea, vomiting, and abdominal cramps. Diarrhea and fever are sometimes present. Symptoms last 24 to 48 hours (287).

Morbidity and mortality: Self-limiting (287).

Toxic dose: Approximately 20 ng of enterotoxin (151).

Vibrio cholerae

Microbiology: A member of the family *Vibrionaceae*, *V. cholerae* is a gram-negative, motile, facultatively anaerobic, curved rod-shaped bacterium (137, 213).

History: Although John Snow deduced in 1854 that cholera was waterborne, it was not until 1883 that Robert Koch first isolated *V. cholerae* (213). Cholera toxin was purified for the first time in 1969 (91). *V. cholerae* has been responsible for seven pandemics since 1817 (213).

Reservoir and geographic distribution: Distribution is worldwide, but the microbe is most prevalent in warmer climates (84, 213). *V. cholerae* can be found in bays and estuaries, especially in warmer climates; it can reach a high concentration in shellfish that grow in contaminated waters (57, 219, 251).

Epidemiology: *V. cholerae* is spread by ingesting contaminated water or food (213). Outbreaks have been traced to contaminated seafood, frozen coconut milk, and home-canned palm fruit (17, 43, 142, 283, 290).

Incubation period: Twenty-four to 72 hours (287).

Symptoms and duration: Symptoms, consisting of profuse, watery diarrhea and vomiting, usually last for 3 to 7 days (287).

Morbidity and mortality: Symptoms are severe and can result in extreme dehydration and death (287). Victims already suffering from other diseases or from malnutrition (such as refugees housed in temporary relief camps) are at an elevated risk of dying from dehydration (202).

Infective dose: Unknown.

Vibrio parahaemolyticus

Microbiology: A member of the family *Vibrionaceae*, *V. parahaemolyticus* is a gram-negative, motile, facultatively anaerobic, halophilic, cold-sensitive, curved rod-shaped bacterium (137).

History: *V. parahaemolyticus* was recognized as a pathogen in the early 1950s (213). A new clone (O3:K6), which emerged in 1996, quickly spread around the world and is credited with being responsible for a pandemic of *V. parahaemolyticus* infections (47, 109, 199, 235, 245, 302, 312).

Reservoir and geographic distribution: *V. parahaemolyticus* is present in coastal marine waters worldwide (68, 199). It is found most readily in warmer climates and in summer months, when the water temperature is more conducive to its survival and growth (309).

Epidemiology: *V. parahaemolyticus* infections are most often contracted by eating contaminated raw or undercooked seafood (68, 97, 199, 245, 309).

Incubation period: Two to 48 hours (287).

Symptoms and duration: Symptoms of *V. parahaemolyticus* gastroenteritis include nausea, vomiting, abdominal cramps, and watery diarrhea and last for 2 to 5 days (287).

Morbidity and mortality: The disease is self-limiting in most cases (309). Individuals who are immunocompromised or who suffer from diseases such as diabetes or liver disease may be at risk for serious complications such as septicemia, carrying a risk of fatal illness (97, 309).

Infective dose: From 10^5 to 10^8 bacteria (101).

Vibrio vulnificus

Microbiology: A member of the family *Vibrionaceae*, *V. vulnificus* is a gram-negative, motile, facultatively anaerobic, curved rod-shaped bacterium (137).

History: *V. vulnificus* was first recognized as a human pathogen in the late 1970s (28, 46).

Reservoir and geographic distribution: Found in coastal marine waters, *V. vulnificus* is more prevalent during summer months or in warmer coastal regions (200, 215). It can also be found in the intestinal contents of fish and shellfish (67). Two biotypes of *V. vulnificus* are known. Biotype 1, a human pathogen, is a normal inhabitant of coastal seawaters; biotype 2 is pathogenic for eels and has also been reported as an opportunistic human pathogen (9, 136).

Epidemiology: Victims become infected by eating contaminated raw or undercooked seafood or by coming into contact with contaminated seawater (200). Handling eels infected with *V. vulnificus* biotype 2 can also result in infection (136).

Incubation period: One to 7 days (287).

Symptoms and duration: Symptoms include abdominal pain, vomiting, diarrhea, bacteremia, and wound infections and typically last for 2 to 6 days (287).

Morbidity and mortality: Immunocompromised individuals and people with liver disease are at greatest risk of fatal septicemia, which has a mortality rate greater than 50% (200, 287, 313).

Infective dose: Unknown.

Yersinia enterocolitica

Microbiology: *Y. enterocolitica* is a member of the family *Enterobacteriaceae*. *Y. enterocolitica*, a coccoid-shaped, gram-negative, facultatively anaerobic bacillus, is cold tolerant and can grow slowly at 32 to 36°F (0 to 2°C) (52, 63, 74).

History: *Y. enterocolitica* was identified as a human pathogen in 1939, but was only recognized as a cause of food-borne disease in the 1970s (34, 74).

Reservoir and geographic distribution: Zoonotic worldwide, *Y. enterocolitica* has been found in many kinds of animals and has also been detected in surface waters. The serotypes most frequently associated with human disease tend to be carried in the mouth and intestinal tract of healthy pigs (74). *Y. enterocolitica* has been detected in meats (beef, lamb, pork, and poultry), milk and other dairy products, environmental waters, and seafood. Many, but not all, of the isolates from food and water are nonpathogenic in humans (70, 74, 80, 132, 229, 258, 286).

Epidemiology: The incidence of *Y. enterocolitica* human infections is highest in cooler climates, notably the cooler regions of Europe and North America (74). Individuals become infected by ingesting food or water that is contaminated with a pathogenic serotype (52). Two U.S. outbreaks involving children were traced to raw chitterlings (raw pork intestines), a traditional winter holiday dish in the black community (41, 153, 154).

Incubation period: One to 11 days (52).

Symptoms and duration: Younger children suffer from enterocolitis, consisting of diarrhea, low-grade fever, and abdominal pain (52). Appendicitis-like symptoms, including diarrhea, vomiting, fever, and abdominal pain, occur mainly in older children and young adults. Symptoms last 1 to 3 weeks (287).

Morbidity and mortality: Most cases are self-limiting, but complications and sequellae can include, among others, appendicitis, arthritis, erythema nodosum, bacteremia, or extraintestinal infections. A case fatality rate of 34 to 50% has been reported for patients suffering from *Y. enterocolitica* bacteremia (52).

Infective dose: Usually 10^9 bacteria (101).

References

1. **Acheson, D.** 2000. Long-term consequences of foodborne disease. *Food Qual.* **7:**29, 31–33.
2. **Ackers, M.-L., N. D. Puhr, R. V. Tauxe, and E. D. Mintz.** 2000. Laboratory-based surveillance of *Salmonella* serotype Typhi infections in the United States. Antimicrobial resistance on the rise. *JAMA* **283:**2668–2673.
3. **Adachi, J. A., J. J. Mathewson, Z.-D. Jiang, C. D. Ericsson, and H. L. DuPont.** 2002. Enteric pathogens in Mexican sauces of popular restaurants in Guadalajara, Mexico, and Houston, Texas. *Ann. Intern. Med.* **136:**884–887.
4. **Albert, M. J., K. Alam, M. Ansaruzzaman, M. M. Islam, A. S. M. H. Rahman, K. Haider, N. A. Bhuiyan, S. Nahar, N. Ryan, J. Montanaro, and M. M. Mathan.** 1992. Pathogenesis of *Providencia alcalifaciens*-induced diarrhea. *Infect. Immun.* **60:**5017–5024.
5. **Albert, M. J., A. S. G. Faruque, and D. Mahalanabis.** 1998. Association of *Providencia alcalifaciens* with diarrhea in children. *J. Clin. Microbiol.* **36:** 1433–1435.
6. **Alfano-Sobsey, E. M., M. L. Eberhard, J. R. Seed, D. J. Weber, K. Y. Won, E. K. Nace, and C. L. Moe.** 2004. Human challenge pilot study with *Cyclospora cayetanensis*. *Emerg. Infect. Dis.* **10:**726–728.
7. **Ali, M. B., K. S. Ghenghesh, R. B. Aissa, A. Abuhelfaia, and M. Dufani.** 2005. Etiology of childhood diarrhea in Zliten, Libya. *Saudi Med. J.* **26:**1759–1765.
8. **Allos, B. M.** 2001. *Campylobacter jejuni* infections: update on emerging issues and trends. *Clin. Infect. Dis.* **32:**1201–1206.
9. **Amaro, C., and E. G. Biosca.** 1996. *Vibrio vulnificus* biotype 2, pathogenic for eels, is also an opportunistic pathogen for humans. *Appl. Environ. Microbiol.* **62:**1454–1457.
10. **Amin, O. M.** 2002. Seasonal prevalence of intestinal parasites in the United States during 2000. *Am. J. Trop. Med. Hyg.* **66:**799–803.
11. **Arroyo, L. G., J. D. Rousseau, H. R. Staempfli, and J. S. Weese.** 2005. Suspected *Clostridium difficile*-associated hemorrhagic diarrhea in a 1-week-old elk calf. *Can. Vet. J.* **46:**1130–1131.
12. **Aulicino, F. A., M. Marranzano, and L. Mauro.** 2005. La contaminazione delle acque superficiali e gli indicatori microbiologici. *Ann. Ist. Super. Sanità* **41:**359–370.
13. **Aureli, P., G. C. Fiorucci, D. Caroli, G. Marchiaro, O. Novara, L. Leone, and S. Salmaso.** 2000. An outbreak of febrile gastroenteritis associated with corn contaminated by *Listeria monocytogenes*. *N. Engl. J. Med.* **342:** 1236–1241.
14. **Avery, S. M., A. Small, C.-A. Reid, and S. Buncic.** 2002. Pulsed-field gel electrophoresis characterization of shiga toxin-producing *Escherichia coli* O157 from hides of cattle at slaughter. *J. Food Prot.* **65:**1172–1176.
15. **Azazy, A. A., and Y. A. Raja'a.** 2003. Malaria and intestinal parasitosis among children presenting to the paediatric centre in Sana'a, Yemen. *East Mediterr. Health J.* **9:**1048–1053.
16. **Baek, S. Y., S.-Y. Lim, D.-H. Lee, K.-H. Min, and C.-M. Kim.** 2000. Incidence and characterization of *Listeria monocytogenes* from domestic and imported foods in Korea. *J. Food Prot.* **63:**186–189.

17. **Bailey, N., M. Louck, D. Hopkins, J. Parker, A. Oglesby, D. Ewert, B. Barrett, K. Laurie, E. Muniz, N. Wade, P. Piercy, B. J. Francis, T. Maxson, M. Russell, and R. Finger.** 1995. Cholera associated with food transported from El Salvador—Indiana, 1994. *Morb. Mortal. Wkly. Rep.* **44:**385–386.

18. **Bal, S. K., and C. Czarnowski.** 2004. A man with fever, cough, diarrhea and a coated tongue. *CMAJ* **170:**1095.

19. **Balayan, M. S.** 1992. Natural hosts of hepatitis A virus. *Vaccine* **10:**S27–S31.

20. **Balfour, A. E., R. Lewis, and S. Ahmed.** 1991. Convalescent excretion of *Salmonella enteritidis* in infants. *J. Infect.* **38:**24–25.

21. **Bartlett, J. G., T. W. Chang, M. Gurwith, S. L. Gorbach, and A. B. Onderdonk.** 1978. Antibiotic-associated pseudomembranous colitis due to toxin-producing clostridia. *N. Engl. J. Med.* **298:**531–534.

22. **Barwick, R. S., A. Uzicanin, S. Lareau, N. Malakmadze, P. Imnadze, M. Iosava, N. Ninashvili, M. Wilson, A. W. Hightower, S. Johnston, H. Bishop, W. A. Petri, Jr., and D. D. Juranek.** 2002. Outbreak of amebiasis in Tbilisi, Republic of Georgia, 1998. *Am. J. Trop. Med. Hyg.* **67:**623–631.

23. **Benoit, V., P. Raiche, M. G. Smith, J. Guthrie, E. F. Donnelly, E. M. Julian, R. Lee, S. DiMaio, M. Rittmann, and B. T. Matyas.** 1994. Foodborne outbreaks of enterotoxigenic *Escherichia coli*—Rhode Island and New Hampshire, 1993. *Morb. Mortal. Wkly. Rep.* **43:**81, 87–89.

24. **Bessesen, M. T., E. Wang, P. Echeverria, and M. J. Blaser.** 1991. Enteroinvasive *Escherichia coli*: a cause of bacteremia in patients with AIDS. *J. Clin. Microbiol.* **29:**2675–2677.

25. **Bidawid, S., J. M. Farber, and S. A. Sattar.** 2000. Contamination of foods by food handlers: experiments on hepatitis A virus transfer to food and its interruption. *Appl. Environ. Microbiol.* **66:**2759–2763.

26. **Biering, G., S. Karlsson, N. C. Clark, K. E. Jónsdóttir, P. Lúdvígsson, and O. Steingrímsson.** 1989. Three cases of neonatal meningitis caused by *Enterobacter sakazakii* in powdered milk. *J. Clin. Microbiol.* **27:**2054–2056.

27. **Blair, D. C.** 1997. A week in the life of a travel clinic. *Clin. Microbiol. Rev.* **10:**650–673.

28. **Blake, P. A., M. H. Merson, R. E. Weaver, D. G. Hollis, and P. C. Heublein.** 1979. Disease caused by a marine *Vibrio*. Clinical characteristics and epidemiology. *N. Engl. J. Med.* **300:**1–5.

29. **Blaser, M. J.** 1997. Epidemiologic and clinical features of *Campylobacter jejuni* infections. *J. Infect. Dis.* **176**(Suppl. 2)**:**S103–S105.

30. **Blaser, M. J., and L. S. Newman.** 1982. A review of human salmonellosis: I. Infective dose. *Rev. Infect. Dis.* **4:**1096–1106.

31. **Blessmann, J., I. K. M. Ali, P. A. T. Nu, B. T. Dinh, T. Q. N. Viet, A. L. Van, C. G. Clark, and E. Tannich.** 2003. Longitudinal study of intestinal *Entamoeba histolytica* infections in asymptomatic adult carriers. *J. Clin. Microbiol.* **41:**4745–4750.

32. **Blessmann, J., A. Le Van, and E. Tannich.** 2006. Epidemiology and treatment of amebiasis in Hue, Vietnam. *Arch. Med. Res.* **37:**270–272.

33. **Bongard, J., R. Savage, R. Dern, H. Bostrum, J. Kazmierczak, S. Keifer, H. Anderson, and J. P. Davis.** 1994. *Cryptosporidium* infections associated

with swimming pools—Dane County, Wisconsin, 1993. *Morb. Mortal. Wkly. Rep.* **43:**561–563.

34. **Bottone, E. J., and T. Robin.** 1977. *Yersinia enterocolitica*: recovery and characterization of two unusual isolates from a case of acute enteritis. *J. Clin. Microbiol.* **5:**341–345.

35. **Bredin, C., J. Margery, L. Bordier, H. Mayaudon, O. Dupuy, B. Vergeau, and B. Bauduceau.** 2004. Diabetes and amoebiasis: a high risk encounter. *Diabetes Metab.* **30:**99–102.

36. **Briley, R. T., J. H. Teel, and J. P. Fowler.** 1994. Investigation and control of a *Shigella sonnei* outbreak in a day care center. *J. Environ. Health* **56:**23–25.

37. **Brock, T. D.** 1999. *Robert Koch. A Life in Medicine and Bacteriology.* ASM Press, Washington, D.C.

38. **Brooks, W. A., A. Hossain, D. Goswami, A. T. Sharmeen, K. Nahar, K. Alam, N. Ahmed, A. Naheed, G. B. Nair, S. Luby, and R. F. Breiman.** 2005. Bacteremic typhoid fever in children in an urban slum, Bangladesh. *Emerg. Infect. Dis.* **11:**326–329.

39. **Buzby, J. C., B. M. Allos, and T. Roberts.** 1997. The economic burden of *Campylobacter*-associated Guillain-Barré syndrome. *J. Infect. Dis.* **176**(Suppl. 2)**:** S192–S197.

40. **Cavagnaro, F., C. Guzman, and P. Harris.** 2006. Hemolytic uremic syndrome associated with *Entamoeba histolytica* intestinal infection. *Pediatr. Nephrol.* **21:**126–128.

41. **Centers for Disease Control and Prevention.** 1990. *Yersinia enterocolitica* infections during the holidays in black families—Georgia. *Morb. Mortal. Wkly. Rep.* **39:**819–820.

42. **Centers for Disease Control and Prevention.** 1993. Preliminary report: foodborne outbreak of *Escherichia coli* O157:H7 infections from hamburgers—Western United States, 1993. *Morb. Mortal. Wkly. Rep.* **42:**85–86.

43. **Centers for Disease Control and Prevention.** 1995. Update: *Vibrio cholerae* O1—Western Hemisphere, 1991–1994, and *V. cholerae* O139—Asia, 1994. *Morb. Mortal. Wkly. Rep.* **44:**215–219.

44. **Centers for Disease Control and Prevention.** 2004. 2003 summary statistics. The total number of foodborne disease outbreaks by etiology. Foodborne and Diarrheal Diseases Branch, Foodborne Outbreak Response and Surveillance Unit, Centers for Disease Control and Prevention. [Online.] http://www.cdc.gov/foodborneoutbreaks/us_outb/fbo2003/summary03.htm. Accessed 18 July 2006.

45. **Chalmers, R. M., A. P. Sturdee, P. Mellors, V. Nicholson, F. Lawlor, F. Kenny, and P. Timpson.** 1997. *Cryptosporidium parvum* in environmental samples in the Sligo area, Republic of Ireland: a preliminary report. *Lett. Appl. Microbiol.* **25:**380–384.

46. **Chong, Y., M. Y. Park, S. Y. Lee, K. S. Kim, and S. I. Lee.** 1982. *Vibrio vulnificus* septicemia in a patient with liver cirrhosis. *Yonsei Med. J.* **23:** 146–152.

47. **Chowdhury, N. R., S. Chakraborty, T. Ramamurthy, M. Nishibuchi, S. Yamasaki, Y. Takeda, and G. B. Nair.** 2000. Molecular evidence of clonal *Vibrio parahaemolyticus* pandemic strains. *Emerg. Infect. Dis.* **6:**631–636.

48. **Claudon, D. G., D. I. Thompson, E. H. Christenson, G. W. Lawton, and E. C. Dick.** 1971. Prolonged *Salmonella* contamination of a recreational lake by runoff waters. *Appl. Microbiol.* **21:**875–877.

49. **Colley, D. G.** 1996. Widespread foodborne cyclosporiasis outbreaks present major challenges. *Emerg. Infect. Dis.* **2:**354–356.

50. **Colomba, C., L. Saporito, L. Infurnari, S. Tumminia, and L. Titone.** 2006. Typhoid fever as a cause of opportunistic infection: case report. *BMC Infect. Dis.* **6:**38.

51. **Connor, B. A., E. Johnson, and R. Soave.** 2001. Reiter syndrome following protracted symptoms of *Cyclospora* infection. *Emerg. Infect. Dis.* **7:**453–454.

52. **Cover, T. L., and R. C. Aber.** 1989. *Yersinia enterocolitica. N. Engl. J. Med.* **321:**16–24.

53. **Cowie, B. C., A. Breschkin, and H. Kelly.** 2005. Hepatitis E virus: overseas epidemics and Victorian travelers. *Med. J. Aust.* **183:**491.

54. **Crabtree, K. D., R. H. Ruskin, S. B. Shaw, and J. B. Rose.** 1996. The detection of *Cryptosporidium* oocysts and *Giardia* cysts in cistern water in the U.S. Virgin Islands. *Water Res.* **30:**208–216.

55. **Crowe, L., W. Lau, L. McLeod, C. M. Anand, B. Ciebin, C. LeBer, S. McCartney, R. Easy, C. Clark, F. Rodgers, A. Ellis, A. Thomas, L. Shields, B. Tate, A. Klappholz, I. LaBerge, R. Reporter, H. Sato, E. Lehnkering, L. Mascola, J. Waddell, S. Waterman, J. Suarez, R. Hammond, R. Hopkins, P. Neves, M. S. Horine, P. Kludt, A. DeMaria, Jr., C. Hedberg, J. Wicklund, J. Besser, D. Boxrud, B. Hubner, M. Osterholm, F. M. Wu, and L. Beuchat.** 1999. Outbreaks of *Shigella sonnei* infection associated with eating fresh parsley—United States and Canada, July–August 1998. *Morb. Mortal. Wkly. Rep.* **48:**285–289.

56. **Crump, J. A., S. P. Luby, and E. D. Mintz.** 2004. The global burden of typhoid fever. *Bull. W. H. O.* **82:**346–353.

57. **Dalsgaard, A., H. H. Huss, A. H. Kittikun, and J. L. Larsen.** 1995. Prevalence of *Vibrio cholerae* and *Salmonella* in a major shrimp production area in Thailand. *Int. J. Food Microbiol.* **28:**101–113.

58. **Danielsson, M.-L., R. Möllby, H. Brag, N. Hansson, P. Jonsson, E. Olsson, and T. Wadström.** 1979. Enterotoxigenic enteric bacteria in foods and outbreaks of food-borne diseases in Sweden. *J. Hyg. Camb.* **83:**33–40.

59. **D'Aoust, J. Y., B. J. Aris, P. Thisdele, A. Durante, N. Brisson, D. Dragon, G. Lachapelle, M. Johnston, and R. Laidley.** 1975. *Salmonella eastbourne* outbreak associated with chocolate. *Can. Inst. Food Sci. Technol. J.* **8:**181–184.

60. **Davies, M., J. Engel, D. Griffin, D. Ginzl, R. Hopkins, C. Blackmore, E. Lawaczec, L. Nathan, C. Levy, G. Briggs, C. Kioski, S. Kreis, J. Keen, L. Durso, J. Schulte, K. Fullerton, C. Long, S. Smith, C. Barton, C. Gleit, M. Joyner, S. Montgomery, C. Braden, B. Goode, D. Chertow, C. O'Reilly, S. Gupta, and J. Dunn.** 2005. Outbreaks of *Escherichia coli* O157: H7 associated with petting zoos—North Carolina, Florida, and Arizona, 2004 and 2005. *Morb. Mortal. Wkly. Rep.* **54:**1277–1280.

61. **Davis, B. D., R. Dulbecco, H. N. Eisen, H. S. Ginsberg, and W. B. Wood, Jr.** 1967. *Microbiology.* Hoeber Medical Division, Harper & Row, Publishers, Inc., New York, N.Y.

62. **Davis, H., J. P. Taylor, J. N. Perdue, G. N. Stelma, Jr., J. M. Humphreys, Jr., R. Rowntree III, and K. D. Greene.** 1988. A shigellosis outbreak traced to commercially distributed shredded lettuce. *Am. J. Epidemiol.* **128:** 1312–1321.

63. **De Boer, E.** 1992. Isolation of *Yersinia enterocolitica* from foods. *Int. J. Food Microbiol.* **17:**75–84.

64. **De Jong, B., Y. Andersson, and K. Ekdahl.** 2005. Effect of regulation and education on reptile-associated salmonellosis. *Emerg. Infect. Dis.* **11:** 398–403.

65. **De Jong, B., and K. Ekdahl.** 2006. The comparative burden of salmonellosis in the European Union member states, associated and candidate countries. *BMC Public Health* **6:**4.

66. **Deneen, V. C., P. A. Belle-Isle, C. M. Taylor, L. L. Gabriel, J. B. Bender, J. H. Wicklund, C. W. Hedberg, and M. T. Osterholm.** 1998. Outbreak of cryptosporidiosis associated with a water sprinkler fountain—Minnesota, 1997. *Morb. Mortal. Wkly. Rep.* **47:**856–860.

67. **DePaola, A., G. M. Capers, and D. Alexander.** 1994. Densities of *Vibrio vulnificus* in the intestines of fish from the U.S. Gulf Coast. *Appl. Environ. Microbiol.* **60:**984–988.

68. **DePaola, A., C. A. Kaysner, J. Bowers, and D. W. Cook.** 2000. Environmental investigations of *Vibrio parahaemolyticus* in oysters after outbreaks in Washington, Texas, and New York (1997 and 1998). *Appl. Environ. Microbiol.* **66:**4649–4654.

69. **De Simon, M., and M. D. Ferrer.** 1998. Initial numbers, serovars and phagevars of *Listeria monocytogenes* isolated in prepared foods in the city of Barcelona (Spain). *Int. J. Food Microbiol.* **44:**141–144.

70. **Desmasures, N., F. Bazin, and M. Guéguen.** 1997. Microbiological composition of raw milk from selected farms in the Camembert region of Normandy. *J. Appl. Microbiol.* **83:**53–58.

71. **Diaz, A. I., R. Z. Rivero, M. A. Bracho, S. M. Castellanos, E. Acurero, L. M. Calchi, and T. R. Atencio.** 2006. [Prevalence of intestinal parasites in children of Yukpa ethnia in Toromo, Zulia State, Venezuela.] [Article in Spanish; English abstract]. *Rev. Med. Chil.* **134:**72–78.

72. **Dixon, T. C., M. Meselson, J. Guillemin, and P. C. Hanna.** 1999. Anthrax. *N. Engl. J. Med.* **341:**815–826.

73. **Donnenberg, M. S., and J. B. Kaper.** 1992. Enteropathogenic *Escherichia coli*. *Infect. Immun.* **60:**3953–3961.

74. **Doyle, M. P.** 1990. Pathogenic *Escherichia coli*, *Yersinia enterocolitica*, and *Vibrio parahaemolyticus*. *Lancet* **336:**1111–1115.

75. **Doyle, M. P., L. R. Beuchat, and T. J. Montville (ed.).** 1997. *Food Microbiology: Fundamentals and Frontiers.* ASM Press, Washington, D.C.

76. **Drake, M. E., A.W. Kitts, M. C. Blanchard, J. D. Farquhar, J. Stokes, Jr., and W. Henle.** 1950. Studies on the agent of infectious hepatitis. II. The disease produced in human volunteers by the agent cultivated in tissue culture or embryonated hen's eggs. *J. Exp. Med.* **92:**283–297.

77. **Dubos, R.** 1998. Chapter 10. Victory over Disease. *In Pasteur and Modern Science.* T. D. Brock (ed.). ASM Press, Washington, D.C.

78. **Eggertson, L.** 2005. *C. difficile* strain 20 times more virulent. *CMAJ* **172:** 1279.

79. **Elder, R. O., J. E. Keen, G. R. Siragusa, G. A. Barkocy-Gallagher, M. Koohmaraie, and W. W. Laegreid.** 2000. Correlation of enterohemorrhagic *Escherichia coli* O157 prevalence in feces, hides, and carcasses of beef cattle during processing. *Proc. Natl. Acad. Sci. USA* **97:**2999–3003.

80. **Erkmen, O.** 1996. Survival of virulent *Yersinia enterocolitica* during the manufacture and storage of Turkish feta cheese. *Int. J. Food Microbiol.* **33:**285–292.

81. **Espinosa-Cantellano, M., and A. Martínez-Palomo.** 2000. Pathogenesis of intestinal amebiasis: from molecules to disease. *Clin. Microbiol. Rev.* **13:** 318–331.

82. **Ewing, W. H.** 1986. *Edwards and Ewing's Identification of Enterobacteriaceae*, 4th ed. Elsevier Science Publishing Co., Inc., New York, N.Y.

83. **Ewing, W. H., and J. L. Gravatti.** 1947. *Shigella* types encountered in the Mediterranean area. *J. Bacteriol.* **53:**191–195.

84. **Falade, A. G., and T. Lawoyin.** 1999. Features of the 1996 cholera epidemic among Nigerian children in Ibadan, Nigeria. *J. Trop. Pediatr.* **45:**59–62.

85. **Falsen, E., B. Kaijser, L. Nehls, B. Nygren, and A. Svedhem.** 1980. *Clostridium difficile* in relation to enteric bacterial pathogens. *J. Clin. Microbiol.* **12:**297–300.

86. **Fankhauser, R. L., S. S. Monroe, J. S. Noel, C. D. Humphrey, J. S. Bresee, U. D. Parashar, T. Ando, and R. I. Glass.** 2002. Epidemiologic and molecular trends of "Norwalk-like viruses" associated with outbreaks of gastroenteritis in the United States. *J. Infect. Dis.* **186:**1–7.

87. **Farmer, J. J., III, M. A. Asbury, F. W. Hickman, D. J. Brenner, and the *Enterobacteriaceae* Study Group.** 1980. *Enterobacter sakazakii*: a new species of "*Enterobacteriaceae*" isolated from clinical specimens. *Int. J. Syst. Bacteriol.* **30:**569–584.

88. **Fasano, A., B. A. Kay, R. G. Russell, D. R. Maneval, Jr., and M. M. Levine.** 1990. Enterotoxin and cytotoxin production by enteroinvasive *Escherichia coli*. *Infect. Immun.* **58:**3717–3723.

89. **Faundez, G., G. Figueroa, M. Troncoso, and F. C. Cabello.** 1988. Characterization of enteroinvasive *Escherichia coli* strains isolated from children with diarrhea in Chile. *J. Clin. Microbiol.* **26:**928–932.

90. **Ferguson, C. M., B. G. Coote, N. J. Ashbolt, and I. M. Stevenson.** 1996. Relationships between indicators, pathogens and water quality in an estuarine system. *Water Res.* **30:**2045–2054.

91. **Finkelstein, R. A.** 2000. Personal reflections on cholera: the impact of serendipity. *ASM News* **66:**663–667.

92. **Fleming, D. W., S. L. Cochi, K. L. MacDonald, J. Brondum, P. S. Hayes, B. D. Plikaytis, M. B. Holmes, A. Audurier, C. V. Broome, and A. L. Reingold.** 1985. Pasteurized milk as a vehicle of infection in an outbreak of listeriosis. *N. Engl. J. Med.* **312:**404–407.

93. **Fordtran, J. S.** 2006. Colitis due to *Clostridium difficile* toxins: underdiagnosed, highly virulent, and nosocomial. *Proc. (Bayl. Univ. Med. Cent.)* **19:**3–12.

94. **Francis, B. J., and J. P. Davis.** 1984. Update: gastrointestinal illness associated with imported semi-soft cheese. *Morb. Mortal. Wkly. Rep.* **33:**16, 22.

95. **Furness, B. W., M. J. Beach, and J. M. Roberts.** 2000. Giardiasis surveillance—United States, 1992–1997. *Morb. Mortal. Wkly. Rep.* **49**(SS07):1–13.

96. **Furtado, C., G. K. Adak, J. M. Stuart, P. G. Wall, H. S. Evans, and D. P. Casemore.** 1998. Outbreaks of waterborne infectious intestinal disease in England and Wales, 1992–5. *Epidemiol. Infect.* **121**:109–119.

97. **Fyfe, M., M. T. Kelly, S. T. Yeung, P. Daly, K. Schallie, S. Buchanan, P. Waller, J. Kobayashi, N. Therien, M. Guichard, S. Lankford, P. Stehr-Green, R. Harsch, E. DeBess, M. Cassidy, T. McGivern, S. Mauvais, D. Fleming, M. Lippmann, L. Pong, R. W. McKay, D. E. Cannon, S. B. Werner, S. Abbott, M. Hernandez, C. Wojee, J. Waddell, S. Waterman, J. Middaugh, D. Sasaki, P. Effler, C. Groves, N. Curtis, D. Dwyer, G. Dowdle, and C. Nichols.** 1998. Outbreak of *Vibrio parahaemolyticus* infections associated with eating raw oysters—Pacific Northwest, 1997. *Morb. Mortal. Wkly. Rep.* **47**:457–462.

98. **Galanis, E., D. M. A. Lo Fo Wong, M. E. Patrick, N. Binsztein, A. Cieslik, T. Chalermchaikit, A. Aidara-Kane, A. Ellis, F. J. Angulo, and H. C. Wegener for World Health Organization Global Salm-Surv.** 2006. Web-based surveillance and global *Salmonella* distribution, 2000–2002. *Emerg. Infect. Dis.* **12**:381–388.

99. **Garcia-Fulgueiras, A., S. Sánchez, J. J. Guillén, B. Marsilla, A. Aladueña, and C. Navarro.** 2001. A large outbreak of *Shigella sonnei* gastroenteritis associated with consumption of fresh pasteurized milk cheese. *Eur. J. Epidemiol.* **17**:533–538.

100. **García-López, H. L., L. E. Rodríguez-Tovar, and C. E. Medina-De la Garza.** 1996. Identification of *Cyclospora* in poultry. *Emerg. Infect. Dis.* **2**: 356–357.

101. **Gascón, J.** 2006. Epidemiology, etiology and pathophysiology of traveler's diarrhea. *Digestion* **73**(S1):102–108.

102. **Gascón, J., M. Vargas, L. Quintó, M. Corachán, M. T. Jimenez de Anta, and J. Vila.** 1998. Enteroaggregative *Escherichia coli* strains as a cause of traveler's diarrhea: a case-control study. *J. Infect. Dis.* **177**:1409–1412.

103. **Gatti, S., G. Swierczynski, F. Robinson, M. Anselmi, J. Corrales, J. Moreira, G. Montalvo, A. Bruno, R. Maserati, Z. Bisoffi, and M. Scaglia.** 2002. Amebic infections due to the *Entamoeba histolytica–Entamoeba dispar* complex: a study of the incidence in a remote rural area of Ecuador. *Am. J. Trop. Med. Hyg.* **67**:123–127.

104. **Geltman, P. L., J. Cochran, and C. Hedgecock.** 2003. Intestinal parasites among African refugees resettled in Massachusetts and the impact of an overseas pre-departure treatment program. *Am. J. Trop. Med. Hyg.* **69**:657–662.

105. **Gilligan, P. H., L. R. McCarthy, and V. M. Genta.** 1981. Relative frequency of *Clostridium difficile* in patients with diarrheal disease. *J. Clin. Microbiol.* **14**:26–31.

106. **Gomes, T. A. T., M. R. F. Toledo, L. R. Trabulsi, P. K. Wood, and J. G. Morris, Jr.** 1987. DNA probes for identification of enteroinvasive *Escherichia coli*. *J. Clin. Microbiol.* **25**:2025–2027.

107. **Gomez-Bautista, M., L. M. Ortega-Mora, E. Tabares, V. Lopez-Rodas, and E. Costas.** 2000. Detection of infectious *Cryptosporidium parvum* oocysts in mussels (*Mytilus galloprovincialis*) and cockles (*Cerastoderma edule*). *Appl. Environ. Microbiol.* **66**:1866–1870.

108. **Gómez-Couso, H., F. Méndez-Hermida, J. A. Castro-Hermida, and E. Ares-Mazás.** 2005. Occurrence of *Giardia* cysts in mussels (*Mytilus galloprovincialis*) destined for human consumption. *J. Food Prot.* **68:**1702–1705.

109. **González-Escalona, N., V. Cachicas, C. Acevedo, M. L. Rioseco, J. A. Vergara, F. Cabello, J. Romero, and R. T. Espejo.** 2005. *Vibrio parahaemolyticus* diarrhea, Chile, 1998 and 2004. *Emerg. Infect. Dis.* **11:**129–131.

110. **Gordillo, M. E., G. R. Reeve, J. Pappas, J. J. Mathewson, H. L. DuPont, and B. E. Murray.** 1992. Molecular characterization of strains of enteroinvasive *Escherichia coli* O143, including isolates from a large outbreak in Houston, Texas. *J. Clin. Microbiol.* **30:**889–893.

111. **Grabowski, D. J., K. J. Tiggs, J. D. Hall, H. W. Senke, A. J. Salas, C. M. Powers, J. A. Knott, L. J. Nims, and C. M. Sewell.** 1989. Common-source outbreak of giardiasis—New Mexico. *Morb. Mortal. Wkly. Rep.* **38:**405–407.

112. **Graczyk, T. K., M. R. Cranfield, R. Fayer, and M. S. Anderson.** 1996. Viability and infectivity of *Cryptosporidium parvum* oocysts are retained upon intestinal passage through a refractory avian host. *Appl. Environ. Microbiol.* **62:**3234–3237.

113. **Graczyk, T. K., R. Fayer, and M. R. Cranfield.** 1996. *Cryptosporidium parvum* is not transmissible to fish, amphibians, or reptiles. *J. Parasitol.* **82:**748–751.

114. **Graczyk, T. K., R. Fayer, J. M. Trout, E. J. Lewis, C. A. Farley, I. Sulaiman, and A. A. Lal.** 1998. *Giardia* sp. cysts and infectious *Cryptosporidium parvum* oocysts in the feces of migratory Canada geese (*Branta canadensis*). *Appl. Environ. Microbiol.* **64:**2736–2738.

115. **Greisman, S. E., R. B. Hornick, F. A. Carozza, Jr., and T. E. Woodward.** 1963. The role of endotoxin during typhoid fever and tularemia in man. I. Acquisition of tolerance to endotoxin. *J. Clin. Invest.* **42:**1064–1075.

116. **Gross, R. J., L. V. Thomas, T. Cheasty, N. P. Day, B. Rowe, M. R. F. Toledo, and L. R. Trabulsi.** 1983. Enterotoxigenic and enteroinvasive *Escherichia coli* strains belonging to a new O group, O167. *J. Clin. Microbiol.* **17:**521–523.

117. **Guerrant, R. L.** 1997. Cryptosporidiosis: an emerging, highly infectious threat. *Emerg. Infect. Dis.* **3:**51–57.

118. **Haas, C. N., and J. B. Rose.** 1996. Distribution of *Cryptosporidium* oocysts in a water supply. *Water Res.* **30:**2251–2254.

119. **Haghighi, A., S. Kobayashi, T. Takeuchi, N. Thammapalerd, and T. Nozaki.** 2003. Geographic diversity among genotypes of *Entamoeba histolytica* field isolates. *J. Clin. Microbiol.* **41:**3748–3756.

120. **Hailemariam, G., A. Kassu, G. Abebe, E. Abate, D. Damte, E. Mekonnen, and F. Ota.** 2004. Intestinal parasitic infections in HIV/AIDS and HIV seronegative individuals in a teaching hospital, Ethiopia. *Jpn. J. Infect. Dis.* **57:**41–43.

121. **Hamilton, J. V., M. J. Lehane, and H. R. Braig.** 2003. Isolation of *Enterobacter sakazakii* from midgut of *Stomoxys calcitrans*. *Emerg. Infect. Dis.* **9:**1355–1356.

122. **Hardie, R. M., P. G. Wall, P. Gott, M. Bardhan, and C. L. R. Bartlett.** 1999. Infectious diarrhea in tourists staying in a resort hotel. *Emerg. Infect. Dis.* **5:**168–171.

123. **Hardy, A.** 2004. *Salmonella*: a continuing problem. *Postgrad. Med. J.* **80:** 541–545.

124. **Harris, J. R., J. Mariano, J. G. Wells, B. J. Payne, H. D. Donnell, and M. L. Cohen.** 1985. Person-to-person transmission in an outbreak of enteroinvasive *Escherichia coli*. *Am. J. Epidemiol.* **122:**245–252.

125. **Hayashi, M., H. Kouzu, M. Nishihara, T. Takahashi, M. Furuhashi, K. Sakamoto, N. Satoh, T. Nishitani, and Y. Shikano.** 2005. Acute renal failure likely due to acute nephritic syndrome associated with typhoid fever. *Intern. Med.* **44:**1074–1077.

126. **Hedberg, C. W., W. C. Levine, K. E. White, R. H. Carlson, D. K. Winsor, D. N. Cameron, K. L. MacDonald, and M. T. Osterholm, for the Investigation Team.** 1992. An international foodborne outbreak of shigellosis associated with a commercial airline. *JAMA* **268:**3208–3212.

127. **Helms, M., P. Vastrup, P. Gerner-Smidt, and K. Mølbak.** 2003. Short and long term mortality associated with foodborne bacterial gastrointestinal infections: registry based study. *Br. Med. J.* **326:**357–361.

128. **Hendriksen, S. W. M., K. Orsel, J. A. Wagenaar, A. Miko, and E. van Duijkeren.** 2004. Animal-to-human transmission of *Salmonella* Typhimurium DT104A variant. *Emerg. Infect. Dis.* **10:**2225–2227.

129. **Henle, W., S. Harris, G. Henle, T. N. Harris, M. E. Drake, F. Mangold, and J. Stokes, Jr.** 1950. Studies on the agent of infectious hepatitis. I. Propagation of the agent in tissue culture and in the embryonated hen's egg. *J. Exp. Med.* **92:**271–281.

130. **Herwaldt, B. L.** 2000. *Cyclospora cayetanensis*: a review, focusing on the outbreaks of cyclosporiasis in the 1990s. *Clin. Infect. Dis.* **31:**1040–1057.

131. **Herwaldt, B. L., M. J. Beach, and the Cyclospora Working Group.** 1999. The return of *Cyclospora* in 1997: another outbreak of cyclosporiasis in North America associated with imported raspberries. *Ann. Intern. Med.* **130:** 210–220.

132. **Highsmith, A. K., J. C. Feeley, P. Skaliy, J. G. Wells, and B. T. Wood.** 1977. Isolation of *Yersinia enterocolitica* from well water and growth in distilled water. *Appl. Environ. Microbiol.* **34:**745–750.

133. **Himelright, I., E. Harris, V. Lorch, M. Anderson, T. Jones, A. Craig, M. Kuehnert, T. Forster, M. Arduino, B. Jensen, and D. Jernigan.** 2002. *Enterobacter sakazakii* infections associated with the use of powdered infant formula—Tennessee, 2001. *Morb. Mortal. Wkly. Rep.* **51:**297–299.

134. **Ho, J. L., K. N. Shands, G. Friedland, P. Eckind, and D. W. Fraser.** 1986. An outbreak of type 4b *Listeria monocytogenes* infection involving patients from eight Boston hospitals. *Arch. Intern. Med.* **146:**520–524.

135. **Hoang, L. M. N., M. Fyfe, C. Ong, J. Harb, S. Champagne, B. Dixon, and J. Isaac-Renton.** 2005. Outbreak of cyclosporiasis in British Columbia associated with imported Thai basil. *Epidemiol. Infect.* **133:**23–27.

136. **Høi, L., I. Dalsgaard, A. DePaola, R. J. Siebeling, and A. Dalsgaard.** 1998. Heterogeneity among isolates of *Vibrio vulnificus* recovered from eels (*Anguilla anguilla*) in Denmark. *Appl. Environ. Microbiol.* **64:**4676–4682.

137. **Holt, J. G., N. R. Krieg, P. H. A. Sneath, J. T. Staley, and S. T. Williams (ed.).** 1994. *Bergey's Manual of Determinative Bacteriology*, 9th ed. Williams & Wilkins, Baltimore, Md.

138. **Hopkins, R. S., R. A. Jajosky, P. A. Hall, D. A. Adams, F. J. Connor, P. Sharp, W. J. Anderson, R. F. Fagan, J. J. Aponte, G. F. Jones,**

D. A. Nitschke, C. A. Worsham, N. Adekoya, and M.-H. Chang. 2005. Summary of notifiable diseases—United States, 2003. *Morb. Mortal. Wkly. Rep.* **52:**1–85.

139. **Hsia, R.-C., P. L. C. Small, and P. M. Bavoil.** 1993. Characterization of virulence genes of enteroinvasive *Escherichia coli* by Tn*phoA* mutagenesis: identification of *invX*, a gene required for entry into HEp-2 cells. *J. Bacteriol.* **175:**4817–4823.

140. **Hsu, B.-M., C. Huang, G.-Y. Jiang, and C.-L. L. Hsu.** 1999. The prevalence of *Giardia* and *Cryptosporidium* in Taiwan water supplies. *J. Toxicol. Environ. Health* Part A **56:**149–160.

141. **Hughes-Hanks, J. M., L. G. Rickard, C. Panuska, J. R. Saucier, T. M. O'Hara, L. Dehn, and R. M. Rolland.** 2005. Prevalence of *Cryptosporidium* spp. and *Giardia* spp. in five marine mammal species. *J. Parasitol.* **91:** 1225–1228.

142. **Huq, A., S. Parveen, F. Qadri, D. A. Sack, and R. R. Colwell.** 1993. Comparison of *Vibrio cholerae* serotype 01 strains isolated from patients and the aquatic environment. *J. Trop. Med. Hyg.* **96:**86–92.

143. **Hutin, Y. J. F., V. Pool, E. H. Cramer, O. V. Nainan, J. Weth, I. T. Williams, S. T. Goldstein, K. F. Gensheimer, B. P. Bell, C. N. Shapiro, M. J. Alter, and H. S. Margolis, for The National Hepatitis A Investigation Team.** 1999. A multistate, foodborne outbreak of hepatitis A. *N. Engl. J. Med.* **340:**595–602.

144. **Inglesby, T. V., T. O'Toole, D. A. Henderson, J. G. Bartlett, M. S. Ascher, E. Eitzen, A. M. Friedlander, J. Gerberding, J. Hauer, J. Hughes, J. McDade, M. T. Osterholm, G. Parker, T. M. Perl, P. K. Russell, and K. Tonat for the Working Group on Civilian Biodefense.** 2002. Anthrax as a biological weapon, 2002. Updated recommendations for management. *JAMA* **287:**2236–2252.

145. **Inian, G., V. Kanagalakshmi, and P. J. Kuruvilla.** 2006. Acute acalculous cholecystitis: a rare complication of typhoid fever. *Singapore Med. J.* **47:**327–328.

146. **Isakbaeva, E. T., M.-A. Widdowson, R. S. Beard, S. N. Bulens, J. Mullins, S. S. Monroe, J. Bresee, P. Sassano, E. H. Cramer, and R. I. Glass.** 2005. Norovirus transmission on cruise ship. *Emerg. Infect. Dis.* **11:**154–157.

147. **Islam, M., J. Morgan, M. P. Doyle, and X. Jiang.** 2004. Fate of *Escherichia coli* O157:H7 in manure compost-amended soil and on carrots and onions grown in an environmentally controlled growth chamber. *J. Food Prot.* **67:**574–578.

148. **Itoh, N., N. Muraoka, J. Kawamata, M. Aoki, and T. Itagaki.** 2006. Prevalence of *Giardia intestinalis* infection in household cats of Tohoku District in Japan. *J. Vet. Med. Sci.* **68:**161–163.

149. **Jackson, S. G., D. A. Yip-Chuck, J. B. Clark, and M. H. Brodsky.** 1986. Diagnostic importance of *Clostridium perfringens* enterotoxin analysis in recurring enteritis among elderly, chronic care psychiatric patients. *J. Clin. Microbiol.* **23:**748–751.

150. **Jackson, T. F., P. G. Sargeaunt, P. S. Visser, V. Gathiram, S. Suparsad, and C. B. Anderson.** 1990. *Entamoeba histolytica*: naturally occurring infections in baboons. *Arch. Invest. Med. (Mex.).* **21**(S1)**:**153–156.

151. **Jay, J. M.** 2000. *Modern Food Microbiology*, 6th ed. Aspen Publishers, Inc., Gaithersburg, Md.

152. **Johnson, S., M. H. Samore, K. A. Farrow, G. E. Killgore, F. C. Tenover, D. Lyras, J. I. Rood, P. DeGirolami, A. L. Baltch, M. E. Rafferty, S. M. Pear, and D. N. Gerding.** 1999. Epidemics of diarrhea caused by a clindamycin-resistant strain of *Clostridium difficile* in four hospitals. *N. Engl. J. Med.* **341:**1645–1651.

153. **Jones, R. C., J. R. Fernandez, S. I. Gerber, W. Paul, L. Williams, R. Turner, and J. T. Watson.** 2003. *Yersinia enterocolitica* gastroenteritis among infants exposed to chitterlings—Chicago, Illinois, 2002. *Morb. Mortal. Wkly. Rep.* **52:**956–958.

154. **Jones, T. F., S. C. Buckingham, C. A. Bopp, E. Ribot, and W. Schaffner.** 2003. From pig to pacifier: chitterling-associated yersiniosis outbreak among black infants. *Emerg. Infect. Dis.* **9:**1007–1009.

155. **Kadhiravan, T., N. Wig, A. Kapil, S. K. Kabra, K. Renuka, and A. Misra.** 2005. Clinical outcomes in typhoid fever: adverse impact of infection with nalidixic acid-resistant *Salmonella typhi. BMC Infect. Dis.* **5:**37.

156. **Kanafani, Z. A., A. Ghossain, A. I. Sharara, J. M. Hatem, and S. S. Kanj.** 2003. Endemic gastrointestinal anthrax in 1960s Lebanon: clinical manifestations and surgical findings. *Emerg. Infect. Dis.* **9:**520–525.

157. **Kandhai, M. C., M. W. Reij, L. G. M. Gorris, O. Guillaume-Gentil, and M. van Schothorst.** 2004. Occurrence of *Enterobacter sakazakii* in food production environments and households. *Lancet* **363:**39–40.

158. **Kang, D.-H., G. A. Barkocy-Gallagher, M. Koohmaraie, and G. R. Siragusa.** 2001. Screening bovine carcass sponge samples for *Escherichia coli* O157 using a short enrichment coupled with immunomagnetic separation and a polymerase chain reaction-based (BAX) detection step. *J. Food Prot.* **64:**1610–1612.

159. **Kapikian, A. Z., R. G. Wyatt, R. Dolin, T. S. Thornhill, A. R. Kalica, and R. M. Chanock.** 1972. Visualization by immune electron microscopy of a 27-nm particle associated with acute infectious nonbacterial gastroenteritis. *J. Virol.* **10:**1075–1081.

160. **Kapperud, G., L. M. Rørvik, V. Hasseltvedt, E. A. Høiby, B. G. Iversen, K. Staveland, G. Johnsen, J. Leitao, H. Herikstad, Y. Andersson, G. Langeland, B. Gondrosen, and J. Lassen.** 1995. Outbreak of *Shigella sonnei* infection traced to imported iceberg lettuce. *J. Clin. Microbiol.* **33:**609–614.

161. **Kapperud, G., E. Skjerve, N. H. Bean, S. M. Ostroff, and J. Lassen.** 1992. Risk factors for sporadic *Campylobacter* infections: results of a case-control study in southeastern Norway. *J. Clin. Microbiol.* **30:**3117–3121.

162. **Karere, G. M., and E. Munene.** 2002. Some gastro-intestinal tract parasites in wild De Brazza's monkeys (*Cercopithecus neglectus*) in Kenya. *Vet. Parasitol.* **110:**153–157.

163. **Kassenborg, H., R. Danila, P. Snippes, M. Wiisanen, M. Sullivan, K. E. Smith, N. Crouch, C. Medus, R. Weber, J. Korlath, T. Ristinen, R. Lynfield, H. F. Hull, J. Pahlen, T. Boldingh, K. Elfering, G. Hoffman, T. Lewis, A. Friedlander, H. Heine, R. Culpepper, E. Henchal, G. Ludwig, C. Rossi, J. Teska, J. Ezzell, and E. Eitzen.** 2000. Human ingestion of *Bacillus anthracis*-contaminated meat—Minnesota, August 2000. *Morb. Mortal. Wkly. Rep.* **49:**813–816.

164. **Katz, D. E., D. Heisey-Grove, M. Beach, R. C. Dicker, and B. T. Matyas.** 2006. Prolonged outbreak of giardiasis with two modes of transmission. *Epidemiol Infect.* **134:**935–941.

165. **Keene, W. E., J. M. McAnulty, F. C. Hoesly, L. P. Williams, Jr., K. Hedberg, G. L. Oxman, T. J. Barrett, M. A. Pfaller, and D. W. Fleming.** 1994. A swimming-associated outbreak of hemorrhagic colitis caused by *Escherichia coli* O157:H7 and *Shigella sonnei*. *N. Engl. J. Med.* **331:**579–584.

166. **Kelly, P., K. S. Baboo, P. Ndubani, M. Nchito, N. P. Okeowo, N. P. Luo, R. A. Feldman, and M. J. G. Farthing.** 1997. Cryptosporidiosis in adults in Lusaka, Zambia, and its relationship to oocyst contamination of drinking water. *J. Infect. Dis.* **176:**1120–1123.

167. **Keskimäki, M., L. Mattila, H. Peltola, and A. Siitonen.** 2000. Prevalence of diarrheagenic *Escherichia coli* in Finns with or without diarrhea during a round-the-world trip. *J. Clin. Microbiol.* **38:**4425–4429.

168. **Ketyi, I.** 1989. Epidemiology of the enteroinvasive *Escherichia coli*. Observations in Hungary. *J. Hyg. Epidemiol. Microbiol. Immunol.* **33:**261–267.

169. **Khodr, M., S. Hill, L. Perkins, S. Stiefel, C. Comer-Morrison, S. Lee, D. R. Patel, D. Peery, C. W. Armstrong, and G. B. Miller, Jr.** 1994. *Bacillus cereus* food poisoning associated with fried rice at two child day care centers—Virginia, 1993. *Morb. Mortal. Wkly. Rep.* **43:**177–178.

170. **Kingsley, D. H., and G. P. Richards.** 2003. Persistence of hepatitis A virus in oysters. *J. Food Prot.* **66:**331–334.

171. **Kofoid, C. A., and E. B. Christiansen.** 1915. On the life-history of *Giardia*. *Proc. Natl. Acad. Sci. USA* **1:**547–552.

172. **Koo, D. T., A. G. Dean, R. W. Slade, C. M. Knowles, D. A. Adams, W. K. Fortune, P. A. Hall, R. F. Fagan, B. Panter-Connah, H. R. Holden, G. F. Jones, and C. L. Maddox.** 1994. MMWR summary of notifiable diseases, United States, 1993. *Morb. Mortal. Wkly. Rep.* **42**(53)**:**1–73.

173. **Koopmans, M., and E. Duizer.** 2004. Foodborne viruses: an emerging problem. *Int. J. Food Microbiol.* **90:**23–41.

174. **Kubota, K., T. J. Barrett, M. L. Ackers, P. S. Brachman, and E. D. Mintz.** 2005. Analysis of *Salmonella enterica* serotype Typhi pulsed-field gel electrophoresis patterns associated with international travel. *J. Clin. Microbiol.* **43:**1205–1209.

175. **Kucik, C. J., G. L. Martin, and B. V. Sortor.** 2004. Common intestinal parasites. *Am. Fam. Physician* **69:**1161–1168.

176. **Kudva, I. T., K. Blanch, and C. J. Hovde.** 1998. Analysis of *Escherichia coli* O157:H7 survival in ovine or bovine manure and manure slurry. *Appl. Environ. Microbiol.* **64:**3166–3174.

177. **Kuzina, L. V., J. J. Peloquin, D. C. Vacek, and T. A. Miller.** 2001. Isolation and identification of bacteria associated with adult laboratory Mexican fruit flies, *Anastrepha ludens* (Diptera: Tephritidae). *Curr. Microbiol.* **42:**290–294.

178. **Labarca, J. A., J. Sturgeon, L. Borenstein, N. Salem, S. M. Harvey, E. Lehnkering, R. Reporter, and L. Mascola.** 2002. *Campylobacter upsaliensis*: another pathogen for consideration in the United States. *Clin. Infect. Dis.* **34:** e59–e60.

179. **Lai, K. K.** 2001. *Enterobacter sakazakii* infections among neonates, infants, children, and adults. Case reports and a review of the literature. *Medicine* **80:**113–122.

180. **Law, D.** 1994. Adhesion and its role in the virulence of enteropathogenic *Escherichia coli*. *Clin. Microbiol. Rev.* **7:**152–173.

181. **Leav, B. A., M. Mackay, and H. D. Ward. 2003.** *Cryptosporidium* species: new insights and old challenges. *Clin. Infect. Dis.* **36:**903–908.

182. **Lee, C.-Y., C.-H. Chiu, Y.-Y. Chuang, L.-H. Su, T.-L. Wu, L.-Y. Chang, Y.-C. Huang, and T.-Y. Lin.** 2002. Multidrug-resistant non-typhoid *Salmonella* infections in a medical center. *J. Microbiol. Immunol. Infect.* **35:** 78–84.

183. **Lefebvre, S. L., D. Waltner-Toews, A. S. Peregrine, R. Reid-Smith, L. Hodge, L. G. Arroyo, and J. S. Weese.** 2006. Prevalence of zoonotic agents in dogs visiting hospitalized people in Ontario: implications for infection control. *J. Hosp. Infect.* **62:**458–466.

184. **Lehner, A., and R. Stephan.** 2004. Microbiological, epidemiological, and food safety aspects of *Enterobacter sakazakii*. *J. Food Prot.* **67:**2850–2857.

185. **Linnan, M. J., L. Mascola, X. D. Lou, V. Goulet, S. May, C. Salminen, D. W. Hird, M. L. Yonekura, P. Hayes, R. Weaver, A. Audurier, B. D. Plikaytis, S. L. Fannin, A. Klees, and C. V. Broome.** 1988. Epidemic listeriosis associated with Mexican-style cheese. *N. Engl. J. Med.* **319:**823–828.

186. **Loncarevic, S., M.-L. Daniolsson-Tham, P. Gerner-Smidt, L. Sahlström, and W. Tham.** 1998. Potential sources of human listeriosis in Sweden. *Food Microbiol.* **15:**65–69.

187. **Loo, V. G., M. D. Libman, M. A. Miller, A.-M. Bourgault, C. H. Frenette, M. Kelly, S. Michaud, T. Nguyen, L. Poirier, A. Vibien, R. Horn, P. J. Laflamme, and P. René.** 2004. *Clostridium difficile*: a formidable foe. *CMAJ* **171:**47–48.

188. **Lopez, A. S., D. R. Dodson, M. J. Arrowood, P. A. Orlandi, Jr., A. J. da Silva, J. W. Bier, S. D. Hanauer, R. L. Kuster, S. Oltman, M. S. Baldwin, K. Y. Won, E. M. Nace, M. L. Eberhard, and B. L. Herwaldt.** 2001. Outbreak of cyclosporiasis associated with basil in Missouri in 1999. *Clin. Infect. Dis.* **32:**1010–1017.

189. **Lopez, D. J., V. K. Abarca, M. P. Paredes, and T. E. Inzunza.** 2006. Intestinal parasites in dogs and cats with gastrointestinal symptoms in Santiago, Chile. [Article in Spanish; English abstract]. *Rev. Med. Chil.* **134:**193–200.

190. **Lung, A. J., C.-M. Lin, J. M. Kim, M. R. Marshall, R. Nordstedt, N. P. Thompson, and C. I. Wei.** 2001. Destruction of *Escherichia coli* O157:H7 and *Salmonella* Enteritidis in cow manure composting. *J. Food Prot.* **64:**1309–1314.

191. **Lyerly, D. M., H. C. Krivan, and T. D. Wilkins.** 1988. *Clostridium difficile*: its disease and toxins. *Clin. Microbiol. Rev.* **1:**1–18.

192. **Lyytikäinen, O., T. Autio, R. Maijala, P. Ruutu, T. Honkanen-Buzalski, M. Miettinen, M. Hatakka, J. Mikkola, V.-J. Anttila, T. Johansson, L. Rantala, T. Aalto, H. Korkeala, and A. Siitonen.** 2000. An outbreak of *Listeria monocytogenes* serotype 3a infections from butter in Finland. *J. Infect. Dis.* **181:**1838–1841.

193. **MacDonald, K. L., and P. M. Griffin.** 1986. Foodborne disease outbreaks, annual summary, 1982. *Morb. Mortal. Wkly. Rep.* **35**(1SS)**:**7ss–10ss.

194. **MacKenzie, W. R., J. J. Kazmierczak, and J. P. Davis.** 1995. An outbreak of cryptosporidiosis associated with a resort swimming pool. *Epidemiol. Infect.* **115:**545–553.

195. **Magner, J. W., and R. D. Coakley.** 1956. Giardiasis: *Giardia (lamblia) intestinalis* as a cause of gastrointestinal disease. *J. Ir. Med. Assoc.* **38:**180–181.

196. **Mahoney, F. J., T. A. Farley, K. Y. Kelso, S. A. Wilson, J. M. Horan, and L. M. McFarland.** 1992. An outbreak of hepatitis A associated with swimming in a public pool. *J. Infect. Dis.* **165:**613–618.

197. **Mak, J. W.** 2004. Important zoonotic intestinal protozoan parasites in Asia. *Trop. Biomed.* **21:**39–50.

198. **Margolles, A., B. Mayo, and C. G. de los Reyes-Gavilán.** 1998. Polymorphism of *Listeria monocytogenes* and *Listeria innocua* strains isolated from short-ripened cheese. *J. Appl. Microbiol.* **84:**255–262.

199. **Martinez-Urtaza, J., L. Simental, D. Velasco, A. DePaola, M. Ishibashi, Y. Nakaguchi, M. Nishibuchi, D. Carrera-Flores, C. Rey-Alvarez, and A. Pousa.** 2005. Pandemic *Vibrio parahaemolyticus* O3:K6, Europe. *Emerg. Infect. Dis.* **11:**1319–1320.

200. **Mascola, L., M. Tormey, D. Dassey, L. Kilman, S. Harvey, A. Medina, A. Tilzer, and S. Waterman.** 1996. *Vibrio vulnificus* infections associated with eating raw oysters—Los Angeles, 1996. *Morb. Mortal. Wkly. Rep.* **45:** 621–624.

201. **Matsushita, S., S. Yamada, A. Kai, and Y. Kudoh.** 1993. Invasive strains of *Escherichia coli* belonging to serotype O121:NM. *J. Clin. Microbiol.* **31:** 3034–3035.

202. **Matthys, F., S. Male, and Z. Labdi.** 1998. Cholera outbreak among Rwandan refugees—Democratic Republic of Congo, April 1997. *Morb. Mortal. Wkly. Rep.* **47:**389–391.

203. **McCarthy, J. S., D. Peacock, K. P. Trown, P. Bade, W. A. Petri, Jr., and B. J. Currie.** 2002. Endemic invasive amoebiasis in northern Australia. *Med. J. Aust.* **177:**570.

204. **McClung, L. S.** 1945. Human food poisoning due to growth of *Clostridium perfringens* (*C. welchii*) in freshly cooked chicken: preliminary note. *J. Bacteriol.* **50:**229–231.

205. **McDonald, L. C., M. Owings, and D. B. Jernigan.** 2006. *Clostridium difficile* infection in patients discharged from US short-stay hospitals, 1996–2003. *Emerg. Infect. Dis.* **12:**409–415.

206. **McKillip, J. L.** 2000. Prevalence and expression of enterotoxins in *Bacillus cereus* and other *Bacillus* spp., a literature review. *Antoine von Leeuwenhoek* **77:**393–399.

207. **Meer, R.** 1993. Dietary risk factors for infant botulism. "Diet and infant botulism." *Dairy Food Environ. Sanit.* **13:**570–573.

208. **Mensah, P., D. Yeboah-Manu, K. Owusu-Darko, and A. Ablordey.** 2002. Street foods in Accra, Ghana: how safe are they? *Bull. W. H. O.* **80:**546–554.

209. **Midura, T. F., S. Snowden, R. M. Wood, and S. S. Arnon.** 1979. Isolation of *Clostridium botulinum* from honey. *J. Clin. Microbiol.* **9:**282–283.

210. **Mintz, E. D., M. Hudson-Wragg, P. Mshar, M. L. Cartter, and J. L. Hadler.** 1993. Foodborne giardiasis in a corporate office setting. *J. Infect. Dis.* **167:**250–253.

211. **Miwa, N., T. Masuda, K. Terai, A. Kawamura, K. Otani, and H. Miyamoto.** 1999. Bacteriological investigation of an outbreak of *Clostridium*

perfringens food poisoning caused by Japanese food without animal protein. *Int. J. Food Microbiol.* **49:**103–106.

212. **Morris, G. K., M. H. Merson, D. A. Sack, J. G. Wells, W. T. Martin, W. E. Dewitt, J. C. Feeley, R. B. Sack, and D. M. Bessudo.** 1976. Laboratory investigation of diarrhea in travelers to Mexico: evaluation of methods for detecting enterotoxigenic *Escherichia coli. J. Clin. Microbiol.* **3:**486–495.

213. **Morris, J. G., Jr., and R. E. Black.** 1985. Cholera and other vibrioses in the United States. *N. Engl. J. Med.* **312:**343–350.

214. **Morris, R. D., E. N. Naumova, and J. K. Griffiths.** 1998. Did Milwaukee experience waterborne cryptosporidiosis before the large documented outbreak in 1993? *Epidemiology* **9:**264–270.

215. **Motes, M. L., A. DePaola, D. W. Cook, J. E. Veazey, J. C. Hunsucker, W. E. Garthright, R. J. Blodgett, and S. J. Chirtel.** 1998. Influence of water temperature and salinity on *Vibrio vulnificus* in northern gulf and Atlantic coast oysters (*Crassostrea virginica*). *Appl. Environ. Microbiol.* **64:**1459–1465.

216. **Mramba, F., A. Broce, and L. Zurek.** 2006. Isolation of *Enterobacter sakazakii* from stable flies, *Stomoxys calcitrans* L. (Diptera: Muscidae). *J. Food Prot.* **69:**671–673.

217. **Mshar, P. A., Z. F. Dembek, M. L. Cartter, J. L. Hadler, T. R. Fiorentino, R. A. Marcus, J. McGuire, M. A. Shiffrin, A. Lewis, J. Feuss, J. Van Dyke, M. Toly, M. Cambridge, J. Guzewich, J. Keithly, D. Dziewulski, E. Braun-Howland, D. Ackman, P. Smith, J. Coates, and J. Ferrara.** 1997. Outbreaks of *Escherichia coli* O157:H7 infection and cryptosporidiosis associated with drinking unpasteurized apple cider—Connecticut and New York, October 1996. *Morb. Mortal. Wkly. Rep.* **46:**4–8.

218. **Murata, T., T. Iida, Y. Shiomi, K. Tagomori, Y. Akeda, I. Yanagihara, S. Mushiake, F. Ishiguro, and T. Honda.** 2001. A large outbreak of foodborne infection attributed to *Providencia alcalifaciens. J. Infect. Dis.* **184:**1050–1055.

219. **Murphree, R. L., and M. L. Tamplin.** 1995. Uptake and retention of *Vibrio cholerae* O1 in the eastern oyster, *Crassostrea virginica. Appl. Environ. Microbiol.* **61:**3656–3660.

220. **Muytjens, H. L., H. Roelofs-Willemse, and G. H. J. Jaspar.** 1988. Quality of powdered substitutes for breast milk with regard to members of the family *Enterobacteriaceae. J. Clin. Microbiol.* **26:**743–746.

221. **Naimi, T. S., J. H. Wicklund, S. J. Olsen, G. Krause, J. G. Wells, J. M. Bartkus, D. J. Boxrud, M. Sullivan, H. Kassenborg, J. M. Besser, E. D. Mintz, M. T. Osterholm, and C. W. Hedberg.** 2003. Concurrent outbreaks of *Shigella sonnei* and enterotoxigenic *Escherichia coli* infections associated with parsley: implications for surveillance and control of foodborne illness. *J. Food Prot.* **66:**535–541.

222. **Nakama, A., M. Terao, Y. Kokubo, T. Itoh, T. Maruyama, C. Kaneuchi, and J. McLauchlin.** 1998. A comparison of *Listeria monocytogenes* serovar 4b isolates of clinical and food origin in Japan by pulsed-field gel electrophoresis. *Int. J. Food Microbiol.* **42:**201–206.

223. **Nassereddin, R. A., and M. I. Yamani.** 2005. Microbiological quality of sous and tamarind, traditional drinks consumed in Jordan. *J. Food Prot.* **68:** 773–777.

224. **Nataro, J. P., and J. B. Kaper.** 1998. Diarrheagenic *Escherichia coli. Clin. Microbiol. Rev.* **11:**142–201.

225. **Nataro, J. P., T. Steiner, and R. L. Guerrant.** 1998. Enteroaggregative *Escherichia coli*. *Emerg. Infect. Dis.* **4:**251–261.

226. **Nazarowec-White, M., and J. M. Farber.** 1997. *Enterobacter sakazakii*: a review. *Int. J. Food Microbiol.* **34:**103–113.

227. **Nocera, D., E. Bannerman, J. Rocourt, K. Jaton-Ogay, and J. Bille.** 1990. Characterization by DNA restriction endonuclease analysis of *Listeria monocytogenes* strains related to the Swiss epidemic of listeriosis. *J. Clin. Microbiol.* **28:**2259–2263.

228. **North, R. A.** 1991. An in-depth investigation into a food poisoning outbreak. *Cater. Health* **2:**25–39.

229. **Nortjé, G. L., S. M. Vorster, R. P. Greebe, and P. L. Steyn.** 1999. Occurrence of *Bacillus cereus* and *Yersinia enterocolitica* in South African retail meats. *Food Microbiol.* **16:**213–217.

230. **Nygård, K., B. Schimmer, Ø. Søbstad, and I. Tveit.** 2004. Waterborne outbreak of giardiasis in Bergen, Norway. *Eur. Surveill. Wkly.* **8.** [Online.] http://www.eurosurveillance.org/ew/2004/041111.asp#2. Accessed 18 July 2006.

231. **O'Hara, C. M., A. G. Steigerwalt, D. Green, M. McDowell, B. C. Hill, D. J. Brenner, and J. M. Miller.** 1999. Isolation of *Providencia heimbachae* from human feces. *J. Clin. Microbiol.* **37:**3048–3050.

232. **Ojeniyi, B., H. C. Wegener, N. E. Jensen, and M. Bisgaard.** 1996. *Listeria monocytogenes* in poultry and poultry products: epidemiological investigations in seven Danish abattoirs. *J. Appl. Bacteriol.* **80:**395–401.

233. **Okeke, I. N., O. Ojo, A. Lamikanra, and J. B. Kaper.** 2003. Etiology of acute diarrhea in adults in southwestern Nigeria. *J. Clin. Microbiol.* **41:**4525–4530.

234. **Okhuysen, P. C.** 2001. Traveler's diarrhea due to intestinal protozoa. *Clin. Infect. Dis.* **33:**110–114.

235. **Okuda, J., M. Ishibashi, E. Hayakawa, T. Nishino, Y. Takeda, A. K. Mukhopadhyay, S. Garg, S. K. Bhattacharya, G. B. Nair, and M. Nishibuchi.** 1997. Emergence of a unique O3:K6 clone of *Vibrio parahaemolyticus* in Calcutta, India, and isolation of strains from the same clonal group from Southeast Asian travelers arriving in Japan. *J. Clin. Microbiol.* **35:**3150–3155.

236. **Olsen, S. J., M. Patrick, S. B. Hunter, V. Reddy, L. Kornstein, W. R. MacKenzie, K. Lane, S. Bidol, G. A. Stoltman, D. M. Frye, I. Lee, S. Hurd, T. F. Jones, T. N. LaPorte, W. Dewitt, L. Graves, M. Wiedmann, D. J. Schoonmaker-Bopp, A. J. Huang, C. Vincent, A. Bugenhagen, J. Corby, E. R. Carloni, M. E. Holcomb, R. F. Woron, S. M. Zansky, G. Dowdle, F. Smith, S. Ahrabi-Fard, A. R. Ong, N. Tucker, N. A. Hynes, and P. Mead.** 2005. Multistate outbreak of *Listeria monocytogenes* infection linked to delicatessen turkey meat. *Clin. Infect. Dis.* **40:**962–967.

237. **Öncü, S., S. Öncü, and S. Sakarya.** 2003. Anthrax—an overview. *Med. Sci. Monit.* **9:**RA276–RA283.

238. **Ong, C., W. Moorehead, A. Ross, and J. Isaac-Renton.** 1996. Studies of *Giardia* spp. and *Cryptosporidium* spp. in two adjacent watersheds. *Appl. Environ. Microbiol.* **62:**2798–2805.

239. **Osterholm, M. T., J. C. Forfang, T. L. Ristinen, A. G. Dean, J. W. Washburn, J. R. Godes, R. A. Rude, and J. G. McCullough.** 1981. An outbreak of foodborne giardiasis. *N. Engl. J. Med.* **304:**24–28.

240. **O'Sullivan, B., V. Delpech, G. Pontivivo, T. Karagiannis, D. Marriott, J. Harkness, and J. M. McAnulty.** 2002. Shigellosis linked to sex venues, Australia. *Emerg. Infect. Dis.* **8:**862–864.

241. **Pagotto, F. J., M. Nazarowec-White, S. Bidawid, and J. M. Farber.** 2003. *Enterobacter sakazakii*: infectivity and enterotoxin production *in vitro* and *in vivo*. *J. Food Prot.* **66:**370–375.

242. **Park, S. K., D.-H. Kim, Y.-K. Deung, H.-J. Kim, E.-J. Yang, S.-J. Lim, Y.-S. Ryang, D. Jin, and K.-J. Lee.** 2004. Status of intestinal parasite infections among children in Bat Dambang, Cambodia. *Korean J. Parasitol.* **42:**201–203.

243. **Porter, J. D., C. Gaffney, D. Heymann, and W. Parkin.** 1990. Food-borne outbreak of *Giardia lamblia*. *Am. J. Public Health* **80:**1259–1260.

244. **Pritchett, R., C. Gossman, V. Radke, J. Moore, E. Busenlehner, K. Fischer, K. Doerr, C. Winkler, M. Franklin-Thomsen, J. Fiander, J. Crowley, E. Peoples, L. Bremby, J. Southard, L. Appleton, D. Bowers, J. Lipsman, H. Callaway, D. Lawrence, R. Gardner, B. Cunanan, R. Snaman, J. Rullan, G. Miller, Jr., S. Henderson, M. Mismas, T. York, J. Pearson, C. Lacey, J. Purvis, N. Curtis, K. Mallet, R. Thompson, D. Portesi, D. M. Dwyer, M. Fletcher, M. Levy, T. Lawford, M. Sabat, and M. Kahn.** 1997. Outbreak of cyclosporiasis—Northern Virginia-Washington, D.C.-Baltimore, Maryland, Metropolitan Area, 1997. *Morb. Mortal. Wkly. Rep.* **46:**689–691.

245. **Quilici, M.-L., A. Robert-Pillot, J. Picart, and J.-M. Fournier.** 2005. Pandemic *Vibrio parahaemolyticus* O3:K6 spread, France. *Emerg. Infect. Dis.* **11:**1148–1149.

246. **Quiroz, E. S., C. Bern, J. R. MacArthur, L. Xiao, M. Fletcher, M. J. Arrowood, D. K. Shay, M. E. Levy, R. I. Glass, and A. Lal.** 2000. An outbreak of cryptosporidiosis linked to a foodhandler. *J. Infect. Dis.* **181:**695–700.

247. **Reed, G. H.** 1994. Foodborne illness (part 3). *Clostridium perfringens* gastroenteritis. *Dairy Food Environ. Sanit.* **14:**16–17.

248. **Reed, G. H.** 1994. Foodborne illness (part 4). *Bacillus cereus* gastroenteritis. *Dairy Food Environ. Sanit.* **14:**87.

249. **Reed, G. H.** 1994. Foodborne illness (part 10). *Listeria monocytogenes*. *Dairy Food Environ. Sanit.* **14:**482–483.

250. **Reed, G. H.** 1994. Foodborne illness (part 12). Shigellosis. *Dairy Food Environ. Sanit.* **14:**591.

251. **Reilly, P. J. A., and D. R. Twiddy.** 1992. *Salmonella* and *Vibrio cholerae* in brackish water cultured tropical prawns. *Int. J. Food Microbiol.* **16:**293–301.

252. **Riedo, F. X., R. W. Pinner, M. de Lourdes Tosca, M. L. Cartter, L. M. Graves, M.W. Reeves, R. E. Weaver, B. D. Plikaytis, and C. V. Broome.** 1994. A point-source foodborne listeriosis outbreak: documented incubation period and possible mild illness. *J. Infect. Dis.* **170:**693–696.

253. **Roels, T. H., M. E. Proctor, L. C. Robinson, K. Hulbert, C. A. Bopp, and J. P. Davis.** 1998. Clinical features of infections due to *Escherichia coli* producing heat-stable toxin during an outbreak in Wisconsin: a rarely suspected cause of diarrhea in the United States. *Clin. Infect. Dis.* **26:**898–902.

254. **Rood, J. I.** 1998. Virulence genes of *Clostridium perfringens*. *Annu. Rev. Microbiol.* **52:**333–360.

255. **Rørvik, L. M., B. Aase, T. Alvestad, and D. A. Caugant.** 2000. Molecular epidemiological survey of *Listeria monocytogenes* in seafoods and seafood-processing plants. *Appl. Environ. Microbiol.* **66:**4779–4784.

256. **Sair, A. I., D. H. D'Souza, and L. A. Jaykus.** 2002. Human enteric viruses as causes of foodborne disease. *Comprehensive Rev. Food Sci. Food Safety* **1:** 73–89.

257. **Salvat, G., M. T. Toquin, Y. Michel, and P. Colin.** 1995. Control of *Listeria monocytogenes* in the delicatessen industries: the lessons of a listeriosis outbreak in France. *Int. J. Food Microbiol.* **25:**75–81.

258. **Sandery, M., T. Stinear, and C. Kaucner.** 1996. Detection of pathogenic *Yersinia enterocolitica* in environmental waters by PCR. *J. Appl. Bacteriol.* **80:**327–332.

259. **Sathyanarayanan, L., and Y. Ortega.** 2004. Effects of pesticides on sporulation of *Cyclospora cayetanensis* and viability of *Cryptosporidium parvum*. *J. Food Prot.* **67:**1044–1049.

260. **Schlech, W. F., III.** 2000. Foodborne listeriosis. *Clin. Infect. Dis.* **31:**770–775.

261. **Schlech, W. F., III, P. M. Lavigne, R. A. Bortolussi, A. C. Allen, E. V. Haldane, A. J. Wort, A. W. Hightower, S. E. Johnson, S. H. King, E. S. Nicholls, and C. V. Broome.** 1983. Epidemic listeriosis—evidence for transmission by food. *N. Engl. J. Med.* **308:**203–206.

262. **Schultsz, C., J. Van Den Ende, F. Cobelens, T. Vervoort, A. Van Gompel, J. C. F. M. Wetsteyn, and J. Dankert.** 2000. Diarrheagenic *Escherichia coli* and acute and persistent diarrhea in returned travelers. *J. Clin. Microbiol.* **38:**3550–3554.

263. **Sealy, D. P., and S. H. Schuman.** 1981. Giardiasis: a common and under-recognized enteric pathogen. *J. Fam. Pract.* **12:**47–54.

264. **Sewell, A. M., and J. M. Farber.** 2001. Foodborne outbreaks in Canada linked to produce. *J. Food Prot.* **64:**1863–1877.

265. **Siegman-Igra, Y., R. Levin, M. Weinberger, Y. Golan, D. Schwartz, Z. Samra, H. Konigsberger, A. Yinnon, G. Rahav, N. Keller, N. Bisharat, J. Karpuch, R. Finkelstein, M. Alkan, Z. Landau, J. Novikov, D. Hassin, C. Rudnicki, R. Kitzes, S. Ovadia, Z. Shimoni, R. Lang, and T. Shohat.** 2002. *Listeria monocytogenes* infection in Israel and review of cases worldwide. *Emerg. Infect. Dis.* **8:**305–310.

266. **Sirisanthana, T., and A. E. Brown.** 2002. Anthrax of the gastrointestinal tract. *Emerg. Infect. Dis.* **8:**649–651.

267. **Smith, J. L.** 1993. *Cryptosporidium* and *Giardia* as agents of foodborne disease. *J. Food Prot.* **56:**451–461.

268. **Smith, J. L.** 1999. Foodborne infections during pregnancy. *J. Food Prot.* **62:**818–829.

269. **Smith, J. L.** 2001. A review of hepatitis E virus. *J. Food Prot.* **64:**572–586.

270. **Smith, J. M., and E. Meerovitch.** 1985. Primates as a source of *Entamoeba histolytica*, their zymodeme status and zoonotic potential. *J. Parasitol.* **71:** 751–756.

271. **Smith, K., D. Boxrud, F. Leano, C. Snider, C. Braden, J. Lockett, S. Montgomery, S. Swanson, and C. O'Reilly.** 2005. Outbreak of multidrug-resistant *Salmonella* Typhimurium associated with rodents purchased at retail pet stores—United States, December 2003–October 2004. *Morb. Mortal. Wkly. Rep.* **54:**429–433.

272. **Smith, L. D. S., and E. O. King.** 1962. Occurrence of *Clostridium difficile* in infections of man. *J. Bacteriol.* **84:**65–67.

273. **Snyder, J. D., J. G. Wells, J. Yashuk, N. Puhr, and P. A. Blake.** 1984. Outbreak of invasive *Escherichia coli* gastroenteritis on a cruise ship. *Am. J. Trop. Med. Hyg.* **33:**281–284.

274. **Sobel, J.** 2005. Botulism. *Clin. Infect. Dis.* **41:**1167–1173.

275. **Sobel, J., D. N. Cameron, J. Ismail, N. Strockbine, M. Williams, P. S. Diaz, B. Westley, M. Rittmann, J. DiCristina, H. Ragazzoni, R. V. Tauxe, and E. D. Mintz.** 1998. A prolonged outbreak of *Shigella sonnei* infections in traditionally observant Jewish communities in North America caused by a molecularly distinct bacterial subtype. *J. Infect. Dis.* **177:**1405–1409.

276. **Solomon, E. B., S. Yaron, and K. R. Matthews.** 2002. Transmission of *Escherichia coli* O157:H7 from contaminated manure and irrigation water to lettuce plant tissue and its subsequent internalization. *Appl. Environ. Microbiol.* **68:**397–400.

277. **Songer, J. G., and F. A. Uzal.** 2005. Clostridial enteric infections in pigs. *J. Vet. Diagn. Invest.* **17:**528–536.

278. **Sonoda, H., M. Abe, T. Sugimoto, Y. Sato, M. Bando, E. Fukui, H. Mizuo, M. Takahashi, T. Nishizawa, and H. Okamoto.** 2004. Prevalence of hepatitis E virus (HEV) infection in wild boars and deer and genetic identification of a genotype 3 HEV from a boar in Japan. *J. Clin. Microbiol.* **42:**5371–5374.

279. **Spitalny, K. C., E. N. Okowitz, and R. L. Vogt.** 1984. Salmonellosis outbreak at a Vermont hospital. *South. Med. J.* **77:**168–172.

280. **Stark, P. L., A. Lee, and B. D. Parsonage.** 1982. Colonization of the large bowel by *Clostridium difficile* in healthy infants: quantitative study. *Infect. Immun.* **35:**895–899.

281. **Stedman, N. L., J. S. Munday, R. Esbeck, and G. S. Visvesvara.** 2003. Gastric amebiasis due to *Entamoeba histolytica* in a Dama wallaby (*Macropus eugenii*). *Vet. Pathol.* **40:**340–342.

282. **Sterling, C. R., and Y. R. Ortega.** 1999. *Cyclospora*: an enigma worth unraveling. *Emerg. Infect. Dis.* **5:**48–53.

283. **Straif-Bourgeois, S., T. Sokol, A. Thomas, R. Ratard, K. D. Greene, E. Mintz, P. Yu, and P. Vranken.** 2006. Two cases of toxigenic *Vibrio cholerae* O1 infection after hurricanes Katrina and Rita—Louisiana, October 2005. *Morb. Mortal. Wkly. Rep.* **55:**31–32.

284. **Stuart, J. M., H. J. Orr, F. G. Warburton, S. Jeyakanth, C. Pugh, I. Morris, J. Sarangi, and G. Nichols.** 2003. Risk factors for sporadic giardiasis: a case-control study in southwestern England. *Emerg. Infect. Dis.* **9:**229–233.

285. **Subekti, D., B. A. Oyofo, P. Tjaniadi, A. L. Corwin, W. Larasati, M. Putri, C. H. Simanjuntak, N. H. Punjabi, J. Taslim, B. Setiawan, A. A. G. S. Djelantik, L. Sriwati, A. Sumardiati, E. Putra, J. R. Campbell, and M. Lesmana.** 2001. *Shigella* spp. surveillance in Indonesia: the emergence or reemergence of *S. dysenteriae*. *Emerg. Infect. Dis.* **7:**137–140.

286. **Tacket, C. O., J. P. Narain, R. Sattin, J. P. Lofgren, C. Konigsberg, Jr., R. C. Rendtorff, A. Rausa, B. R. Davis, and M. L. Cohen.** 1984. A multistate outbreak of infections caused by *Yersinia enterocolitica* transmitted by pasteurized milk. *JAMA* **251:**483–486.

287. **Tan, L. J., J. Lyznicki, P. M. Adcock, E. Dunne, J. Smith, E. Parish, A. Miller, H. Seltzer, and R. Etzel.** 2001. Diagnosis and management of foodborne illnesses: a primer for physicians. *Morb. Mortal. Wkly. Rep.* **50**(RR02):1–69.

288. **Tanyuksel, M., and W. A. Petri, Jr.** 2003. Laboratory diagnosis of amebiasis. *Clin. Microbiol. Rev.* **16:**713–729.

289. **Taylor, D. N., P. Echeverria, O. Scthabutr, C. Pitarangsi, U. Leksomboon, N. R. Blacklow, B. Rowe, R. Gross, and J. Cross.** 1988. Clinical and microbiologic features of *Shigella* and enteroinvasive *Escherichia coli* infections detected by DNA hybridization. *J. Clin. Microbiol.* **26:**1362–1366.

290. **Taylor, J. L., J. Tuttle, T. Pramukul, K. O'Brien, T. J. Barrett, B. Jolbitado, Y. L. Lim, D. Vugia, J. G. Morris, Jr., R. V. Tauxe, and D. M. Dwyer.** 1993. An outbreak of cholera in Maryland associated with imported commercial frozen fresh coconut milk. *J. Infect. Dis.* **167:**1330–1335.

291. **Teng, L.-J., P.-R. Hsueh, S.-J. Liaw, S.-W. Ho, and J.-C. Tsai.** 2004. Genetic detection of diarrheagenic *Escherichia coli* isolated from children with sporadic diarrhea. *J. Microbiol. Immunol. Infect.* **37:**327–334.

292. **Thurman, R., B. Faulkner, D. Veal, G. Cramer, and M. Meiklejohn.** 1998. Water quality in rural Australia. *J. Appl. Microbiol.* **84:**627–632.

293. **Tutenel, A. V., D. Pierard, J. Van Hoof, and L. De Zutter.** 2003. Molecular characterization of *Escherichia coli* O157 contamination routes in a cattle slaughterhouse. *J. Food Prot.* **66:**1564–1569.

294. **Ustun, S., H. Dagci, U. Aksoy, Y. Guruz, and G. Ersoz.** 2003. Prevalence of amebiasis in inflammatory bowel disease in Turkey. *World J. Gastroenterol.* **9:**1834–1835.

295. **Uyeyama, R. R., S. B. Werner, S. Chin, S. F. Pearce, C. L. Kollip, L. P. Williams, and J. A. Googins.** 1982. Isolation of *E. coli* O157:H7 from sporadic cases of hemorrhagic colitis—United States. *Morb. Mortal. Wkly. Rep.* **31:**580, 585.

296. **Vaishnavi, C., R. Kochhar, G. Singh, S. Kumar, S. Singh, and K. Singh.** 2005. Epidemiology of typhoid carriers among blood donors and patients with biliary, gastrointestinal and other related diseases. *Microbiol. Immunol.* **49:**107–112.

297. **Van Acker, J., F. de Smet, G. Muyldermans, A. Bougatef, A. Naessens, and S. Lauwers.** 2001. Outbreak of necrotizing enterocolitis associated with *Enterobacter sakazakii* in powdered milk formula. *J. Clin. Microbiol.* **39:**293–297.

298. **Van Houten, R., D. Farberman, J. Norton, J. Ellison, J. Kiehlbauch, T. Morris, and P. Smith.** 1998. *Plesiomonas shigelloides* and *Salmonella* serotype Hartford infections associated with a contaminated water supply—Livingston County, New York, 1996. *Morb. Mortal. Wkly. Rep.* **47:**394–396.

299. **Van Immerseel, F., F. Pasmans, J. De Buck, I. Rychlik, H. Hradecka, J.-M. Collard, C. Wildemauwe, M. Heyndrickx, R. Ducatelle, and F. Haesebrouck.** 2004. Cats as a risk for transmission of antimicrobial drug-resistant *Salmonella*. *Emerg. Infect. Dis.* **10:**2169–2174.

300. **Vieira, A. B. R., I. H. J. Koh, and B. E. C. Guth.** 2003. *Providencia alcalifaciens* strains translocate from the gastrointestinal tract and are resistant to lytic activity of serum complement. *J. Med. Microbiol.* **52:**633–636.

301. **Vollaard, A. M., S. Ali, H. A. G. H. van Asten, S. Widjaja, L. G. Visser, C. Surjadi, and J. T. van Dissel.** 2004. Risk factors for typhoid and paratyphoid fever in Jakarta, Indonesia. *JAMA* **291:**2607–2615.

302. **Vuddhakul, V., A. Chowdhury, V. Laohaprertthisan, P. Pungrasamee, N. Patararungrong, P. Thianmontri, M. Ishibashi, C. Matsumoto,**

and M. Nishibuchi. 2000. Isolation of a pandemic O3:K6 clone of a *Vibrio parahaemolyticus* strain from environmental and clinical sources in Thailand. *Appl. Environ. Microbiol.* **66:**2685–2689.

303. **Vugia, D., A. Cronquist, J. Hadler, M. Tobin-D'Angelo, D. Blythe, K. Smith, K. Thornton, D. Morse, P. Cieslak, T. Jones, K. Holt, J. Guzewich, O. Henao, E. Scallan, F. Angulo, P. Griffin, R. Tauxe, and E. Barzilay.** 2006. Preliminary FoodNet data on the incidence of infection with pathogens transmitted commonly through food—10 states, United States, 2005. *Morb. Mortal. Wkly. Rep.* **55:**392–395.

304. **Wachi, K., K. Tateda, Y. Yamashiro, M. Takahashi, T. Matsumoto, N. Fuyura, Y. Ishii, Y. Akasaka, K. Yamaguchi, and K. Uchida.** 2005. Sepsis caused by food-borne infection with *Escherichia coli. Intern. Med.* **44:**1316–1319.

305. **Wada, A., Y. Masuda, M. Fukayama, T. Hatakeyama, Y. Yanagawa, H. Watanabe, and T. Inamatsu.** 1996. Nosocomial diarrhoea in the elderly due to enterotoxigenic *Clostridium perfringens. Microbiol. Immunol.* **40:**767–771.

306. **Wallis, P. M., S. L. Erlandsen, J. L. Isaac-Renton, M. E. Olson, W. J. Robertson, and H. Van Keulen.** 1996. Prevalence of *Giardia* cysts and *Cryptosporidium* oocysts and characterization of *Giardia* spp. isolated from drinking water in Canada. *Appl. Environ. Microbiol.* **62:**2789–2797.

307. **Wang, L., and H. Zhuang.** 2004. Hepatitis E: an overview and recent advances in vaccine research. *World J. Gastroenterol.* **10:**2157–2162.

308. **Warny, M., J. Pepin, A. Fang, G. Killgore, A. Thompson, J. Brazier, E. Frost, and L. C. McDonald.** 2005. Toxin production by an emerging strain of *Clostridium difficile* associated with outbreaks of severe disease in North America and Europe. *Lancet* **366:**1079–1084.

309. **Wechsler, E., C. D'Aleo, V. A. Hill, J. Hopper, D. Myers-Wiley, E. O'Keeffe, J. Jacobs, F. Guido, A. Huang, S. N. Dodt, B. Rowan, M. Sherman, A. Greenberg, D. Schneider, B. Noone, L. Fanella, B. R. Williamson, E. Dinda, M. Mayer, M. Backer, A. Agasan, L. Kornstein, F. Stavinsky, B. Neal, D. Edwards, M. Haroon, D. Hurley, L. Colbert, J. Miller, B. Mojica, E. Carloni, B. Devine, M. Cambridge, T. Root, D. Schoonmaker, M. Shayegani, W. Hastback, B. Wallace, S. Kondracki, P. Smith, S. Matiuck, K. Pilot, M. Acharya, G. Wolf, W. Manley, C. Genese, J. Brooks, Z. Dembek, and J. Hadler.** 1999. Outbreak of *Vibrio parahaemolyticus* infection associated with eating raw oysters and clams harvested from Long Island Sound—Connecticut, New Jersey, and New York, 1998. *Morb. Mortal. Wkly. Rep.* **48:**48–51.

310. **Whipp, S. C., H. W. Moon, and N. C. Lyon.** 1975. Heat-stable *Escherichia coli* enterotoxin production *in vivo. Infect. Immun.* **12:**240–244.

311. **Widdowson, M.-A., A. Sulka, S. N. Bulens, R. S. Beard, S. S. Chaves, R. Hammond, E. D. P. Salehi, E. Swanson, J. Totaro, R. Woron, P. S. Mead, J. S. Bresee, S. S. Monroe, and R. I. Glass.** 2005. Norovirus and foodborne disease, United States, 1991–2000. *Emerg. Infect. Dis.* **11:**95–102.

312. **Wong, H.-C., S.-H. Liu, T.-K. Wang, C.-L. Lee, C.-S. Chiou, D.-P. Liu, M. Nishibuchi, and B.-K. Lee.** 2000. Characteristics of *Vibrio parahaemolyticus* O3:K6 from Asia. *Appl. Environ. Microbiol.* **66:**3981–3986.

313. **Wright, A. C., R. T. Hill, J. A. Johnson, M.-C. Roghman, R. R. Colwell, and J. G. Morris, Jr.** 1996. Distribution of *Vibrio vulnificus* in the Chesapeake Bay. *Appl. Environ. Microbiol.* **62:**717–724.

314. **Xiao, L., R. Fayer, U. Ryan, and S. J. Upton.** 2004. *Cryptosporidium* taxonomy: recent advances and implications for public health. *Clin. Microbiol. Rev.* **17**:72–97.

315. **Yoh, M., J. Matsuyama, M. Ohnishi, K. Takagi, H. Miyagi, K. Mori, K. S. Park, T. Ono, and T. Honda.** 2005. Importance of *Providencia* species as a major cause of travellers' diarrhoea. *J. Med. Microbiol.* **54**:1077–1082.

316. **Yoo, D., P. Willson, Y. Pei, M. A. Hayes, A. Deckert, C. E. Dewey, R. M. Friendship, Y. Yoon, M. Gottschalk, C. Yason, and A. Giulivi.** 2001. Prevalence of hepatitis E virus antibodies in Canadian swine herds and identification of a novel variant of swine hepatitis E virus. *Clin. Diagn. Lab. Immunol.* **8**:1213–1219.

appendix B

Glossary

Aerobe A microorganism that requires molecular oxygen (O_2) for growth. See also **Anaerobe** and **Microaerophile**.

Agglomeration A process of reintroducing a small amount of moisture to particles (e.g., powdered milk) to cause them to clump together, rendering them more easily soluble. See also **Instantizing**.

Albumen Egg white.

Anaerobe A microorganism that grows in the absence of molecular oxygen (O_2). Those microorganisms that only grow in the absence of O_2 are known as "strict anaerobes"; those that are able to grow either in the presence or absence of O_2 are known as "facultative anaerobes."

Antibiotic/antimicrobial agent Biologically active substance that interferes with the metabolism, growth, or replication of microorganisms. The term "antibiotic" refers specifically to an antibacterial agent; an antimicrobial agent may target bacteria, fungi, protozoa, or even some viruses.

Antibiotic resistance The ability of a bacterium to survive and grow in the presence of what should be an effective dose of an antibiotic. Bacteria vary in their degree of resistance to an antibiotic. Inappropriate use of antibiotics usually results in selective growth of the more resistant bacteria.

Antibody A protein molecule produced by an organism in response to the presence of foreign bodies or molecules.

Antigen A substance that, when introduced into an organism, can stimulate the organism to produce a specific antibody against that substance.

Appertization Preservation of food by canning, heating, sealing, and cooling. The contents of the sealed can contract as they cool, resulting in the production of a partial vacuum. Named for its developer, Nicholas Appert.

Attenuated culture A culture that has been modified, usually through repeated culturing or passage through a host, to lessen or eliminate its ability to cause disease. Attenuated cultures may be used as vaccines.

Autosomal mutation A mutation that takes place in the DNA of a germ cell (egg or sperm) and that can be passed down to succeeding generations.

Avirulent Benign. Without virulence (see **Virulence**).

Bacteremia Presence of bacteria in the bloodstream.

Bacteriophage A virus that infects and replicates itself within bacterial cells.

Bacterium (*pl.*, bacteria) Single-celled microorganism capable of self-replication. Bacterial cells are surrounded by a cell wall but lack an organized nucleus.

Biotype Varieties within a species of bacteria or fungus that can be differentiated by their physiological or metabolic characteristics (e.g., ability to ferment a specific sugar or to produce a particular metabolite).

Blog Short for "web log." A blog is an informal Internet posting, usually by an individual. A blog can take the form of a personal daily journal, a news article, a commentary, or simply one or more links to articles on a subject of interest to the writer (known as a "blogger") appearing elsewhere on the Internet.

Brine A salt solution.

Cecum A pouch located at the junction between the small and large intestines.

Chalaza Twisted "ropes" that attach an egg yolk to the two ends of the eggshell.

Cholecystitis, acalculous Inflammation of the gallbladder without the appearance of calculi or gallstones.

Cloaca In egg-laying animals and birds, the opening shared by the oviduct and the intestine. This opening is used for laying eggs and for excretion.

Colitis, pseudomembranous Inflammation of the colon lining, accompanied by the formation of a layer of exudate that resembles a membrane.

Colon The main part of the large intestine.

Competitive exclusion The process of preventing the intestinal tract of an animal from becoming colonized with undesirable bacteria (e.g., *Salmonella*) by feeding the animal massive doses of competing, benign bacteria.

Conching An intermediate step in the production of chocolate consisting of a combination of heating, milling, and mixing.

Cross-contamination Transfer of microbes from a contaminated food item, utensil, or working surface to a previously uncontaminated food, utensil, or surface.

Cytotoxin A toxin that is damaging to living cells.

Disease syndrome A collection of symptoms that characterize a disease.

Disinfectant A chemical agent that kills or inactivates microorganisms.

Duodenum The first part of the small intestine, located between the stomach and the jejunum.

Dysentery A disease syndrome resulting from infection with some species of *Shigella*.

***E. coli* biotype 1** The biotype of *Escherichia coli* that is most characteristic of the species. Referred to by journalists, certain government agencies (including the U.S. Department of Agriculture), and some scientists as "generic *E. coli*" to distinguish it from *E. coli* O157:H7.

Emetic An inducer of vomiting.

Endemic disease A disease that is embedded in a population or a geographic area.

Enteritis An inflammation of the small intestine, usually as a result of infection or the ingestion of a microbial toxin. The most common symptom of enteritis is diarrhea.

Enterotoxin A toxin that acts on or is produced in the intestinal tract.

Enzyme A molecule (usually, but not always, a protein) produced by an organism that catalyzes or facilitates a biochemical reaction.

Epidemic A widespread disease outbreak.

Epidemiology Branch of medical science that studies the incidence, possible sources, patterns of spread, and potential control of diseases and disease outbreaks.

Epithelial cells Cells that form the outermost layer of skin; epithelial cells also line the intestinal tract and the internal surface of some other organs.

Erythema nodosum An eruption of tender, red bumps, usually found on the shins.

Erythrocyte Red blood cell.

Etiology Cause of a disease.

Eviscerate Remove internal organs from a carcass.

Fecal coliforms Coliform bacteria usually associated with, or found in, feces. These are gram-negative, rod-shaped bacteria that are able to ferment lactose when incubated at 44.5°C (112.1°F). Most *E. coli* strains are fecal coliforms, as are members of several other genera and species of bacteria.

Fimbriae Threadlike protein structures that protrude from a bacterial cell wall.

Fomites *(sing., fomes)* Inanimate disease reservoirs, such as dust particles, pens, or doorknobs.

Gastroenteritis A disease syndrome that affects the stomach and/or intestines.

Germination The initiation of growth from a previously dormant spore.

Hemorrhagic Accompanied by or causing bleeding (hemorrhage).

Humanized monoclonal antibody A monoclonal antibody that has been produced from a cell culture formed by fusing together a mouse cell with a human cell. Humanized monoclonal antibodies are less likely than conventional monoclonal antibodies to provoke a tissue rejection reaction when used to treat human illness.

Iatrogenic disease A disease that is transmitted as the result of a medical procedure (e.g., by the presence of infectious material on a medical instrument or by transfusion with contaminated blood). See also **Nosocomial infection**.

Ileum A portion of the small intestine located between the jejunum and cecum.

Immunocompromised Made more susceptible to infectious disease due to damage or weakening of the immune system. An individual's immunity can be affected, for example, by other illness, by certain medications, or by pregnancy.

Immunoglobulin Part of the immune system. A collection of protein molecules that function as antibodies.

Immunohistochemistry (IHC) A procedure that consists of fixing a thin section of a tissue to a microscope slide, flooding the tissue with a labeled antibody preparation (usually the label is a fluorescent molecule or a dye), incubating to allow the antibodies to react to their corresponding antigen molecules that may be located in the tissue, washing the stained preparation, and examining the slide microscopically. IHC is useful to detect the presence and the location of foreign antigens (e.g., pathogens or prions) in a tissue sample.

Incubation period The period of time between infection and the onset of symptoms.

Index case The first documented case in a disease outbreak.

Infectious dose A minimum quantity of a specific strain of living microbial cells or virus particles that is needed to establish an infection in an individual. Not all those who become infected with a pathogen will develop symptoms of the disease. The infectious dose can vary, depending on the individual strain and the susceptibility of the host.

Instantizing Treating a powdered food to improve its solubility in water. A common step in the production of powdered milk.

Intestinal mucosa The internal lining of the small intestine.

Jaundice A symptom of hepatitis, manifesting as a yellowing of the skin and the whites of the eyes.

Jejunum A portion of the small intestine located between the duodenum and the ileum.

Lesion Wound or injury to a tissue or organ.

Meningitis Inflammation of the membranes surrounding the brain or spinal cord.

Microaerophile A microorganism that requires a minute amount of molecular oxygen (O_2) for growth but that cannot grow in the presence of the normal concentration of O_2 present in the atmosphere.

Microbe/microorganism An organism, usually single celled, that is too small to be seen without the aid of a microscope.

Monoclonal antibodies Antibodies produced from a cell culture (typically from a mouse cell line) that originated from a single antibody-producing cell (i.e., from a clone of antibody-producing cells). All of the antibodies produced from a single clone should react to the same antigen.

Neurotoxin A toxin that attacks the nervous system.

Nosocomial infection Infection acquired in a hospital setting, either as a result of a medical procedure (see **Iatrogenic disease**), person-to-person contact, contamination in the hospital environment, or ingestion of contaminated food or water.

Offal Internal organs and entrails of a slaughtered animal. Some offal may or may not be fit for human consumption, depending on its condition and on cultural customs.

Oocyst A cyst containing the fertilized egg of a protozoan.

Ostomy bag A pouch used to collect bodily waste. Surgical removal of all or part of the intestines often includes ileostomy, the creation of an opening in the abdominal wall through which waste can be emptied directly from the small intestine into an ostomy bag, which must be removed and emptied periodically by the patient.

Outbreak A cluster of illnesses with a common source and cause.

Parasite An organism that lives in (or on), and obtains its nourishment from, another living organism (host).

Pasteurization Heat treatment (usually of a liquid) that is adequate to kill most vegetative bacterial cells, but not spores. Named for its developer, Louis Pasteur.

Pathogen A microorganism that causes disease. Pathogens can be viruses, bacteria, fungi, or protozoa and may infect other microorganisms, plants, or animals (including humans).

Perineum The area of skin located between the anus and the scrotum (in the male) or vulva (in the female).

Prions Infectious protein particles that are believed to cause transmissible spongiform encephalopathies, including mad cow disease.

Probiotic A microorganism that, when ingested in sufficient quantity, has a beneficial effect on the digestion process.

Protozoa (*sing.*, protozoan) Single-celled animals; members of the phylum Protista.

Pseudopodium (*pl.*, pseudopodia) Literally, a "false foot." A temporary protrusion of the cell, the pseudopodium is used by amoebae for locomotion and feeding.

Quorum sensing The ability of bacteria to sense and communicate with other bacteria through the release and detection of particular chemicals.

Rendering Melting down fat, or heat-processing portions of an animal carcass to extract usable components.

Retort A pressure cooker used to heat-process canned foods.

Ruminant animal An animal that chews its cud. Examples of ruminants include cattle, sheep, goats, and camels.

Salmonellosis Disease syndrome produced by infection with *Salmonella.*

Sanitizer A chemical that, when used correctly, can reduce the microbial load of the surface to which it is applied.

Sepsis Presence of bacteria in a tissue, such as a wound. Can also refer to the generalized presence of bacteria in organs and tissues throughout the body.

Serotypes/serovars Varieties within a species (usually of bacteria) that can be differentiated based on the antigens presented on their exterior surfaces.

Shiga toxin Toxin produced by *Shigella.*

Shiga-like toxin A toxin closely resembling the toxin produced by *Shigella* that is produced by some strains of *Escherichia coli.*

Shigellosis Disease syndrome produced by infection with *Shigella.*

Somatic mutation A mutation that takes place in the nucleus of a somatic cell and is, therefore, not passed down to subsequent generations.

Sorbitol A sugar alcohol. Sorbitol can be metabolized by most strains of *E. coli*, but *E. coli* O157:H7 is usually unable to ferment sorbitol.

Spore Some species of bacteria produce spores. Spores are a survival mechanism. They are far more tolerant of adverse environmental conditions (dryness, heat, cold, ionizing radiation, etc.) than the vegetative cells in which they are produced. Mold spores, by contrast, are more than just a survival mechanism; they are the primary means of reproduction for these microorganisms.

Sporozoite Usually infectious, the sporozoite is a motile stage in the life cycle of some protozoa.

Superchlorination Dosing with an exceptionally high concentration of chlorine to achieve disinfection or to decontaminate a contaminated water supply.

Surveillance Monitoring of illness reports to detect disease outbreaks or changes in the incidence of particular diseases.

Syndrome A group of symptoms that, when present, are characteristic of a particular disease.

Tenesmus A constant feeling of needing to pass feces, usually accompanied by pain or abdominal cramps.

Thrombocytopenia Reduced level of platelets in the blood; results in bruising and slow clotting of blood after an injury.

Thrombosis Blood clot.

Toxin Poisonous substance.

Toxoid A toxin that has been inactivated to serve as a vaccine.

Vaccine A microbe (usually either an attenuated or killed preparation) or toxoid that, when inoculated into an animal or human, will stimulate an immune response.

Vector A living organism of whatever size or classification that is able to transmit a disease is referred to as a vector for that disease. For example, the mosquito is a vector for malaria.

Virulence The degree of severity of a disease.

Virus A submicroscopic particle that consists largely of protein and nucleic acids. Viruses cannot replicate themselves; they hijack the host cell's metabolism in order to achieve their own replication. Viral replication and release often, but not always, results in the death of the host cell. Viruses are usually host-specific, and even organ-specific; that is, they are only able to invade and replicate within the cells of a particular species or group of species, and only within the cells of certain organs.

Vitelline membrane A membrane that envelops the yolk in an egg.

Water activity (a_w) A measure of the amount of free water (i.e., water that is not bound up by solutes such as salts or sugars) in a food.

Western blot A test used to characterize proteins by their movement in an electric field (electrophoresis). Western blot analysis is one of the confirming tests for the presence of the bovine spongiform encephalopathy prion.

Zoonosis A disease that can be transmitted from animals to humans; a disease for which the reservoir is one or more nonhuman hosts.

appendix C

Abbreviations and Acronyms

Agencies

ADA American Dietetic Association.

CDC Centers for Disease Control and Prevention (originally Center for Disease Control).

CFIA Canadian Food Inspection Agency.

EPA Environmental Protection Agency.

FAO Food and Agricultural Organization of the United Nations.

FDA U.S. Food and Drug Administration.

FSIS Food Safety and Inspection Service (part of USDA).

GAO Government Accountability Office (formerly General Accounting Office).

HPB Health Protection Branch, Health and Welfare Canada.

MDPH Massachusetts Department of Public Health.

MOE Ministry of the Environment (Ontario, Canada).

MOSPL Microbial Outbreaks and Special Projects Laboratory (part of USDA).

NACMCF National Advisory Committee for Microbiological Criteria for Foods.

NASA National Aeronautics and Space Administration.

NFPA National Food Processors Association (now Food Products Association).

NIH National Institutes of Health (U.S.).

NRC National Research Council (U.S.).

OIG Office of Inspector General of USDA.

PUC Public Utilities Commission (Walkerton, Ontario, Canada).

USDA U.S. Department of Agriculture.

WHO World Health Organization.

Miscellaneous Abbreviations

AWSC Andrew & Williamson Sales Company.

BARF Biologically Appropriate Raw Food diet.

BSE Bovine spongiform encephalopathy.

CDAD *Clostridium difficile*-associated diarrhea.

CJD; vCJD Creutzfeldt-Jakob disease; new variant Creutzfeldt-Jakob disease.

CUSTA Canada–United States Free Trade Agreement.

HACCP Hazard Analysis and Critical Control Points.

HAV Hepatitis A virus.

HUS Hemolytic uremic syndrome.

IFT Institute of Food Technologists.

IHC Immunohistochemistry.

MBM Meat and bone meal.

MSU Michigan State University.

NAFTA North American Free Trade Agreement.

NICU Neonatal intensive care unit.

SBO Specified bovine offals.

TSE Transmissible spongiform encephalopathy.

TTP Thrombotic thrombocytopenic purpura.

VSP Vessel Sanitation Program.

Index

F

G

J

K

L

T

Y

Z